Lecture Notes in Physics

The Editorial Policy for Proceedings

The series Lecture Notes in Physics reports new developments in physical research and teaching – quickly, informally, and at a high level. The proceedings to be considered for publication in this series should be limited to only a few areas of research, and these should be closely related to each other. The contributions should be of a high standard and should avoid lengthy redraftings of papers already published or about to be published elsewhere. As a whole, the proceedings should aim for a balanced presentation of the theme of the conference including a description of the techniques used and enough motivation for a broad readership. It should not be assumed that the published proceedings must reflect the conference in its entirety. (A listing or abstracts of papers presented at the meeting but not included in the proceedings could be added as an appendix.)

When applying for publication in the series Lecture Notes in Physics the volume's editor(s) should submit sufficient material to enable the series editors and their referees to make a fairly accurate evaluation (e.g. a complete list of speakers and titles of papers to be presented and abstracts). If, based on this information, the proceedings are (tentatively) accepted, the volume's editor(s), whose name(s) will appear on the title pages, should select the papers suitable for publication and have them refereed (as for a journal) when appropriate. As a rule discussions will not be accepted. The series editors and Springer-Verlag will normally not interfere with the detailed editing except in fairly obvious cases or on technical matters.

Final acceptance is expressed by the series editor in charge, in consultation with Springer-Verlag only after receiving the complete manuscript. It might help to send a copy of the authors' manuscripts in advance to the editor in charge to discuss possible revisions with him. As a general rule, the series editor will confirm his tentative acceptance if the final manuscript corresponds to the original concept discussed, if the quality of the contribution meets the requirements of the series, and if the final size of the manuscript does not greatly exceed the number of pages originally agreed upon.

The manuscript should be forwarded to Springer-Verlag shortly after the meeting. In cases of extreme delay (more than six months after the conference) the series editors will check once more the timeliness of the papers. Therefore, the volume's editor(s) should establish strict deadlines, or collect the articles during the conference and have them revised on the spot. If a delay is unavoidable, one should encourage the authors to update their contributions if appropriate. The editors of proceedings are strongly advised to inform contributors about these points at an early stage.

The final manuscript should contain a table of contents and an informative introduction accessible also to readers not particularly familiar with the topic of the conference. The contributions should be in English. The volume's editor(s) should check the contributions for the correct use of language. At Springer-Verlag only the prefaces will be checked by a copy-editor for language and style. Grave linguistic or technical shortcomings may lead to the rejection of contributions by the series editors.

A conference report should not exceed a total of 500 pages. Keeping the size within this bound should be achieved by a stricter selection of articles and not by imposing an upper limit to the length of the individual papers.

Editors receive jointly 30 complimentary copies of their book. They are entitled to purchase further copies of their book at a reduced rate. As a rule no reprints of individual contributions can be supplied. No royalty is paid on Lecture Notes in Physics volumes. Commitment to publish is made by letter of interest rather than by signing a formal contract. Springer-Verlag secures the copyright for each volume.

The Production Process

The books are hardbound, and quality paper appropriate to the needs of the authors is used. Publication time is about ten weeks. More than twenty years of experience guarantee authors the best possible service. To reach the goal of rapid publication at a low price the technique of photographic reproduction from a camera-ready manuscript was chosen. This process shifts the main responsibility for the technical quality considerably from the publisher to the authors. We therefore urge all authors and editors of proceedings to observe very carefully the essentials for the preparation of camera-ready manuscripts, which we will supply on request. This applies especially to the quality of figures and halftones submitted for publication. In addition, it might be useful to look at some of the volumes already published.

As a special service, we offer free of charge LATEX and TEX macro packages to format the text according to Springer-Verlag's quality requirements. We strongly recommend that you make use of this offer, since the result will be a book of considerably improved technical quality.

To avoid mistakes and time-consuming correspondence during the production period the conference editors should request special instructions from the publisher well before the beginning of the conference. Manuscripts not meeting the technical standard of the series will have to be returned for improvement.

For further information please contact Springer-Verlag, Physics Editorial Department V, Tiergartenstrasse 17, W-6900 Heidelberg, FRG

M. Remoissenet   M. Peyrard (Eds.)

Nonlinear Coherent Structures in Physics and Biology

Proceedings of the 7th Interdisciplinary Workshop
Held at Dijon, France, 4-6 June 1991

Springer-Verlag Berlin Heidelberg GmbH

Editors

M. Remoissenet
M. Peyrard
Université de Bourgogne
Laboratoire Ondes et Structures Cohérentes (OSC)
6 Bd Gabriel, F-21100 Dijon, France

ISBN 978-3-662-13838-0 ISBN 978-3-540-46458-7 (eBook)
DOI 10.1007/978-3-540-46458-7

Originally published by Springer-Verlag Berlin Heidelberg New York in 1991
Softcover reprint of the hardcover 1st edition 1991

58/3140-543210 - Printed on acid-free paper

This Book is Dedicated to the Memory of Stephanos Pnevmatikos

PREFACE

This volume contains the text of most of the contributions presented at the 7th Interdisciplinary Workshop on "Nonlinear Coherent Structures in Physics and Biology", which was held on the campus of the Université de Bourgogne, Dijon, France, from June 4 to 6, 1991, with about 80 participants.

As with earlier workshops in this series, the purpose of this workshop was to bring together scientists concerned with recent developments and various aspects of nonlinear structures and to provide a forum to stimulate the exchange of ideas among scientists of different backgrounds, including physicists, mathematicians, biologists and engineers.

Nature provides many examples of coherent nonlinear structures and waves, and these have been observed and studied in various fields, ranging from fluids and plasmas through solid state physics to chemistry and biology. Among these beautiful nonlinear phenomena, localized wave packets, solitary waves and solitons, which propagate without dispersing, are the simplest structures, and these provide a continuing source of fascination for the student of nonlinearity. In fact, many real systems sharing the same underlying nonlinear phenomenon can be modeled by the same basic equations, leading to an understanding of their dynamic properties.This correctly indicates the importance of maintaining the interdisciplinary feature of nonlinear science.

The proceedings reflect the remarkable progress in understanding and modeling nonlinear phenomena in various systems, and these new developments show that the study of nonlinear coherent structures is in a state of healthy growth. Experimental, numerical and theoretical activities are interacting in various studies, which we present according to the following classification :

- magnetic and optical systems
- biosystems and molecular systems
- lattice excitations and localized modes
- two-dimensional structures
- theoretical physics
- mathematical methods

We gratefully acknowledge the *Centre National de la Recherche Scientifique*, the *Region Bourgogne* and the *Université de Bourgogne*, which contributed to the opportunity of gathering in Dijon leading scientists in both experimental and theoretical nonlinear science by providing the workshop with financial support .

We are grateful to Mrs A.Levy, D.Arnoux and Y.Boiteux for their active collaboration in the meeting, and to all our colleagues who helped us in many ways.

Dijon , June 1991

M. REMOISSENET

M. PEYRARD

CONTENTS

PART I

MAGNETIC AND OPTICAL SYSTEMS

EQUATIONS OF MOTION FOR VORTICES IN 2-D EASY-PLANE MAGNETS

G.M. Wysin,* F.G. Mertens†
*Kansas State University, Manhattan, KS 66506 USA
†Physics Institute, University of Bayreuth,
D-8580 Bayreuth, Germany

The dynamics of individual and pairs of vortices in a classical easy-plane Heisenberg spin model is studied. There are two types of vortices possible: in-plane, with small out-of-plane spin components present only at nonzero velocity, and out-of-plane, with large out-of-plane spin components even when at rest. As a result, the two types are governed by different equations of motion when in the presence of neighboring vortices. We review the static spin configurations and the changes due to nonzero velocity. An equation of motion introduced by Thiele and used by Huber will be re-examined. However, that equation may be inadequate to describe vortices in the XY model, due to their zero gyrovector. An alternative dynamical equation is developed, and effective mass and dissipation tensors are defined. These are relevant for models with spatially anisotropic coupling in combination with easy-plane spin exchange.

INTRODUCTION

A model for the dynamic correlations of vortices in easy-plane two-dimensional magnets has been presented, that uses the idea of an ideal gas of weakly interacting vortices.[1] Assuming a Boltmann velocity distribution, and if the velocity-dependent spin field of the vortices is known, then the dynamic structure function $S^{\alpha\alpha}(\vec{q},\omega)$ can be determined. At the microscopic level we would like to investigate the time-dependent motion of a single vortex, to understand how the neighboring vortices cause forces and accelerations, and to have a clear picture of how equilibrium is achieved.

Huber[2,3] has done such an analysis for diffusive motion of so-called "out-of-plane" vortices, ones that possess large out-of-easy-plane spin components. However, it is now realized that there are two type of vortices possible,[4,5] depending on the strength of the easy-plane anisotropy.[6,7] The stable vortices of the XY model, for example, are so-called "planar" vortices that only have small out-of-plane spin components. In that case the equation of motion that was used[3,8] is found to be inapplicable because these planar vortices have a zero gyrovector, to be discussed below. Here we propose an alternative dynamic equation of motion that applies to both types of vortices.

We begin by summarizing the properties of the two types of vortices allowed in the easy-plane anisotropic ferromagnetic Heisenberg model. The derivation of the equation of motion introduced by Thiele,[8] in terms of conserved force densities, will be sketched out, and the breakdown for planar vortices will be discussed. An alternative formalism using a canonical momentum for the vortex is developed. The new equation of motion includes the effects of vortex shape changes that are the result of acceleration. This leads to definition of an effective mass tensor, and, the gyrovector also re-appears. The new equation allows for a consistent description of both types of vortices.

Anisotropic Heisenberg Ferromagnet

The model system is the nearest neighbor 2D Heisenberg ferromagnet with easy-plane anisotropic exchange, characterized by a parameter $0 \leq \lambda < 1$; the Hamiltonian is

$$\mathbf{H} = -J\sum_{ij} \left(S_i^x S_j^x + S_i^y S_j^y + \lambda S_i^z S_j^z\right) . \tag{1}$$

J is an energy scale and the $\vec{S}_i$ are classical spins with fixed length. The XY model is given when $\lambda=0$, the Heisenberg model when $\lambda=1$. The spin dynamics is described by the Landau-Lifshitz equation,[9,10]

$$\dot{\vec{S}}_i = \{\vec{S}_i, H\} - \alpha\, \vec{S}_i \times \dot{\vec{S}}_i = \vec{S}_i \times \left(\vec{H}_i - \alpha \dot{\vec{S}}_i\right) \tag{2a}$$

$$\vec{H}_i = J\sum_{(ij)} \left(S_j^x \hat{x} + S_j^y \hat{y} + \lambda S_j^z \hat{z}\right) \tag{2b}$$

The sum is only over the neighbors of site i. A Landau-Gilbert damping term of strength α has been included. At any given time each spin is instantaneously precessing about the effective field $(\vec{H}_i - \alpha\dot{\vec{S}}_i)$. Initially the vortices will be described in the absence of damping, $\alpha=0$, which can be later re-introduced at the phenomenological level.

Static Vortices

The spins are parametrized in terms of an in-plane angle $\phi(\vec{x},t)$ and an out-of-plane angle $\theta(\vec{x},t)$ (or we use $S^z=S\sin\theta$),

$$\vec{S}(\vec{x},t) = S(\cos\theta\cos\phi,\ \cos\theta\sin\phi,\ \sin\theta) . \tag{3}$$

Then in a continuum limit including terms up to 2nd order in gradients the equations of motion are[4,5,7]

$$\dot{\theta} = JS\,(\cos\theta\,\nabla^2\phi - 2\sin\theta\,\nabla\theta\cdot\nabla\phi) \tag{4a}$$

$$\dot{\phi}\cos\theta = -\frac{1}{2}\,JS\left\{\left[\delta\left(|\vec{\nabla}\theta|^2-4\right) + |\vec{\nabla}\phi|^2\right]\sin2\theta + 2\left(1-\delta\cos^2\theta\right)\nabla^2\theta\right\} \tag{4b}$$

where $\delta\equiv1-\lambda$. Using polar coordinates (r,φ), and assuming a spatially isotropic solution, $\phi=\phi(\varphi)$, while $\theta=\theta(r)$, static vortices always have an in-plane angle satisfying Laplace's equation,

$$\phi(\vec{x}) = q\varphi + \phi_0 = q\tan^{-1}\left(\frac{y-y_0}{x-x_0}\right) + \phi_0 . \qquad (5)$$

The charge q is an integer, and ϕ_0 is a constant of integration.

The two types of vortices are distinguished by the out-of-plane angle θ. The static planar vortices have an out-of-plane angle that is zero, $\theta=0$. This solution exists independent of the anisotropy λ. However, when placed on a lattice,[6,7] it is found to be unstable for $\lambda>\lambda_c$, where λ_c is a critical anisotropy depending on the lattice ($\lambda_c=0.72$ for a square lattice). Planar vortices are stable only for $\lambda<\lambda_c$.

Out-of-plane vortices have a nonzero out-of-plane angle with asymptotic behavior[5,7]

$$\sin\theta \sim \begin{cases} p(1-ar^2) & \text{as } r\to 0 \\ \sqrt{\dfrac{r_v}{r}}\, e^{-r/r_v} & \text{as } r\to\infty \end{cases} \qquad (6a)$$

$$r_v = \frac{1}{2}\sqrt{\frac{\lambda}{1-\lambda}}, \qquad (6b)$$

where r_v is a characteristic vortex radius and a is a constant. The charge p is ±1 which determines whether the spin at the vortex center points along $+\hat{z}$ or $-\hat{z}$. When this solution is placed on a lattice,[6,7] it is found to be unstable for $\lambda<\lambda_c$ and stable only for $\lambda>\lambda_c$. Thus we have a situation where either planar ($\lambda<\lambda_c$) or out-of-plane ($\lambda>\lambda_c$) vortices are stable, and we expect that the dynamics may also reflect this as a crossover point in other quantities.

Dynamic Vortices

The equilibrium correlations between vortices can be found in an ideal gas phenomenology using the known spin profiles given above.[1] However, for correlations of the z spin components, for $\lambda<\lambda_c$, static planar vortices can contribute nothing to $S^{zz}(\vec{q},\omega)$. Then the lowest order vortex contribution must come from moving planar vortices, which do have nonzero S^z components. One can determine the perturbation due to a constant velocity $\vec{v}$ by assuming a solution $\vec{S}(\vec{x}-\vec{v}t)$. For planar vortices, with $\lambda<<1$, to first order in $\vec{v}$ we have no change in ϕ. The change in θ is given by[6,7]

$$\sin\theta = \frac{-\vec{v}\cdot\vec{\nabla}\phi}{JS(4\delta-|\nabla\phi|^2)} = \frac{q}{4JSr^2}(\vec{v}\times\vec{r})_z \tag{7}$$

in the asymptotic $r\to\infty$ regime, and $\vec{r}$ is measured from the instantaneous center position of the vortex. A similar change in $\sin\theta$ occurs for moving out-of-plane vortices, but it is small compared to the large out-of-plane profile already present in the static out-of-plane vortex.

Thiele's Equation of Motion

We review Thiele's vortex equation of motion[8] and the definition of the gyrovector, which vanishes for planar vortices. The equation is based on an interesting force-density interpretation of the Landau-Lifshitz equation, first rewritten in equivalent form,

$$\vec{S}\times\vec{H}_{net} = 0, \tag{8a}$$

$$\vec{H}_{net} = \vec{S}\times\dot{\vec{S}} + \vec{H} - \alpha\,\dot{\vec{S}}\,. \tag{8b}$$

$\vec{H}$ is analogous to that in Eq.(2b), representing the effective local field from neighboring spins. The other terms in $\vec{H}_{net}$ are dynamic and damping terms, respectively. In this notation the dynamics is "simple," in that each spin remains parallel to its instantaneous local field $\vec{H}_{net}$. Thus we could write $\vec{S}=\beta\vec{H}_{net}$ where $\beta(\vec{x},t) = S^2/(\vec{S}\cdot\vec{H})$. Combinations of $\vec{H}_{net}$ with gradients of $\vec{S}$ have dimensions of force per unit volume. Applying the operator $\cdot\,\partial_j\vec{S}\,\hat{e}_j = \cdot\vec{\nabla}\vec{S}$, (sum over repeated indices j=1,2) and realizing $\vec{S}\cdot\vec{\nabla}\vec{S} = 0$, there results the statement of conserved force density,

$$\vec{H}_{net}\cdot\vec{\nabla}\vec{S} = \left(\vec{H} + \vec{S}\times\dot{\vec{S}} - \alpha\dot{\vec{S}}\right)\cdot\vec{\nabla}\vec{S} = 0 \tag{9}$$

where the contraction is over spin components.

To apply this to a vortex we assume a travelling wave $\vec{S}(\vec{x} - \vec{v}t)$, and rewrite time derivatives using $\dot{\vec{S}} = -v_k\partial_k\vec{S}$.

There results

$$\vec{H}\cdot\vec{\nabla}\vec{S} + \vec{S}\cdot\left(\partial_j\vec{S}\times\partial_k\vec{S}\right)\hat{e}_j v_k + \alpha\,(\partial_j\vec{S})\cdot(\partial_k\vec{S})\,\hat{e}_j v_k = 0\,. \tag{10}$$

This then motivates the definition of the gyrovector $\vec{G}$,

$$G_i = -\frac{1}{2}\,\epsilon_{ijk}G_{jk}, \tag{11a}$$

$$G_{jk} \equiv -\int d^2x\,\vec{S}\cdot\left(\partial_j\vec{S}\times\partial_k\vec{S}\right) \tag{11b}$$

and the symmetric dissipation tensor $\tilde{D}$

$$D_{jk} = -\int d^2x\,\alpha\,(\partial_j\cdot\vec{S})\cdot(\partial_k\vec{S})\;. \tag{12}$$

The gyrovector is derived from an antisymmetric tensor G_{jk}. An equivalent expression for $\vec{G}$ is

$$\vec{G} = S^2\int d^2x\,\vec{\nabla}\Phi\times\vec{\nabla}S^z. \tag{13}$$

The remaining term concerns reversible effects. It is taken to give the effective force acting on the vortex,

$$\vec{F} = -\int d^2x\,\vec{H}\cdot\vec{\nabla}\vec{S}. \tag{14}$$

Then the Thiele equation of motion is

$$\vec{F} + \vec{G} \times \vec{V} + \tilde{D} \cdot \vec{V} = 0 . \qquad (15)$$

This equation can be used to describe the motion of out-of-plane vortices, for example, interacting in pairs, with a force $F=2\pi JS^2 q_1 q_2/r_{12}$. The gyrovector is found to be $\hat{G}=2\pi pq\hat{z}$. In the absence of damping, the pair will move in a circle if the gyrovector are parallel ($p_1q_1=p_2q_2$) or they will have a parallel translational motion if the gyrovectors are antiparallel ($p_1q_1=-p_2q_2$).

A problem arises for planar vortices. The gyrovector for static and moving planar vortices is <u>zero</u>.[11] This is due to the asymmetry of the S^z component about the direction of motion. The Thiele equation of motion becomes singular because the dynamic term $\vec{G}\times\vec{v}$ vanishes. This is most obvious with no damping, then the equation reads $\vec{F}=0$, which is not necessarily true for a vortex in the field of its neighbors. This leads to a conceptual difficulty in making a vortex ideal gas description for $\lambda<\lambda_c$.

Vortex Momentum

The problem seems to be that the analysis above does not allow for shape changes of a vortex in response to external fields as will occur for an accelerated vortex. That is, S^z for a planar vortex depends approximately linearly on its velocity.[6,7] If it accelerates it changes shape by developing more spin tilting out of the easy plane. On the other hand, the out-of-plane vortices have a large S^z component even at zero velocity so velocity-dependent changes in S^z may have a lesser effect.

An alternative viewpoint is to define a canonical momentum[12] $\vec{P}$ for the vortex, conjugate to its position $\vec{r}$, and then use the equation of motion $\dot{\vec{P}} = -\partial H/\partial \vec{r} = \vec{F}$. A Lagrangian that gives the correct spin-dynamics equations of motion is

$$\mathcal{L} = \int d^2x\, S^z \dot{\phi} - H \tag{16}$$

because S^z is the field momentum conjugate to ϕ. This then suggests that we take the definition of the vortex momentum to be

$$\vec{P} = -\int d^2x\, S^z\, \vec{\nabla}\phi \tag{17}$$

and then $\mathcal{L}=\vec{v}\cdot\vec{P}-H$ for a vortex of velocity $\vec{v}$. This definition is analogous to the canonical momentum developed for describing solitons in 1-D magnets[12,13,14] (generator of translations). For a planar vortex we get

$$\vec{P} = \frac{\pi}{4JS}\, \ln(R/a_o)\ \vec{v} \tag{18}$$

where R is a large distance cutoff (system size), and a_o is a short distance cutoff (≈ lattice constant). The effective mass seen here is identical to that found from the kinetic energy of the planar vortex.[7]

An equation of motion results by conserving momentum,[15] $\dot{\vec{P}} = \vec{F}$, and using

$$\dot{\vec{P}} = -\int d^2x \left(\dot{S}^z\, \vec{\nabla}\phi + S^z\, \vec{\nabla}\dot{\phi}\right) . \tag{19}$$

The vortex shape is determined by its velocity $\vec{V} = \dot{\vec{r}}(t)$ as well as its position $\vec{r}(t)$, so we assume $\vec{S} = \vec{S}(\vec{x} - \vec{r}(t), \vec{v}(t))$. Therefore the time derivative is replaced by space ($\partial_j \equiv \partial/\partial x_j$) and velocity ($\tilde{\partial}_j \equiv \partial/\partial v_j$) gradients,

$$\frac{d}{dt} = -\, v_j \partial_j + a_j \tilde{\partial}_j . \tag{20}$$

In the absence of damping the total rate of change of momentum is

$$\dot{\vec{P}} = - \hat{e}_j K_{jk} v_k + \hat{e}_j M_{jk} a_k = - \tilde{K}\cdot\vec{v} + \tilde{M}\cdot\vec{a} \tag{21}$$

where $\tilde{K}$ and $\tilde{M}$ are an effective gyrovector and mass tensor,

$$K_{jk} = - \int d^2x \, \partial_k (S^z \, \partial_j \phi) \tag{22a}$$

$$M_{jk} = - \int d^2x \, \tilde{\partial}_k (S^z \, \partial_j \phi) \; . \tag{22b}$$

The generalized gyrovector $\tilde{K}$ is related to Thiele's gyrotensor $\tilde{G}$. We separate $\tilde{K}$ into a symmetric and an antisymmetric part,

$$K_{jk} = g_{jk} + L_{jk}, \tag{23a}$$

$$g_{jk} = \frac{1}{2} (K_{jk} - K_{kj}), \tag{23b}$$

$$L_{jk} = \frac{1}{2} (K_{jk} + K_{kj}) \; . \tag{23c}$$

Then the equation of motion now becomes

$$\vec{F} = - \vec{g} \times \vec{v} - \tilde{L}\cdot\vec{v} + \tilde{M}\cdot\vec{a} \tag{24a}$$

where

$$\vec{g} = - \frac{1}{2} \epsilon_{ijk} \hat{e}_i g_{jk} = \frac{1}{2} \vec{G} \tag{24b}$$

$\vec{g}$, is the gyrovector, smaller by a factor of 1/2 from equation 11. Note that the origin must be excluded from the integrals giving $\tilde{K}$ and $\tilde{M}$. The symmetric tensor $\tilde{L}$ has no effect for vortices, it is zero for moving planar and out-of-plane vortices. The mass tensor $\tilde{M}$, determined by mixed space and velocity derivatives, exhibits the dynamic effect that changes in velocity cause changes in the internal structure of the vortex spin field. The force need not be parallel to $\vec{g} \times \vec{v}$ or $\vec{a}$.

For <u>planar</u> vortices, $\vec{g}=0$, $\tilde{L}=0$, $M_{ij}=(\pi/4JS)\ln(R/a_o)\,\delta_{ij}$. Then the equation of motion is Newtonian,

$$\vec{F} = M\vec{a} \tag{25}$$

stating that a pair of interacting vortices move straight away or straight toward each other in the absence of damping. This behavior is seen in simulations. This is not qualitately different from Thiele's equation with damping, but this new approach remains applicable even when the damping is removed. In this way we now have a microscopic dynamics for the vortices in the XY model, or, whenever $\lambda<\lambda_c$.

For out-of-plane vortices, $\vec{g}=\pi pq\hat{z}$, $\tilde{L}=0$, and the mass is equivalent to the planar vortex mass. Then the equation of motion is

$$\vec{F} + \vec{g} \times \vec{v} = M\vec{a} \ . \tag{26}$$

We can consider the interaction of a pair of out-of-plane vortices, with force $F=A/r_{12}$, $A=2\pi JS^2 q_1 q_2$. If the gyrovectors are antiparallel ($p_1q_1=-p_2q_2$), then the motion is unaccelerated parallel translation as mentioned earlier. However, if the gyrovectors are parallel ($p_1q_1=p_2q_2$), the motion can be circular, but the angular frequency depends on whether the pair's interaction is repulsive ($A>0$) or attractive ($A<0$). For small mass ($AM/G^2 r^2{}_{12}<<1$) the frequency is found to be

$$\omega = \frac{A}{gr_{12}^2}\left(1 + \frac{AM}{g^2 r_{12}^2}\right) \ . \tag{27}$$

The repulsive interaction gives the larger frequency, and has the angular velocity $\vec{\omega}$ parallel to $\vec{g}$. The attractive interaction, with smaller frequency, has $\vec{\omega}$ antiparallel to $\vec{g}$. This new effect is related to a competition between an intrinsic vortex momentum (similar to $\vec{g}$) and the orbital angular momentum of the pair. For a given size of force transverse to the direction of motion, a larger acceleration occurs when the force is parallel to $\vec{g} \times \vec{v}$. The path of the vortex is more easily bent in the $\vec{g} \times \vec{v}$ direction than in the $-\vec{g} \times \vec{v}$ direction (i.e., left turns are easier than right turns for an "up" vortex, $\vec{g}=g\hat{z}$).

Conclusion

An equation of motion developed by Thiele[8] and Huber[2,3] is limited to cases where the vortex shape is fixed as a function of velocity, excluding application to the important XY model. An alternative equation of motion was developed here, based on finding the time rate of change of vortex momentum. The new equation alleviates the difficulty encountered when $\vec{G}=0$ (for planar vortices) and indicates new behavior for interacting out-of-plane vortices. The differences with this new equation of motion are summarized as follows: 1) an equation of motion exists for planar and out-of-plane vortices even for zero damping; 2) the new mass term accounts for velocity-dependent shape changes that result from acceleration; 3) a pair of out-of-plane vortices with parallel gyrovectors and $q_1=q_2$ (vortex-vortex) orbits faster than a pair with $q_1=-q_2$ (vortex-antivortex), all other things being equal. It should have important applications for microscopic vortex dynamics, especially for models with spatial anisotropy, whose vortices will possess an anisotropic mass tensor.

REFERENCES

1. F.G. Mertens, A.R. Bishop, G.M. Wysin and C. Kawabata, Phys. Rev. Lett. 59, 117 (1987); Phys. Rev. B39, 591 (1989).
2. D.L. Huber, Phys. Lett. 76A, 406 (1980).
3. D.L. Huber, Phys. Rev. B26, 3758 (1982).
4. S. Takeno and S. Homma, Prog. Theor. Phys. 64, 1193 (1980); 65, 172 (1980).
5. S. Hikami and T. Tsuneto, Prog. Theor. Phys. 63, 387 (1980).

6. G.M. Wysin, M.E. Gouvêa, A.R. Bishop and F.G. Mertens, in *Computer Simulation Studies in Condensed Matter Physics*, D.P. Landau, K.K. Mon and H.-B. Schüttler, eds., Springer-Verlag 1988.

7. M.E. Gouvêa, G.M. Wysin, A.R. Bishop and F.G. Mertens, Phys. Rev. B39, 11,840 (1989).

8. A.A. Thiele, Phys. Rev. Lett. 30, 230 (1973); J. Appl. Phys. 45, 377 (1974).

9. L.D. Landau and E.M. Lifshitz, Phys. Z. Sowjet 8, 153 (1935).

10. F.H. de Leeuw, R. van den Poel and U. Enz, Rep. Prog. Phys. 43, 44 (1980).

11. A.R. Völkel, F.G. Mertens, A.R. Bishop and G.M. Wysin, Phys. Rev. B43, 5992 (1991).

12. J. Tjon and J. Wright, Phys. Rev. B15, 3470 (1977).

13. E.K. Sklyanin, Sov. Phys. Dokl. 24, 107 (1979).

14. H.C. Fogedby, in *Theoretical Aspects of Mainly Low-Dimensional Magnetic Systems*, Springer-Verlag (1980).

15. The effects of damping in this formalism are somewhat complicated and will be given elsewhere.

CENTRAL PEAK SIGNATURES FROM VORTICES IN 2D EASY-PLANE ANTIFERROMAGNETS

F.G. Mertens, A. Völkel, G.M. Wysin*, A.R. Bishop+
University of Bayreuth, Germany, *Kansas State University, Manhattan, USA,
+Los Alamos National Laboratory, USA

We investigate the dynamics of a classical, anisotropic Heisenberg model. Assuming a dilute gas of ballistically moving vortices above the Kosterlitz-Thouless transition temperature, we calculate the dynamic form factors $S(\vec{q},\omega)$ and test them by combined Monte Carlo-molecular dynamics simulations. For both in-plane and out-of-plane correlations we predict and observe central peaks (CP) which are, however, produced by quite different mechanisms, depending on whether the correlations are globally or locally sensitive to the presence of vortices. The positions of the peaks in q-space depend on the type of interaction and on the velocity dependence of the vortex structure. For a ferromagnet both CP's are centered at $q = 0$; for an antiferromagnet the static vortex structure is responsible for a CP at the Bragg points, while deviations from it due to the vortex motion produce a CP at $q = 0$. By fitting the CP's to the simulation data we obtain the correlation length and the mean vortex velocity.

1. INTRODUCTION

A wide class of quasi-two-dimensional magnetic materials (e.g. layered magnets[1], graphite intercalation compounds[2], magnetic lipid layers[3]) can be described approximately by the classical anisotropic Heisenberg model

$$H = J \sum_{\langle ij \rangle} (S_i^x S_j^x + S_i^y S_j^y + \lambda\, S_i^z S_j^z) \qquad (1.1)$$

with $0 \le \lambda < 1$. The spins $\vec{S}_i$ are coupled only to nearest neighbors on a square lattice and tend to be aligned in the xy-plane (easy-plane).

Below the Kosterlitz-Thouless transition temperature T_c bound vortex-antivortex pairs are thermally excited, but do not move. Above T_c the pairs begin to unbind and the single vortices move around due to the interaction with the other vortices.

For the static spin correlations many exact results from the Kosterlitz-Thouless theory are known. However, for the <u>dynamic</u> correlations only a phenomenological treatment has been possible so far. Assuming a dilute gas of ballistically moving vortices above T_c, the dynamic form factors $S(\vec{q},\omega)$ have been calculated for the case of a ferromagnetic coupling[4,5]. The typical signatures of moving vortices are central peaks (CP's), which have been observed both in combined Monte Carlo-molecular dynamics simulations[4] and in inelastic neutron scattering experiments[1,2].

In this paper we will shortly review these results and discuss the characteristic changes of the CP's for the case of an antiferromagnetic coupling.

2. ANTIFERROMAGNETIC VORTICES

We parametrize the classical spins by four angles[6]

$$S_n^x = (-1)^n S \cos[\Phi_n + (-1)^n \phi_n] \cos[\theta_n + (-1)^n \vartheta_n]$$

$$S_n^y = (-1)^n S \sin[\Phi_n + (-1)^n \phi_n] \cos[\theta_n + (-1)^n \vartheta_n] \qquad (2.1)$$

$$S_n^z = (-1)^n S \sin[\theta_n + (-1)^n \vartheta_n] ,$$

where the even n denote one sublattice, the odd n the other one. In the continuum approximation the capital angles $\Phi(\vec{r})$ and $\theta(\vec{r})$ describe a perfect local antiferromagnetic structure, while the small angles $\phi(\vec{r})$ and $\vartheta(\vec{r})$ describe deviations from this structure.

In a previous paper[7], we have derived the four equations of motion and we have found the following continuum limit vortex solutions (to first order in the vortex velocity u)

$$\Phi(\vec{r}) = \pm \arctan(y/x) , \quad \phi(\vec{r}) = 0 , \quad \theta(\vec{r}) = 0 \qquad (2.2)$$

$$\vartheta(\vec{r}) \sim \frac{u}{4(1+\lambda)JS} \frac{\sin(\varphi-\varepsilon)}{r} \quad \text{for large } r . \qquad (2.3)$$

Here r, φ are the polar coordinates of $\vec{r}$, and ε is the angle between $\vec{u}$ and the x-axis.

When placed on a lattice, this solution is stable only for $\lambda < \lambda_c \approx 0.72$ and is called "in-plane vortex" because the out-of-plane components of the spins are small, or even zero in the static limit. In this paper we do not discuss the case $\lambda > \lambda_c$ for which "out-of-plane vortices" are stable, here θ is nonzero independent of the velocity.

3. IN-PLANE CORRELATION FUNCTION

Inserting the vortex solution (2.2/3) in the definition $S^{xx}(\vec{r},t) = \langle S^x(\vec{r},t)\, S^x(\vec{0},0)\rangle$ we obtain

$$S^{xx} = S^2 \langle e^{i\vec{K}\vec{r}} \cos \Phi(\vec{r},t) \cos[e^{i\vec{K}\vec{r}}\, \vartheta(\vec{r},t)] \cos \Phi(\vec{0},0) \cos \vartheta(\vec{0},0)\rangle. \qquad (3.1)$$

Here $\exp(i\vec{K}\vec{r}) = \pm 1$, depending on the sublattice to which $\vec{r}$ belongs; $\vec{K} = (\pi,\pi)$ restricting ourselves to the first Brioullin zone.

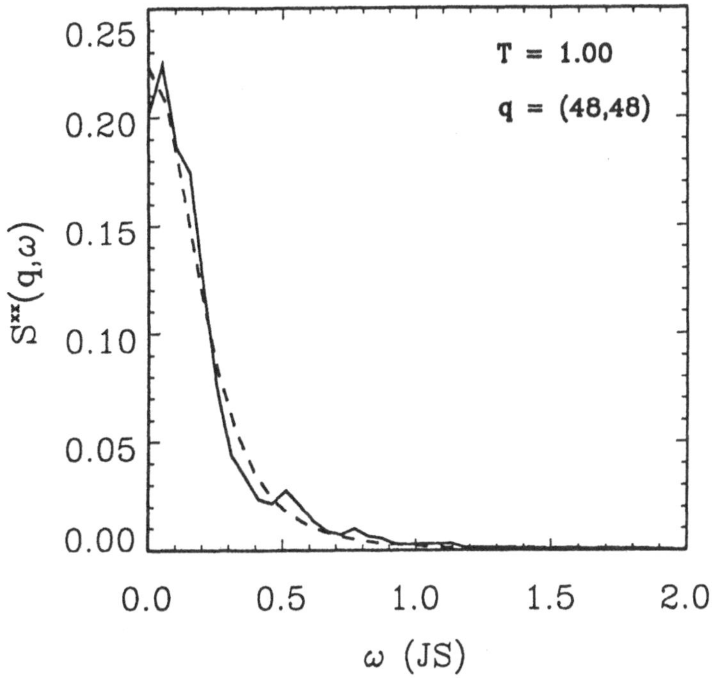

Fig. 1. In-plane dynamic form factor for $\lambda = 0$ (XY-model), $\vec{q}$ in units of $2\pi/L$ with lattice size $L = 100$ a. Solid line: simulation data; dashed line: fit of squared Lorentzian (3.3).

Similar to the ferromagnet[4], for large r the main effect of a vortex passing a lattice site is to flip the spin at this site, i.e. $\cos \Phi$ changes its sign. Compared to this big effect the small angle ϑ can be neglected and we obtain

$$S^{xx}(\vec{r},t) = S^2 e^{i\vec{K}\vec{r}} \langle \cos^2\Phi\rangle\langle(-1)^{N(\vec{r},t)}\rangle . \qquad (3.2)$$

The first average is a static one and gives 1/2; the dynamics is contained in the second average: $N(\vec{r},t)$ is the number of vortices passing an arbitrary, nonintersecting contour between $(\vec{0},0)$ and $(\vec{r},t)$. The calculation proceeds in the same way as for the ferromagnet[4] and yields a squared Lorentzian CP for the dynamic form factor

$$S^{xx}(\vec{q},\omega) = \frac{S^2}{2\pi} \frac{\gamma^3\xi^2}{\{\omega^2+\gamma^2[1+(\xi q^*)^2]\}^2} , \qquad (3.3)$$

the only difference is the appearance of $\vec{q}^* = \vec{K} - \vec{q}$ instead of $\vec{q}$. Here $\gamma = \sqrt{\pi}\,\bar{u}/(2\xi)$, with rms vortex velocity $\bar{u}$, and 2ξ is the average distance between two vortices (according to the Kosterlitz-Thouless theory, ξ is also the static correlation length).

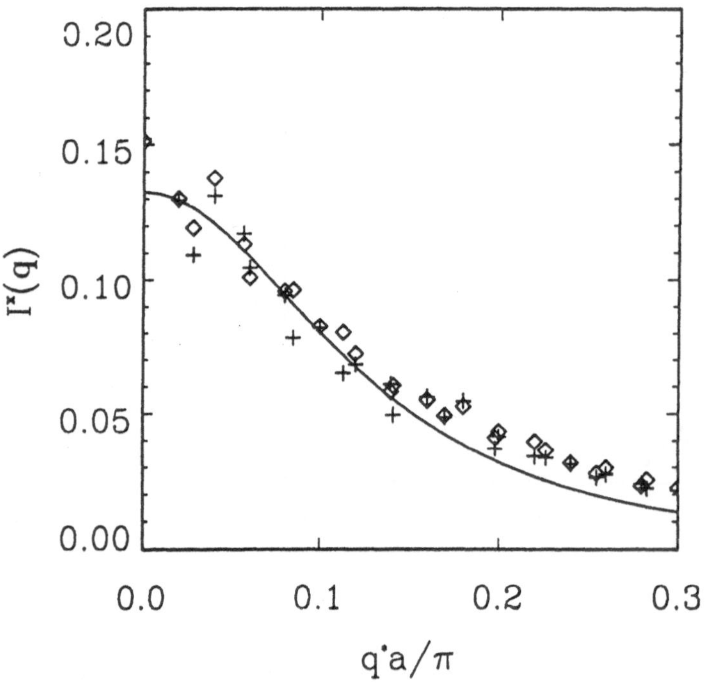

Fig 2. Integrated intensity of in-plane dynamic form factor for T = 1.0. +: intensity of fitted central peak (3.3); ◇: intensity from simulation data; solid line: fit to (3.4) for small q^*.

Since (3.3) contains no parameters characterizing the vortex structure, S^{xx} is only globally sensitive to the presence of vortices. In fact, the moving vortices disturb the local antiferromagnetic order and thus diminish the correlations. For this reason the integrated intensity

$$I^x(q^*) = \frac{S^2}{4\pi} \frac{\xi^2}{[1+(q^*\xi)^2]^{3/2}} \tag{3.4}$$

is inversely proportional to the density $n_v \simeq (2\xi)^{-2}$ of free vortices.

The prediction (3.3) can be well fitted to the CP which we observe in simulations on a 100 x 100 lattice (Fig.1). The resulting data for the width Γ^x and intensity I^x are displayed in Figs. 2 and 3. Moreover, we have fitted these data for small q^* (because of our large-r approximation) to (3.4) and to

$$\Gamma^x(q^*) = \tfrac{1}{2}\sqrt{\pi(\sqrt{2}-1)}\,\frac{\bar{u}}{\xi}\sqrt{1+(\xi q^*)^2} \tag{3.5}$$

which represents the half width of (3.3). This fit gives us $\bar{u}$ and ξ. Just above $T_c \simeq 0.79$ we obtain $\bar{u} \approx 1$ (in dimensionless units) and this value increases slightly with the reduced temperature $\tau = (T - T_c)/T_c$. The values for ξ agree rather well with the Kosterlitz-Thouless prediction $\xi(T) \simeq \exp(b/\sqrt{\tau})$ choosing $b \simeq 0.5$.

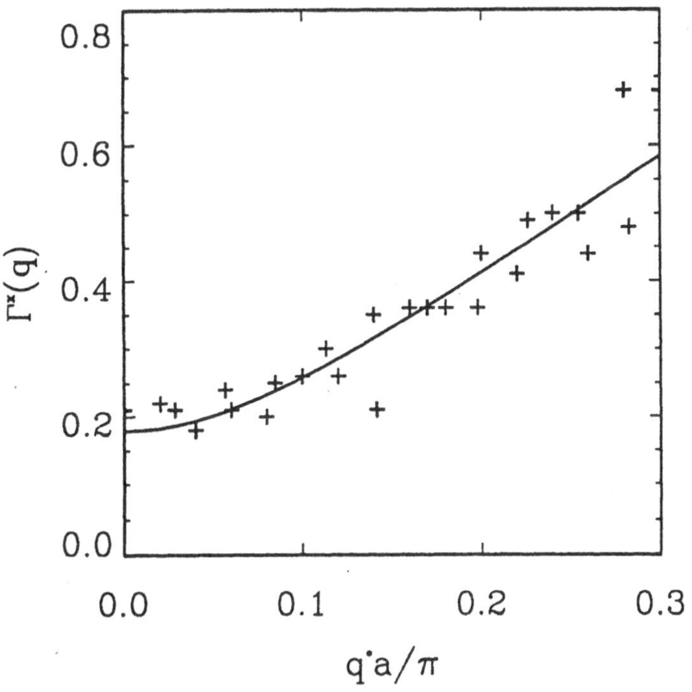

Fig 3. Width of in-plane dynamic form factor for T = 1.0. +: width of fitted central peak (3.3); solid line: fit to (3.5) for small q^*.

4. OUT-OF-PLANE CORRELATION FUNCTION

Contrary to the in-plane structure, the out-of-plane structure (2.3) of a vortex is localized. Therefore we expect that $S^{zz}(\vec{r},t)=\langle S^z(\vec{r},t)S^z(\vec{0},0)\rangle$ depends on the shape of this structure. We are mostly interested in small wave vectors and therefore we evaluate

$$S^{zz} = \langle e^{i\vec{K}\vec{r}} \sin [e^{i\vec{K}\vec{r}} \vartheta(\vec{r},t)] \cdot \sin \vartheta(\vec{0},0)\rangle \tag{4.1}$$

for large r where ϑ is small, and obtain $\langle\vartheta(\vec{r},t)\vartheta(\vec{0},0)\rangle$. Using (2.3) we obtain in a vortex-gas approximation, similarly to the ferromagnetic case[5], a Gaussian CP

$$S^{zz}(q,\omega) = \frac{n_v \bar{u}}{32(1+\lambda)^2 J^2 \sqrt{\pi}\, q^3} \exp \{-[\omega/(\bar{u}q)]^2\} . \tag{4.2}$$

Here $1 + \lambda$ appears, instead of $1 - \lambda$ for the ferromagnet. However, otherwise the formula is the same, i.e. in both cases there is a CP with a maximum at $\vec{q} = (0,0)$ and $\omega = 0$. The reason is that the z-components of neighboring spins on the two sublattices have the same sign, thus the out-of-plane structure of an antiferromagnetic vortex shows a local ferromagnetic order.

Because (4.2) is proportinal to the density n_v of free vortices, the CP is very small, cf. the different scales in the Figs. 1 and 4. Moreover, the stiffness constant of the out-of-plane spin waves suffers no sudden jump to zero at T_c, contrary to the in-plane magnons[8]. Therefore magnon peaks show up in the out-of-plane dynamic form factor also for $T > T_c$, though they are much broader than for $T < T_c$. Unfortunately there is an "acoustic" branch that interferes with the CP (4.2) for small q, where chances are best to observe the CP. Therefore CP and magnon contributions here cannot be distinguished clearly.

Fig. 4 shows simulation data for small q, together with a plot of (4.2) using the values for $\bar{u}$ and n_v obtained from the in-plane correlations. From such figures for different temperatures we can verify that the CP-width is consistent with the prediction $\Gamma^z = \bar{u}q$ resulting from (4.2). However, the intensity

$$I^z(q) = \frac{n_v \bar{u}}{32(1+\lambda)^2 J^2 \sqrt{\pi}\, q^2} \qquad (4.3)$$

cannot be tested because of the difficulty to separate the magnon contributions. (Remark: The singularity in (4.3) for $q \to 0$ can be avoided by introducing a cut-off in the order of ξ in order to account for the finite size of the vortices due to the presence of the other vortices[5].)

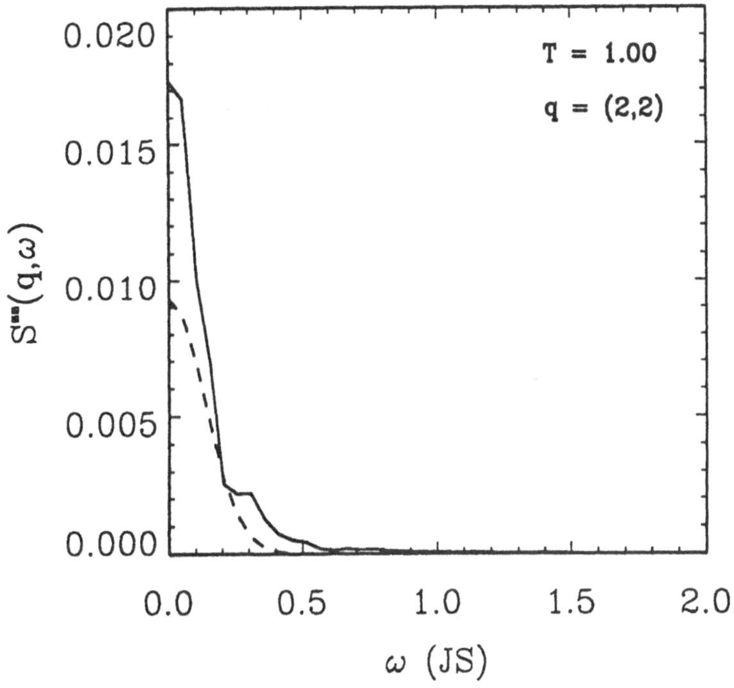

Fig 4. Out-of-plane dynamic form factor for $\lambda = 0$ (XY-model). Solid line: simulation data; dashed line: Gaussian (4.2) with $\bar{u}$ and ξ from the in-plane correlations.

5. CONCLUSION

Single vortices in easy-plane antiferromagnets with $\lambda < \lambda_c$ have a static structure with a locally perfect antiferromagnetic order. Therefore the in-plane dynamic form factor $S^{xx}(\vec{q},\omega)$, which is only globally sensitive to the presence of vortices, shows a squared Lorentzian CP at the Bragg points.

The finite velocity of the vortices causes small deviations from the above structure, producing a localized out-of-plane structure with a ferromagnetic character. Therefore the out-of-plane dynamic form factor $S^{zz}(\vec{q},\omega)$ shows a Gaussian CP at the center of the Brioullin zone. The q-dependence of S^{zz} reflects the shape of the ferromagnetic out-of-plane structure and the ω-dependence reflects the velocity distribution of the vortices.

Both CP's are observed in computer simulations which allows to fit the phenomenological parameters ξ and $\bar{u}$.

ACKNOWLEDGEMENTS

This work was supported by Deutsche Forschungsgemeinschaft (SFB 213, project C 19), by NATO (Collaborative Research Grant 0013/89), and by the United States Department of Energy.

REFERENCES

1. L.P. Regnault, J.P. Boucher, J. Rossat-Mignot, J. Bouillot, R. Pynn, J.Y. Henry, J.P. Renard, Physica 136 B, 329 (1986);
   S.T. Bramwell, M.T. Hutchings, J. Norman, R. Pynn, P. Day, J. Phys. C 8, 1435 (1988)
2. D.G. Wiesler, H. Zabel, S.M. Shapiro, Physica B 156 + 157, 292 (1989)
3. M. Pomerantz, Surface Science 142, 556 (1984);
   D.I. Head, B.H. Blott, D. Melville, J. Phys. C 8, 1649 (1988)
4. F.G. Mertens, A.R. Bishop, G.M. Wysin, C. Kawabata, Phys. Rev. B 39, 591 (1989)
5. M.E. Gouvêa, G.M. Wysin, A.R. Bishop, F.G. Mertens, Phys. Rev. B 39, 11840 (1989)
6. H.J. Mikeska, J. Phys. C 13, 2913 (1980)
7. A.R. Völkel, F.G. Mertens, A.R. Bishop, G.M. Wysin, Phys. Rev. B 43, 5992 (1991)
8. D.R. Nelson, J.M. Kosterlitz, Phys. Rev. Lett. 39, 1201 (1977)

Free Vortices in the Quasi-Two Dimensional XY Antiferromagnet $BaNi_2(PO_4)_2$?

P. Gaveau, J.P. Boucher, A. Bouvet, L.P. Regnault and Y. Henry
Département de Recherche Fondamentale, Service de Physique
Centre d'Etudes Nucléaires de Grenoble, 85 X, 38041 Grenoble cedex, France.

Non-linear excitations are expected to play an important role in low dimensional magnetic systems. The existence of soliton excitations in magnetic chains is now well established theoretically and experimentally. In two dimensions non-linear excitations have also been predicted to be important, at least for the so-called XY model where the spins are mainly confined within the magnetic plane. As shown by Kosterlitz and Thouless (KT) and Berezinskii [1], a topological ordering, associated with the pairing of vortex excitations, is expected to occur at a finite temperarure T_{KT}. Above T_{KT}, the unbinding of the vortex pairs gives rise to a gas of freely moving vortices. Recently, it was shown that the vortex motion results in a flipping process of the ordered spins [2]. In the fluctuation spectrum, this flipping process yields a characteristic central peak —the flipping mode. As shown for the case of solitons in antiferromagnetic (AF) chains where a similar effect was observed, the nuclear magnetic resonance (NMR) technique is well suited to probe such a narrow peak, centered at $\omega \approx 0$ [3].

In real systems, the KT transition and the topological order can never be observed. This is due to the coupling between planes which, always, induces a three-dimensional (3D) order at a temperature $T_N > T_{KT}$. However, for a very small interplane coupling, T_N and T_{KT} can be very close and vortex excitations, if they do exist, can only be observed above T_N. The application of an external magnetic field H may change the properties and the dynamics of the vortices. This latter point has been discussed recently for the case of ferromagnets [4]. For planar antiferromagnets, one expects the effect of a field H to be small, if it is applied perpendicular to the magnetic plane ($H_\perp$). In that case, at least for moderate field values ($H \ll J$) the effect of a field is essentially to reinforce the planar character of the spin system.

In the present work, we report on nuclear spin-lattice relaxation time (T_1) measurements performed on the compound $BaNi_2(PO_4)_2$ [5]. As shown on fig. 1, a spectacular diverging

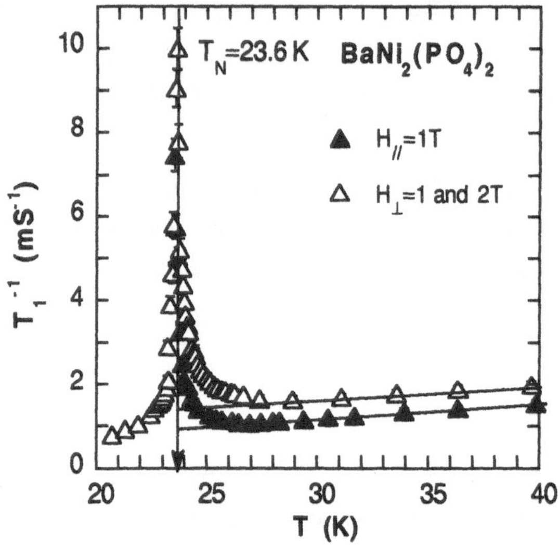

Fig.1 : Nuclear spin-lattice relaxation rate $1/T_1$ of ^{31}P in $BaNi_2(PO_4)_2$ for H perpendicular and parallel to the XY magnetic plane.

behaviour is observed for $1/T_1$ as the 3D transition ($T_N \approx 23.6K$) is approached. The data for $T>T_N$ are tentatively analysed within a model of freely moving vortices. The extrapolation of this model to the case where the field is parallel to the plane ($H_{//}$) is also discussed.

The compound $BaNi_2(PO_4)_2$ is a good quasi-2D planar antiferromagnet [6]. Its magnetic structure consists of 2D layers of Ni ions (with spin S=1), located on a honeycomb lattice of parameter a=2.8 Å. The in-plane exchange coupling is antiferromagnetic with the (average) value $J \approx 11K$. The planar character observed at low temperature results from a large single ion anisotropy $D \approx 7.3K$. The 3D magnetic ordering occuring at $T_N \approx 23.6K$ is attributed to a very small interplane coupling J' ($J'/J \approx 10^{-3} — 10^{-4}$). As shown by Regnault et al. [6], just above T_N ($T-T_N<10K$), the short range order is mainly two dimensional and can be analysed in the model of the KT transition. In that case, it was deduced that T_N should be very close to T_{KT} : $T_{KT} \approx 0.95—0.98\ T_N$ [6].

Our T_1 measurements have been performed on the ^{31}P ions, as a function of the temperature and for different values and orientations of the external field. In all cases, the same kind of divergence for $1/T_1$ is observed as $T \rightarrow T_N$. However, while the data are the same for $H_{\perp}$=1T and 2T, they appear to be strongly field dependent for $H_{//}$. Since we want to focus on the diverging part of $1/T_1$, we consider only the contributions above the lines drawn on fig. 1. The data to be

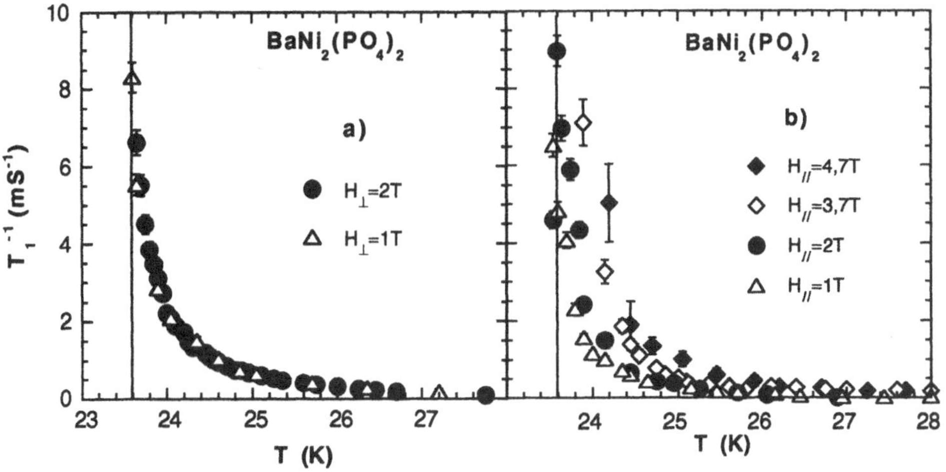

Fig.2 : Contributions from the expected critical fluctuations to $1/T_1$ for H perpendicular (a) and H parallel (b) to the XY plane.

analysed are shown in fig. 2. These values are expected to describe mainly the critical fluctuations.

For the phosphorus nuclear spins, the relaxation rate $1/T_1$ is given by $1/T_1=\Sigma_{\alpha q}A_q^{\alpha}S^{\alpha}(q,\omega_N)$ where the A_q^{α} are geometrical coefficients associated with the hyperfine coupling and the $S^{\alpha}(q,\omega)$ are the dynamical structure factors which describe the magnetic fluctuations. ω_N is the nuclear Larmor frequency (ω_N<80MHz) and q is a wavevector of the Brillouin zone. For such a 2D system, α=xy and α=z refer to the in-plane and out-of-plane fluctuations, respectively. According to the model of freely moving vortices proposed by Mertens et al. [2], one expects the main contribution to $1/T_1$ to be given by the flipping mode of the in-plane fluctuations. In reciprocal space, this flipping mode is limited to a narrow domain of the order of $\Delta q\approx\xi^{-1}$ where ξ is the 2D correlation length. Over this domain, A_q^{xy} is practically q independent ($A_q^{xy}\approx A^{xy}$) and $1/T_1$ can be written: $1/T_1\approx A^{xy}\Sigma_q S^{xy}(q,\omega_N)= A^{xy}S^{xy}(\omega_N)$ where $S^{xy}(\omega)$ is the spectrum of local fluctuations. It has a simple Lorentzian shape with a characteristic width γ which measures the rate of the spin flippings induced by the moving vortices. Finally :

$$1/T_1=A^{xy}.\gamma \ / \ (\gamma^2+\omega_N^2) \approx A^{xy}/\gamma \qquad (1)$$

for $\omega_N<<\gamma$. The flipping rate is evaluated to be :

$$\gamma=\sqrt{\pi n_v}.U \tag{2}$$

where

$$n_v=1/(2\xi)^2 \tag{3}$$

is the density of the free vortices [7]

$$n_v=(2\xi_0)^{-2}\exp(-2b/\sqrt{\tau})$$

and U the vortex velocity [8]:

$$U=\sqrt{(\pi/2)(JS^2a^2/h)^2 n_v \,Ln\{k_B T_{KT}/(JS^2 n_v a^2)\}} \tag{4}$$

In these expressions $\tau= T/T_{KT}-1$ and $\xi_0\approx$ a. The parameter b is not known accurately. The KT theory predicts $b= \pi/2 \approx 1.57$ [1] ; however recent numerical simulations yield much smaller values: $b\approx 0.5$—0.3 [4,9]. The data of figure 2 are analysed with Eq.1, where A^{xy}, b and T_{KT}/T_N are considered as adjustable parameters. Fig.3 gives a few exemples of resulting fits yielding

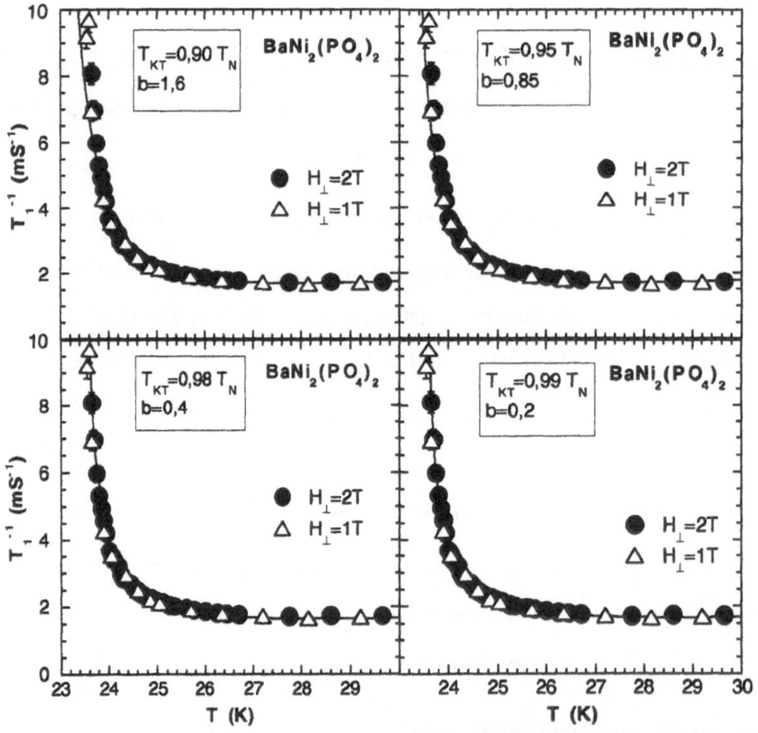

Fig.3 : Comparisons between theory (Eq. 1) and experimental data for different values of A^{xy}, b and T_{KT}/T_N

T_{KT} (K)	T_{KT}/T_N	b	A^{xy} (rad^2S^{-2})	γ (GHz) pour T=24 K
21,2	0,90	1,6	6 10^{11}	0,027
22,4	0,95	0,85	5 10^{12}	0,23
23,1	0,98	0,4	7 10^{13}	3,22
23,3	0,99	0,2	3 10^{14}	13

Table I : Values of the fitting parameters A^{xy}, b and T_{KT}/T_N corresponding to the different fits of fig. 3.

the values given in table I. The values for the flipping rate γ at 24 K are also reported in table I. A comparison with a recent neutron investigation [10] ($\gamma \approx 10$ GHz at 24 K) allows us to conclude that the values to be retained for T_{KT}/T_N, b and A^{xy} range certainly in between the values reported in the two last lines of table I. For the geometrical coefficient A^{xy} the corresponding values agree also with the evaluation we have made from the frequency shift of the NMR line : $A^{xy} \approx 10^{14}$ rad^2S^{-2}.

For $H_{//}$, a divergence is also observed for $1/T_1$, which is seen to be field dependent (fig. 2b). At low field ($H_{//} \approx 1T$), the vortex model defined above remains valid with similar values for T_{KT}/T_N and b. For larger field values, the observed increase of the divergence can be interpreted as essentially due to a slowing down of the flipping process: γ is reduced by the field $H_{//}$ (see Eq.2). As shown on fig.4, the 3D ordering temperature [10] is observed to increase slightly with $H_{//}$: for $H_{//}$ = 6 T, $\Delta T_N/T_N \approx 1.6\%$. Since T_N is proportional to the square of the 2D correlation lentgh ξ this effect corresponds to an increase of ξ. In the vortex model (see Eq. 3), this means that the vortex density n_V is decreasing with $H_{//}$: for $H_{//}$=6 T, the relative decrease of n_V is very small ($\Delta n_V/n_V$=-1.6%). Therefore, if the vortex density does not change very much, it is the vortex velocity which is reduced by the field. From the $1/T_1$ data, we can deduce the field dependence of U. The corresponding values are shown on fig.5, where they are compared to

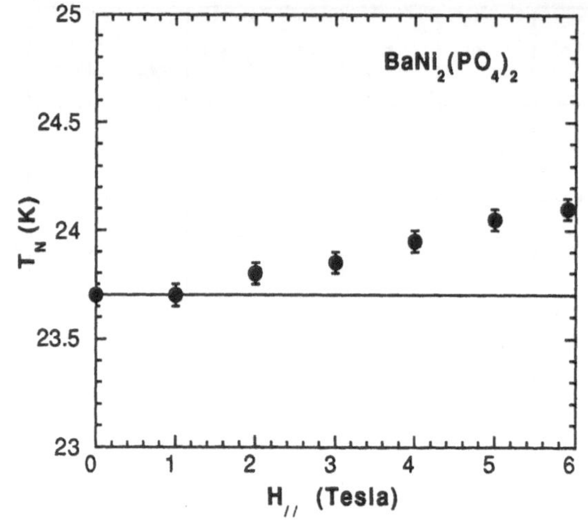

Fig. 4 : The 3D ordering temperature T_N as a function of $H_{//}$, observed by neutron scattering measurements [10].

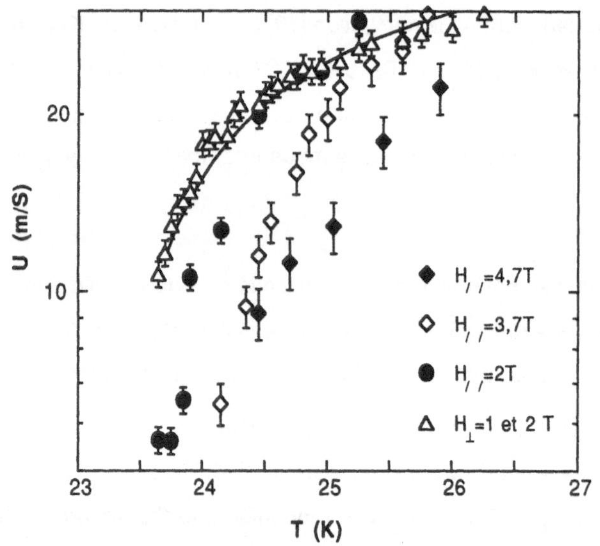

Fig. 5 : Vortex velocity U deduced from the $1/T_1$ data of fig.2. The full line represents the theoretical expression given by Huber.

the velocity given by Eq.4. In antiferromagnets, a field $H_{//}$ appears to reduce appreciably the vortex velocity. This conclusion differs from that of the ferromagnetic case [4].

The divergence of $1/T_1$ observed in $BaNi_2(PO_4)_2$ when approaching TN is well explained by a model of free moving vortices. The values obtained for b (b=0.4—0.2) is in reasonable agreement with the recent numerical simulations of Mertens et al. [4;9]. For H=0, the vortex velocity is in accord with Eq.4 given by Huber [8]. However, it is observed to be strongly field dependent.

References :

[1] J.M. Kosterlitz and D. Thouless, J.Phys. C : Solid State Phys.**6** (1973) 1186.

V.L. Berezinskii, Sov. Phys. JETP **34** (1972) 610.

[2] F.G. Mertens, A.R. Bishop, G.M. Wysin and C. Kawabata, Phys. Rev. B **39** (1989) 591.

[3] J.P. Boucher, Hyperfine Interactions **49** (1989) 423.

[4] M.E. Gouvêa, F.G. Mertens, A.R. Bishop and G.M. Wysin J.Phys: Condens. Matter **2** (1990) 1853.

[5] P. Gaveau, J.P. Boucher, L.P. Regnault and Y. Henry in *Nonlinear Coherent Structures* , Eds. M. Barthes and J. Leon, p.141 (Springer-Verlag,1990).

P. Gaveau, J.P Boucher, L.P. Regnault and J.Y.Henry, to be published in J. Appl. Phys.

[6] L.P. Regnault and J. Rossat-Mignod, in *Magnetic Properties of Layered Transition Metal Compounds,* Ed. L.J De Jongh, p.271 (Kluwer Academic Publisher, 1990).

[7] D.J. Bishop and J.D. Reppy, Phys. Rev. Lett. **40** 1727 (1978)

[8] D.L. Huber, Phys. Rev. B **26** 3758 (1982)

[9] A.R. Völkel, G.M. Wysin, A.R. Bishop and F.G. Mertens, preprint.

[10] A. Bouvet et al., to be published.

THE ELECTRIC RESISTIVITY OF A MAGNETIC SEMICONDUCTOR WITH EASY-AXIS OF ANISOTROPY POPULATED BY MAGNON SOLITONS

M.V.Satarić[(1)] and J.A.Tuszynski[(2)]

1) Faculty of Technical Sciences, 21000 Novi Sad, Serbia, Yugoslavia

2) Department of Physics, The University of Alberta, Edmonton, Canada T6G 2J1

1. INTRODUCTION

We examine the possibilities of the formation of solitonlike magnon excitations in "quasi-one-dimensional (QOD) ferromagnets with "easy-axis" anisotropy (EAA). Such a possibility is indicated[(1)] under the constraint that the magnetic anisotropy energy is small compared to the direct exchange energy between magnetic ions. It seems us that good candidates which can support the existence of such solitons may be found among ferromagnetic semiconductors (FMS), first of all europium oxide (EuO).

Not a single FMS had been discovered until 1960; moreover an opinion had been voiced that ferromagnetic and semiconducting properties were incompatible. After discovering first FMS ($CsBr_3$) their number was growing and at present is close to 100. Nagaev[(2)] states the data showing that, for example EuO crystal possesses properties which allow it to be mapped approximately onto EAA chain. The EuO crystals have cubic, of NaCl type, structure. Their importante feature is the absence of an orbital angular momentum of the electrons in the partially filled f-shells of the Eu^{++} ions. The ground state of those f-shells is the $^8S_{7/2}$ with L=0, S=7/2. The magnetic dipole--dipole energy of ferromagnetically ordered spins in that cubic lattice is zero. Those are the reasons for the crystallographic anisotropy of such crystals to be relatively very small (the anisotropy field for EuO is equal to 0,02 T while the effective exchange field is of the order of 10^2T). On the other hand, such crystals appear to be almost ideal Heisenberg magnets[(2)].

The second, and main stage of our work is dedicated to examination of collisions of conduction electrons with magnon solitons in a FMS with EAA. It is well known that the existence of localized spin moments that couple to the conduction electrons (CE) has important consequences on the electrical conductivity of the corresponding crystals. The magnetic ions act as scattering centers so that at sufficiently low temperatures the scattering they cause will be the primary source of electrical resistance (ER).

Vonsovsky[(3)] first realized that an important contribution to ER could occur in ferromagnets as a result of the exchange interaction among CEs and the localized magnetic ions (LMI), often called the s-d or s-f interaction. Nagaev emphasizes[2] that the model proposed by Vonsovsky represents the most adequate description of FMSs. In his model the electrons in LMI's d or f shells interact with one another via the Heisenberg nearest-neighbor exchange mechanism while an entirely distinct subsystem exists which is composed of quasifree CEs in Bloch states of the conduction band (s). Since in this model the localized d and f electrons can be analized using a virtually identical treatment we will use the symbol "l" (localized) for both.

2. THE VONSOVSKY MODEL

Let us construct the s-l model Hamiltonian due to Vonsovsky

$$H_V = H_s + H_l + H_{sl} \ . \tag{1}$$

The first term on the right-hand side represents the noninteracting CEs described by the operators $a_{k\sigma}$, $a^+_{k\sigma}$ which destroy and create an electron with the wave number k and with up (↑) and down (↓) spin projection ($\sigma=\pm 1/2$), respectively

$$H_s = \sum_{k,\sigma} E_{k\sigma} a^+_{k\sigma} a_{k\sigma} \ . \tag{2}$$

The normalized energy of CEs, induding the Zeman energy due to an applied external magnetic field h (AEMF), and the zone shift caused by s-l interaction has the form

$$E_{k\sigma} = \frac{\hbar^2 k^2}{2m^*} + \sigma g_s m_B h - SW_{kk}\delta_{\sigma\downarrow} \tag{3}$$

where m* represents the effective mass of a CE; g_s is the Lande factor; m_B is the Bohr magneton; S is the spin of a single LMI; W_{kk} is the interaction energy of Vonsovsky type which arises from thirth term on the right hand side of Eq. (1).

The compounds of europium may serve as an example with a wide conduction band. In them the effective mass of CEs is of the order of free-electron mass m_0. On the basis of this, and taking into account that the lattice constant is of the order $R_0 \simeq 3\text{-}5\cdot 10^{-10}$ m, we may estimate that the bandwidth A in Eu chalogenides is of the order $A \simeq 3\text{-}5$ eV.

The second term on the right-hand side of Eq. (1) represents the interionic magnetic interaction which is of direct or indirect exchange type

$$H_l = -g_s m_B h \sum_n S_n^z - \frac{J_0}{4} \sum_n (S_n^+ S_{n+1}^- + S_n^- S_{n+1}^+) - \frac{J_0^z}{2} \sum_n S_n^z S_{n+1}^z \ , \tag{4}$$

where the spin ladder operators are $S_n^{\pm} = S_n^x \pm i S_n^y$ and the spin projection operator in the directon of AEMF is S_n^z. The energy parameters introduced above are the exchange interaction constant J_0 and anisotropic exchange interaction constant J_0^z.

The last term on the right-hand side of Eq. (1) describes the exchange coupling between the CEs and the spins of LMIs. In the single-band approximation this can be expressed in the direct space representation as follows

$$H_{s-l} = -\sum_{n,\sigma,\sigma'} W_{n\sigma\sigma'}(S_n \cdot s)_{\sigma\sigma'} a^+_{n\sigma'} a_{n\sigma} , \tag{5}$$

where the exchange energy $W_{n\sigma\sigma'}$ is of short-range type; $s_{\sigma\sigma'}$ are Pauli matrices, and $a^+_{n\sigma}$, $a_{n\sigma}$ are Fourier transforms of the operators a^+_k, a_k. The EuO has rather strong constant of interaction of the order W~0,1eV.

3. THE SOLITONLIKE BOUND STATE OF *N* MAGNONS

The formation of bound states of a few magnons in QOD ferromagnet with EAA was experimentally observed for the first time by Torrance and Tinkham(4). The basic theoretical discussion about necessary conditions for the formation of such bound states in the same crystals is given by Ivanov and Kosevich(1). The starting point in this discussion is the assumption about smallness of anisotropy, i.e. the relative magnitude of the energy $J^z_o - J_o$ in comparison with the exchange energy J_o. For EuO this condition is fairly fulfiled due to $(J^z_o - J_o)J^{-1}_o \simeq 10^{-3}$.

We used a quantum-mechanical method formulated in the space of coherent states(5) and our main results are congruent with those obtained in Ref. 1 in terms of a classical approach.

We start from Hamiltonian (4). The Holstein-Primakoff representation, which is adequate description due to S=7/2, allows us to go over from the spin operators to the magnon annihilation and creation operators, B_q, B^+_q. In such a way obtained Hamiltonian of the "magnon gas" in the k representation is given by the approximate expression

$$H_1 = \sum_q \varepsilon_q B^+_q B_q - N^{-1/2}(J^z_o - J_o) \sum_{q,q_1,q_2} B^+_{q+q_1} B^+_{q_2-q_1} B_q B_{q_2} \tag{6}$$

The energy of a one magnon state is defined in terms of the corresponding dispersion relation

$$\varepsilon_q = g_s m_B h + 2SJ^z_o - 2J_o S\cos(R_o q) . \tag{7}$$

The second term on the right-hand side of Eq. (6) represents the magnon-magnon attractive interaction. Even in the long-wave limit ($R_o q \to 0$) the mentioned interaction remains nonzero as a result of a non-zero spin-wave collision amplitude.

We formed the ansatz trial function as a product of Glauber's coherent states and after straightforward procedure by virtue of usuage the continuum approximation, Schrödinger equation with cubic nonlinearity (NLSE) appears.

Assuming that a fixed number of magnons N is involved in clusterization in terms of attractive magnon-magnon forces, we solved NLSE including the normalization condition $\int_{-\infty}^{+\infty} |\psi(z,t)|^2 dz = N$. The corresponding envelope of the clusterized "magnon drop" moves along the chain with velocity v in the form of a bell shaped soliton. After performing the Fourier transform of the bell shaped solution $\psi(z,t)$ we express the density of magnons with the wave vector q involved in a cluster as follows

$$\langle N_q \rangle = \frac{dN}{dq} = |\psi_\varepsilon(t)|^2 = N \frac{\pi^2}{2\mu} \operatorname{sech}^2\{\frac{\pi R_o}{2\mu}(k_s - q)\} \; . \tag{8}$$

The corresponding parameters have the following meanings: k_s is the solitonic quasi-wave number; μ represents the inverse solitonic domain of localization and $\hbar\omega_s$ is the energy of a cluster.

$$k_s = N \frac{\hbar v}{2J_o S R_o^2} \; ; \; \mu = N \frac{J_o^z - J_o}{2J_o S} \; ; \; \hbar\omega_s = E_o + \frac{1}{2} M^* v^2 \; . \tag{9}$$

Where the static energy of a cluster E_o and its effective mass M^* are expressed by

$$E_o = N\varepsilon_o \{1 - \frac{N^2 (J_o^z - J_o)^2}{12 J_o S \varepsilon_o}\} \; ; \; \varepsilon_o = \varepsilon_{q=o} \; ; \; M^* = N \frac{\hbar^2}{2SJ_o R_o^2} \; . \tag{10}$$

The expression (8) has basic importance because it will play the role of magnon population in the expressions containing the corresponding correlators in the theory which follows in the next section.

4. THE SCATTERING PROCESSES BETWEEN CEs AND MAGNON SOLITONS AND THEIR INFLUENCE ON FMS's CONDUCTIVITY

The method is based on the formulation of nonequilibrium density matrix first proposed by Zubarev[6]

$$\hat{\rho} = \{1 + \int_0^1 d\tau \{\exp(-\hat{M}\tau)\delta\hat{M} \exp(\hat{M}\tau)\}\}\hat{\rho}_{eq} \tag{11}$$

where operator $\hat{M}$ is proportional to the diagonal parts of Hamiltonian (2) and (3) including the thermodinamical forces

$$\hat{M} = \sum_q \gamma_q B_q^+ B_q + \sum_{k\sigma} \tilde{\gamma}_{k\sigma} a_{k\sigma}^+ a_{k\sigma} \; ; \; \gamma_q = \beta_2(\varepsilon_q - \mu_2) \; ; \; \tilde{\gamma}_{k\sigma} = \beta_1(E_{k\sigma} - \mu_1) \tag{12}$$

where μ_1 is the Fermi level of the conduction band, while μ_2 is the chemical potential of the magnon subsystem; τ is dimensionless parameter. The important assumption is that in a nonequilibrium state where the magnon population is generated by an external alternative field the subsystems are at slightly different temperatures[7]

$\beta_1=(k_BT_1)^{-1}\neq\beta_2=(k_BT_2)^{-1}$; T_1 correspones to the free electron gas while T_2 characterizes the magnon's gas. Operator $\delta\hat{M}$ in Eq. (11) involves the Vonsovsky interaction between subsystems

$$\delta\hat{M} = (\beta_2-\beta_1)\int_{-\infty}^{0} \exp(\theta t)\dot{H}_s(t)dt \; ; \; \dot{H}_s = \frac{1}{i\hbar}\left[H_s, H_v\right] \; ; \; \theta<<1 \; . \quad (13)$$

At last, $\hat{\rho}_{eq}$ represents the equilibrium statistical operator (NSO)

$$\hat{\rho}_{eq} = \exp(-\hat{M})\{Tr\{\exp(-\hat{M})\}\}^{-1} \; . \quad (14)$$

The main step in the procedure is to find the average value of the energy current between subsystems taking the average of expression (13) with respect to NSO

$$<\dot{H}_s> = Tr(\hat{\rho}\dot{H}_s) = (\beta_1-\beta_2)L_{12} \quad (15)$$

where L_{12} represents the kinetic coefficient of the scattering process

$$L_{12} = \int_{-\infty}^{0} dt\exp(\theta t)\int_{0}^{1} d\tau\{Tr\{\dot{H}_s(0)\exp(-\hat{M}\tau)\dot{H}_s(t)\exp(\hat{M}\tau)\hat{\rho}_{eq}\}\} \; . \quad (16)$$

Performing very cumbersome calculations with the simplifications enabled by the estimations of different parameters, and averaging the kinetic coefficient over all alowed velocities of "ideal gas" of solitons (IGS) we finally get the mean relaxation time for collisions of conducting electrons with IGS as a function of constant AEMF as follows

$$<\tau(h)> = \frac{<H_s^2>}{<L_{12}>} = \frac{a(\beta_1)+b(\beta_1)h+c(\beta_1)h^2}{m(\beta_1,\beta_2,\Delta T)+n(\beta_1,\Delta T)h} \quad (17)$$

where we used the symbols as follows

$$a(\beta_1) = f_1(\beta_1)+SWf_2(\beta_1)+4S^2W^2f_3(\beta_1)$$

$$b(\beta_1) = \frac{1}{4}\, g_s m_B f_2(\beta_1)+2SWg_s m_B f_3(\beta_1) \; ; \; C(\beta_1) = \frac{1}{4}\, g_s^2 m_B^2 f_3(\beta_s)$$

$$f_1(\beta_1) = \frac{3}{8}\frac{\eta}{\beta_1^2} \; ; \; f_2(\beta_1) = \frac{1}{4}\frac{\eta}{\beta_1} \; ; \; f_3(\beta_1) = \frac{1}{8}\,\eta \; ; \; \eta = \frac{R_o}{\sqrt{\pi}}\left(\frac{2m^*}{\beta_1\hbar^2}\right)^{1/2} \quad (18)$$

$$m(\beta_1,\beta_2\Delta T) = D\left(\frac{\hbar^4 N^2}{8J_o^2S^2R_o^4m^*M^*\beta_2} - SW\right) \; ; \; n(\beta_1,\Delta T) = \frac{1}{2}\, g_s m_B D$$

$$D = \frac{\pi^2SW^2R_oNk_BT_1^2}{\mu\mu_1\Delta T\hbar^2}\left(\frac{m^*M^*\beta_2}{\pi^2\beta_1}\right)^{1/2} x\,\mathrm{erf}\left(\frac{\pi^2\hbar^2\beta_1}{8m^*R_o^2}\right) \; ; \; \Delta T = T_2-T_1<<T_2 \; .$$

We make a semiquantitative estimation of the relaxation time by using the following set of available parameters: $R_o\sim3\cdot10^{-10}$ m; $m^*\sim10^{-30}$ kg, $T_1=10$ K, $\Delta T\sim1$K; $S=7/2$, $W\sim10^{-21}$J, $M^*\sim10^{-30}$ kg, $J_o\sim10^{-20}$ J, $N\sim5$, $\mu_1\sim\mu_2\sim10^{-19}$ J, $\mu\sim10^{-3}$.

On the basis of aforementioned estimations we obtain the following AEMF dependence of relaxation time

$$\langle\tau(h)\rangle \simeq \frac{1}{4}\left(\frac{6+3h+h^2}{1+h}\right)\cdot 10^{-8}\ \mathrm{s} \qquad (19)$$

which yealds that for $h \simeq 0{,}7$ T the minimum relaxation time has the value $\langle\tau\rangle_{min} \sim 1\cdot 10^{-8}$s, Fig. (1). The possible conclusions could be as follows. Supppose that by the method of parametric resonance by an alternating magnetic field[8] at radio frequencies the

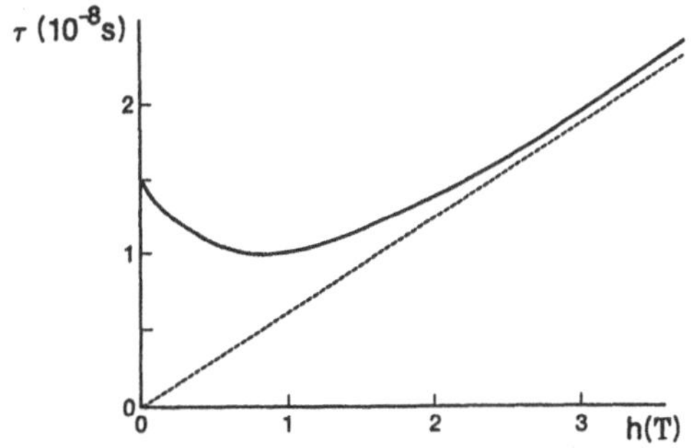

FIG. 1. The field dependence of the relaxation time corresponding to the conduction electron IGS at T_1=10 K and $\Delta T \sim 1$ K.

generation of coherent magnons leads to soliton formation. Having in mind that the scattering of CEs with phonons in the aforementioned temperature range is negligible, we expect that measurments of electrical resistivity of MS as a function of AEMF will reveal the features predicated by law presented by formula (17).

REFERENCES

1° B. Ivanov and A.M. Kosevich, Zh. Eksp. Teor. Fiz 72, 2000 (1977) |Sov. Phys. JETP 45, 1050 (1977)|
2° E.L. Nagaev, Physics of Magnetic Semiconductors (MiR Moscow, 1983) English translation
3° S.V. Vonsovski, Zh. Eksp. Teor. Fiz 18, 219 (1948)
4° J. Torrance and M. Tinkham, J. Appl. Phys. 39, 822 (1968)
5° M. Satarić and J.A. Tuszynski, Phys. Rev. B, 43, 8450 (1991)
6° D.N. Zubarev, Nonequilibrium Statistical Thermodynamics (Nauka, Moscow, 1971) in Russian
7° Yu.A. Izumov and M.V. Medvedev, Zh. Eksp. Teor. Fiz 59, 553 (1970) |Sov. Phys. JETP 32, 302 (1971)|
8° V.S. L'vov, in Solitons in Modern Problems in Condensea Matter Sciences (North--Holand, Amsterdam 1986), vol 17, Chap 5.

THERMODYNAMICS of QUANTUM SPIN CHAINS

Alessandro Cuccoli[a], *Valerio Tognetti*[a], *Ruggero Vaia*[b], and *Paola Verrucchi*[a]
[a] Dipartimento di Fisica, Universitá di Firenze, Largo E. Fermi 2, 50125 Firenze, Italy.
[b] Istituto di Elettronica Quantistica C.N.R., Via Panciatichi 56/30, 50127 Firenze, Italy.

1. Introduction and summary.

In this paper we consider the ferromagnetic chain with easy-plane anisotropy D and in-plane applied Zeeman field H. Its quantum hamiltonian reads:

$$\hat{\mathcal{H}} = \sum_{i=1}^{N} \left[-J\hat{\mathbf{S}}_i \cdot \hat{\mathbf{S}}_{i+1} + D\big(\hat{S}_i^z\big)^2 - g\mu_B H \hat{S}_i^x \right] \quad , \tag{1.1}$$

with $\left|\hat{\mathbf{S}}\right|^2 = S(S+1)$. This hamiltonian is believed to have a physical realization in the real system $CsNiF_3$, with $S = 1$, $J = 23.6$ K, $D = 9$ K, $g = 2.4$.

For strong planar anisotropy, the classical counterpart of (1.1) can be approximated by the so-called planar model, which in the continuum limit becomes the sine-Gordon (SG) model. The latter admits soliton excitations contributing to the thermodynamics. This has lead to the search for soliton-like behavior in the easy-plane ferromagnetic chain, both theoretically [1,2] and experimentally [3–5]. Indeed, experimental measurements [4] of the magnetic contribution to the specific heat in $CsNiF_3$ have risen many attempts to relate the peak in the *excess* specific heat (i.e. the difference between the specific heat with and without applied field H) with the existence of SG-like nonlinear excitations [5,6].

However, as it has been clearly shown by Pini and Rettori [7], the thermodynamics of $CsNiF_3$ requires a quantum mechanical treatment, especially in order to account for the relevant quantum character of the out-of-plane fluctuations [5–10]. Quantum corrections to the classical SG thermodynamics have been introduced [5,8,9], and modified SG-like theories have been developed [6,10] in order to include a better description of the out-of-plane part.

In this paper we put forward a new approach. We start by observing that the quantum hamiltonian (1.1) can be expressed in terms of canonical operators by means of the Villain transformation [11]. Then, in section 2 we show how to approximate the general path-integral formula for the partition function [12,13] by using a nonlocal quadratic "trial" action functional, containing a number of variational parameters which are functions defined in phase-space. These parameters can be at best determined after a variational principle, and the approximated path-integral can be put into the classical form of a phase-space integral, allowing us to define an effective hamiltonian. In order to get simple and useful expressions, we then introduce an approximation of "low quantum coupling", which, however, preserves an exact description of the quantum behavior of the harmonic part — even at lowest temperatures, where the results of the selfconsistent gaussian approximation are actually recovered —, as well as of the classical nonlinear behavior — with quantum renormalizations which, in the high temperature limit, do agree with the results of the Wigner method —.

In section 3 we apply the above framework to the (Villain transformed) hamiltonian (1.1). We show that the corresponding effective hamiltonian can be recast, in terms of classical spin variables, into the same form of the original hamiltonian, with suitably renormalized parameters, but for the appearance of an exchange anisotropy term. Then we use the effective hamiltonian in numerical transfer matrix calculations, and report some quantum results for the specific heat, showing their fairly good quantitative agreement with the available experimental data for $CsNiF_3$, as well as with computational data by other authors [14]. In addition, we also compare with the results of the quantum planar and SG models, for which a simplified version of the variational method leads to the corresponding effective potentials [9,10], showing that the nonlinearity of the exchange term (which is neglected in SG) plays a relevant role. We conclude that the nonlinear thermodynamic behavior of the spin chain only qualitatively can be explained in terms of SG solitons.

2. The effective hamiltonian.

Let $\hat{\mathcal{H}}(\hat{p},\hat{q})$, be the hamiltonian operator of a quantum mechanical system with N degrees of freedom. We use the matrix notation $\hat{p} = \{\hat{p}_i\}_{i=1,\dots,N}$ and $\hat{q} = \{\hat{q}_i\}_{i=1,\dots,N}$ for its momentum and coordinate operators, which satisfy canonical commutation relations $[\hat{p}_i,\hat{q}_j] = i\hbar\delta_{ij}$.

The equilibrium partition function $\mathcal{Z}$ of this system at the temperature $T = \beta^{-1}$ can be expressed as a functional integral over paths $\{p(u), q(u)\}$ in phase-space:

$$\mathcal{Z} \equiv e^{-\beta F} = \int \mathcal{D}[p(u)] \int_{q(0)=q(\beta)} \mathcal{D}[q(u)]\; e^{\,S[p(u),q(u)]} \quad , \tag{2.1}$$

with the action functional

$$S[p(u),q(u)] = \int_0^{\beta\hbar} \frac{du}{\hbar} \left[ip^t\dot{q} - \mathcal{H}(p,q)\right] \quad . \tag{2.2}$$

In order to give a mathematical meaning [12] to the above formulas, one has to define the function $\mathcal{H}(p,q)$, which is associated with the hamiltonian operator $\hat{\mathcal{H}}(\hat{p},\hat{q})$ by some ordering rule (e.g. the p-q order [12,13], obtained by moving the $\hat{p}_i$ on the left of the $\hat{q}_i$), and an additional condition on the $p^t\dot{q}$ term (which, in the case of p-q order, reads $p^t\dot{q} \to p^t(u)\dot{q}(u+0)$).

If we are able to calculate path-integrals with a trial action $S_0[p(u), q(u)]$, containing a proper set of variational parameters [13], it is possible to minimize the r.h.s. of the so-called Feynman-Jensen inequality:

$$F \leq F_0 + \beta^{-1}\langle S_0 - S\rangle_{S_0} \quad , \tag{2.3}$$

in order to approximate (2.1). The functional average $\langle \cdot \rangle_{S_0}$ appearing in (2.3) corresponds to the path integral (2.2) with the integrand functional e^{S_0} multiplied by $S_0 - S$, divided by $\mathcal{Z}_0 = \exp(-\beta F_0)$.

The trial action we will use has the following general form:

$$S_0[p(u), q(u)] = \int_0^{\beta\hbar} \frac{du}{\hbar} \left[i p^t \dot{q} - w - \frac{1}{2}\delta p^t \, \mathbf{A}^2 \, \delta p - \delta p^t \, \mathbf{X} \, \delta q - \frac{1}{2}\delta q^t \, \mathbf{B}^2 \, \delta q \right] \quad , \tag{2.4}$$

where $\delta p \equiv p(u) - \bar{p}$ and $\delta q \equiv q(u) - \bar{q}$, and the functionals

$$(\bar{p}, \bar{q}) = \int_0^{\beta\hbar} \frac{du}{\beta\hbar} \, (p(u), q(u)) \tag{2.5}$$

represent the ***average point*** of each path $(p(u), q(u))$. Moreover, the real c-number w and the real $N \times N$ matrices **A**, **X** and **B** are functions of $(\bar{p}, \bar{q})$, and are considered as the variational parameters of the trial action. Of course, they are subjected to proper constraints (for instance, **A** and **B** are symmetric). In addition, the matrix $\mathbf{\Sigma} \equiv \mathbf{A}^{-1}\mathbf{X}\mathbf{A}$ is constrained to commute with the matrix **B**, which makes the path integrals with S_0 appearing in (2.3) of easy evaluation. The final result for $\mathcal{Z}_0$ can be put in the classical form of a phase space integral over $\bar{p}$ and $\bar{q}$ (in the forthcoming formulas we will suppress the bars).

The calculation leads to the formal diagonalization problem of the matrices $\mathbf{\Sigma}$ and $\mathbf{\Omega}^2 \equiv \mathbf{A}\mathbf{B}^2\mathbf{A} - \mathbf{\Sigma}^t\mathbf{\Sigma}$, which are diagonalized by the orthogonal matrix $\mathbf{U}(\bar{p}, \bar{q})$:

$$\sigma_k \, \delta_{kl} = \left(\mathbf{U} \, \mathbf{\Sigma} \, \mathbf{U}^t\right)_{kl} \quad , \qquad \omega_k^2 \, \delta_{kl} = \left(\mathbf{U} \, \mathbf{\Omega}^2 \, \mathbf{U}^t\right)_{kl} \quad , \tag{2.6}$$

and we can equivalently consider as variational parameters to be determined by (2.3) the independent components of the matrices $\mathbf{A}(p,q)$ and $\mathbf{U}(p,q)$, the sets of eigenvalues $\{\omega_k^2(p,q)\}$ and $\{\sigma_k(p,q)\}$, and the function $w(p,q)$. The minimization of the r.h.s. of (2.3) with respect to w yields the vanishing of $\langle S - S_0\rangle_{S_0}$, so that we are allowed to define an effective hamiltonian $\mathcal{H}_{\rm eff}(p,q)$ by which the approximated partition function can be written in the classical form

$$\mathcal{Z}_0 \equiv e^{-\beta F_0} = \int \frac{dp \, dq}{2\pi\hbar} \, e^{-\beta \mathcal{H}_{\rm eff}(p,q)} \quad , \tag{2.7}$$

$$\mathcal{H}_{\rm eff}(p,q) = e^{\langle\!\langle \phi^2 \rangle\!\rangle} \, \mathcal{H}(p,q) - \sum_k \omega_k^2 \, \alpha_k + \frac{1}{\beta} \sum_k \ln \frac{\sinh f_k}{f_k} \quad , \tag{2.8}$$

where

$$f_k(p,q) = \frac{1}{2}\beta\hbar\omega_k \, , \qquad \alpha_k(p,q) = \frac{\hbar}{2\omega_k}\left(\coth f_k - f_k^{-1}\right) \quad . \tag{2.9}$$

In eq. (2.8) $\mathcal{H}(p,q)$ represents the function associated with $\hat{\mathcal{H}}$ by Weyl ordering [12], also called the "Weyl symbol for $\hat{\mathcal{H}}$":

$$\mathcal{H}(p,q) = \int ds \Big\langle q - \frac{s}{2} \Big| \hat{\mathcal{H}}(\hat{p},\hat{q}) \Big| q + \frac{s}{2} \Big\rangle \, e^{ip^t s/\hbar} \quad , \tag{2.10}$$

and the differentiation operator $\not{\partial} \equiv (\eta^t \partial_p + \xi^t \partial_q)/\sqrt{2}$ acts on the arguments of $\mathcal{H}(p,q)$. The gaussian average $\langle\!\langle \cdot \rangle\!\rangle$, which operates on the fluctuation variables $\{\eta_i\} = \eta = \mathbf{A}^{-1}\mathbf{U}^t\tilde{\eta}$ and $\{\xi_i\} = \xi = \mathbf{A}\mathbf{U}^t\tilde{\xi}$, is uniquely defined by the second moments of their transformed components $\{\tilde{\eta}_k\}$ and $\{\tilde{\xi}_k\}$:

$$\langle\!\langle \tilde{\eta}_k^2 \rangle\!\rangle = (\omega_k^2 + \sigma_k^2)\alpha_k \quad , \qquad \langle\!\langle \tilde{\eta}_k \tilde{\xi}_k \rangle\!\rangle = -\sigma_k \alpha_k \quad , \qquad \langle\!\langle \tilde{\xi}_k^2 \rangle\!\rangle = \alpha_k \quad . \tag{2.11}$$

The further minimization of F_0 with respect to the remaining parameters gives their determination as

$$\begin{aligned} \mathbf{A}_{ij}^2 &= e^{\langle\!\langle \not{\partial}^2 \rangle\!\rangle} \, \mathcal{H}_{p_i p_j}(p,q) \quad , \\ (\omega_k^2 + \sigma_k^2)\, \delta_{kl} &= \sum_{ij} (\mathbf{UA})_{ki} \, (e^{\langle\!\langle \not{\partial}^2 \rangle\!\rangle} \, \mathcal{H}_{q_i q_j}(p,q)) \, (\mathbf{UA})_{lj} \quad , \\ \sigma_k \, \delta_{kl} &= \sum_{ij} (\mathbf{UA}^{-1})_{ki} \, (e^{\langle\!\langle \not{\partial}^2 \rangle\!\rangle} \, \mathcal{H}_{p_i q_j}(p,q)) \, (\mathbf{UA})_{lj} \quad . \end{aligned} \tag{2.12}$$

These are coupled self-consistent secular equations, the first of which determines the "reciprocal mass" matrix $\mathbf{A}^2$. The subscripts of $\mathcal{H}$ denote the corresponding derivatives.

The above formalism, when applied to a quadratic hamiltonian (with the "mixed" term satisfying the above mentioned constraint of commutativity), gives the exact partition function. Indeed the variational parameters turn out to be constant, and equal to the corresponding coefficients in the starting hamiltonian, so that the partition function turns out to be the fully quantum one, $\mathcal{Z} = \sum_k \left(2\sinh f_k\right)^{-1}$, thanks to the logarithmic term appearing in $\mathcal{H}_{\text{eff}}$.

However, it is generally very hard to solve the above self-consistent equations for any phase-space point, a task that presumably will numerically last as a heavy quantum Monte Carlo simulation. Therefore a further simplification is in order. If the operator $e^{\langle\!\langle \not{\partial}^2 \rangle\!\rangle}$ only slightly affects the Weyl hamiltonian, we can rederive the "low coupling approximation" (LCA) that we have extensively used in the standard case of the effective potential [9,13] .

The LCA consists in expanding the variational parameters around their values in the self-consistent minimum (p_0, q_0) of $\mathcal{H}_{\text{eff}}(p,q)$, in such a way that $\mathcal{H}_{\text{eff}}$ is correct within terms of order α_k^2. The simplest form of the LCA occurs in the case of translationally invariant systems, for which the matrix $\mathbf{U}(p_0, q_0)$ is nothing else but a standard real Fourier transformation, which also diagonalizes the reciprocal mass matrix:

$$(\mathbf{U}\mathbf{A}^2\mathbf{U}^t)_{kl} = m_k^{-1} \, \delta_{kl} \quad . \tag{2.13}$$

Here and in the following the dependence on (p_0, q_0) is understood. The LCA effective hamiltonian then reads

$$\mathcal{H}_{\rm eff}(p,q) = e^{\langle\!\langle \not{\partial}^2 \rangle\!\rangle}\, \mathcal{H}(p,q) - \langle\!\langle \not{\partial}^2 \rangle\!\rangle\, e^{\langle\!\langle \not{\partial}^2 \rangle\!\rangle}\, \mathcal{H}(p_0,q_0) + \frac{1}{\beta}\sum_k \ln\frac{\sinh f_k}{f_k} \quad , \tag{2.14}$$

and the variational parameters can be obtained by Fourier transformig the second derivatives of $\mathcal{H}_{\rm eff}$

$$\begin{aligned} m_k^{-1} &= \sum_{ij} U_{ki}U_{kj}\partial_{p_i}\partial_{p_j}\mathcal{H}_{\rm eff}(p,q)\Big|_{p_0,q_0} \quad ,\\ \sigma_k &= \sum_{ij} U_{ki}U_{kj}\partial_{p_i}\partial_{q_j}\mathcal{H}_{\rm eff}(p,q)\Big|_{p_0,q_0} \quad ,\\ m_k\left(\omega_k^2+\sigma_k^2\right) &= \sum_{ij} U_{ki}U_{kj}\partial_{q_i}\partial_{q_j}\mathcal{H}_{\rm eff}(p,q)\Big|_{p_0,q_0} \quad . \end{aligned} \tag{2.15}$$

It turns out that the first term of the effective hamiltonian is the result of a broadening of the Weyl hamiltonian, on a scale given by eqs.(2.11) for each "normal mode". Eqs.(2.11) represent the pure quantum contributions (i.e. the quantum ones minus the corresponding classical) to the square fluctuations of the canonical coordinates for a hamiltonian corresponding to the quadratic approximation to $\mathcal{H}_{\rm eff}$. Finally, we note that the high temperture limit of $\mathcal{H}_{\rm eff}$ concides with the Wigner effective hamiltonian, so that the Weyl ordered hamiltonian $\mathcal{H}$ represents the *well-done* classical limit.

3. Effective hamiltonian of the spin chain.

Exploiting the easy-plane character of the system (1.1), we perform, for each spin operator $\hat{\mathbf{S}}$, the quantum Villain transformation [11] to canonical coordinate operators $\hat{\mathbf{S}} \longrightarrow \{\hat{S}^z, \hat{\varphi}\}$, satisfying $[\hat{\varphi}, \hat{S}^z] = i$:

$$\hat{S}^+(\hat{S}^z,\hat{\varphi}) = e^{i\hat{\varphi}}\sqrt{S(S+1)-\hat{S}^z(\hat{S}^z+1)} \quad , \qquad \hat{S}^- = (\hat{S}^+)^\dagger \quad . \tag{3.1}$$

The Villain transformation allows us to use the variational method described in the preceding section. The Weyl symbol for $\hat{S}^\pm(\hat{S}^z,\hat{\varphi})$ is easily found to be:

$$S^\pm(S^z,\varphi) = \sqrt{\left(S+\frac{1}{2}\right)^2 - (S^z)^2}\; e^{\pm i\varphi} \quad , \tag{3.2}$$

It follows that it is natural to scale the "momenta" $\hat{S}_i^z$ with $\widetilde{S} = S + \frac{1}{2}$, defining $\hat{p} = \hat{S}^z/\widetilde{S}$. Then $[\hat{\varphi}, \hat{p}] = i/\widetilde{S}$, and $\widetilde{S}^{-1}$ plays the role of $\hbar$. Eventually, the Weyl symbol for $\hat{\mathcal{H}}(\hat{p},\hat{\varphi})$ reads

$$\mathcal{H}(p,\varphi)= J\widetilde{S}^2 \sum_i \left[-p_ip_{i+1}-\sqrt{1-p_i^2}\sqrt{1-p_{i+1}^2}\cos\left(\varphi_i-\varphi_{i+1}\right) + \frac{p_i^2}{2\gamma}-h\sqrt{1-p_i^2}\cos\varphi_i\right] , \tag{3.3}$$

where $\gamma = J/(2D)$ and $h = g\mu_B H/(J\widetilde{S})$. This hamiltonian is a "classical counterpart" of (1.1), with the unusual correspondence of the quantum spin operators $\hat{\mathbf{S}}$ to classical unit vectors of length $\widetilde{S} = S+1/2$. For low values of the spin S this is, in our opinion, a sensible classical counterpart of (1.1), at difference with the usual *naïve* procedure [3,7].

In the present case the "skew" variational parameters $\sigma_k = 0$, and only the following three quantum renormalization parameters turn out to be relevant

$$\begin{aligned} D_p &\equiv \langle\!\langle \eta_i^2 \rangle\!\rangle = \frac{\gamma\lambda}{N}\sum_k \frac{b_k}{2a_k}\left(\coth f_k - \frac{1}{f_k}\right) \quad , \\ D_\varphi &\equiv \langle\!\langle \xi_i^2 \rangle\!\rangle = \frac{\lambda}{N}\sum_k \frac{a_k}{2b_k}\left(\coth f_k - \frac{1}{f_k}\right) \quad , \\ D_{\delta\varphi} &\equiv \langle\!\langle \left(\xi_i-\xi_{i-1}\right)^2 \rangle\!\rangle = \frac{\lambda}{N}\sum_k 4\sin^2\frac{k}{2}\,\frac{a_k}{2b_k}\left(\coth f_k - \frac{1}{f_k}\right) \quad . \end{aligned} \tag{3.4}$$

The usual field theoretic definition of the coupling constant is $\lambda = \left(\gamma\widetilde{S}^2\right)^{-1/2}$, and $f_k = \lambda a_k b_k/(2t)$, where $t = T/\left(J\widetilde{S}^2\right)$ is the dimensionless reduced temperature and

$$a_k^2 = 1+\gamma\left[\widetilde{h} - 2\left(1-e^{-D_{\delta\varphi}/2}\right) + 4\sin^2\frac{k}{2}\right] , \quad b_k^2 = \delta^2\left[\widetilde{h} + 4\tau\sin^2\frac{k}{2}\right] \quad . \tag{3.5}$$

Moreover, $\widetilde{h} = h\, e^{-D_\varphi/2}$, $\delta^2 = 1 - D_p/2$ and $\tau = \delta^2\, e^{-D_{\delta\varphi}/2}$.

If we neglect, in view of the easy-plane character and according to the low-coupling approximation, terms of the order D_p^2 , the resulting effective hamiltonian can be written for classical spin variables $\mathbf{s}_i$ $\left(\left|\mathbf{s}_i\right|^2 = 1 \ , \ s_i^z = p_i/\delta\right)$ in the same form of (1.1), but for the appearance of a further anisotropy term in the exchange:

$$\begin{aligned} \frac{\mathcal{H}_{\rm eff}}{J\widetilde{S}^2} &= \delta^2 \sum_{i=1}^N \left[-s_i^z s_{i+1}^z - \tau\left(s_i^x s_{i+1}^x + s_i^y s_{i+1}^y\right) + \frac{\left(s_i^z\right)^2}{2\gamma} - \widetilde{h} s_i^x\right] + t\sum_k \ln\frac{\sinh f_k}{f_k} + \Delta, \\ \Delta &= \frac{N}{2}\left[\widetilde{h}\left(D_\varphi + D_p - D_p D_\varphi/2\right) + e^{-D_{\delta\varphi}/2}\left(D_{\delta\varphi} + 2D_p - D_p D_{\delta\varphi}\right)\right] + t\ln\delta \ . \end{aligned} \tag{3.6}$$

If we let $\lambda \to 0$, keeping γ fixed (which corresponds to the limit $S \to \infty$), the Weyl symbol $\mathcal{H}$ is recovered.

The model parameters characteristic of $CsNiF_3$ correspond to the value of $\gamma = 1.3$. The low value of the spin $S = 1$ yields a rather high value of the coupling parameter, $\lambda = 0.58$, which is at the limit of reliability of the above framework. The evaluation of the thermodynamic quantities of $CsNiF_3$ has been done by the classical transfer matrix method, using the effective hamiltonian (3.6). The convergence of the method at lowest temperatures has been checked and connected with the corresponding self-consistent gaussian approximation.

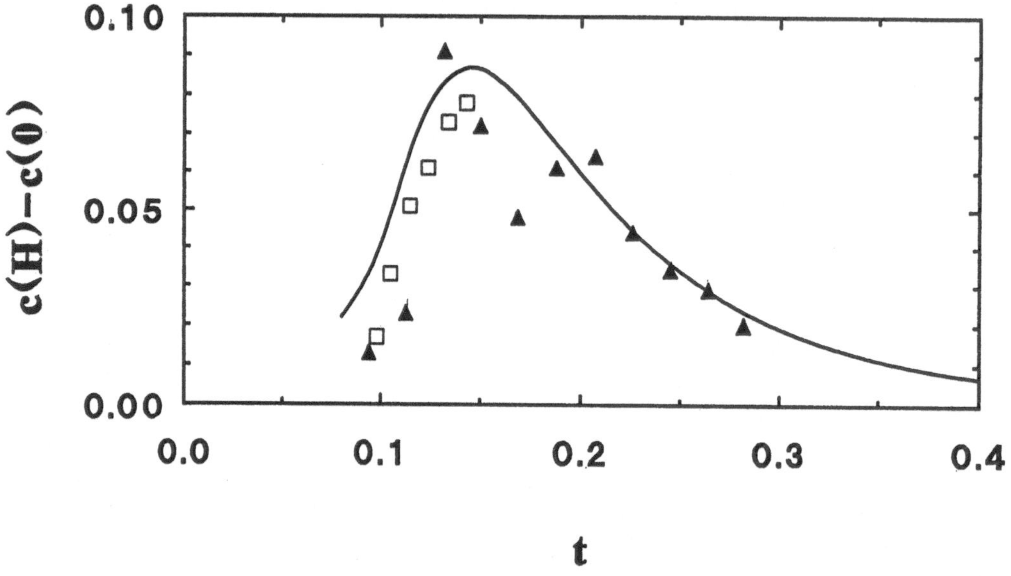

Figure 1 — *Excess specific heat per site (in units of* k_B*) versus the reduced temperature* $t = T/(J\widetilde{S}^2)$. *Model parameters characteristic of* $CsNiF_3$, *to which the values of* $\gamma = 1.3$, $\lambda = 0.58$ *do correspond. Field:* $H = 5$ kG ($h = 0.022$). *Solid line: quantum result. Squares: experimental data (ref.* 4). *Triangles: quantum Monte Carlo data (ref.* 14).

In the figures we report data for the *excess* specific heat per site, i.e. the difference between the specific heat and its value for zero Zeeman field. Indeed, the experimental measurements in $CsNiF_3$ made by Ramirez and Wolf [4] do report this difference, since it is not affected by the lattice contribution. In figure 1 the results of our calculation by use of the effective hamiltonian (3.6) are reported versus the reduced temperature $t = T/(J\widetilde{S}^2)$, and compared with the experimental data and with quantum Monte Carlo results by Wysin and Bishop [14]. The good agreement witnesses both the usefulness of the method and the correctness of the interaction parameters used for $CsNiF_3$.

In figure 2 the same results of the variational method are compared with the quantum results of the planar model and of the sine-Gordon model. Although the latter were calculated previously [10], we note that here we have used the correct procedure of approximating the *quantum* (Weyl ordered) hamiltonian to these two limits models, which gives rise to the same coupling parameter $\lambda = 0.58$ reported above. The quantum calculation has been performed within the framework of the variational method, using the corresponding effective potential. It is apparent that the anharmonicity of the in-plane exchange, which is retained by the planar model, is important, but is not enough to explain the full nonlinear contribution to the thermodynamics of the easy-plane ferromagnetic chain. The necessity of using a realistic model, which takes into account together out-of-plane fluctuations and anharmonic exchange is apparent.

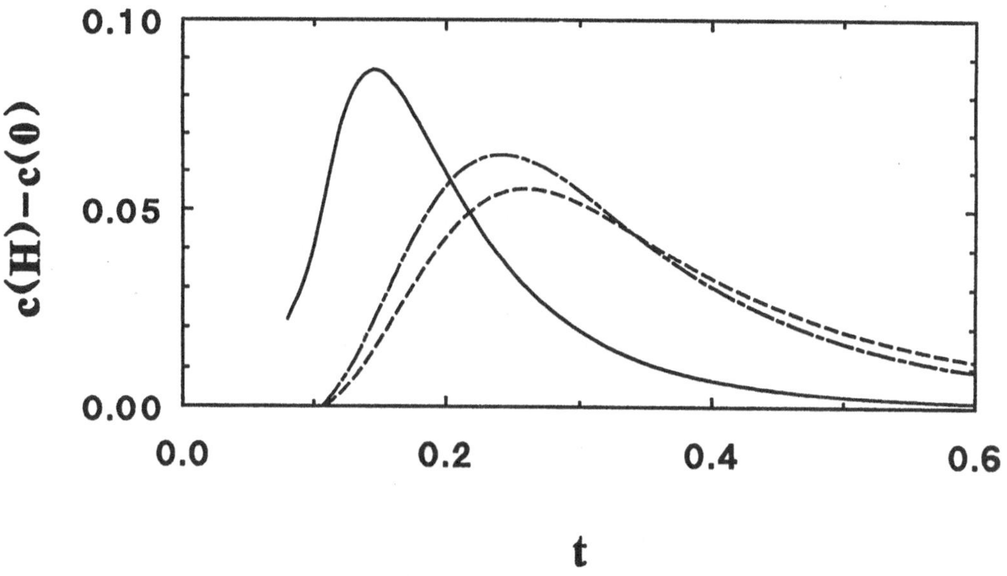

Figure 2 — *Quantum results for the excess specific heat. Units and model parameters as in figure 1. Dashed line: sine-Gordon approximation. Dash-dotted line: planar approximation. Solid line: full effective hamiltonian.*

References.

1. *H.J. Mikeska*, J. Phys. **C11**, L29 (1978).
2. *P.S. Riseborough* and *S.E. Trullinger*, Phys. Rev. **B22**, 4389, (1980).
3. *J.K. Kjems* and *M. Steiner*, Phys. Rev. Lett. **41**, 1137 (1978).
4. *A.P. Ramirez* and *W.P. Wolf*, Phys. Rev. Lett. **49**, 227 (1982); Phys. Rev. **B32**, 1639 (1985).
5. *M.D. Johnson* and *N.F. Wright*, Phys. Rev. **B32**, 5798 (1985).
6. *H.J. Mikeska* and *H. Frahm*, J. Phys. **C19**, 3203 (1986).
7. *M.G. Pini* and *A. Rettori*, Phys. Rev. **B29**, 5246 (1984).
8. *R. Giachetti* and *V. Tognetti*, Phys. Rev. Lett. **55**, 912 (1985).
9. *R. Giachetti*, *V. Tognetti* and *R. Vaia*, Phys. Rev. **A37**, 2165 (1988)
10. *R. Giachetti* and *V. Tognetti*, Phys. Rev. **B37**, 5512 (1987).
11. *J. Villain*, J. de Physique **35**, 27 (1974).
12. *F.A. Berezin*, Sov. Phys. Usp. **23**, 763 (1980).
13. *R. Giachetti*, *V. Tognetti* and *R. Vaia*, in *Path integrals from meV to MeV*, eds. *V. Sa-yakanit* et. al., (Word Scientific, Singapore, 1989).
14. *G.M. Wysin* and *A.R. Bishop*, Phys. Rev. **B34**, 3377 (1986).

DISSIPATIVE SUPERLUMINOUS BRILLOUIN SOLITONS IN AN OPTICAL-FIBER RING CAVITY

Carlos Montes, Eric Picholle, Jean Botineau
Olivier Legrand, and Claude Leycuras

Laboratoire de Physique de la Matière Condensée, (U.R.A. - C.N.R.S. N° 190)
Université de Nice-Sophia Antipolis, Parc Valrose, F-06034 Nice CEDEX, France.

Abstract : Generation of large-scale spatio-temporal coherent structures caused by stimulated Brillouin backscattering of a narrow-band laser wave in a large-gain one-dimensional nonlinear medium is studied by comparing the numerical simulations and the analytical asymptotics of the three-wave resonant model to actual experiments in a single-mode optical-fiber. This comparison recently allowed us to predict [1] and to perform the first experimental observation [2] of the "superluminous" Brillouin soliton backward propagating with respect to the cw pump in an optical-fiber ring-cavity.

I Introduction

Stimulated Brillouin scattering (SBS) is the dominant stimulated scattering process in many optical media, and particularly in optical fibers, pumped by single-frequency lasers [3] [4]. Considered as a detrimental effect for optical communications and above all for laser-plasma fusion experiments, where it is responsible of reflecting a large fraction of the laser energy, an important aim has been to limit its efficiency [5]. However, it allows also to achieve efficient pulse amplification and compression since 1968 [6].

The spatio-temporal SBS kinetics is well described by a three-wave nonlinear resonant interaction [7]. Our purpose has been to better understand the time-dependent SBS generation of large-scale coherent structures by comparing the numerical simulations and the analytical asymptotics of the three-wave coherent model to actual experiments in an optical-fiber.

In the one-dimensional problem, the forward-propagating pump wave (at frequency $\omega_p = k_p c/n_0$) couples with the thermal phonon fluctuations of the material medium (at frequency $\omega_a \simeq 2c_a\omega_p n_0/c$, where c_a is the acoustic velocity and n_0 the unperturbed refractive index) and stimulates a counterpropagating Stokes wave (at frequency $\omega_S = \omega_p - \omega_a$) downshifted by the acoustic frequency (ω_a). The acoustic wave is in turn amplified by electrostriction. The resonant condition for the three-wave coherent interaction ($\omega_p = \omega_S + \omega_a$) provides maximum power transfer when the wave-vector mismatch is zero ($\mathbf{k}_p = \mathbf{k}_S + \mathbf{k}_a$; $\Rightarrow k_a = k_p + k_S \simeq 2k_p$). Thus, assuming slowly varying envelopes for the waves, neglecting optical dispersion, and respectively denoting the complex amplitude E_p for the pump wave, E_S for the counterpropagating Stokes wave and E_a for the acoustic wave, the three-wave equations read as follows in a coherent dimensionless form [8] [9] :

$$(\partial_t + \partial_x + \mu_e)E_p = -E_S E_a$$
$$(\partial_t - \partial_x + \mu_e)E_S = E_p E_a^* \quad (1)$$
$$(\partial_t + \mu_a)E_a = E_p E_S^*,$$

where the acoustic velocity has been neglected, due to its smallness relative to light velocity (normalized to unity), and where μ_e (μ_a) is the damping coefficient for the optical (acoustic) wave. Moreover, additive terms in Eqs.(1) allow our numerical simulations to account for phase modulation due to optical Kerr contribution [9].

We have carried out numerical and asymptotical studies of the coherent space-time dependent kinetics, with actual experiments in an optical-fiber ring-cavity, which lead us to consider two main SBS time dependent regimes : (i) Stokes pulse amplification and compression under nonstationary conditions [8] [9], and (ii) generation of a backscattered "superluminous" Stokes soliton accompanied by self-induced transparency for the pump [1] [2]. Let us mention two previous papers [10] [11], closely related to the theoretical and numerical models, though some confusion appears between regimes (i) and (ii).

The compression regime (i) takes place when two separated pump and Stokes pulses interact in a counter-propagating motion. The Stokes pulse behaves as a shock-wave, the amplification of which depleting the pump wave. The interaction amplifies the acoustic wave by electrostriction starting from the Stokes leading edge and propagating with the Stokes pulse in its backward traveling motion. The Stokes amplification strongly depends on the shape of its leading edge, but for long enough times it takes the shape of a "π-pulse" self-similar profile [12] whose leading maximum amplitude grows linearly in time while its width decreases as the inverse of time [8]. It is difficult to identify this asymptotic stage in an actual experiment, since we must control the initial Stokes pulse. A stable train of compressed Stokes pulses (to $\sim$ 10 ns) has been obtained in a stimulated Brillouin fiber ring laser (of length L = 83 m), similar to that shown in figure 1, but where the acousto-optic modulator (AOM) was put inside the ring-cavity,

Fig. 1 *Experimental setup.*

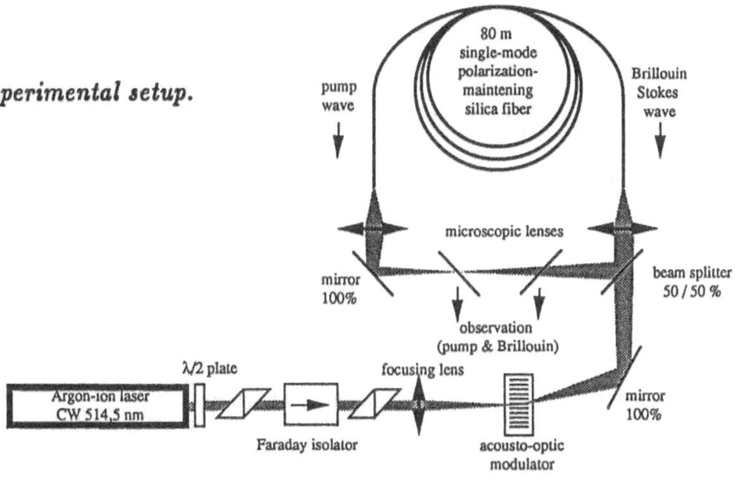

and the Ar-ion cw pump beam, at $\lambda_p = 514.5$ nm, was coupled into the fiber through the first Bragg order of the AOM [9]. By periodically interrupting the pump beam and the Stokes leading edge at the round-trip flight time $\Delta t_L = Ln_0/c \simeq 420$ ns, we periodically repeated the compression experience. The experimental time scale for the generated Stokes pulse width ($\simeq 10$ ns) is comparable to the spontaneous acoustic decay time in silica ($\simeq 7$ ns deduced from a spectral Brillouin bandwidth of $\Delta\nu = 150$ MHz at a pump wavelength $\lambda_p = 514.5$ nm) [13] and therefore absolutely calls for the use of the coherent description given by Eqs.(1). These studies have allowed us to better precise the experimental configuration for obtaining the second regime, which is the object of this paper.

II. Dissipative Brillouin soliton

Let us look for soliton solutions of Eqs.(1) which involve the coherent dynamics of a self-similar backward-propagating three-wave (pump-Stokes-acoustic) structure. For the three-wave interaction (1) in the nondissipative case (i.e. $\mu_e = \mu_a = 0$), specific traveling wave solutions have been studied in the forward-scattering case [14] - [16]. Corresponding solutions in the SBS case display new interesting features that persist in the dissipative case [10]. These solutions are special cases of a broader class of solitons previously described through inverse scattering transform [17] [18].

However, in the case of silica optical fibers, it is not possible to neglect the dampings ($\mu_i \neq 0$). Analytic traveling-wave solutions are still available if the pump attenuation is neglected, which is locally legitimate as long as $\mu_e/\mu_a \ll 1$; it is indeed the case in our experiment. Starting from Eqs.(1) we first perform the change of frame moving in the backscattered direction $x \longrightarrow x + vt$, $t \longrightarrow t$ which yields :

$$\begin{aligned}
[\partial_t + (1+v)\partial_x]E_p &= -E_S E_a - \mu_e E_p \\
[\partial_t + (v-1)\partial_x]E_S &= E_p E_a^* - \mu_e E_S \\
[\partial_t + v\partial_x]E_a &= E_p E_S^* - \mu_a E_a.
\end{aligned} \tag{2}$$

Then, by defining the A_i's fields as

$$A_1 = |1+v|^{1/2} E_p \; ; \quad A_2 = |v-1|^{1/2} E_S \; ; \quad A_3 = |v|^{1/2} E_a \; ; \tag{3}$$

and looking for stationary solutions in the new frame, we have :

$$\begin{aligned}
\partial_X A_1 &= -s_1 A_2 A_3 - s_1 \rho_1 A_1 \\
\partial_X A_2 &= s_2 A_1 A_3^* - s_2 \rho_2 A_2 \\
\partial_X A_3 &= s_3 A_1 A_2^* - s_3 \rho_3 A_3
\end{aligned} \tag{4}$$

where $X = x/|(v-1)(1+v)v|^{1/2}$; $\rho_1 = \mu_e |v-1|^{1/2} |v|^{1/2} / |1+v|^{1/2}$; $\rho_2 = \mu_e |1+v|^{1/2} |v|^{1/2} / |v-1|^{1/2}$; $\rho_3 = \mu_a |1+v|^{1/2} |v-1|^{1/2} / |v|^{1/2}$; and $s_1 = sgn(1+v)$, $s_2 = sgn(v-1)$, $s_3 = sgn(v)$. Taking into account stability arguments [18], we shall be concerned only with the solutions for which $v > 1$, i.e. a structure moving at a velocity larger than velocity of light (the $v < -1$ case only intervenes in a transient stage of the nonlinear interaction and is asymptotically unstable). Since $\mu_e \ll \mu_a$ and $v > 1$ we also have $\rho_1 \ll \rho_3$ but ρ_2 and ρ_3 are of the same order, i.e. $\mu_e \sim |v-1|\mu_a$. Thus, for $\rho_1 = 0$ and $s_1 = s_2 = s_3 = 1$, we obtain the following solution

$$A_1 = -A \tanh AX + A_0$$

$$A_2 = A_3 = A \operatorname{sech} AX \tag{5}$$

with $A_0 = \rho_2 = \rho_3$. Therefore

$$\frac{\mu_e}{\mu_a} = \left|\frac{v-1}{v}\right| \quad \Rightarrow \quad v = \frac{1}{1-\mu_e/\mu_a}. \tag{6}$$

From (3) we obtain the expressions for the initial amplitudes of the waves

$$E_p = P_0 - P \tanh\big[(x+vt)/\Delta\big],$$

$$E_S = S \operatorname{sech}\big[(x+vt)/\Delta\big], \tag{7}$$

$$E_a = SP\big[2/(S^2+P^2)\big]^{1/2} \operatorname{sech}\big[(x+vt)/\Delta\big],$$

with the extra relationships

$$P_0 = (\mu_e\mu_a)^{1/2}, \quad S/P = (2\mu_a/\mu_e - 1)^{1/2}, \tag{8}$$

the width of this backward complex structure being given by

$$\Delta = \frac{(\mu_e/\mu_a)^{1/2}}{P(1-\mu_e/\mu_a)}. \tag{9}$$

There is only one free parameter left which will be taken as the pump amplitude at the far left end of the structure shown in figure 2, namely $P_0 + P = E_p(x+vt \ll -\Delta) = 1$.

This superluminous self-similar Stokes pulse does not contradict by any means the special theory of relativity. Its motion can be viewed as the result of an amplification of the leading edges of the Stokes and acoustic pulses while, at the same time, their rears are depleted, the pump wave being partially restored after the interaction, in some cases with an opposite phase. This behavior is quite reminiscent of the self-induced transparency effect encountered in the coherent pulse propagation in a two-level medium [19]. No transportation of information can be obtained via this process which can only occur if a sufficiently extended background of Stokes light is available so that the superluminous self-similar amplification takes place. If at a given time (t=200 in fig.2) we perturb the soliton profile by "engraving" a signal, this one moves at the velocity of light (i.e. remains at rest in the backward moving frame of fig.2) and the Brillouin soliton runs away at the "superluminous" velocity. Figure 2 also shows the elasticity of the soliton structure. An additive interesting feature obtained by the numerical treatment of Eqs. (1) is that, when it is built, the Stokes pulse is no longer perturbed by pump phase fluctuations, which are compensated by the acoustic wave (cf. fig. 3).

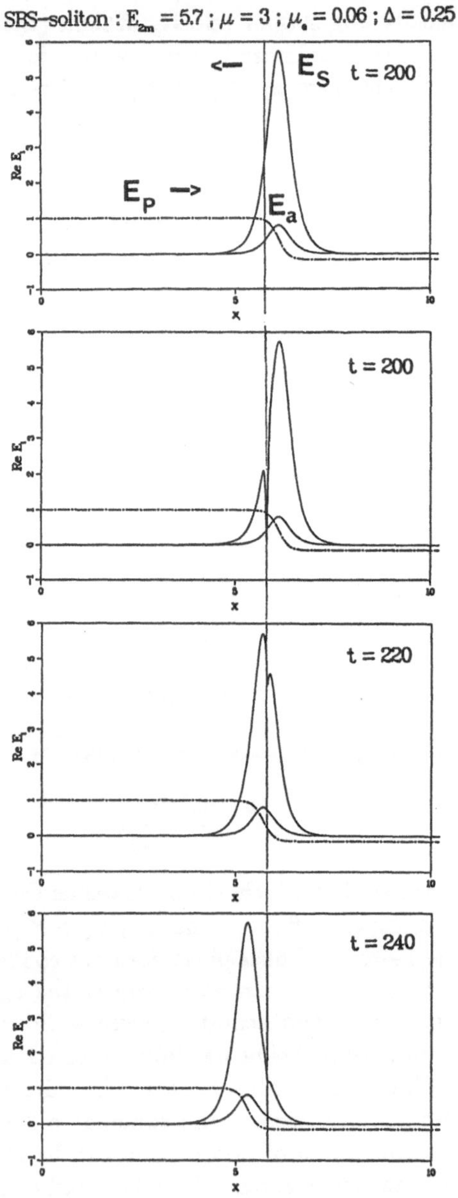

Fig. 2 *Dissipative superluminous Brillouin soliton given by formulas (7), for $\mu_a = 3$, $\mu_e = 6 \times 10^{-2}$ in the frame moving backward at the velocity of light (Stokes frame). Note that the pump is partially restored after the interaction. At time t=200 a signal is "engraved" in the soliton profile. The signal remains at rest in this backscattered frame and the Brillouin soliton runs backward away at the "superluminous" velocity. It cannot transport information.*

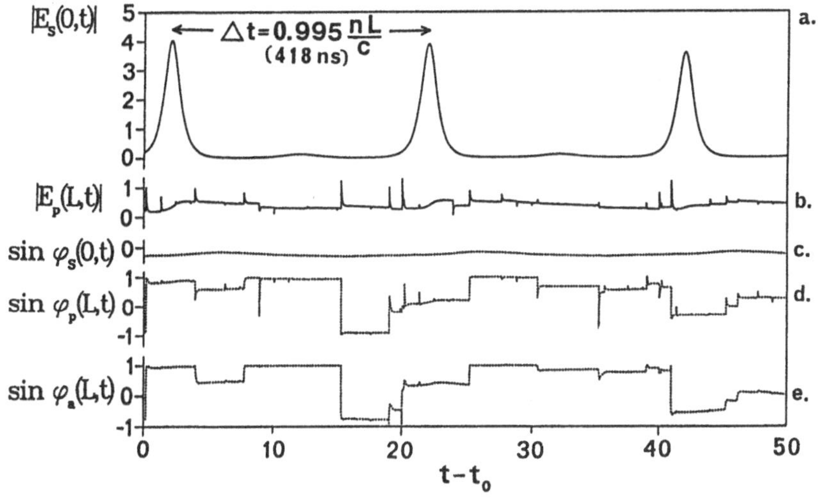

Fig. 3 *Numerical computation of Eqs.(1) for the ring cavity with* $\mu_a = 10$, $\mu_e = 10^{-2}$ *and* $L/\Lambda = 20$, *where* Λ *is the SBS characteristic length (Ref. [9]). Curve* (**a**), *train of backscattered solitons [output Stokes component* $E_S(0,t)$*]; curve* (**b**), *transmitted pump amplitude* $E_p(L,t)$ *showing spikes generated by random phase shifts; curve* (**c**), *Stokes phase, almost unsensitive to the pump phase jumps; curve* (**d**), *strongly modulated pump phase; curve* (**d**), *acoustic phase, following the pump phase jumps.*

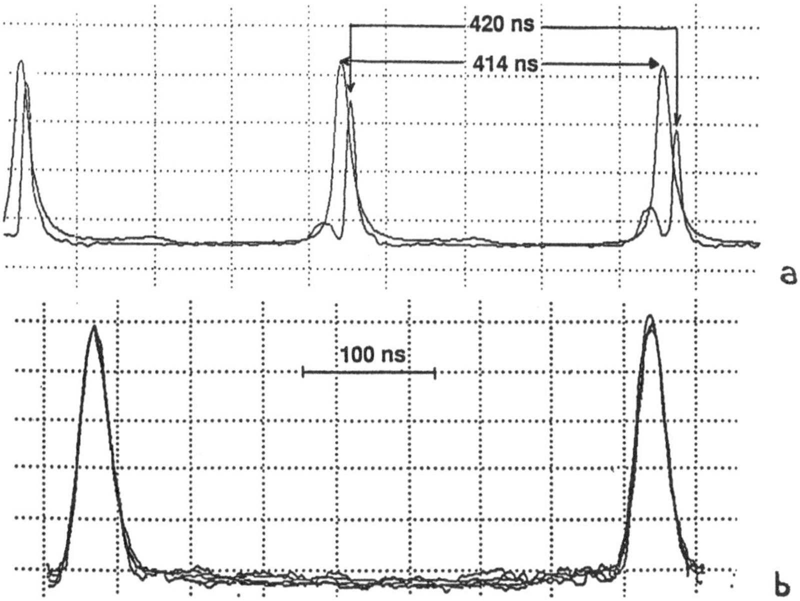

Fig. 4 (**a**) *Experimental superluminous Brillouin Stokes solitons superimposed to the evolutive pulses spaced by* Δt_L.
(**b**) *Experimental stability test of the Brillouin fiber-ring laser: six superimposed uncorrelated couples of pulses.*

III. Numerical and experimental results

In order to observe the spatial extended soliton envelopes in actual experimental situations, we must choose the interaction length as long as possible by allowing the Stokes envelope to be always in interaction with the pump in order to avoid the initial stage of the shock-wave regime (i), i.e. whithout cutting the Stokes leading edge. This has been achieved in the optical-fiber ring-cavity shown in figure 1, where the 50/50% beam splitter ensures the coupling of the cw pump wave into the cavity and the continuously recoupling of the backscattered Stokes wave. This configuration allows the Stokes wave to spread along the whole ring cavity.

However, we must also prevent the asymptotic steady state regime of the so-called *Brillouin mirror* : In the case of an ideal monochromatic cw pump wave, the numerical treatment of the coherent three-wave equations (1), with the feedback boundary conditions due to the cavity configuration, shows that, after some pulsed transients, the parametric SBS instability is saturated by the large depletion of the pump inside the medium, due to the accumulation of the phonons at the fiber input end, and gives rise to its reflection into a backscattered continuous Stokes wave [9]. This ideal case is indeed obtained in the experiment for high enough pump powers (beyond 250 mW coupled into the fiber), but for small cw input powers (below 100 mW) the experiments show pulsed nonstationary regimes exhibiting soliton-like profiles for the Stokes envelopes and self-induced transparency for the pump. These regimes are unstable and present sometimes several Stokes pulses in a round trip. But the stabilization of the pulsed regime, even at high intensity level, may be obtained by periodically interrupting the pump wave with the AOM during a time longer than the spontaneous acoustic damping time μ_a^{-1}. The modulation is achieved outside the ring cavity, so that the Stokes pulses may develop their wings through the whole ring cavity. For a good efficiency, the modulation frequency must roughly correspond to the round-trip flight time in the cavity. In this configuration the Brillouin mirror regime is totally prohibited and a stable soliton-like regime may be reached. We have obtained by this way sequences of well-shaped Stokes pulses for a few milliseconds [2]. In figure 4 are superimposed two couples of experimental Stokes pulses recorded at the output of the cavity during two consecutive sequences of qualitatively different dynamical behaviors. The first one is associated with the building phase [compressional regime (i)] of Stokes fine evolutive structures separated by 420ns, which is, within a precision of 2 ns, the round-trip flight time $\Delta t_L = Ln_0/c$ in the cavity at the group velocity of light in the fiber. The second one corresponds to a very stable sequence of hyperbolic-secant-like fitted pulses which repeat themselves every 414ns. The stability of these soliton-like pulses is quite remarkable as can be verified on figure 4(b), which shows six superimposed uncorrelated couples of Stokes pulses. We also verify that a non-zero level of pump intensity is transmitted at the other end of the fiber. Superluminous velocity and partial self-induced transparency are certainly the most undisputable proofs of the physical relevance of the dissipative Brillouin solitons. It is interesting to note that, by using such an externally modulated pump wave, a numerical treatment of Eqs.(1) shows the birth of the solitonlike profile by starting from completely stochastic amplitude and phase acoustic fluctuations.

It should nevertheless be kept in mind that the asymptotic stage is not reached. What we observe is an incompleted growing quasisoliton. In fact, in the experiment, a quasisoliton pulse is amplified and accelerated at each round trip in the cavity, but is also prevented from completing its growth every time it reaches the end of the fiber before being recoupled at the other end with a lower amplitude due to recoupling losses pertaining to the ring setup.

In conclusion, dissipative superluminous quasisolitons, observed for the first time in a Brillouin optical-fiber ring laser [2], can account for nonstationary dynamic behaviors of the backscattered Stokes wave and self-induced transparency on the transmitted pump wave. The coherent Brillouin soliton-like behaviour contributes to stabilize the Stokes output. It may be responsible of the high spectral coherence for the backscattered Stokes wave (much greater than the pump coherence) observed in recent experiments [20] [21].

[1] O. Legrand and C. Montes, J. Phys. Colloq. France, **50**, C3-147 (1989).
[2] E. Picholle, C. Montes, C. Leycuras, O. Legrand, and J. Botineau, Phys. Rev. Lett. **66**, 1454 (1991).
[3] E.P. Ippen and R.H. Stolen, Appl. Phys. Lett. **21**, 539 (1972).
[4] D. Cotter, J. Opt. Commun. **4**, 10 (1983).
[5] C. Montes, Phys. Rev. A **31**, 2366 (1985) and references therein.
[6] M. Maier, Phys. Rev. **166**, 113 (1968).
[7] W. Kaiser and M. Maier, in *Laser Handbook 2*, T.F. Arecchi and F.O. Schulz-Dubois, eds. (North-Holland, Amsterdam, 1972) p. 1077.
[8] J. Coste and C. Montes, Phys. Rev. A **34**, 3940 (1986)
[9] J. Botineau, C. Leycuras, C. Montes, and E. Picholle, J. Opt. Soc. Am. B **6**, 300 (1989).
[10] S.F. Morozov, L.V. Piskunova, M.M. Sushchik, and G.I. Freidman, Sov. J. Quant. Electron. **8**, 576 (1978).
[11] V.A. Gorbunov, Sov. J. Quant. Electron. **14**, 1066 (1984).
[12] G.L. Lamb, Jr., Phys. Lett. **29 A**, 507 (1969); Rev. Mod. Phys. **43**, 99 (1971).
[13] J. Pelous and R. Vacher, Solid State Commun. **16**, 279 (1975).
[14] J.A. Armstrong, S.J. Sudhanshu, and N.S. Shiren, J. Quant. Elect. **QE-6**, 591 (1974).
[15] K. Nozaki and T. Taniuti, J. Phys. Soc. Jpn. **34**, 796 (1973); Y. Oshawa and K. Nozaki, J. Phys. Soc. Jpn. **36**, 591 (1974).
[16] C.J. McKinstrie, Phys. Fluids **31**, 288 (1988)
[17] D.J. Kaup, Stud. Appl. Math. **55**, 9 (1976).
[18] S.C. Chiu, J. Math. Phys. **19**, 168 (1978).
[19] S.L. McCall and E.L. Hahn, Phys. Rev. **183**, 457 (1969).
[20] M. Douay, P. Bernage, and P. Niay, Opt. Comm. **81**, 231 (1991).
[21] S.P. Smith, F. Zarinetchi, and S.E. Ezekiel, Opt. Lett; **16**, 393 (1991).

POLARISATION FLUCTUATIONS IN NONLINEAR OPTICAL FIBRES

A.P. Mayer and D.F. Parker

Department of Mathematics, University of Edinburgh,
King's Buildings, Edinburgh EH9 3JZ, Scotland

1. INTRODUCTION

Nonlinear optical fibres are systems in which nonlinearity and dispersion give rise to the existence of stable solitary envelope pulses associated with light at a given carrier frequency [1]. The direct observability of the formation and propagation of these solitary pulses and the prospect of their application in long-distance telecommunication systems has stimulated a large number of theoretical investigations in this field [2]. Mathematically, pulse propagation in the monomode regime of an axisymmetric optical fibre can be described by a pair of nonlinearly coupled nonlinear Schrödinger equations (NLS). The two complex variables in these equations play the role of slowly varying amplitudes of two degenerate fibre modes (i.e. fibre modes having the same propagation constant k for given frequency ω). Unitary transformations of the complex two-component vector of these amplitudes leave the intensity unchanged but affect phase and polarisation of a pulse. If the coupled NLS equations are invariant under U(2)-transformations, then the system is integrable via the inverse scattering transform [3]. It possesses a multi-parameter family of pulse solutions which have been termed vector solitons [4]. For real material coefficients, the nonlinear terms in the coupled equations are not invariant under U(2)-transformations and the system is not integrable [5]. It does however possess a multi-parameter family of solitary wave solutions [6] which may be regarded as generalisations of vector solitons and are, for special polarisations, identical to them. It will be shown in detail that material inhomogeneities lead to extra terms in the evolution equations that generate unitary transformations of the two-component vector of complex amplitudes in the linear limit. If the system of coupled NLS equations is integrable, the effect of these terms is trivial, while in the general case, they give rise to interesting behaviour of pulse propagation in which the aforementioned solitary pulse solutions play an important role.

2. VARIATIONS OF THE LINEAR REFRACTIVE INDEX

To derive evolution equations for the propagation of envelope pulses in nonlinear optical fibres, we start from the wave equation

$$\nabla\times(\nabla\times\underset{\sim}{E}) + \frac{\partial^2}{\partial t^2}(\epsilon+N|\underset{\sim}{E}|^2)\underset{\sim}{E} = 0 \qquad (2.1)$$

for the electric field $\underset{\sim}{E}$, which we expand in powers of a small parameter ν,

$$\underset{\sim}{E}(\underset{\sim}{x},t) = e^{i(kz-\omega t)}\sum_{n=1}^{\infty} \nu^n \underset{\sim}{E}_n(\underset{\sim}{x},t) + c.c. \qquad (2.2)$$

We also introduce stretched coordinates $Z_n=\nu^n z$, $T_n=\nu^n t$, where z is the coordinate in the fibre direction. In axisymmetric fibres which are homogeneous along the z-direction, the linear dielectric constant ϵ and the Kerr coefficient N depend on the radial coordinate r only and the linearised system allows $\underset{\sim}{E}$ to be a linear combination of the modal fields $\underset{\sim}{E}_{\pm}(r,\theta)$ of two degenerate fibre modes which we choose to be of circular polarisation, so that

$$\underset{\sim}{E}_1(\underset{\sim}{x},t) = \sum_{\sigma=\pm} \underset{\sim}{E}_\sigma(r,\theta)\; A_\sigma(Z_1,T_1,Z_2,..) \; . \qquad (2.3)$$

In the following, we also take into account small and gradual variations of ϵ that may violate the axial symmetry, by decomposing $\epsilon(\underset{\sim}{x})=\epsilon_1(r)+\epsilon_2(r,\theta,z)$ and scaling $\epsilon_2=\nu^m\delta(r,\theta,Z_2)$. For m=2, the variations of the dielectric constant due to inhomogeneities are of the same order in ν as its variations induced by the Kerr nonlinearity. Following the standard procedure of multiple scales, simplified by the use of compatibility conditions for the equations of third order in ν [7], evolution equations are obtained for the amplitudes $A_{\pm}$ which, after rescaling, take the form

$$i\frac{\partial}{\partial z}A_{\pm} = \frac{\partial^2}{\partial t^2}A_{\pm} + \sum_{\sigma=\pm} \Lambda_{\pm\sigma}(z)A_\sigma + (|A_{\pm}|^2 + h|A_{\mp}|^2)A_{\pm} \quad , \qquad (2.4)$$

where we now use the symbols z and t for a length proportional to Z_2 and a retarded time variable proportional to T_1-Z_1/S, S being the group velocity of the linear degenerate fibre modes with $\varepsilon_2 \equiv 0$. For h=1, the system (2.4) is integrable [3]. However, for weakly guiding fibres, the material coefficient h takes values close to 2, and the system is not integrable [5]. The components of the Hermitian matrix $\underset{\sim}{\Lambda}$ are proportional to overlap integrals of the function $\delta(r,\theta,Z_2)$ with products of two modal fields, of the form

$$\Lambda_{++}=\Lambda_{--} \sim \int_0^\infty f_0(r) \int_0^{2\pi} \delta(r,\theta,Z_2)\, d\theta\, dr \tag{2.5}$$

$$\Lambda_{+-}=\Lambda^*_{-+} \sim \int_0^\infty f_1(r) \int_0^{2\pi} e^{-2i\theta} \delta(r,\theta,Z_2)\, d\theta\, dr \, . \tag{2.6}$$

Due to the symmetry of the problem, the diagonal elements are equal. They can be different if δ is allowed to be a tensor. In the absence of nonlinearity, the matrix $\underset{\sim}{\Lambda}$ causes the two-component vector (A_+,A_-) to undergo continuous unitary transformations. In the integrable case h=1, this matrix may be eliminated from the evolution equations by transformation to the variables $B_\pm(z,t)=\sum_{\sigma=\pm} R_{\sigma\pm}(z)A_\sigma(z,t)$ with the unitary matrix $\underset{\sim}{R}$ satisfying the ordinary differential equation

$$-i\frac{d}{dz}\underset{\sim}{R} = \underset{\sim}{\Lambda}^*\underset{\sim}{R} \quad . \tag{2.7}$$

If $h \neq 1$, this is no longer possible. The diagonal elements of $\underset{\sim}{\Lambda}$ (if they are equal) may however still be absorbed by a redefinition of phase, so that we may assume without loss of generality that $\Lambda_{++}=\Lambda_{--}=0$.

3. SCATTERING OF A PULSE AT A BIREFRINGENCE DEFECT

We now investigate the effect on pulse propagation of an irregularity in ε causing Λ_{+-} to be significantly non-zero only in a finite interval. We shall call such a localised irregularity a birefringence defect. The simple functional form $\Lambda_{+-}(z)=iS\exp\{-(z-z_0)^2/w^2\}$ with $z_0 \gg w$ is chosen, and for z=0, we assume a pulse of the form $A_+(0,t)=a(0,t)=\sqrt{2}\,\mathrm{sech}(t)$, $A_-(0,t)=0$, where a(z,t) is a soliton solution of the single NLS equation describing a circularly polarised mode. In the integrable case the solution takes exactly the form $(A_+(z,t),\ A_-(z,t))= (\cos\alpha,\ \sin\alpha)\ a(z,t)$, where $\alpha=\alpha(z)$ has the limit $-\sqrt{\pi}Sw$ for $z \gg z_0+w$. The effect of a birefringence defect thus consists in transforming a vector soliton into another one with altered polarisation which depends only on the product Sw.

Although the system is not integrable for $h\neq1$, it possesses solitary wave solutions corresponding to simple pulses [6],

$$A_{\pm}(z,t) = \eta F_{\pm}(\eta t;\alpha) \exp(-i\eta^{2}\beta_{\pm}z) . \qquad (3.1)$$

The real functions $F_{\pm}(\tau;\alpha)$ are solutions of the ordinary differential equations

$$\frac{d^2}{d\tau^2}F_{\pm} = (\beta_{\pm}-F_{\pm}^{2}-hF_{\mp}^{2})F_{\pm} \qquad (3.2)$$

decaying to zero exponentially as $\tau\to\pm\infty$. The parameter α plays the role of a polarisation angle if we define $\tan\alpha=F_{-}(0)/F_{+}(0)$ and $F_{\pm}$ are even functions. Numerical integrations have been carried out with the above "initial conditions" at z=0. Their results suggest that, after passing the birefringence defect, the pulse evolves into a solution of the coupled NLS equations being predominantly of the form (3.1), (3.2). This has been tested by using a consistency relation between $\beta_{\pm}$ and $F_{\pm}(0)$ following from the nonlinear eigenvalue problem (3.2). In addition, superimposed oscillations and continuous output of radiation has been observed for sufficiently large S. The numerical integrations have been performed for different combinations of S and w, but with fixed product Sw=0.2. For h=1, the limiting behaviour of $A_{\pm}$ for $z\gg z_0+w$ is then identical for all these combinations.

Different behaviour is found for h=2. Here, two regimes may be distinguished. For large S and small width w on the length scale on which nonlinearity and dispersion are effective, the main effect of the birefringence irregularity is to reset the initial conditions for the evolution of $A_{\pm}$. This is illustrated in Fig. 1, which shows the behaviour of the maxima of $|A_{\pm}(t)|$ as functions of z. This resetting of initial conditions, and also the subsequent evolution, will only depend on the product Sw. (It should be noted that exchange of intensity between the two modes can take place only if Λ_{+-} is nonzero.) With increasing w and decreasing S, less intensity is converted from the first (+) to the second (-) mode and finally, A_{-} is nonzero only in the neighbourhood of z_0. In other words, the pulse regains its initial polarisation after having passed the defect, in strong contrast to the integrable case. This qualitatively different behaviour can be understood in the framework of soliton perturbation theory (recently applied to similar problems in optical fibres in refs. [8-11]) based on the assumption that, if w is large on the scale on

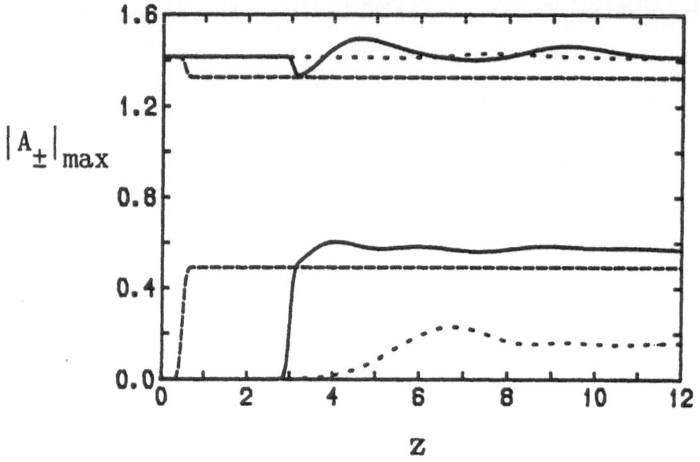

Fig. 1. Maxima of $|A_\pm(t)|$ as functions of z. In all three cases Sw=0.2. dashed: h=1, w=0.1; solid: h=2, w=0.1; dotted: h=2, w=1.6.

which nonlinear evolution takes place and S is small, the fields $A_\pm$ can be approximated by the functional form

$$A_\pm(z,t) = \eta(z)\ F_\pm(\eta(z)t;\alpha(z))\ \exp\{-i[\phi(z)\pm\Psi(z)/2]\} \qquad (3.3)$$

with parameters η, α, ϕ and Ψ varying slowly along the fibre. By inserting this Ansatz into the action integral for $A_\pm$ and taking the variation with respect to these parameters, coupled equations of the form

$$\frac{d}{dz}\alpha = \lambda\ K_1(\alpha)\ \sin(\Psi+\gamma)\ , \qquad \frac{d}{dz}\Psi = K_2(\alpha) + \lambda\ K_3(\alpha)\ \cos(\Psi+\gamma) \qquad (3.4)$$

are obtained, where λ and γ are the modulus and argument of Λ_{+-}. In the integrable case, the term K_2 in (3.4) is absent. For a qualitative discussion, we confine ourselves to the case $|h-1|<<1$. Then, η is approximately a constant and $K_2(\alpha)\approx(4/3)(1-h)\eta^2\cos(2\alpha)$. After a transformation to linearly polarised modes, coupled equations are obtained of the form derived earlier for $\gamma=0$ and λ constant [8,10]. These may be linearised around the initial pulse parameters to yield driven harmonic oscillator equations

$$\frac{d^2}{dz^2}q_\pm + \Omega^2 q_\pm = f_\pm \qquad (3.5)$$

for variables $q_\pm$ connected with the polarisation angle α via $\alpha^2 \approx q_+^2 + q_-^2$. The driving forces $f_\pm$ are linear combinations of the real and imaginary parts of Λ_{+-} and their derivatives with respect to z, and the "frequency" is

$\Omega=(4/3)|h-1|\eta^2$. The deviation of h from 1 thus gives rise to a restoring force that causes the polarisation angle to return to its initial value 0 after the pulse has passed the birefringence defect. This behaviour does not seem to occur for initially linearly polarised pulses with $A_+=A_-$.

With the initial conditions $a(0,t)=2\sqrt{2}\mathrm{sech}(t)$ a two-soliton bound state of the NLS equation evolves. In distinction to the single soliton case, the effect of a birefringence defect now depends on its location z_0 relative to the stage of periodic internal oscillation of the two-soliton bound state. Results of a numerical integration for h=2 and a defect with width w=0.1 and strength S=2 at $z_0=0.78$ indicate that as the pulse encounters the defect, it strongly distorts producing a large amount of radiation. In a second phase, an extensive reshaping of the pulse takes place with lesser generation of radiation. The internal oscillations gradually disappear and a sharp central peak emerges with two broad wings detaching from it (Fig.2). Analysis of

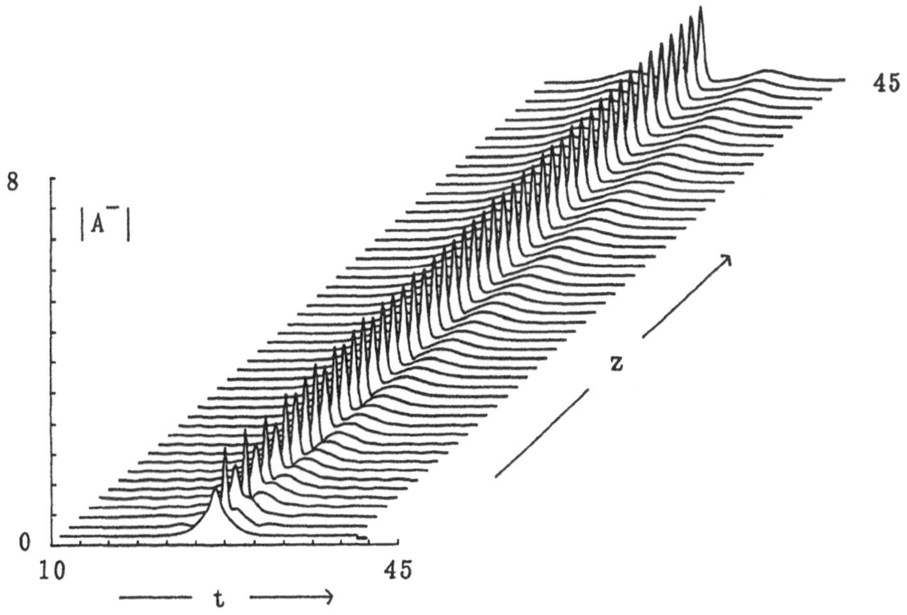

Fig. 2. Evolution of $|A_-|$ of a pulse initially corresponding to a two-soliton bound state scattered at a birefringence defect at position $z_0=0.78$ marked by a little bar.

the moduli and arguments of $A_\pm$ at maximum suggests that the central peak corresponds to a solitary wave solution of the type (3.1) with (3.2). These findings compare with earlier numerical results by Blow et al. [12] for a fibre with constant birefringence, who found that a two-soliton bound state evolves into a highly compressed single pulse.

4. STRONG BIREFRINGENCE

The evolution equations (2.4) have been obtained with the assumption that the inhomogeneous part ε_2 of the permittivity is of second order in ν (m=2). We now consider briefly the case m=1. In order to keep the multiple scales expansion uniform [13], we introduce coordinates $\zeta_j = f_j(Z_2)/\nu$ with $f_1'(Z_2) = \lambda(Z_2)$ and $f_2'(Z_2) = \Lambda_{++}(Z_2)$. We then perform the transformation

$$A_\pm(\zeta_1, \zeta_2, Z_2, ..) = \sum_{\sigma=\pm} R_{\pm\sigma}(\zeta_1, \zeta_2)\, B_\sigma(Z_2, ..) \,, \tag{4.1}$$

where the unitary matrix $\underset{\sim}{R}$ has elements

$$R_{\sigma\sigma'} = \exp\{i(\Psi_{\sigma\sigma'} - \zeta_2)\}/\sqrt{2} \,, \quad \Psi_{++} = \gamma - \zeta_1 = -\Psi_{--} + \pi \,, \quad \Psi_{-+} = -\zeta_1 = -\Psi_{+-} \,. \tag{4.2}$$

Following now the procedure of multiple scales, a pair of evolution equations emerges for $B_\pm$, from which all fast oscillations on the scales of $\zeta_{1,2}$ are eliminated. After rescaling in the same way as in the derivation of (2.4), they take the form

$$i\frac{\partial}{\partial z}B_\pm = \Pi_\pm(z)B_\pm + iM_\pm(z)\frac{\partial}{\partial t}B_\pm + \frac{\partial^2}{\partial t^2}B_\pm + \frac{1}{2}(1+h)\{|B_\pm|^2 + \frac{2}{1+h}|B_\mp|^2\}B_\pm \,, \tag{4.3}$$

where $\Pi_\pm$ and $M_\pm$ are real functions. The quantities $M_\pm$ may be expressed in terms of the $\Lambda_{\sigma\sigma'}$ and their first derivatives with respect to the propagation constant k (prior to rescaling). By simple transformations, the first term on the right-hand side of (4.3) can be eliminated and the group velocity shifts $M_\pm(z)$ can be arranged to have equal modulus and opposite signs. For the case of homogeneous fibres, equations of this type have been obtained by Menyuk [14,15] and pulse propagation in this case has been studied numerically [15,16] and analytically [11] on the basis of such equations. For constant $M_\pm$ the second term on the right-hand side of (4.3) can be eliminated by a simple phase transformation, and the resulting pair of coupled NLS equations has solitary pulse solutions of the form (3.1) with altered coefficients in the nonlinear terms of (3.2).

In the preceding discussions, we have considered symmetry-breaking of the refractive index which varies gradually (on the scale of Z_2) along the fibre. If the refractive index fluctuates on length scales considerably shorter than the characteristic length ℓ for the pulse evolution due to nonlinearity and dispersion, the effect of such fluctuations is primarily to introduce effective coefficients in the evolution equations for slowly varying amplitudes as averages of the fluctuations over lengths smaller than ℓ. In an approximation, these averages may then be calculated as averages over an ensemble of different realisations [17]. Gaussian statistical properties of the fluctuating part ϵ_2 of the dielectric constant translate into a Gaussian ensemble for the matrix $\underset{\sim}{\Lambda}$. For systems in which ϵ_2 is of first order in ν and preserves the axial symmetry <u>on average</u>, evolution equations have been derived for slowly varying amplitudes $B_\pm$ which are related to $A_\pm$ by a unitary transformation of the form (4.1) [18]. These evolution equations are of the same form as those in the absence of fluctuations. However, the coefficients in front of the nonlinear terms acquire corrections determined by the correlation function $\langle\Lambda_{+-}(z)\Lambda_{-+}(z')\rangle$.

REFERENCES

1. A. Hasegawa and F. Tappert, Appl. Phys. Lett. 23, 142 (1973).
2. For a recent review see: G.P. Agrawal, "Nonlinear Fiber Optics", (Academic Press, 1989) and references therein.
3. S.V. Manakov, Zh. Eksp. Teor. Fiz. 65, 505 (1973) [Sov. Phys. -JETP 38, 248 (1974)].
4. D.N. Christodoulides and R.I. Joseph, Opt. Lett. 13, 53 (1988).
5. V.E. Zakharov and E.I. Schulman, Physica 4D, 270 (1982).
6. D.F. Parker and G.K. Newboult, J. Phys. Colloq. 50, C3-137 (1989).
7. G.K. Newboult, D.F. Parker and T.R. Faulkner, J. Math. Phys. 30, 930 (1989).
8. D.J. Muraki and W.L. Kath, Phys. Lett. A 139, 379 (1989).
9. M. Lisak, A. Höök and D. Anderson, J. Opt. Soc. Am. B 7, 810 (1990).
10. B.A. Malomed, Phys. Rev. A 43, 410 (1991).
11. D. Anderson, Yu.S. Kivshar and M. Lisak, Physica Scripta 43, 273 (1991).
12. K.J. Blow, N.J. Doran and D. Wood, Opt. Lett. 12, 202 (1987).
13. A.H. Nayfeh, "Perturbation Methods", (Wiley, 1973), Chapter 6.
14. C.R. Menyuk, IEEE J. Quantum Electron. QE-23, 174 (1987).
15. C.R. Menyuk, J. Opt. Soc. Am. B 5, 392 (1988).
16. L.F. Mollenauer, K. Smith, J.P. Gordon and C.R. Menyuk, Opt. Lett. 14, 1219 (1989).
17. S. Solimeno, B. Crosignani and P. di Porto, "Guiding, Diffraction, and Confinement of Optical Radiation", (Academic Press, 1986), Chapter 8.
18. A.P. Mayer and D.F. Parker, in preparation.

STOCHASTIC DYNAMICS OF SPATIAL SOLITONS ON THE PERIODIC INTERFACE OF TWO NONLINEAR MEDIA

F.Kh.Abdullaev, B.A.Umarov

Thermal Physics Department of the Uzbek Academy of Sciences ,Katartal Str.28,Chilanzar,Tashkent 700135,USSR

1.Introduction

Currently a problem of nonlinear electromagnetic wave propagation through a interface separating two dielectric media attracts much attention. It is caused by the possibilities to construct optical switchers ,scanners and other optical devices based on the applications of the beam and surface wave properties on the interface of nonlinear media.

In [1-4] a problem of interaction of plane self-focusing light channel in a self-focusing medium with cubic nonlinearity (of *spatial* soliton) with interface has been numerically and analytically considered.The main results of these papers are: there exist some regimes of complete internal reflection, trapping and transformation of a beam into nonlinear surface wave and beam passing depending upon the incident beam parameters.

The purpose of this work is to investigate nonlinear surface waves taking into account the periodic modulation of the interface.

2.Basic equations

Let us briefly consider a conclusion of the basic equations. The wave equation describing a propagation of monochromatic electromagnetic TE field in (x-z) plane takes the following form

$$\frac{\partial^2 E}{\partial z^2} + \frac{\partial^2 E}{\partial x^2} = -n^2 k^2 E \, , \qquad (1)$$

where $n^2 = n_l^2 + \alpha|E|^2 = \frac{\omega^2}{c^2k^2}[1 + 4\pi\chi^{(1)} + 4\pi\chi^{(3)}|E|^2]$

is the nonlinear refractive index of the medium, ω is the frequency of the electromagnetic field, βk is the wave number of the incident field in z, β is the waveguide mode, and $k = \omega/c$. $\chi^{(1)}, \chi^{(3)}$ are the linear and nonlinear responses respectively, characterizing the medium properties. We will consider a self-focusing nonlinearities when $\chi^{(3)} > 0$. The electromagnetic field is assumed to be almost monochromatic in x and z. Let

$$E = F(x,z)\, exp(i\beta kz). \tag{2}$$

Substituting (2) into (1), making variable change $x' = kx$, $z' = kz$ and omitting a prime, one obtains

$$2i\beta \frac{\partial F}{\partial z} + \frac{\partial^2 F}{\partial x^2} - (\beta^2 - n^2)F = 0 \quad , \tag{3}$$

where refractive index $n^2 = n_i^2 + \alpha_i|F|^2$ is discontinuous at the interface. Here the index $i = 0$ at $x < 0$ and $i = 1$ at $x > 0$. Following [1] in this work we assume that

$$\Delta = n_0^2 - n_1^2 > 0 \text{ , } \alpha = \alpha_0/\alpha_1 \leqslant 1 \text{ .} \tag{4}$$

Let a wave packet falls on the interface at the right side ($x > 0$). Make a variable change

$$F(x,z) = (2/\alpha_1)A(x,\tau)\, exp[i(\beta^2 - n_1^2)z/2\beta] \text{ ,} \tau = \beta k. \tag{5}$$

Substituting (5) into (3) one obtains

$$i\frac{\partial A}{\partial \tau} + \frac{\partial^2 A}{\partial x^2} + 2|A|^2 A = VA \quad , \tag{6}$$

$$V = \begin{cases} 0 & , \quad x>0 \\ -\Delta - 2(\alpha - 1)|A|^2, & x<0 \end{cases} .$$

The equation (6) is the nonlinear Schrodinger equation (NLS) with a perturbing term $V(x)A$ .At the interface absence $\alpha=1,\Delta=0$ it follows that $V = 0$. In this case the equation is exactly integrable and its general solution consists of solitons and nonsoliton component, which parameters are defined by an initial condition.

A self-focused wave channel at $x > 0$ is described by a solitonic solution which takes the following form when the perturbation is absent.

$$A(x,\tau) = 2\eta_1 \; sech[2\eta_1(x-\bar{x})] \; exp[i(\frac{vx}{2} + 2\sigma)] \quad . \tag{7}$$

It is clear, that in general case the equation (6) solution is a rather cumbersome problem but nevertheless, if the medium parameter changes are assumed to be small,i.e $\alpha^{-1} - 1 \ll 1$, $\Delta \ll 1$, the interface influence can be taken into account, considering that the soliton parameters, velocity v and amplitude η_1, change slowly with τ changing. The equation defining the soliton parameter dependence upon τ can be derived by application of the conservation laws or the perturbation theory for solitons [1-2,4]. The soliton amplitude appeared on adiabatic approximation not to depend on τ, i.e. $d\eta/d\tau = 0$, and the velocity changes are described by the following equation

$$\frac{d^2\bar{x}}{d\tau^2} = - \frac{dU}{d\bar{x}} \quad , \tag{8}$$

where

$$U = \Delta(1 - S_1^{-1}) tanh s + (\Delta/3S_1) \; tanh^3 s \; . \tag{9}$$

$$s=2\eta_1\bar{x} \; . \; S_1 = [8(\eta_1/\Delta)^2(\alpha-1)]^{-1} \; ,$$

As seen from (8) the soliton center is moving as a particle in anharmonic potential U, and in this case the equation (8) is the Newton one for a particle, i.e we obtain a quasiparticle analog. While investigating the potential U properties one can completely describe the beam dynamics. In this case it is necessary to take into account the velocity v corresponding to the propagation angle Φ, $v = d\bar{x}/d\tau = 2\eta_1 \sin\Phi$,i.e. while considering trajectories with various initial velocities on a phase plane, indeed we investigate a problem of a beam on the at various fall angels on the interface.

3. Description of quasiparticle motion

Let us consider a potential U more detaily.First of all we start with critical point $\bar{x}_0$ where $dU/d\bar{x}=0$. With a help of calculation we find

$$2\eta_1\bar{x}_0 = \tanh^{-1}[(1-S_1)^{1/2}] ,$$

i.e. $\bar{x}_0$ exists only at $S_1 < 1$.
Let us calculate the second derivative from the potential in a point x_0,

$$\left.\frac{d^2U}{d\bar{x}^2}\right|_{\bar{x}=\bar{x}_0} = 8\Delta\eta_1^2 S_1 (1-S_1)^{1/2} > 0 ,$$

i.e. the point $\bar{x}_0$ is one of potential minimum. Hence, we have a conclusion on existence of stable stationary surface wave, which center is located in distance of $\bar{x}_0$ from the interface. At the initial velocities being not equal to 0, the soliton center will be a periodic function τ. A frequency of small oscillations w can be easily calculated by potential $U(\bar{x})$ expansion near the point x_0 into degrees ($\bar{x} - \bar{x}_0$) and preserving quadratic terms, we obtain

$$U(\bar{x}) \approx U(\bar{x}_0) + \frac{1}{2}\left.\frac{d^2U}{d\bar{x}^2}\right|_{\bar{x}=\bar{x}_0} (\bar{x}-\bar{x}_0)^2 .$$

If one denotes $\bar{x}-\bar{x}_o = y$, then the equation for particle small oscillations near a minimum of the potential well is written in the form

$$\frac{d^2y}{d\tau^2} + w^2 y = 0 , \qquad (10)$$

where $w^2 = 4\Delta\eta_1^2 S_1^2 (1-S_1)^{1/2}$.

If the quasiparticle initial velocity is rather different from zero the oscillations become anharmonic ones. Let us consider a complete equation. In this case, having integrated the equation (8) we obtain

$$\frac{1}{2}\left(\frac{d\bar{x}}{d\tau}\right)^2 = E - U(\bar{x}) . \qquad (11)$$

Let us investigate the potential $U(\bar{x})$. The quasiparticle can be trapped by a potential and make oscillating motion or reflect from it and go away at $+\infty$ depending on the initial energy. On the phase plane there also exists a separatrix trajectory which separates oscillating and reflective trajectories, and the velocity on a separatrix v_S tends to zero in infinity

$$v_s \Rightarrow 0 \text{ ,if } \bar{x} \Rightarrow \infty .$$

The equation (11) is integrated in quadrature

$$\tau(\bar{x}) = \int_{\bar{x}_1}^{\bar{x}} [2(E - U(\bar{x}))]^{-1/2} dx . \qquad (12)$$

From (12) one can obtain a period of oscillations T

$$T = \int_{\bar{x}_m}^{\bar{x}_n} [2(E - U(\bar{x}))]^{-1/2} dx ,$$

where $\bar{x}_m$ and $\bar{x}_n$ are the points in reflection at the oscillating motion which depend upon the initial velocity.

Let us investigate a separatrix trajectory. We write the equation (11) in the form

$$\frac{1}{2}v^2(\tau) = E - U(\bar{x}),$$

where

$$U(\bar{x}) = a \tanh s + b \tanh^3 s ,$$

$$a = \Delta(1 - S_1^{-1}),\ b = \Delta/3S_1 .$$

On the separatrix $v \Rightarrow 0$ at $x \Rightarrow \infty$. We define a value of the constant E from these condition

$$E = a + b . \tag{13}$$

Substituting (13) into (12) we obtain the equation of motion on the separatrix

$$\tau = \int_{\bar{x}_0}^{\bar{x}} [2b(1 - \tanh s)(\tanh s - m)(\tanh s - n)]^{-1/2}\, dx ,$$

where

$$m = -1/2 + (3/2)(1 - 4S_1/3)^{1/2} ,$$

$$n = -1/2 - (3/2)(1 - 4S_1/3)^{1/2} ,$$

A qualitative motion of the particle on the separatrix can be described by the following way: in a moment $\tau = -\infty$ the particle is located at the point $x = -\infty$ with the velocity $v = 0$, then it moves to the left and in a moment $\tau = 0$ it is reflected at the point $s = \tanh^{-1} m$, then it goes far away at $\bar{x} = +\infty$. The velocity on the separatrix is an odd function τ , a coordinate is an even function from τ .

4. Soliton motion along modulated interface

Let us describe soliton propagation in the case when the interface is periodically modulated. In this case the perturbation potential takes the form

$$V_1 = \theta(-x + \varepsilon\, sin\Omega\tau)[-\Delta + 2(\alpha - 1)|A|^2] ,$$

where ε is the amplitude of the modulation, Ω is the frequency of modulation. The equation of soliton motion in a potential takes the following form

$$\frac{d^2\bar{x}}{d\tau^2} = -(2\Delta\eta_1 sech^2 s' + 16\eta_1^3(\alpha - 1)\, sech^4 s') , \qquad (14)$$

$$s' = 2\eta_1(\bar{x} - \varepsilon\, sin\Omega\tau).$$

Let us assume that $\varepsilon \ll 1$ and expand a function of the right hand of the equation (14) in the series in small parameters and preserve the first order terms . In this case we obtain the following equation for $\bar{x}$

$$\frac{d^2\bar{x}}{d\tau^2} = - \frac{dU}{d\bar{x}} + \varepsilon f(\bar{x})\, sin\, \Omega\tau , \qquad (15)$$

where $f(\bar{x}) = - \frac{1}{2}(d^2U/d\bar{x}^2) = 4\Delta\eta_1^2 sech^2 s\; tanh s(1- 2S_1 sech^2 s).$

At $\varepsilon = 0$ the equation (15) is reduced to equation having been investigated in [1]. At $\varepsilon \ll 1$ the second term from the right hand (15) can be taken into account as a small periodic perturbation. As known, during the periodic perturbation influence upon the particle moving in anharmonic potential there arises a whole number of new physical phenomena, such as higher harmonic appearance, nonlinear resonances, phase oscillations, and under certain conditions, nonlinear resonance interaction and chaotic motion. An estimate of chaotic layer width near the separatrix can be found applying the

Melnikov function [5], that in our case takes the form

$$D(\tau_0) = \varepsilon \int_{-\infty}^{+\infty} \sin[\Omega(\tau - \tau_0)] f(\bar{x})\, v(\tau)\, d\tau ,$$

where $\bar{x} = \bar{x}(\tau)$ is the separatrix trajectory. Applying $v(-\tau) = -v(\tau)$, $\bar{x}(-\tau) = \bar{x}(\tau)$, and expanding $\sin[\Omega(\tau - \tau_0)]$ we can write for $D(\tau_0)$:

$$D(\tau_0) = 2\varepsilon \cos(\Omega\tau_0) \int_{0}^{+\infty} \sin(\Omega\tau) f(\bar{x})\, v(\tau) d\tau .$$

Let us make variable changes $v(\tau)d\tau = d\bar{x}$.

Then

$$D(\tau_0) = 2\varepsilon \cos(\Omega\tau_0) \int_{\bar{x}_m}^{+\infty} \sin[\Omega\tau(\bar{x})] f(\bar{x})\, d\bar{x} ,$$

$$\bar{x}_m = (2\eta_1)^{-1} \tanh^{-1} m.$$

As known, the chaotic motion arises if the Melnikov function has infinite sets of zeros and this condition is realized in our case. The coefficient

$$|D| = 2\varepsilon \int_{\bar{x}_m}^{+\infty} \sin[\Omega\tau(\bar{x})] f(\bar{x})\, d\bar{x},$$

defines a stochastic layer width. In Fig. 1 the dependence $|D|$ on Ω for specific values is presented.

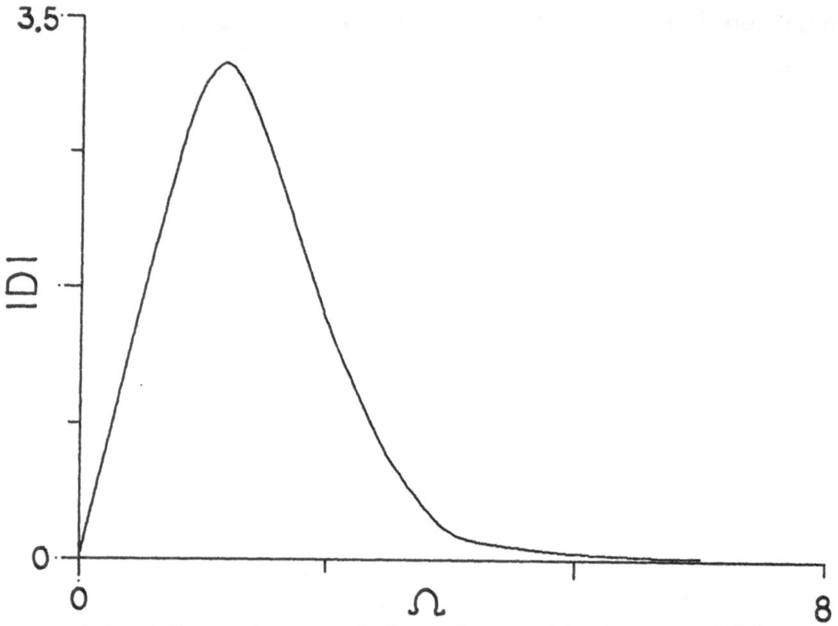

Fig.1. Dependence of the stochastic layer width versus modulation frequency at $\Delta = 0.1$, $\alpha = 0.5$, $2\eta_1 = 1.264$, $\varepsilon = 0.05$.

As seen, $|D|$ is rapidly decreasing with increase of Ω and has a maximum at the frequency $\Omega = 0.55$. If the angle of the incident wave is located within the stochastic layer, then a ray is firstly trapped by the interface, slides along the interface for some distance, and the beam center is nonlinearly oscillating, then it reaches a region of chaos and reflects from the interface.

REFERENCES

1. Aceves A.V., Moloney J.V., Newell A.C., Phys.Lett.A **129**, 231 (1988).
2. Nesterov L.A., Opt. Spektrosk. **64**, 1166 (1988) [Opt. Spektrosc. (USSR) **64**, 694 (1988)].
3. Akhmediev N.N., Korneev V.I., Kuz'menko Y.V., Sov.Phys.JETP **61** (1), 62 (1985).
4. Kivshar Yu.S., Kosevich A.M., Chubykalo O.A., Phys. Rev. A **41**, 1667 (1990).
5. Lichtenberg A.G., Liberman M.A., Regular and Stochastic Motion. (Springer-Verlag, 1984).

CONVERSION OF ULTRASHORT OPTICAL SOLITONS IN THE FIBRE-OPTICAL LOOP

Dzhakhangir V. Khaidarov
Thermal Physics Department, Uzbek Academy of Sciences
Chilanzar "C", ul. Katartal 28, Tashkent, 700135, USSR

1. INTRODUCTION

Different types of the fibre-optical elements have been proposed and designed recently for ultra-short pulses selfswitching and ultra-fast all-optical light control. One of such constructions is so-called fibre-optical loop mirror [1] - all-fibre Sagnac interferometer based on the single-mode directional coupler with two output ports linked by long (in comparison with coupler length) single-mode fibre. In this device some effects were observed: self-switching of the ultra-short pulses in the soliton [2] and non-soliton [3] regimes of propagation; pulse shaping [4] and cross-switching [5] in the non-soliton regime.

But actually fundamental solitons (with energy and duration connected by fixed relation) haven't been described in the fibre-optical loop. Other authors varied energy of the pulses at the input of the loop without changing of the pulse duration. Here some possibilities of the conversion of the fundamental soliton in the fibre-optical loop are demonstrated: fundamental soliton selfswitching, its filtration from the background, and ultrashort pulse generation from the CW radiation due to cross-phase modulation (XPM) of this radiation by the fundamental soliton in the loop.

2. LOOP CHARACTERIZATION

Fiber-optical loop is the two-beam interferometer, in which radiation passes through the coupler then, after splitting, propagates along opposing directions in the loop fibre and at last returns back to the coupler and interferes there. One part of the radiation passes to the output of the loop the other comes back to the input fibre. The main advantage of this configuration is the separation of the process of the nonlinear interaction in the long optical fibre from the interference in the short single-mode directional coupler. It gives a lot of possibilities for varying of the working characteristics of the elements in the wide region.

After the travelling around the loop the radiations have the phase shift consisted of two parts. One part - the common linear phase shift, the same for both radiations because their optical paths are the same. The other - nonreciprocal phase shift which could exist due to the nonlinearity. One could take into account only the phase difference $\delta\varphi$ and receive the normalized transmittivity of the loop [6]:

$$\beta = \beta_0 + (1-\beta_0)\cdot SIN^2(\delta\varphi/2) \qquad (1)$$

Where $\beta_0 = (2\alpha-1)^2$ - transmittivity of the loop when one could neglect the phase difference $\delta\varphi$, α - the coupling coefficient of the coupler.

As it follows from equation (1), loop transmittivity takes on a maximal value $\beta_{max} = 1$ when $\delta\varphi = (2m+1)\cdot\pi$, and minimal $\beta_{min} = \beta_0$ when $\delta\varphi = 2m\pi$, where $m = 0, \pm1, \ldots$; and switching between these values is possible under varying of $\delta\varphi$.

This analysis is valid for short squared pulses with constant amplitude. For the real bell-like pulses, transmittivity varies along pulse due to varying of the $\delta\varphi$, and pulse could be broken-up or not switched fully [4]. Since optical solitons [7] are the most convenient carrier of the information in such devices due to their quasi particle propagation with constant phase along whole pulse, which makes high contrast and avoidance of the pulse break-up under switching.

Soliton passage through the loop we have described in detail in [6]. Here we concentrate our consideration on soliton filtration from the background.

There are two possibilities for the arising of the phase difference $\delta\varphi$ in the loop due to nonlinearity. The first - self-phase modulation of the nonequaled parts of the pulse propagating in the opposite direction, and the second - different phase shift of the equal counter propagated parts of the CW or quasi-CW radiation due to XPM from co propagated nonequal parts of the short pulse. Consequently at first we will discuss single soliton passage and self-switching in loop and then soliton interaction with CW or quasi CW radiation at the different wavelengths.

3. SOLITON PASSAGE THROUGH THE LOOP

Let's suppose that soliton amplitude at the input port of the coupler is described by the expression: $q_0(\tau)=SECH(\tau)$ where $\tau = t/\tau_0$ - normalized on the soliton duration τ_0 time t. After the coupler and sufficiently long fiber (L - fiber length is much greater than dispersive length of the initial soliton $L_d = \tau_0^2/(\partial^2 k/\partial\omega^2)$) the new perturbed soliton with the formfactor $æ_1 = (2\sqrt{\alpha}-1)$ is formed [8]:

$$q(\tau) = æ_1 \cdot SECH(æ_1\tau) \cdot EXP(iæ_1^2\xi/2), \tag{2}$$

where $\xi = L/L_d$. This expression is valid only if $\alpha>0.25$ $(æ_1>0)$. otherwise the soliton is not formed and the waveform is dispersed.

As one could see, two different cases are possible. The first: $\alpha > 0.75$ or $\alpha < 0.25$ when only one soliton is formed in the loop. And the second: $0.25 < \alpha < 0.75$ when two counterpropagating solitons are formed and switching take place. Consequently we'll consider this two different cases separately.

3.1. Soliton filtering from the background ($\alpha > 0.75$)

In this case the soliton forms, after the coupler, another perturbed soliton, propagating only in one direction in the loop and switching is not the case. At the loop output, immediately after the coupler, the pulse is formed: $q'(\tau) = æ_1 \cdot \sqrt{\alpha} \cdot SECH(æ_1\tau)$. Then, after sufficiently long fibre, the soliton $q(\tau)$ with formfactor $æ=(2\sqrt{\alpha}-1)^2$ is formed. The duration T and the energy E of this soliton are connected with initial ones: $T/\tau_0 = æ^{-1}$; $E/E_0 = \beta = æ$. In the calculations we neglected the nonsoliton part of the radiation. It is right, when $L \gg L_d$ and the intensity of the nonsoliton radiation is critically weaker than soliton intensity.

When the α value is unsufficient for the soliton formation in "α"-channel, then soliton is formed in "1-α"-channel. So, in this case, we can exchange in all formulas α with $1-\alpha$. Besides, it is clear that for the loop all dependencies with α are symmetrical relatively to the point $\alpha = 0.5$.

It should be noted that all this formulas are independent from the loop length. It is due to the fact that soliton doesn't changes its shape and energy under propagation in the sufficiently long loop.

The dependence of the loop transmittivity on α in the linear case and for the fundamental solitons is shown in Figure 1. The linear loop transmittivity β_0 is represented by parabola. The correlation $\beta > \beta_0$ shows that the soliton transmittivity is greater than linear one even without switching. Consequently the loop filtrates valid solitons from nonsoliton background.

Figure 2 presents the intensity profile of the initial soliton on the continuous background (0.1 of the soliton amplitude) and normalized intensity of the soliton formed after the loop in the "far field" at $\alpha = 0.75$. It is also shown pulse, formed immediately after the loop. One could see that relation between intensities of the background and soliton approximately equals to initial one. But relation between energies - really registered values, is in two times greater due to increasing of soliton width. So in this case (without switching) it is possible to make contrast soliton - background higher in two times.

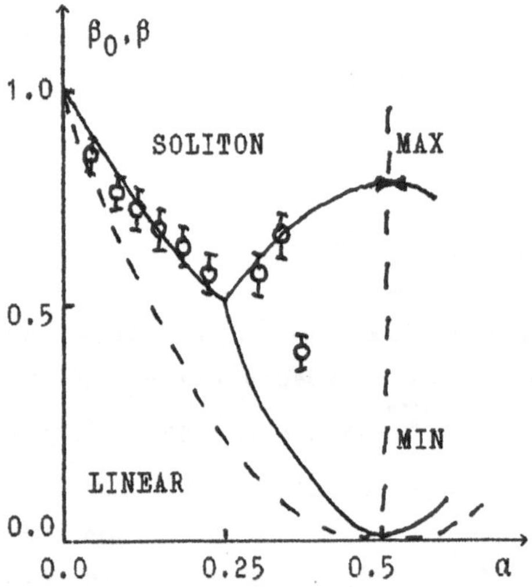

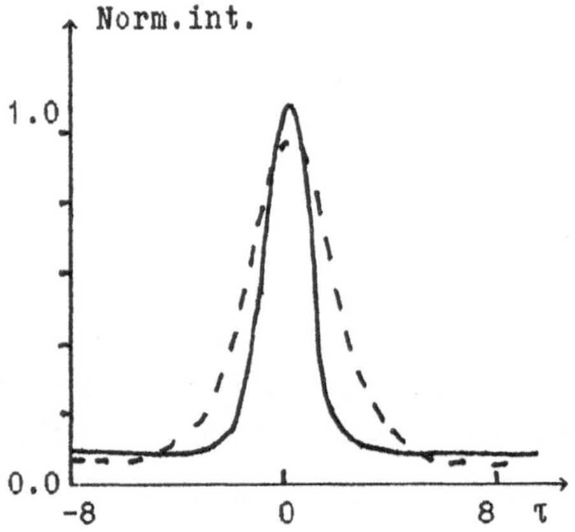

Fig.1: Loop transmittivity vs α.

Fig.2: Soliton envelope before and after the loop (α=0.75).

3.2. Fundamental soliton switching in the loop ($0.25 < \alpha < 0.75$)

In this case the counterpropagating solitons are formed in the loop and interferes in the coupler at the output. When $\alpha \neq 0.5$ the solitons amplitudes in the loop are different and the solitons phase difference exists and we have to take into account this phase difference. Solitons phase difference $\delta\varphi$ is expressed as (see (2)):

$$\delta\varphi = 2\cdot(2\alpha - \sqrt{\alpha} + \sqrt{1-\alpha} - 1)\cdot\xi \qquad (3)$$

The dependence $\beta(\alpha)$ is nonmonotonous and more complicated then in the "nonswitching" regions due to the length dependence of the $\delta\varphi$. The maximal and minimal values of the β may be calculated (with taking into account conditions for the $\delta\varphi$) by the expression:

$$\beta(\alpha) = E/E_0 = \alpha\cdot(2\sqrt{\alpha}-1) + (1-\alpha)\cdot(2\sqrt{1-\alpha}-1) \pm$$
$$\pm \sqrt{\alpha(1-\alpha)}\cdot(2\sqrt{\alpha}-1)\cdot(2\sqrt{1-\alpha}-1)\cdot I \qquad (4)$$

where $I = \int [SECH((2\sqrt{\alpha}-1)\tau')\cdot SECH((2\sqrt{1-\alpha}-1)\tau')]d\tau'$. Upper sign in (4) corresponds to maximal β, lower - to minimal. The results of calculations are presented at Figure 1 at $0.25<\alpha<0.75$ and at Figure 3.

It should be noted that there is one singular point at β_{max} curve. When α exactly equals to 0.5 the same portions of the initial pulse energy propagated in both directions, their nonlinear phase shifts equals and $\delta\varphi=0$. Consequently $\beta_{max}=\beta_{min}=0$.

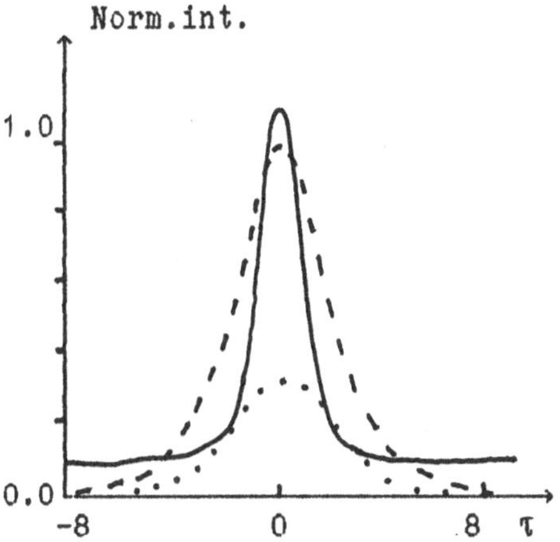

Fig.3 Soliton envelop before and after the switching (α=0.45)

The loop transmittivity depends on the phase difference $\delta\varphi$ which varies with the input solitons energy variation. The energy and time duration of the fundamental soliton are connected by the relation: $ET = CONST$ [7]. When the energy varies, the time duration and dispersive length of the soliton varies too. So, under fundamental soliton energy variation at the input of the loop, the normalized length of the loop $\xi = L/L_d$ and phase difference $\delta\varphi$ varies too. But when $\delta\varphi$ varies the loop transmittivity changes between maximal and minimal value and i.e. the self switching could take place.

The ratio of the maximal and minimal β gives the switching contrast $K = \beta_{max}/\beta_{min}$ which could be achieved at given α. The contrast K is maximal near the point $\alpha = 0.5$. In this region the parameters of the solitons, formed in the loop, are close to each other. So the envelops of the solitons are practically the same and switching with high contrast could be achieved.

When α=0.45 the maximal switching contrast equals to $K\approx30$. To observe at this α the soliton selfswitching between m-th maximum and minimum the normalized loop length must equal $\xi \approx 30\cdot m\cdot\pi$. In this case the selfswitching would be observed for the solitons with energy relation $K_0 \approx 1 + 1/(4m)$. These relations are received from (3) and (4). When m=1 then $\xi \approx 100$ and $K_0 \approx 1.25$, and when m = 5 then $\xi \approx 500$

and $K_0 \approx 1.05$. So, when the loop length is equal $L = 500L_d$, one can sort with contrast $K \approx 30$ solitons with initial energy difference only about 5%. The real loop length L is about 100 ÷ 200 meters for the solitons with duration of about 100 fs.

We have done some experiments on the fundamental soliton transmission through the fibre-optical loop based on the fused tapered coupler [6]. At Figure 1 experimentally measured transmittivity and at Figure 4 the dependence of the time duration of the formed after the loop solitons are shown . Good agreement between experiment and numerical estimations confirms correctness of the theoretical model.

4.SOLITON INTERACTION WITH CW RADIATION IN THE LOOP

When $\alpha = 0.5$ (the symmetrical coupler) the amplitudes of the counter propagated radiation equals each other and $\beta = 0$, i.e. the radiation returns back to the input fiber. Nonreciprocal phase shift could arises under interaction of CW radiation (wavelength λ_1) with the short pulse (λ_0), propagating in the loop in one direction. Due to XPM this pulse leads to additional phase shift $\delta\varphi$ to the CW radiation propagating in the same direction. The optimal pulse for modulation is soliton due to its constant form.

If soliton duration is shorter than loop round trip time the phase distortion in the counter - propagating wave will be negligible. If the loop length is sufficiently short, then phase modulation of the CW radiation will not transform to the amplitude one. In this case to calculate output signal we can take into account only phase correlation. This situation could be realized by using spectral-selective coupler with coupling coefficient varying with wavelength: $\alpha_0(\lambda_0) = 1$, $\alpha_1(\lambda_1) = 0.5$. Under these conditions pulse is formed from the CW radiation and its intensity envelop is described by the following expression:

$$I(\tau) = I_1 \cdot SIN^2(\delta\varphi(\tau)/2) \qquad (5)$$

where I_1 - initial intensity of the CW radiation; and

$$\delta\varphi(\tau) = (\eta\cdot\gamma)\cdot\int_{\tau-\gamma}^{\tau} SECH^2(\tau')\,d\tau' \qquad (6)$$

where $\eta=4\pi n_2 I_0 L/\lambda_1$; $\gamma=\Delta L/\tau_0$; n_2 - nonlinear refracted index; I_0 - peak intensity of the soliton, L - loop length; $\Delta=(1/u_0-1/u_1)$; $u_{0,1}$ - group velocities at the wavelengths λ_0 and λ_1. The formed pulse is of minimal duration and maximal intensity at the conditions: $\eta=\pi$ and $\gamma=0$:

$$I(\tau) = I_1\cdot SIN^2(\pi\cdot SECH^2(\tau/2)) \qquad (7)$$

These conditions are satisfied when $u_0 = u_1$ and $L = L_d\cdot\pi\cdot(\lambda_1/\lambda_0)/2$. When the parameter η grows, formed pulse is splitted to some peaks. When $\gamma\neq0$ then pulse amplitude decreases and duration grows due to soliton sliding with respect to CW radiation. Asymptotical (at high γ) dependence of the peak intensity I and FWHM time duration T are connected with initial ones by the following expressions ($\eta = \pi$):

$$I/I_1 \simeq (\pi/\gamma)^2 \;;\; T/\tau_0 = \gamma - 0.884$$

Optimal conditions of the pulse formation ($\gamma = 0$) could be obtained under interaction of the soliton in the negative dispersion region with the CW radiation in the positive dispersion region of the single-mode fiber. Let us suppose that soliton with duration 200 fs at the wavelength 1.6 μm ($L_d \simeq 80$ cm). Let's suppose that group velocities of the soliton and CW radiation at 1.064 μm are the same. In this case modulation is optimal ($\eta = \pi$) at the $L \simeq 80$ cm.

Analytical approximation is good for estimations but is not valid when loop length is not sufficiently short and one have to take into account dispersive evolution of the CW radiation due to the XPM phase. To receive more realistic results computer simulation have been done. Figure 4 shows results of this simulation.

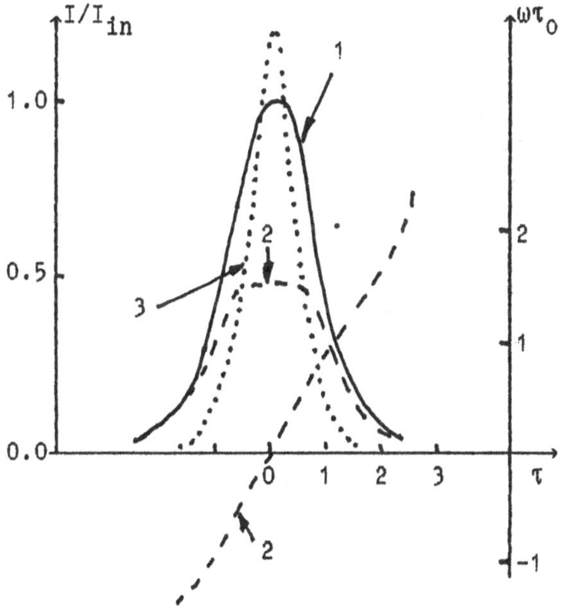

Fig.4 Soliton (1), generated pulse (2), and compressed pulse (3) at $\gamma=0$, $\eta\simeq\pi$. Dashed line - pulse chirp.

Very interesting feature of the generated pulse is the practically linear chirp within all pulse at the loop lengths $L/L_d\simeq\pi/4$. At the longer length the chirp becomes lower and curved in the central part of the pulse. When the chirp is linear through all pulse, it could be effectively compressed without any wings [9]. Compressed pulse at the loop length $L \simeq L_d\cdot\pi/4$ is shown at Figure 4.

When the intensity of the CW radiation is greater than soliton intensity it is necessary to take into account XPM by CW radiation to soliton. But at the loop length about $L/L_d=\pi/4$ the parameters of the formed pulses are approximately the same as at low intensity radiation.

5.CONCLUSIONS

All the results mentioned above are connected with the fiber-optical loop mirror - Sagnac interferometer based on a single coupler. By using two or more couplers one could receive more complicated device in which could be realized another types of the fundamental soliton conversion and radiation handling. In Mach-Zehnder interferometer [1], for example, there are two couplers and two fiber arms and it is possible to form dark or programmable ultrashort pulses from the CW radiation under soliton XPM.

REFERENCES

1. N.J.Doran and D.Wood,
   Optics Letters, 13, 56 (1988)
2. K.J.Blow, N.J.Doran and B.K.Nayar,
   Optics Letters, 14, 754 (1989)
3. N.J.Doran, D.S.Forrester and B.K.Nayar
   Electronics letters, 25, 267 (1989)
4. K.J.Smith, N.J.Doran and P.G.J.Wigley
   Optics Letters, 15, 1294 (1990)
5. M.C.Farries and D.N.Payne
   Appl. Phys. Letters, 55, 25 (1989)
6. E.A.Zakhidov, F.M.Mirtadzhiev, D.V.Khaidarov,
   A.V.Kuznetsov and O.G.Okhotnikov
   Kvantovaya Electronika (Moscow), 18, 333, (1991)
7. A.Hasegava and F.Tappert
   Appl. Phys. Letters, 23, 142 (1973)
8. J.Satsuma and N.Yajima
   Progr. Theor. Phys. Suppl., 55, 284 (1984)
9. W.J.Tomlinson, R.H.Stolen and C.V.Shank
   JOSA B 1, 139, (1984)

PART II

BIOSYSTEMS AND MOLECULAR SYSTEMS

DYNAMICS OF BREATHER MODES IN A NONLINEAR "HELICOIDAL" MODEL OF DNA

T. DAUXOIS and M. PEYRARD
Physique Non-Linéaire: Ondes et Structures Cohérentes
6 Bd Gabriel, 21000 Dijon, France.

ABSTRACT: Via a recent model with an additional helicoidal coupling, the dynamics of breathers modes in DNA are studied analytically and with the use of numerical simulations. It is shown that these excitations are longlived and can match experimentally observed fluctuational openings.

1. INTRODUCTION

Biological macromolecules undergo a complex dynamics and the knowledge of their motions provides insights into biological phenomena. Recently, attention was focused on dynamics of large amplitude localized excitations in DNA [1–4], in which the double helix fluctuates between an open state and its equilibrium structure. These oscillatory states, also called breathing modes [5] or fluctuational openings, are expected to be precursor states for the local denaturation observed during DNA transcription or thermal denaturation. In these studies, the molecule is modeled by two parallel chains of nucleotides, linked by nearest neighbor harmonic interactions along the chains and the strands are coupled to each other by Morse potentials which represent the bonding inside one base pair. Such a model does not include the helical geometry of the molecule.

But, one of the consequences of the helical structure is that nucleotides which are far apart in the one-dimensional model can be close enough in the three-dimensional structure to be connected by hydrogen-bonded water filaments. These strong water filaments has been suggested by indirect experiments [6] and results of Monte Carlo simulations [7]. They connect a phosphate group P_n at one side of the major groove with an another phosphate group $P_{n\pm4}$ at the opposite side. Therefore, in order to take into account the presence of this dynamically stable filament, the model must include a coupling between the nth nucleotide on one strand and the $n+h$ one on the other ($h = 4$ according to the experiments). Such an extension was carried out by Gaeta [8], but he considered only its consequences on the dispersion curves of the small amplitude excitations of the molecule. We consider here the nonlinear excitations in the extended model and show how the additional coupling increases the ability of the molecule to bear rather broad and sufficiently large amplitude breatherlike modes, which propagate easily along the molecule.

2. MODEL

In our model we consider a simplified geometry for the DNA chain in which, we have neglected the assymetry of the molecule and we represent each strand by a set of point masses wich correspond to the nucleotides. The characteristics of the model are the following:

(i) Like Peyrard et al [1], we only take into account transversal motions. The displacement from equilibrium of the nth nucleotide is denoted u_n (respectively v_n) for the chain C_1 (resp. C_2).

(ii) Two neighboring nucleotides of the same strand are connected by harmonic potential because we assume that the displacements due to the bubbles change only gradually from one site to the next. On the contrary, the bonds connecting the two bases belonging to different strands are extremely stretched when the double helix open locally: their nonlinearity must not be ignored. We use a Morse potential to represent not only the hydrogen bonds, but the repulsive interactions of the phosphate, and the surrounding solvent action.

(iii) Finally, we add to the model introduced by Peyrard and Bishop, a harmonic coupling which takes account of the helical geometry discussed above. It connects the nth mass on the chain C_1 to both the $(n+h)$th and $(n-h)$th masses on chain C_2.

Therefore the Hamiltonian is written as

$$H = \sum_n \Bigg[\tfrac{1}{2}m(\dot{u}_n^2 + \dot{v}_n^2) \quad + \tfrac{1}{2}k\Big[(u_n - u_{n-1})^2 + (v_n - v_{n-1})^2\Big] + D\Big[e^{-a(u_n - v_n)} - 1\Big]^2 \quad + \tfrac{1}{2}K\Big[(u_n - v_{n+h})^2 + (u_n - v_{n-h})^2\Big]\Bigg] \qquad (1)$$

where the four terms are respectively the kinetic energy of transverse vibrations, the potential energy of the longitudinal, transverse (analog to a substrate potential) and helicoidal connections. Here k (respectively K) is the harmonic constant of the longitudinal (resp. helicoidal) spring, m the nucleotide mass and D (resp. a) the depth (resp. width) of the Morse potential.

Using the variables $x_n = (u_n + v_n)/\sqrt{2}$ and $y_n = (u_n - v_n)/\sqrt{2}$, which represent the in-phase and out-of-phase motions respectively, the dynamical equations are then:

$$\begin{cases} m\ddot{x}_n = k(x_{n+1} + x_{n-1} - 2x_n) \quad +K(x_{n+h} + x_{n-h} - 2x_n) & (2) \\ m\ddot{y}_n = k(y_{n+1} + y_{n-1} - 2y_n) \quad -K(y_{n+h} + y_{n-h} + 2y_n) \\ \qquad +2\sqrt{2}aD(e^{-a\sqrt{2}y_n} - 1)e^{-a\sqrt{2}y_n} & (3) \end{cases}$$

The two equations decouple exactly and we find two linear dispersion relations (an acoustical and an optical branch). The introduction of the new coupling affects the spectrum [9], by increasing the frequencies and introducing oscillations in agreement with Gaeta's results [8].

3. BREATHER IN THE SEMI-DISCRETE APPROXIMATION

Let us focus our attention on the nonlinear equation (3), which includes the only degree of freedom interesting for the local denaturation: the stretching y_n between two nucleotides of different strands.

We are interested in collective oscillations which are large enough to be strongly anharmonic, but still much smaller than the motions which result in permanently open states, where the nucleotides reach the plateau of the Morse potential. In this hypothesis, the atoms oscillates near the bottom of the potential well, so that we assume $y = \varepsilon\phi$ (where $\varepsilon \ll 1$) and expand the substrate potential to fourth order terms in $\varepsilon\phi$. The equation of motion is then:

$$\begin{aligned} \ddot{\phi}_n = \ & \tfrac{k}{m}(\phi_{n+1} + \phi_{n-1} - 2\phi_n) \quad -\tfrac{K}{m}(\phi_{n+h} + \phi_{n-h} + 2\phi_n) \\ & -\omega_g^2(\phi_n + \alpha\phi_n^2\varepsilon + \beta\phi_n^3\varepsilon^2) \qquad (4) \end{aligned}$$

by setting $\omega_g^2 = 4a^2D/m$, $\alpha = -3a/\sqrt{2}$ and $\beta = 7a^2/3$.

According to the experimental results, the problem implies two times-scale: one corresponds to the vibration of the particle around its equilibrium position and the second, much larger, to the propagation of a collective coherent stucture along the chain.

So we will use the reductive perturbation method in which we expand in the small parameter ε and, using $\theta_n = qn\ell - \omega t$ (where ω is the optical frequency of the linear approximation and ℓ the distance between adjacent nucleotides on the same strand), we substitute

$$\phi_n(t) = \left[\varepsilon\left[F_1(\varepsilon n\ell,\varepsilon t)e^{i\theta_n} + \varepsilon^2\left(F_0(\varepsilon n\ell,\varepsilon t) + F_2(\varepsilon n\ell,\varepsilon t)e^{i2\theta_n}\right)\right] + cc + O(\varepsilon^3)\right. \tag{5}$$

in (4) by using the semi-discrete approximation [10] (the complete continuum limit would be too restrictive for DNA, where discreteness effects may be important).
Indeed, as we limit ourselves to large enough width excitations, we can determine the envelope in the continuum limit, as function of the slow variables $Z = \varepsilon z$ et $T = \varepsilon t$, while the fast oscillations of the quasiharmonic carrier, inside the envelope, are treated exactly. Equating the coefficients of ε for each harmonic, we get $F_0 = \mu\,|F_1|^2$ and $F_2 = \delta\,F_1^2$, and finally obtain the Nonlinear Schrödinger (NLS) equation for the envelope function F_1:

$$i\frac{\partial F_1}{\partial \tau} + P\frac{\partial^2 F_1}{\partial S^2} + Q|F_1|^2 F_1 = 0 \tag{6}$$

where we have made the transformation $\tau = \varepsilon T$ and $S = Z - V_g T$, with the linear group velocity $V_g = \ell\left[k\sin(q\ell) - Kh\sin(qh\ell)\right]/m\omega$, the dispersion coefficient $P = \left[\ell^2\left(k\cos(q\ell) - Kh^2\cos(qh\ell)\right)/m - V_g^2\right]/2\omega$ and the nonlinear one $Q = -\omega_g^2[2\alpha(\mu+\delta) + 3\beta]/2\omega$.

We will briefly discuss the stability of analytic solutions of NLS, which depends on the signs of PQ. However, to simplify this study, we expand these quantities to first order terms in q (a numerical study shows that the results of the stability discussion are almost unaffected by this expansion), since in next section we limit ourselves to large width bubbles, ie $q \ll 1$.

In this limit, P has the sign of $(k - Kh^2)$ and Q of $(1 - 7K/8a^2D)$. Therefore the solutions changes qualitatively, depending on the value of K. PQ is negative for $k/h^2 \le K \le 8a^2D/7$; in this case, the solution of (6) is a finite amplitude plane wave with a dip near $S - u_e\tau \simeq 0$, called a dark-soliton (or a enveloppe hole), which does not correspond to the small amplitude limit of breather modes. For $0 \le K \le k/h^2$ (this case includes the usual model without helicoidal coupling) and $8a^2D/7 \le K$, PQ is positive; we have plane waves solutions, unstable because of modulational (or Benjamin-Feir) instability, and a localised envelope solution, with a vanishing amplitude at $|z| \to \infty$: such a solution has the appropriate shape to represent breathing modes in DNA.

Therefore, the solution of (6) is then:

$$F_1(S,\tau) = A\,\text{sech}[\frac{1}{L_e}(S-u_e\tau)]\exp[i\frac{u_e}{2P}(S-u_c\tau)] \tag{7}$$

with u_e and u_c the velocities of the envelope and carrier waves, the amplitude
$A = \sqrt{(u_e^2 - 2u_eu_c)/2PQ}$ and the width $L_e = 2P/\sqrt{u_e^2 - 2u_eu_c}$. The envelope soliton, solution of (4), is a plane wave with a frequency corrected for the nonlinearity, an amplitude modulated by a sech-type envelope modified by the second harmonic and the non-oscillating components. By setting $V_e = V_g + \varepsilon u_e$, $\Theta = q + \varepsilon u_e/2P$ and $\Omega = \omega + (V_g + u_c\varepsilon)\varepsilon u_e/2P$, it reads:

$$y_n(t) = 2\varepsilon A\,\text{sech}[\varepsilon(n\ell - V_e t)/L_e]\Big[\cos(\Theta n\ell - \Omega t)$$

$$+\varepsilon A\,\text{sech}[\varepsilon(n\ell - V_e t)/L_e] \times \Big(\mu/2 + \delta\cos[2(\Theta n\ell - \Omega t)]\Big)\Big] \quad +O(\varepsilon^3)$$

When $K < k/h^2$, we obtain a very narrow pulse, almost identical to those found in the model without helicoidal interactions [4] (because K approaches 0). On the contrary, when $K > 8a^2D/7$, the solution is much broader and has a larger amplitude so that it could provide a better representation of the fluctuational openings of DNA. We have investigated its stability numerically.

4. NUMERICAL RESULTS

The lifetime of the solutions determined above is an important parameter, because only long-lived excitations can be detected experimentally. First we discuss briefly the numerical technique, and then we compare the numerical and theoretical results.

Basically, we perform the simulation by using a continuum breather as an initial condition in the discrete lattice, with the complete Morse potential. Then, we simulate the ensuing propagation of the pulse, solving the Newtonian equations of motion with a fourth-order Runge-Kutta method. The timestep Δt is chosen so that the total energy of the system is conserved to a relative accuracy better than 10^{-3}.

The question of the choice of parameters for this model is still a controversial topic, as shown by the debate over these values in the literature [11]. We have chosen a dissociation energy $D = 0.1\ eV$, $a = 2\ \text{\AA}^{-1}$, coupling constants $k = 1.5\ eV/\text{\AA}^2$ and $K = 0.5\ eV/\text{\AA}^2$, a distance between base pairs $\ell = 3.4$ Å and a mass of 300 m.u. for each nucleotide. To generate the bubbles, we choose a small value for the wave vector ($q = 0.01$ Å^{-1}) and therefore th wavelength of the carrier wave is in the range of the envelope width: the solution is similar then to a local opening which oscillates.

As long as the amplitude remains in the region where the Taylor's development is justified (typically where y is lower than the Morse potential inflection point), our approximations are valid so that the solution can be expected to be stable. In order to describe the large amplitude fluctuational openings observed in DNA, we must however consider initial conditions with a larger amplitude.

The figure 1 shows the motion of a breather with an initial amplitude of 1 Å and a half-width of 18 nucleotides. We can see that, when the motion begins, the amplitude adapts to the real substrate potential. The figure exhibits an amplitude modulation not explained by the calculations performed in the limit of small displacements, ie in the bottom of the Morse well.

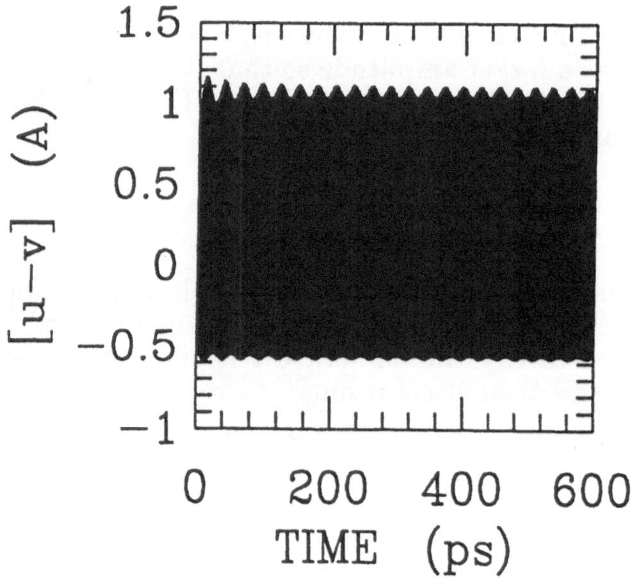

Fig 1: $(u - v)$ vs time for the center of the breather ($\varepsilon = 0.007$, $u_e = 10^3$ Å$/ps$ and $u_c = 0$ Å$/ps$). The figure contains about 1000 oscillations of the breather.

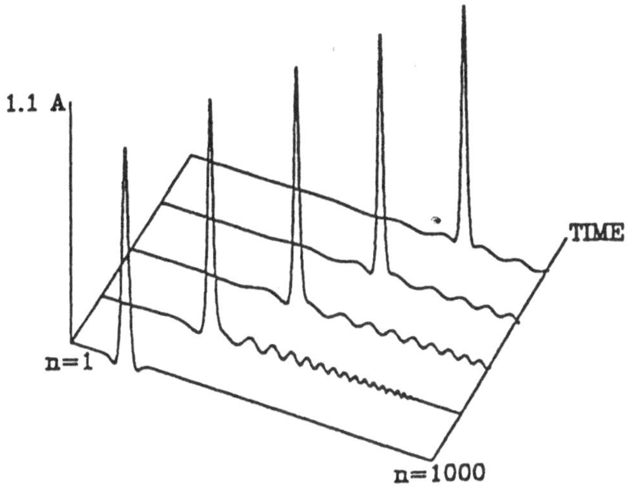

Fig 2: Propagation of the breather along the chain (only 1000 nucleotides are represented). The transverse stretchings are shown every 250 oscillations, when the position of the breather center is at its maximun. Note the assymetry of the backward and forward radiation patterns.

Figure 2 shows these excitations to be very long-lived, although some radiation is emitted by the breather. In spite of this radiation, it should be noticed that the decrease in amplitude is only very weak.

In order to analyse the emitted waves, we have studied the amplitude of the stretching at a distance of 100 particles away from the center of the breather. After the first burst due to adaptation, the radiative rate decreases, and finally corresponds to a permanent emission of resonant phonons. Indeed, a temporal Fourier transform of the same simulation data, started at $t \simeq 400$ ps, shows that the frequency of the breathing oscillation $\omega_B = 11.20$ ps^{-1} is about 1 % higher than the analytical value; the frequency of the radiated phonons is $\omega_P = 10.97$ ps^{-1}, which coincides with ω within 0.2 % and attests the coupling mechanism of the breatherlike motion to phonons radiation.

The position of the frequency in the bottom of the dispersion relation ($V_g \simeq 0$), explains the slow speed of the radiations packets, compaired to the speed of the burst due to adaptation. Besides, the propagation speed of the breather $V_e = 3.7$ $Å/ps$, is about 20% less than the theoretical value, because of the discreteness effects which usually tends to slow down the motion.

5. CONCLUSION

Our primary aim was to construct a new extended model for the coherent dynamics of bubbles in DNA. We considered, on one hand, first-neighbour harmonic longitudinal and nonlinear transverse interaction and, on the other hand, an harmonic helicoidal coupling, due to transgroove hydrogenbonded water filaments. Then envelope solitons, solutions of the NLS equation were obtained using a perturbation approach and simulation results were used to show the coupling mechanism between the motion of the breather and phonons radiation. Note that the addition of the helicoidal term, introducing modifications in P and Q, has created a special zone without breather modes. We emphasize that this model can have large amplitude broad oscillations which better correspond to the fluctuational openings of DNA, whereas the previous model with similar parameters cannot.

Nevertheless, it is obvious that before obtaining a suitable description of DNA, we have to take into account the local assymetry of the two helices, as well as the second principal source of nonlinearity which appears as DNA chains unwind: the bistability of the sugar ring, which allows sugar puckering modes.

REFERENCES:

1. M.Peyrard, A.Bishop, *Phys. Rev. Lett.* 62 (1989) 2755
2. M.Techera et al, *Phys. Rev. A* **41** (1990) 4543
3. V.Muto, A.C.Scott,P.L.Christiansen *Phys. Rev. A* **136** (1989) 33
4. M.Peyrard,T.Dauxois, H.Hoyet, in proceedings of NATO ARW *Coherent and Emergent Phenomena in Biomolecular Systems*, Tucson, USA, January 1991
5. E.W.Prohofsky et al, *Phys. Lett. A* **70** (1979) 492
6. U.Dahlborg, A.Rupprecht, *Biopolymers* **10** (1971) 849
7. G.Corongiu,E.Clementi, *Biopolymers* **20** (1981) 551
8. G.Gaeta, *Phys. Lett. A* **143** (1990) 227
9. T.Dauxois, submitted to *Phys. Lett. A*
10. M.Remoissenet, *Phys. Rev. B* **33** (1986) 2386
11. M.Techera et al, *Phys. Rev. A* **42** (1990) 5033

Equilibrium and Nonequilibrium Statistical Mechanics of a Nonlinear Model of DNA

Mario Techera
Max Planck Inst. For Biophysical Chemistry,
Dept. of Molecular Biology,
W-3400 Göttingen, Germany

L.L.Daemen
Theoretical Division, Los Alamos National Laboratory,
Los Alamos, New Mexico 87545

E.W.Prohofsky
Department of Physics, Purdue University,
West Lafayette, Indiana 47907

Experimental and theoretical studies indicate that the hydrogen bond stretch mode dominates DNA dynamics close to denaturation temperatures. We analyze a simplified model for DNA which retains only this (nonlinear) degree of freedom. The dynamics and thermodynamics of the system are discussed. In particular, the analytical and numerical results do not exhibit a melting transition but instead a state of pseudo-equilibrium distinct from the state expected from equilibrium thermodynamics. Finally numerical results show that energy transport is unlikely at biological temperatures.

I. THE MODEL

Deoxyribonucleic acid, or DNA, is conceivably the most important biomolecule. Its double stranded helical structure is of particular interest since the four bases (Adenine, Thymine, Guanine and Cytosine or A,T,G and C), whose sequence determines the genetic code, are projected inward toward the helix axis. On the outside of the double helix is found the backbone formed by two strands consisting of alternating phosphate groups and deoxyribose sugars. An excellent overview of DNA structure and function can be found in Saenger.[1] As such, the geometry of the double helix requires that the two complementary strands come apart in order for the base sequence to be read by other molecules. This melting, or denaturation of DNA has been the study of intensive experimental and theoretical investigation because of its biological importance.[2] In the present paper we present the motivation for a very simple model of DNA along with an analysis of its thermodynamics.

The infrared transmission spectrum of DNA has shown the existence of soft mode around 85 cm^{-1}.[3] This mode is seen to drop in frequency as the melting temperature is approached. Using normal mode analysis Awati was able to characterize this mode as a collective motion of the bases that stretch the interbase hydrogen bonds (HBs).[4] With the use of the Modified Self-Consistent Phonon Approximation, MSPA, he was able to predict the temperature dependence of this mode along with the fact that it also gained further HB stretch character as the melting temperature is approached.[4]

The following simplified geometry is considered for DNA; the molecule is first untwisted and each strand is then represented by a set of point masses (the nucleotides) connected by linear springs. The intrastrand interactions (*i.e.* the HBs between base pairs) are modeled by a Morse potential. Schematically this can be represented as in Fig.1(a). The displacement from equilibrium of the nth mass point is denoted by $u_n(v_n)$ in the top (bottom) chain respectively. Only transverse motions are considered. The equations of motion for u_n and v_n:

$$m\ddot{u}_n = k(u_{n+1} + u_{n-1} - 2u_n) - \frac{\partial \phi}{\partial(u_n - v_n)}, \tag{1}$$

$$m\ddot{v}_n = k(v_{n+1} + v_{n-1} - 2v_n) + \frac{\partial \phi}{\partial(u_n - v_n)}, \tag{2}$$

where ϕ is the non-linear potential describing the HB interaction. At this point, it should be emphasized again that the source of the nonlinearity in the model lies in the coupling between the strands not between adjacent particles on the same strand.

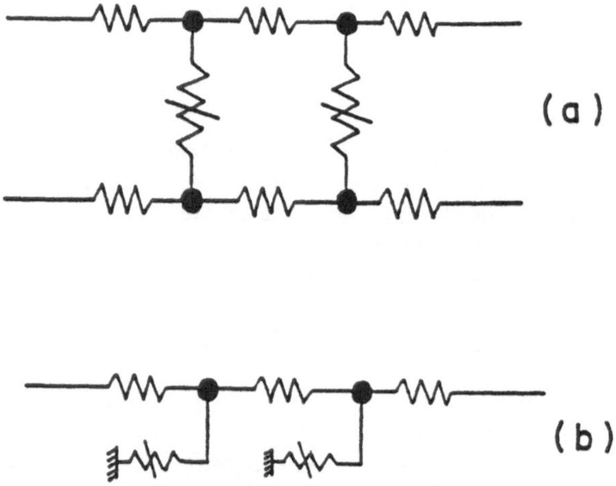

Figure 1

Since the individual masses of the four different nucleotides differ by at most 13%, the masses of the particles on each strand have been made equal in the above equations.[1] Furthermore, for the sake of simplicity, it has been assumed that the force constants k are the same throughout the chain. These assumptions permit a transformation to center of mass coordinates:

$$x_n = \frac{1}{\sqrt{2}}(u_n + v_n), \tag{3}$$

$$y_n = \frac{1}{\sqrt{2}}(u_n - v_n), \tag{4}$$

Eqns.(1) and (2) become:

$$m\ddot{x}_n = k(x_{n+1} + x_{n-1} - 2x_n), \tag{5}$$

$$m\ddot{y}_n = k(y_{n+1} + y_{n-1} - 2y_n) - \frac{\partial \phi}{\partial y_n}. \tag{6}$$

x_n is the motion of the center of mass and y_n describes the motion about the center of mass (a positive y_n represents a stretch). The potential energy ϕ is chosen to model the nonlinear HB interaction between base pairs. This is typically done by using the Morse potential:

$$\phi_M(y_n) = V_\infty(1 - e^{-\sqrt{2}ay_n})^2, \tag{7}$$

which, with suitably chosen parameters, can provide a good description of HB potentials in DNA.These equations may be obtained from the following Hamiltonian:

$$\begin{aligned} H &= \sum_n \frac{p_n^2}{2m} + \frac{k}{2}(x_{n+1} - x_n)^2 + \frac{q_n^2}{2m} + \frac{k}{2}(y_{n+1} - y_n)^2 + V_\infty[1 - \exp(-\sqrt{2}ay_n)]^2 \\ &\equiv H_x + H_y. \end{aligned} \tag{8}$$

We note here for future reference, that H_y can be viewed as an ensemble of Morse oscillators with nearest-neighbour harmonic coupling in the displacements and that in the limit $k \to 0$ we have an ensemble of independent Morse oscillators. The equations of motion are now uncoupled. Eqn.(5) represents a pure harmonic lattice with plane wave solutions. In what follows, all the attention will be focused on the motion about the center of mass Eqn.(6). This equation can also be viewed as describing longitudinal displacements in the one-dimensional chain shown in Fig.1(b). An analysis of the dynamics of this model has been presented elsewhere.[5]

II. EQUILIBRIUM THERMODYNAMICS

An initial attempt at calculating the thermodynamics for this model was presented by Peyrard and Bishop in 1989.[6] Their idea was to apply the transfer integral method to evaluate the partition function in the canonical ensemble.[7] Here we will show that strictly speaking this approach is incorrect due to the fact that the Morse potential is bounded for large stretches. To make this statement as clear as possible we consider first an ensemble of independent Morse oscillators i.e. H_y of Eqn.(8), in the limit $k \to 0$. H_y then becomes:

$$H_y = \sum_n \frac{q_n^2}{2m} + V_\infty[1 - \exp(-\sqrt{2}ay_n)]^2. \tag{9}$$

In order to calculate the thermodynamic properties in the canonical ensemble, the usual procedure is to calculate the partition function Z_y: [8]

$$Z_y = \frac{1}{h^N} \int_{-\infty}^{+\infty} \prod_{n=1}^{N} dy_n dq_n \exp(-\beta H_y), \tag{10}$$

$$= \frac{1}{h^N} \left(\frac{2\pi m}{\beta}\right)^{\frac{N}{2}} I^N(-\infty, +\infty), \tag{11}$$

where

$$I(-\infty, +\infty) \equiv \int_{-\infty}^{+\infty} dy_n \exp\left\{-\beta V_\infty[1 - \exp(-\sqrt{2}ay_n)]^2\right\}, \tag{12}$$

and where N is the number of particles, $\beta = 1/k_B T$ and the integration is performed over all the available phase space *i.e.* $y_n \in (-\infty, +\infty)$. Everything appears to be in order until one notices that for $y_n \to +\infty$ the integrand in Eqn.(12) is *bounded.* In other words,*the integral* $I(-\infty, +\infty)$ *and consequently the partition function* Z_y *are divergent*! Furthermore, if one considers

$$\lim_{y_u \to +\infty} I(-\infty, y_u), \tag{13}$$

it can be shown that the divergence is linear in y_u, meaning that the integral diverges linearly with volume. This result *per se* may not be distressing since the partition function has no direct physical meaning, however on further examination the consequences become clear. Consider the average displacement from the Morse well , $< y_n >$ (which for nonzero values of k in Eqn.(8) represents the average HB stretch of a base pair in the molecule), for a given particle in the ensemble:

$$\begin{aligned} < y_n > &= \frac{1}{Z_y h^N} \int_{-\infty}^{+\infty} \prod_{n=1}^{N} dy_n dq_n \, y_n \exp(-\beta H_y) \\ &= \left[\frac{\int_{-\infty}^{+\infty} dy_n \, y_n \exp\left\{-\beta V_\infty [1 - \exp(-\sqrt{2} a y_n)]^2\right\}}{I(-\infty, +\infty)} \right]^N . \end{aligned} \tag{14}$$

The numerator in the above expression is also infinite but diverges as the square of the volume and the denominator has the linear volume divergence mentioned previously. The average position of a particle in the ensemble is then $< y_n >= +\infty$. *Thus the particles are at equilibrium only when they have escaped the well.* This interpretation can be made more rigorous by considering finite upper boundaries on the integrals and taking the limit of the boundary to $+\infty$ as in Eqn.(13). As long as the boundary remains finite, so does the quantity $< y_n >$, but this quantity diverges in the limit of a boundary at infinity. We thus conclude that a meaningful physical treatment of such Hamiltonians in the canonical ensemble requires the explicit introduction of a cutoff in the integral which of course one hopes can be interpreted physically.

The Morse oscillator illustrates a much more general feature of systems of particles evolving under the influence of long-range, unscreened forces which asymptotically tend to a finite value. Considerable care must be exerted in applying the methods of statistical mechanics to these systems.[9] In particular, concepts such as thermodynamical equilibrium, statistical ensemble, thermodynamic limit, and the meaning of averages of various thermodynamic quantities must be thoroughly examined. In particular, several conceptual difficulties arise when the methods of statistical mechanics are applied in the canonical ensemble. Other physical systems of interest also involve long-range forces. Recently, an extensive study of the statistical mechanics of gravitating systems was performed by Padmanabhan.[10] Many of the conclusions he drew for gravitating systems are similar to our results for the Morse potential.[11]

One can now go back to the full Hamiltonian H_y as defined in Eqn.(8) and show that it exhibits the same problem. The partition function is given by:

$$Z_y = \frac{1}{h^N} \left(\frac{2\pi m}{\beta} \right)^{N/2} \int_{-\infty}^{+\infty} \prod_{n=1}^{N} dy_n \exp\left\{ -\beta \left[\frac{k}{2} (y_{n+1} - y_n)^2 + V_\infty [1 - \exp(-\sqrt{2} a y_n)]^2 \right] \right\} \tag{15}$$

which again is not finite, since the integrand does not vanish at infinity. In particular, one can easily verify this from the above equation by considering the line $y_1 = y_2 = \ldots = y_n$ as $y_n \to +\infty$. A dramatic consequence of this divergence, from the point of view of the model is that the the average stretch of a base pair is infinite at all nonzero temperatures at equilibrium. The implication is that the thermal equilibrium state of the the chain, is one with all the bonds stretched to infinity. This is a direct consequence of the boundedness of the Morse potential as stretches become large (even infinite). This straightforward analysis does not, however, give us an idea of the time it takes for the stretches to become infinite, a question which will also be analyzed further in the next section.

For the sake of completion, we present in Fig.2 the results for the average HB stretch, $< y_n >$, as a function of temperature, in the thermodynamic limit for various values of the cutoff, y_u, i.e. the Morse potential is valid for $y_n < y_u$ and there is an infinite barier for stretches greater than y_u. These results were calculated by solving numerically a transfer integral equation.[11] We stress that the transfer integral method is valid for this calculation only because we have added an explicit cutoff. The depth of the well, V_∞ is set to 0.2eV, $k = 0.3eV/\AA^2$ and $a = 2.77\AA^{-1}$ (this puts the inflexion point of the Morse at about $0.2\AA$) which is a typical parametrization of the Morse potential.[12]

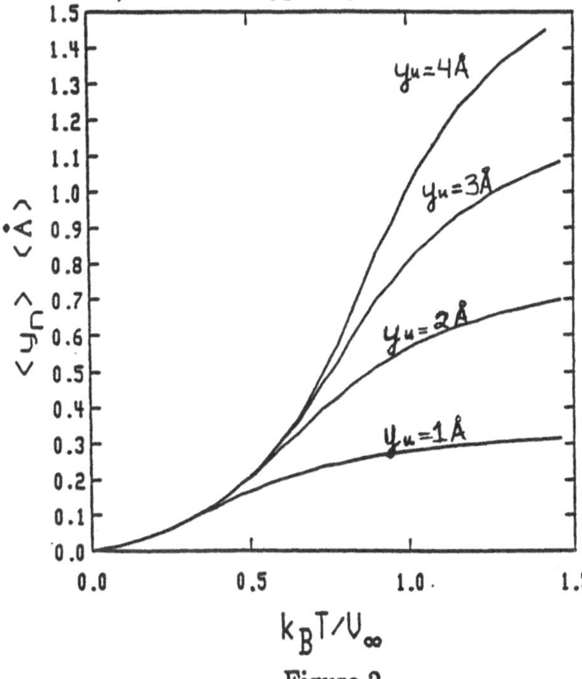

Figure 2

III. NONEQUILIBRIUM THERMODYNAMICS

Based on the aforementioned results in the canonical ensemble and as a consequence of treating the whole Morse potential with no cutoff, one can then ask the following question: *If we don't restrict the Morse potential with a cutoff how much time is required for this divergence in the average HB stretch to occur at some temperature T, given some generic initial condition of the chain?* Of course the obvious answer is an infinite amount of time since the particle must travel to $+\infty$. One can, nevertheless, define some finite stretch as representing dissociation, in which case the required time is finite. In the present model, this stretch can be identified with the stretch required for imino proton exchange in DNA to occur. By the argument presented in the last section, the above question is one of nonequilibrium thermodynamics.

In order to answer this question, one approach consists in using the Langevin equations.[13] The equations of motion are modified to include a heat bath with which the system is in contact and which is at a constant temperature (this can be associated to the solution in which the molecule is typically found). We add two additional terms to simulate stochastic collisions and dynamical friction of the molecule with the environment.[14] Eqn.(6) now becomes a system of stochastic differential equations given by:

$$m\ddot{y}_n = k(y_{n+1} + y_{n-1} - 2y_n) - \frac{\partial \phi}{\partial y_n} - m\gamma\dot{y}_n + \eta_n(t). \tag{16}$$

Characteristic of such equations, the influence of the surrounding medium is split into two parts: 1) a systematic term $-m\gamma\dot{y}_n$ which causes the friction and 2) a fluctuating part $\eta_n(t)$ which represent the 'random' collisions of the constituents of the medium with the system in question. This fluctuating term will be assumed to posess certain properties, namely:

$$< \eta_n(t) > = 0 \tag{17}$$

$$< \eta_n(t_1)\eta_m(t_2) > = q\delta(t_1 - t_2)\delta_{nm}, \tag{18}$$

where $q = 2mk_BT\gamma$.[15] The details of how this system is solved numerically along with further results are presented elsewhere. [11] The lattice consists of $N = 125$ particles with circular boundary conditions *i.e.* $y_1 = y_{126}$. The dissipative constants are assumed to be equal at all the lattice sites with a value of $\gamma^{-1} = 20\tau$, where τ is the small amplitude period of the Morse oscillators. Thus, the system will be underdamped and not dominated by friction, otherwise, the parameters will be taken to be similar to those used in the previous section.

The procedure to be followed will be to start the lattice at its dynamical equilibrium position *i.e.* with all particles at the bottom of their respective Morse wells, with no kinetic energy, and then to evolve the system according to Eqns.(16). In all runs, the average kinetic energy of a base pair on the lattice, $< K_n >$, always thermalized to $\frac{1}{2}k_BT$ as expected (after about $50ps$). For the above parametrization of the Morse potential, a temperature of $T_c = 4642K$ would correspond to the average kinetic energy being equal to the depth of the Morse well, nevertheless, at temperatures well below T_c it was possible to observe the system denature in a brief period of time as seen in Figure 3(a) which plots the average HB stretch vs time for a $T = 3000K$. The vertical scale is important since the inflexion point of the Morse is at about $y_n = 0.2\AA$ as mentioned above. Our calculations have shown that the lower the temperature the longer the required time for the divergence discussed in the previous section to become evident. In fact, at $T = 300K$, room temperature, we performed several runs of $10ns$ without ever observing any such divergence. Figure 3(b) shows a portion of such a run. Once again notice the vertical scale. We can therefore see that even though the system is not in equilibrium, that at low temperatures the system behaves as if it were in a *pseudo-equilibrium* because its average kinetic energy is at the right value but other quantities, such as the average HB stretch $< y_n >$, are not.

Finally it should be mentioned that the Langevin approach was used to study the effects of thermalization on the dynamics of this model that had been observed previously, particularly on the quasi-solitonic modes that exist.[5] The essential effect was that for realistic Morse parameters, nonlinear wave propagation was strongly hindered at temperatures above $10K$.[11]

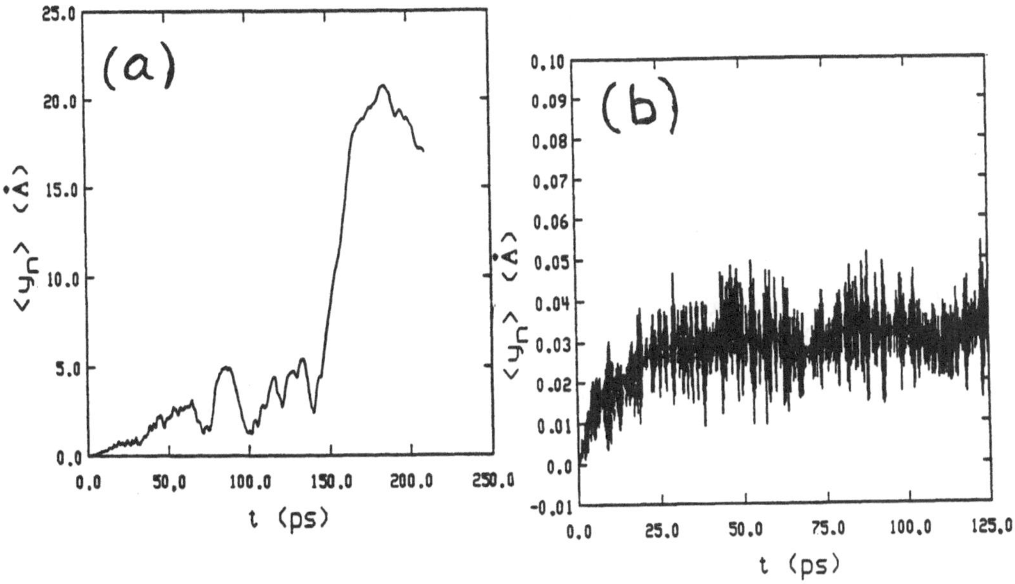

Figure 3

IV. CONCLUSION.

We have shown that the thermodynamics of this simple DNA model can be treated meaningfully in the canonical ensemble, even though the partition function diverges - one just has to interpret carefully the nature of the thermodynamic equilibrium state. Furthermore the denaturation predicted for all temperatures has been shown to be directly related to the boundedness of the Morse potential which represents the HB interaction. This problem has been encountered recently in the study of Coulomb gases[9] and the statistical mechanics of gravitating systems.[10] We have presented two ways of circumventing this problem: 1) avoid the divergences by putting a cutoff to the Morse potential i.e. allow only for a certain maximum stretch 2) accept the divergence as a final equilibrium state in the average stretch for all T and study how the system in a nonequilibrium state evolves to this final state. In the latter case, we presented briefly results on the dynamical approach to equilibrium and showed that the time required for the bases to separate can be very long for low temperatures.

DNA is known to undergo a structural phase transition during thermal denaturation[1], but this model showed no such clear cut behaviour at any particular temperature. We can therefore conclude that more physics needs to be incorporated such as next-to-nearest neighbour coupling and the helical twist. The HBs alone seem to account for energy localization due to nonlinearity and that might be an important precursor to thermal denaturation, but the transition itself requires that more phenomenology be added to such a model.

---

[1]W.Saenger,'Principles of Nucleic Acid Structure'(S pringer-Verlag,New York,1983).

[2]V.Muto,J.Halding,P.L.Christiansen and A.C.Scott, J.Biom.Struct.Dyn. 5,873 (1988);V.Muto, A.C.Scott and P.L. Christiansen, Phys.Let.A 136,33 (1989); S.Yomosa,Phys.Rev.A 27,2120 (1983);S.Takeno,Prog.Theor.Phys.71,395 (1984); C.Zhang,Phys.Rev.A 35, 886 (1987).

[3]J.W.Powell,G.S.Edwards,L.Genzel,F.Kremer,A.Wittlin, W.Kubasek and W.Peticolas, Phys.Rev.A 35, 9 (1987).

[4]K.Awati, Ph.D. Thesis, Purdue University, West Lafayette, Indiana, USA (1989).

[5]M.Techera, L.L.Daemen and E.W.Prohofsky, Phys.Rev.A 40, 6636 (1989);M.Techera, L.L.Daemen and E.W.Prohofsky, Phys.Rev.A 41, 4543 (1990);M.Techera, L.L.Daemen and E.W.Prohofsky, Phys.Rev.A 42, 1008 (1990).

[6]M.Peyrard and A.R.Bishop, Phys.Rev.Let. 62, 2755 (1989).

[7]D.J.Scalapino,M.Sears,and R.A.Ferrel, Phys.Rev.B 11,3535(1975).

[8]K.Huang, 'Statistical Mechanics', 2nd Edition (Wiley,Chichester,1987).

[9]M.K.H.Kiessling, J.Stat.Phys.59, 1157, (1990).

[10]T.Padmanabhan, Phys.Reports, 188, 285 (1990).

[11]M.Techera, Ph.D. Thesis, Purdue University, West Lafayette, Indiana, USA (1991).

[12]R.M.Wartell and A.S.Benight, Phys.Rep.126, 67 (1985).

[13]J.McLennan,'Introduction to Non-Equilibrium Statistical Mechanics' (Prentice Hall,New Jersey,1989).

[14]S.Chandrasekhar, Rev.Mod.Phys. 15,1 (1943).

[15]T.Schneider and E.Stoll, Phys.Rev.B 17, 1302 (1978).

A simple model of DNA dynamics

Giuseppe Gaeta
Centre de Physique Theorique
Ecole Polytechnique, F-91128 Palaiseau (France)

In 1989 professor Yakushevich (of the Institute of Biological Physics of the Academy of Sciences of the URSS) proposed a simple model for DNA (torsion) dynamics [1], from now on called Y model. Further details on this model, as well as on many other "theoretical physics" issues in biology and molecular biology are also contained in the book [2].

This model is schematically illustrated in Figs.1 and 2; the (longitudinal) springs joining discs on the same chain give only a torque, while the (transversal) ones joining discs on different chains work as shown in Fig.2; see [1] for further details. It is quite clear that this model does not take into account the helical structure of DNA.

A consequence of the double helix structure of DNA and a simple way to take this into account is that, even if we consider only next neighbour interactions among the bases, this leads to interaction (also called non-covalent interactions) among bases which are far apart along the bases (as the two which are pointed out schematically in Fig. 4), since they are actually neighbouring in space.

A proposal to modify Yakushevich model in this direction [3], see Fig.3, was indeed presented shortly after the the appearance of [1] (see also [4] for a related discussion on topological features), and leads to qualitatively new phenomena. We refer to [1],[3] for a full discussion of the model, while here we just discuss the aspects which are relevant to the topic of the conference.

In Y model, the degrees of freedom correspond to torsion of the bases, so that to each base is associated a single scalar variable, $\phi_k^i \equiv \phi_k^i(t)$, where $i = \pm 1$ identifies the chain of the double helix to which the base belongs, while k identifies the site along the helix.

The Hamiltonian will be written as

$$\mathcal{H} = T + V^{(\ell)} + V^{(t)} \tag{1}$$

Contribution to the workshop *Nonlinear Coherent Structures in Physics and Biology* (Dijon, 4-6 June 1991) where T is the kinetic energy,

$$T = \sum_{i,n} \frac{1}{2} I (\dot{\phi}_n^{(i)})^2 \tag{2}$$

and $V^{(\ell)}, V^{(t)}$ are the potential energy terms corresponding to longitudinal and transversal interactions respectively:

$$V^{(l)} = \sum_{i,n} \frac{1}{2} K_L a^2 (\phi_{n+1}^{(i)} - \phi_n^{(i)})^2 \tag{3}$$

$$V^{(t)} = \sum_{n} \frac{1}{2} K_T (\Delta l_n)^2 \tag{4}$$

Here K_L, K_T are "elastic constants"; I is the moment of inertia of the discs, and Δl_n is, after ref. [1],

$$\Delta l_n = [(2r + l_0 - r\cos\phi_n^{(1)} - r\cos\phi_n^{(2)})^2 + (r\sin\phi_n^{(1)} - r\sin\phi_n^{(2)})^2\,]^{1/2} - l_0 \equiv l_n - l_0 \tag{5}$$

where a is the distance among the bases, and r their radius.

Notice that we have assumed that the discs are all equal, that the coupling constants for interactions of the same type are equal, and that the longitudinal elastic constants along the two chains are equal.

In the "helicoidal" version of Y model [3], from now on referred to as modified Y model, one adds a term

$$V^{(h)} = \sum_{i,n} \frac{1}{2} K_H d^2 (\phi_{n+h}^{(i)} - \phi_n^{(-i)})^2 \tag{6}$$

in the Hamiltonian, with K_H another elastic constant and d the distance in space among bases interacting via the $V^{(h)}$ term, to take into account the helicoidal (i.e. non-covalent) interaction of the kind illustrated in Fig.4; the original Y model corresponds then to $K_H = 0$.

In the continuum limit, we have z the coordinate along the chains, and we are left with two scalar fields $\phi^i \equiv \phi^i(z,t)$, $i = \pm 1$.

We want to consider the case $\ell_0 \simeq 0$, i.e. $\Delta\ell_n \simeq \ell_n$; moreover we are interested in long wavelenght solutions, which justifies passing to the continuum limit. When all these approximations are made, one is left with the equations

$$I\phi_{tt}^{(i)} = K_L a^4 \phi_{zz}^{(i)} - K_T r^2 [2\sin\phi^{(i)} - \sin(\phi^{(i)} + \phi^{(-i)})] + K_H d^2 [2(\phi^{(-i)} - \phi^{(i)}) + \phi_{zz}^{(i)} w^2] \tag{7}$$

where w is the lenght of an half-wind of the helix in the z coordinate (i.e. along the helix itself).

From our point of view here, this model presents two remarkable features:

Soliton solutions

There are special ansatzes which lead to well known equations and in turn to soliton solutions. Indeed, for $\phi^{\mu} = 0, \phi^{(1)} \equiv \phi$, we get

$$I\phi_{tt} = K_L a^4 \phi_{zz} - K_T r^2 \sin\phi - K_H d^2 \phi \tag{8}$$

Similarly, for $\phi^{(-1)} = -\phi^{(1)} \equiv \phi$, we get

$$I\phi_{tt} = (K_L a^4 - K_H d^2 w^2)\phi_{zz} - 2K_T r^2 \sin\phi - 4K_H d^2 \phi \tag{9}$$

I.e., we get sine-Gordon type equations; the K_H coupling is responsible for the appearance of a mass term.

As for the $\phi^{(-1)} = \phi^{(1)} \equiv \phi$ case, we have no qualitative difference due to the introduction of the helicoidal interaction: indeed, we get

$$I\phi_{tt} = (K_L a^4 + K_H d^2 w^2)\phi_{zz} - 2K_T r^2 \sin\phi + K_T r^2 \sin 2\phi \tag{10}$$

We remark that this kind of ansatzes fits well in the framework of "conditional symmetries" [5,6] in the group theoretical approach to differential equations [7,8,9]

Dispersion relations

By linearizing eq. (7), we get

$$I\phi_{tt}^{(i)} = K_L a^4 \phi_{zz}^{(i)} - K_T r^2 [\phi^{(i)} - \phi^{(-i)}] + K_H d^2 [2(\phi^{(-i)} - \phi^{(i)}) + \phi_{zz}^{(i)} w^2] \tag{11}$$

If now we look at travelling wave solutions,

$$\begin{aligned} \phi^{(1)}(z,t) &= \alpha e^{i(qz-\omega t)} \\ \phi^{(-1)}(z,t) &= \beta e^{i(qz-\omega t)} \end{aligned} \tag{12}$$

we get the relation

$$(\sigma - I\omega^2)^2 = \mu^2 \tag{13}$$

where

$$\sigma = q^2(K_L a^4) + (K_T r^2 + 2K_H d^2) \tag{13'}$$

$$\mu = K_T r^2 + 2K_H d^2(1 - w^2 q^2) \tag{13"}$$

so that the spectrum of the model consists of an acoustical and an optical branch,

$$\omega^2 = \frac{\sigma \pm \mu}{I} \tag{14}$$

given explicitely by

$$\begin{aligned} I\omega_a^2 &= (K_L a^4 + K_H d^2 w^2) q^2 \\ I\omega_o^2 &= (K_L a^4 - K_H d^2 w^2) q^2 + (2K_T R^2 + 4K_H d^2) \end{aligned} \tag{15}$$

In this case, the introduction of an helicoidal coupling is responsible for quite relevant qualitative changes in the behavious of the model.

Indeed, for $K_H = 0$, the two branches are at constant distance (see Fig.5),

$$\omega_0^2 - \omega_a^2 = \frac{2K_T}{I} r^2 \tag{16}$$

while for $K_H \neq 0$ they can cross, as depicted in Fig.6. The corresponding wavelenght λ_{cr} is given by

$$\frac{\lambda_{cr}}{w} = 2\pi \left(\frac{K_H}{K_T r^2 + 2K_H d^2} \right)^{1/2} \equiv 2\pi \left(\frac{1}{2+\chi} \right)^{1/2} \tag{17}$$

Conditions for such a crossing are shortly discussed in [3]; here we notice that a similar crossing of bands in the spectrum of (simple) molecules is well known in laser spectroscopy, and usually corresponds to the appearance of complex phenomena (also called "quantum bifurcations") [10,11,12]; it can be considered, from our point of view, as the signal of appearance of a complicate, possibly chaotic, dynamics.

References

[1] L.V. Yakushevich; *Phys. Lett. A* **136** (1989), 413

[1] L.V. Yakushevich; "Methods of theoretical physics and their applications to biopolymer sciences" (in russian - english translation to appear), Pushchino 1990

[3] G. Gaeta; *Phys. Lett. A* **143** (1990), 227

[4] G. Gaeta; *Phys. Lett. A* **151** (1990), 62

[5] D. Levi and P. Winternitz; *J. Phys. A* **22** (1989), 2915

[6] G. Gaeta; *J. Phys. A* **23** (1990), 3643

[7] P.J. Olver; "Applications of Lie groups to differential equations"; Springer (N.Y.) 1986

[8] G.W. Bluman and S. Kumei; "Symmetries and differential equations", Springer (N.Y.) 1989

[9] H. Stephani; "Differential equations. Their solution using symmetries", Cambridge 1989

[10] I.M. Pavlichenkov and B.I. Zhilinskii; *Ann. Phys.* (N.Y.) **184** (1988), 1

[11] V.B. Pavlov-Verevkin, D.A. Sadovskii and B.I. Zhilinskii; *Europhys. Lett.* **6** (1988), 573

[12] G. Pierre, D.A. Sadovskii and B.I. Zhilinskii; *Europhys. Lett.* **10** (1989), 409

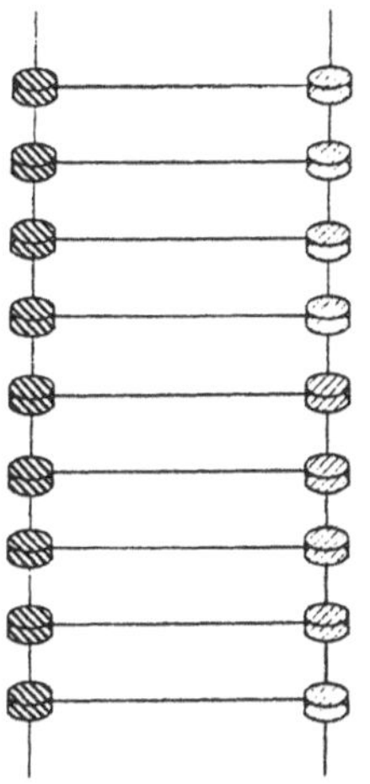

Figure 1 - The Yakushevich model

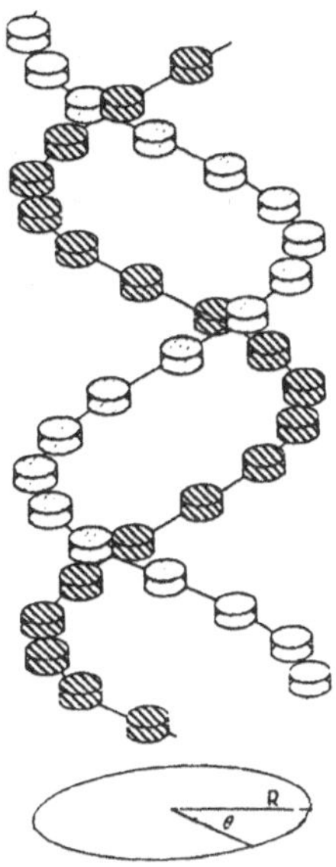

Figure 3 - The modified Y model
(interactions are not indicated)

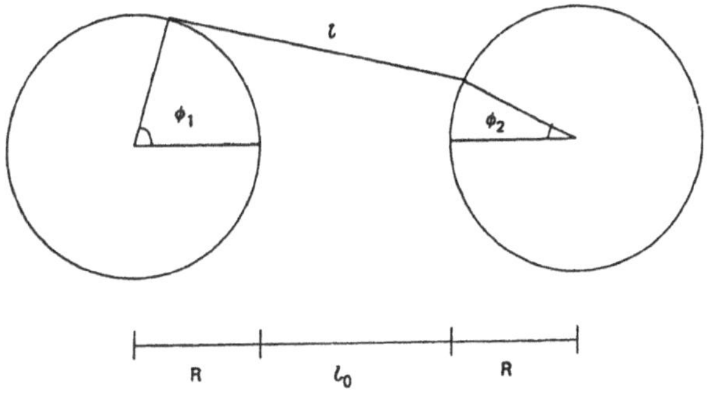

Figure 2 - Detail of the "transversal" interaction in the Y model (from [1])

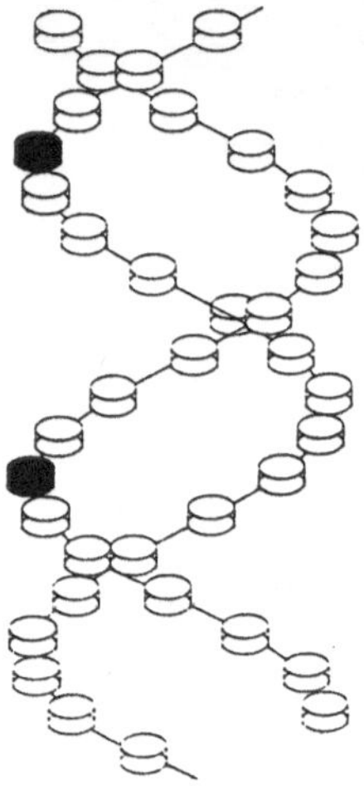

Figure 4 - Two bases interacting via the "helicoidal" term

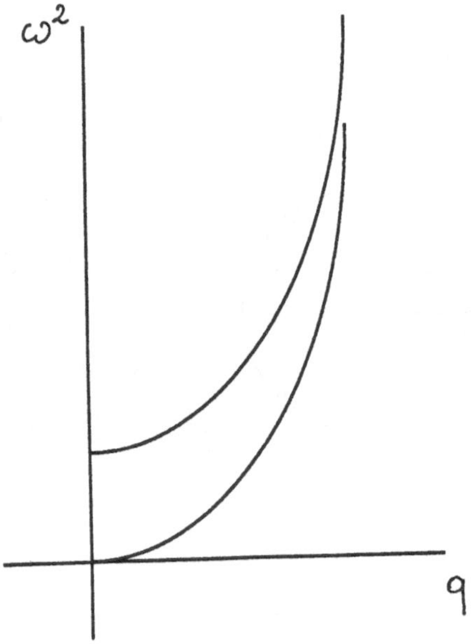

Figure 5 - Dispersion relation for the Y model

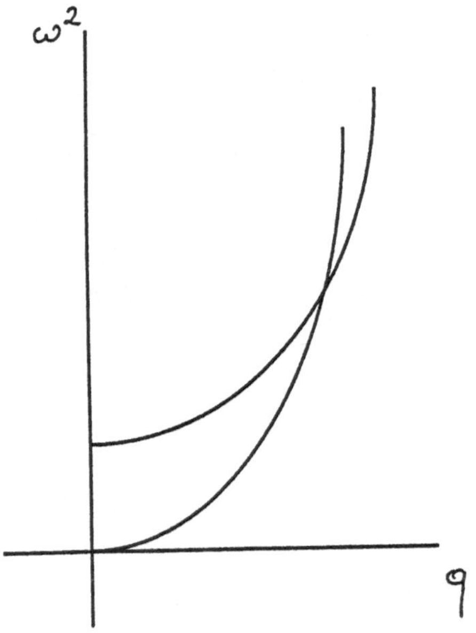

Figure 6 - Dispersion relation for the modified Y model

ANOMALOUS VIBRATIONAL MODES IN ACETANILIDE :

a F.D.S. incoherent inelastic neutron scattering study

Mariette BARTHES*, Juergen ECKERT+, Susanna W. JOHNSON+, Jacques MORET*, Basil I. SWANSON≈ and Clifford J. UNKEFER≈.

*Groupe de Dynamique des Phases Condensées , Université des Sciences , 34095 Montpellier , France

+LANSCE , and ≈INC4 ,Los Alamos National Laboratory , Los Alamos ,NM 87545 , USA

** *** **

Extended Abstract

The origin of the anomalous infra-red and Raman modes in acetanilide ($C_6H_5NHCOCH_3$, or ACN)(1) , remains a subject of considerable controversy. One family of theoretical models involves Davydov-like solitons (2) nonlinear vibrational coupling (3), or "polaronic" localized modes (4)(5). An alternative interpretation of the extra-bands in terms of a Fermi resonance was proposed (6) and recently the existence of slightly non-degenerate hydrogen atom configurations (7) in the H-bond was suggested as an explanation for the anomalies.

In this paper we report some new results on the anomalous vibrational modes in ACN that were obtained by inelastic incoherent neutron scattering (INS) . Comparisons of the spectra of ACN and of five deuterated derivatives,greatly facilitates assignments of the main features .

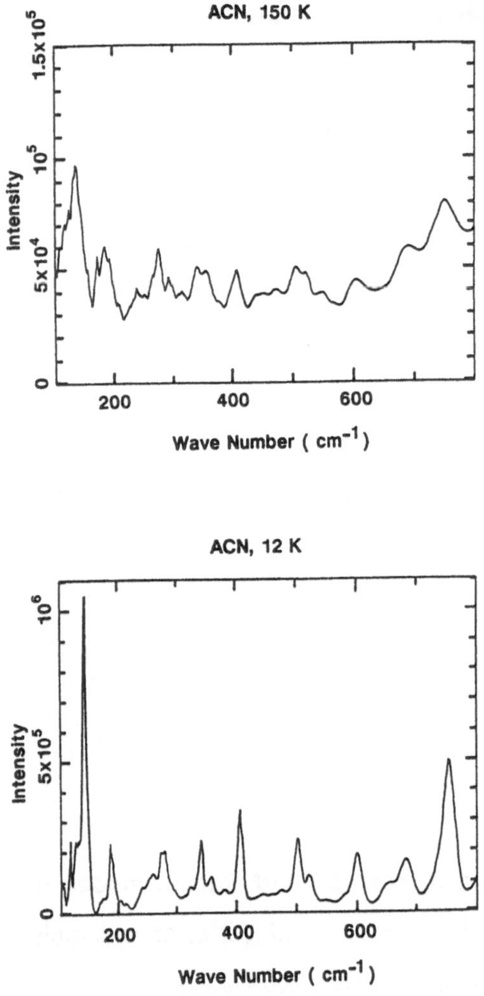

- Fig.1 - INS spectra of ACN

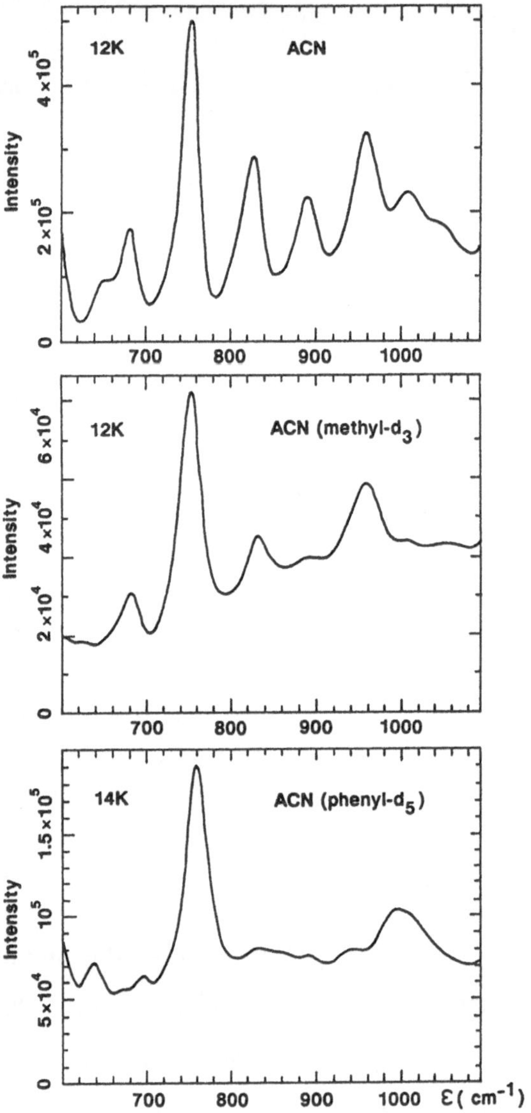

- Fig.2 - INS spectra of the γ-NH mode

The mutual influence of the different chemical groups (amide, methyl group and phenyl ring) are emphasized by means of selective deuterations. The possibility of vibrational coupling between them is discussed.

In addition, the temperature dependence of the γ(NH) modes (out-of-plane bend of the H-bond) and of the methyl torsional modes are described.

These data are then compared to the predictions of the different theoretical models that attempt to explain the anomalies.

The INS data were collected on theFilter Difference Spectrometer (FDS), at Manuel Lujan Jr. Neutron Scattering Center of Los Alamos National Laboratory, between lOK and 3OOK and deconvoluted ,to determine the frequency distribution . The useful range of energy transfers , was 1OO cm^{-1}-16OOcm^{-1} (12meV-2OOmeV) with a relative resolution (ΔE/E) of about 2% .

The five deuterated derivatives are :

$C_6H_5NHCOCD_3$: ACN -d_3 ; $C_6H_5NDCOCD_3$: ACN -d_4 ; $C_6D_5NHCOCH_3$: ACN -d_5 ; $C_6D_5NDCOCH_3$: ACN -d_6 ; $C_6D_5NDCOCD_3$: ACN-d_9 .

Comparison of the INS spectra of the various ACN isotopomers is a powerful tool for determining the origin of the modes and their assignments . In addition,direct subtraction ("differential spectroscopy") can be used to highlight the modes of a particular molecular group.

The low temperature spectrum of ACN is displayed in fig. 1 from 1OO to 8OOcm^{-1}. Fig. 2 shows the INS spectrum in the region of γ(NH) at 12K for the three samples that have a protonated amide group. Tentative assignments of the bands are given in table I. These are in agreement with available previous data from Raman and IR spectroscopy .However , the INS do provide new informations : for example ,in amide deuterated derivatives some small peaks at 750 cm-1 may be attributed to out- of-plane C-H bending of the phenyl . In ACN these bands are screened by the N-H bending , and never before identified .

Fig 3 shows the methyl torsional modes at 12 K in three samples. Despite the limited resolution, the methyl mode appears to be split , in agreement with the Raman scattering spectra (8)(9). In the single crystal polarized Raman study (9) with a z(xy)z configuration the methyl torsion was found to gradually split as the temperature is lowered below about 200K . The frequencies of the three

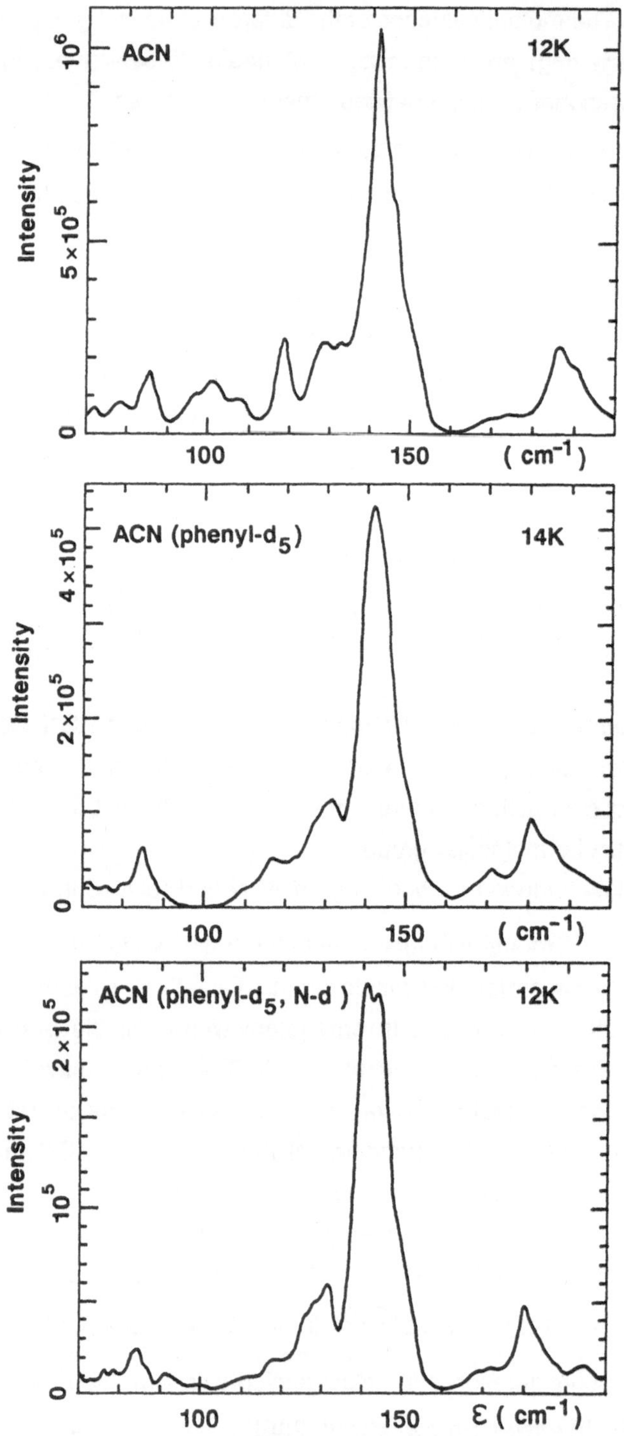

- Fig.3 - Methyl torsional modes

components as well as the splittings all increase with decreasing temperature . INS studies [10] previously identified the methyl torsion in ACN but the splitting was not observed . In the present work at least three components may be identified in ACN , namely at 142, 146 and 152cm-1 . The effect of deuteration of the phenyl ring is a significant change in the shape and of width of the methyl librational band. In ACN-d_6 the splitting is even more evident, and the relative intensities of the component have changed. These observations clearly demonstrate the influence on the dynamics of the amide and methyl groups on each other.

The frequency shift of the methyl torsions as a function of temperature obtained from our data ,fig.4 , is consistent with the measurements of Johnston et al.[9] . The γ (NH) mode at 754 cm^{-1} is also found to be affected by deuteration of the other groups as shown in Fig.2. This mode undergoes a change in shape and width upon deuteration of the methyl or phenyl group. Similar changes in shape were observed in the corresponding IR absorption bands [11].

Another important feature of these spectra is the temperature dependence of the γ (NH) mode, at 750-770cm^{-1}. The intensity of this mode was observed to increase strongly with decreasing temperature (Fig.1) . Moreover if one considers the two spectra of ACN-d3 shown in fig 5 , in which there is no longer any contribution from the methyl group, it is apparent that the intensity of the γ(NH) peak increases much more than other peaks with decreasing temperature. So, the unusual intensity increase of the γ(NH) with decreasing temperature is therefore an intrinsic property of this mode .

These observations are in agreement with our IR spectra [11] which indicate an total increase of intensity of the γ(NH) with decreasing temperature of about 20% between 300K and 15K. Similar observations have also been made by Raman scattering in ACN[12].

Data from the different techniques are thus in agreement about this important result , namely that γ(NH) also has an anomalous temperature dependence. So , ACN exhibits not only an anomalous amide mode at 1650cm^{-1} but also another one at 750cm^{-1}.

Among these new results, two of them seem to be especially relevant to the problem of the anomalous modes in ACN :

- observation of the interdependence of the chemical groups, by selective deuteration.

- Anomalous increase of the intensity of γ(NH) with decreasing temperature.

The band corresponding to the methyl torsional transitions is split in three components and is sensitive to the deuteration of other parts of the molecule.This splitting (also observed in the polarized Raman scattering[(9)]) may be rationalized by a small departure from three-fold symetry for the rotational potential. This could be induced by the low symetry of the steric environment of the methyl group which in turn could be caused by differences in the positions of the hydrogen bond protons . The methyl group would as a consequence occupy energetically inequivalent positions and therefore have different librational frequencies . The change in shape and width of the librational components when the amide or phenyl ring are deuterated could result either from direct vibrational coupling of the methyl torsion with motions of the other groups (or with low frequency phonons) or from a modification of the rotational potentiel by changing the local environment of CH_3.

The former model is consistent with the hypothesis of "polaronic" local modes or solitons [(5)], while the latter would be related to the assumption of multiple conformations of the molecular chain[(7)].

The width and shape changes of the bands corresponding to the N-H bending mode upon deuteration of other groups of the molecule could be accounted for by the same mechanisms , i.e. either direct vibrational coupling, or a multiple-well potential for the amide proton .In this latter case , changes in the steric environment affects the shape of this potential , and thus the frequencies and intensities of the transitions to the excited levels.

At this point it is not possible to decide which of these possible explanations is the relevant one.

The second result, which has now been observed by INS , IR , and in Raman scattering is the anomalous thermal behaviour of γ (NH) . It may be expected to provide further important input into the determination of the origin of these anomalies.For example,in the case of localized modes (polarons, solitons or coupled oscillators) the temperature dependence of the anomalous intensity should obey a characteristic law $I(T)/I(0) \approx \exp(-T^2/\Theta^2)$ [(4)] while the hypothesis of non degenerate substates for the amide proton would mean either a temperature independent global intensity , or one governed by the Boltzman population of each level.

Our former infrared data [11] indicate that the intensity of the γ(NH) mode of ACN obeys the $\exp(-T^2/_{\Theta}2)$ which would favor the family of models of localized nonlinear excitations.However , former theories of the ACN problem only took into account the anomaly at 1650 cm-1, and the recent observations of new anomalies suggest that the self-trapping mechanism in ACN may be more complex . Further analysis is underway.

Acknowledgments : This work has benefitted from the use of facilities at the Manuel Lujan Jr. Neutron Scattering Center, Los Alamos . This work is supported by NATO under grant n° 910281 .

REFERENCES

1 - G.Careri, U.Buontempo , F.Galluzi , A.C.Scott , E.Gratton and E.Shyamsunder - Phys.Rev.B 30, 4689, 1984.

2 - J.C.Eilbeck , P.S.Lomdahl and A.C.Scott - Phys.Rev. B 30, 4703, 1984.

3 - S. Takeno - Prog. Theor. Phys. 75,1,1986.

4 - D.M.Alexander and J.A.Krumhansl - Phys. Rev.B 33, 7172, 1986.

5 - A.C. Scott , I. J.Bigio and C.T. Johnston - Phys. Rev. B 39,12883 , 1989

6 - C.T.Johnston , B.J.Swanson - Chem. Phys. Lett. 114, 547, 1985.

7 - W. Fann , L. Rothberg , M. Roberson , S. Benson , J. Madey, S. Etemad ,and R. Austin - Phys. Rev. Lett. 64 , 607 , 1990

8 - J.L.Sauvajol,J.Moret,R.Almairac and M. Barthes. - J.Ram. Spect. 20,517,1989.

9 - C.T.Johnston,B.Swanson,J.Eckert and C.J.Unkefer- J.Phys.Chem.(in press 1991)

10 - M. Barthes ,R. Almairac, J.L. Sauvajol ,J. Moret,R. Currat ,J. Dianoux - Phys.Rev.B 43 , 5223, 1991 .

11 - M . Barthes - in " Self-Trapping of Vibrational Energy in Protein " - "Davydov's Soliton Revisited" A.C. Scott Ed. - Plenum Press 1990 -

12 - C.T. Johnston and B.I. SWANSON (unpublished)

TABLE I

TENTATIVE ASSIGNMENT OF THE INS SPECTRUM OF ACETANILIDE

FREQUENCY(cm^{-1})	ASSIGNMENT	OBSERVATIONS
0-33	Acoustic modes	
35-100	External modes librations.	IR and Raman spectroscopy
	FDS	
130-140	Methyl torsional transitions	(10), (9), sensitive to the deuteration of other group
171		shoulder disappearing at 15K
186	External mode	progressive energy shift with deuteration.
275	Phenyl modes	
345	Methyl modes	
406	(C-C-C) out-of-plane deformation	
502	Methyl modes	
521	(C-C-C) out-of-plane deformation	
600	Methyl modes	
646	Methyl	
683	Phenyl modes ,C-C-H deformation	
754	γ(NH) (out-of -plane bending mode) and γ(CH),Phenyl .	Anomalous modes
829	γ(CH) , Phenyl	
890	Combination band	decreases in ACN-d3 and in $ACNd_5$
959	γ(CH) Phenyl	Breathing modes (1)
1020	Methyl rock	
1140	δ(CH) Phenyl	

Notes: Vibrational modes that involve predominantly methyl group motions appear to be heavily coupled with other modes (see Fig.3) .Their precise assignment requires a normal coordinate analysis . γ and δ refer , respectively , to out-of-plane and in-plane bends .

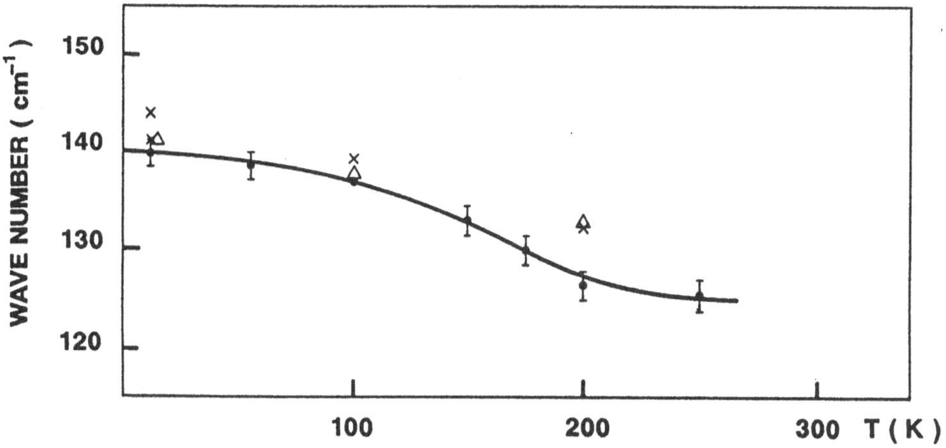

- Fig.4 - Frequency shift of the methyl torsions .

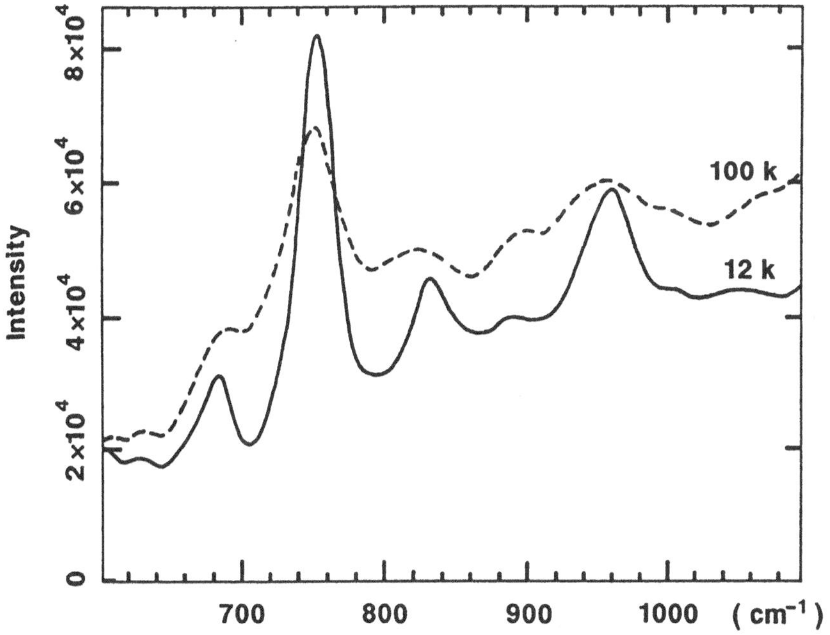

- Fig.5 - The γ-NH mode at 12K and 100K in ACN-d3

Nonlinear Excitations in a Quantum Dimer

Lisa J. Bernstein
Department of Chemistry, B-040 and the Institute for Nonlinear Science
University of California, San Diego, La Jolla, CA 92093 USA

1. Introduction

Nonlinear dynamical systems provide us with a relatively simple way to describ rich collection of physical phenomena. Taking advantage of the descriptive power nonlinear equations requires special justification, however, if the dynamical system interest is a molecule. For any physical system, a nonlinear dynamical model is approximation of the exact quantum-mechanical dynamics that comes from Schrödinge linear wave equation. For molecular systems, we must ask what errors are involved approximating Schrödinger's equation by a simpler but nonlinear model. We should a understand how interesting nonlinear phenomena are manifested in quantum-mechani energy spectra, in order to test the predictions of nonlinear models against experimer evidence.

One classically-nonlinear phenomenon that is relevant to molecular systems is tendency for the vibrational energy in a system of coupled oscillators to become localiz When the forces in coupled-oscillator system are nonlinear, it can be energetically fav able for vibrational energy to be concentrated on a single oscillator instead of distribu evenly over the system. Such localization of vibrational energy is observed experim tally in, for example, the C-H stretch vibrations of benzene [1], and is known in the ph cal chemistry literature as a local mode. Satisfactory quantum theories for local mode small molecules have been developed [2,3].

The work presented here is motivated by a related theory for localized and possi mobile excitations in another type of molecular system: alpha helix protein. In 19 Davydov proposed that interactions (which are nonlinear when described in terms of c sical physics) between Amide-I (C==O) bond oscillations and the vibrations of adjac hydrogen bonds could lead to stable coherent excitations in alpha helix sections of biol ical protein [4]. Although much theoretical work has been done since Davydov's origi paper, most calculations are based on nonlinear approximations of the true quant dynamics, for which precise estimates of the approximation errors are not known.

I am studying a model for the same classically-nonlinear interactions involved in Davydov's protein theory, but which is composed of only four coupled oscillators instead of the hundreds of bonds included in realistic protein models. For my system, it is possible to calculate the exact quantum dynamics by numerically solving the Schrödinger equation. These exact solutions can be used to evaluate the accuracy of nonlinear approximation methods. Here I will present some exact wavefunctions for this model and make a preliminary evaluation of the Discrete Self-Trapping Equation as a nonlinear approximation.

2. The Fröhlich-Einstein Dimer

The Fröhlich-Einstein Dimer (FED) is a system of four coupled oscillators, two with frequency Ω_0, representing the Amide-I vibrations in a protein model, and two Einstein oscillators with frequency ω_0, representing deformations of a surrounding molecular structure such as the stretching of hydrogen bonds in an alpha helix. The FED is shown schematically in Figure 1.

Figure 1:

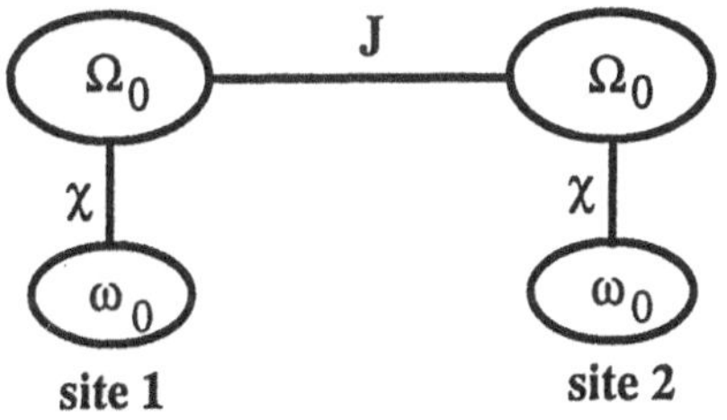

This model is defined by a two-site version of the Fröhlich Hamiltonian used in Davydov's protein theory. For the case where a single quantum of Amide-I energy is present in the system, the FED energy operator can be written

$$H_{FED} = \sum_i [\hbar\, \omega_0 b_i^\dagger b_i + \chi \sqrt{\frac{\hbar}{2M\,\omega_0}} (b_i + b_i^\dagger) B_i^\dagger B_i] + J(B_1^\dagger B_2 + B_2^\dagger B_1),$$

where b_i ($b_i^\dagger$) is a lowering (raising) operator for excitations of the ith Einstein oscillator and B_i ($B_i^\dagger$) is a lowering (raising) operator for excitations of the ith Amide-I bond. The Einstein oscillators have a characteristic mass M and a spring constant w such that $\omega_0 = \sqrt{w/M}$. The displacement operator Y_i for the ith Einstein oscillator is

$$Y_i = \sqrt{\frac{\hbar}{2M\,\omega_0}} (b_i + b_i^\dagger).$$

The strength of the intersite coupling between the two Amide-I bonds is given by the real parameter J. The magnitude of intrasite interactions between the Amide-I bond and the Einstein oscillator at each site is given by the real parameter χ.

The values for the FED parameters considered here are those appropriate for the one-dimensional version of Davydov's protein model [5], with the values for χ and w taken to be the midpoints of the appropriate ranges of values. These parameter values are listed in Table I.

Table I: Parameters		
J	1.55×10^{-22}	Joules
χ	48.5×10^{-12}	Newtons
w	48.75	Newtons/meter
M	5.7×10^{-25}	kilograms

The calculations presented here for the parameters of Table I complement recent results which concentrate on systems with lower Einstein oscillator frequency (smaller ω_0) [6] and stronger intrasite coupling (larger χ) [7].

3. FED Quantum Dynamics

Three different descriptions of the FED quantum dynamics are presented here. First, for comparative purposes, the exact quantum theory for the classically-linear case $\chi = 0$ is described. For $\chi \neq 0$, we consider both the exact quantum theory and a nonlinear approximation called the Discrete Self-Trapping Equation. We look for the FED version of a Davydov soliton - a quantum-mechanical wavefunction for which there is initially a greater probability of finding the quantum of Amide-I energy on one of the two sites. Without loss of generality, we consider local modes on site 1.

3.1. The Zero-Interaction Limit ($\chi = 0$)

If the value of χ is set to zero, the two stationary quantum states of lowest energy for the FED are

$$\psi_\pm(t) = \frac{1}{\sqrt{2}}(|1>|0> \pm |0>|1>)|0(Y_1)>|0(Y_2)>\exp(\frac{iE_\pm t}{\hbar})$$

where $|1>|0>$ and $|0>|1>$ are harmonic-oscillator wavefunctions for the two Amide-I bonds, with the single quantum of vibrational energy localized on site 1 and site 2, respectively. The factor $|0(Y_i)>$ is a harmonic-oscillator ground state wavefunction for the ith Einstein oscillator, with equilibrium position $Y_i = 0$. The corresponding energies are $E_{\pm} = \pm J$.

3.2. The Discrete Self-Trapping Approximation

Most of the nonlinear approximations used on Davydov's model are based on assumptions about the form of the quantum wavefunctions. One commonly-used trial wavefunction is

$$\psi(t) = (c_1|1>|0> + c_2|0>|1>)|0(Y_1 - \beta_1)>|0(Y_2 - \beta_2)> \exp(\frac{iEt}{\hbar}) \qquad (1)$$

when written specifically for the FED. Here, $|1>|0>$ and $|0>|1>$ are the Amide-I states discussed in the previous section, and c_1, c_2, β_1 and β_2 are time-dependent parameters. The Einstein oscillator harmonic-oscillator states have time-dependent equilibrium positions $\beta_i(t)$. The complex amplitudes c_1 and c_2 have the property that $|c_i(t)|^2$ gives the probability of finding an Amide-I excitation on site i at time t. In the jargon of Davydov soliton fanatics, this is the "D2 Ansatz" for the FED.

With the additional assumption that the Einstein oscillators respond instantaneously to changes in the Amide-I probabilities, the equations of motion for the parameters in the trial state (1) are

$$i\hbar\, \dot{c}_1 = -Jc_2 - \frac{\chi^2}{w}|c_1|^2 c_1 \qquad (2a)$$

$$i\hbar\, \dot{c}_2 = -Jc_1 - \frac{\chi^2}{w}|c_2|^2 c_2 \qquad (2b)$$

and

$$\beta_i = -\frac{\chi}{w}|c_i|^2. \qquad (3)$$

The system (2a,b) is known as the Discrete Self-Trapping Equation (DST) [8]. The time-evolution of a local mode is obtained by integrating (2a,b) from an initial condition such that $|c_1|>|c_2|$. Because the trial state (1) is not general enough to include all the possible FED wavefunctions, (2a,b) will generally not give the exact quantum dynamics [9].

The solutions of (2a,b) are known for arbitrary initial conditions in terms of Jacobi elliptic functions [10,11]. For $c_1(0) = 1$ and $c_2(0) = 0$, the site 1 excitation probability evolves as

$$P_1(t) = |c_1(t)|^2 = \tfrac{1}{2}[1 + \mathrm{cn}(u\,|\,k)]$$

where $\mathrm{cn}(u\,|\,k)$ is the Jacobi elliptic function cn with argument $u = 2.94\,t$ and modulus $k = .078$. This elliptic function oscillates between 1 and -1 with a frequency $\omega = 2.93\ \mathrm{psec}^{-1}$. Assumption (3) about the Einstein oscillators implies

$$<Y_1(t)> = A\,[1 + \mathrm{cn}(u\,|\,k)],$$

where, for the Table I parameters, $A = -.497$.

3.3. Exact Quantum States of the FED

Because the two sites in the FED are identical, we can choose all stationary quantum states to be either symmetric or antisymmetric with respect to site exchanges. Using the coordinates

$$\delta = \frac{1}{\sqrt{2}}(Y_1 - Y_2) \text{ and } \sigma = \frac{1}{\sqrt{2}}(Y_1 + Y_2),$$

the exact stationary states of the FED system can be written

$$\psi^{\pm}(t) = [f^{\pm}(\delta)|1>|0> \pm f^{\pm}(-\delta)|0>|1>]\,|m(\sigma + \frac{\chi}{\sqrt{2}w})> \exp(\frac{iE^{\pm}t}{\hbar}),$$

where $|m(\sigma + \chi/\sqrt{2}w)>$ is a harmonic oscillator state for the coordinate σ with equilibrium position $\sigma = -\chi/\sqrt{2}w$, and the $\pm$ sign indicates whether the state $\psi^{\pm}(t)$ is symmetric or anti-symmetric. A particular function $f^{+}(\delta)$ ($f^{-}(\delta)$) corresponds to each symmetric (anti-symmetric) stationary state.

Shore [12] showed that a straightforward numerical calculation can be used to obtain the functions $f^{\pm}(\delta)$ in terms of the coefficients in the expansion

$$f^{\pm}(\delta) = \sum_{j=0}^{\infty} c_j^{\pm}\,|j(\delta)>, \tag{4}$$

where $|j(\delta)>$ is the jth harmonic oscillator number state in the coordinate δ. The coefficients $c_j^{\pm}$ are found by diagonalizing the matrices H^{+} and H^{-} with elements

$$H_{jj}^{\pm} = \pm(-1)^{j+1} J + \hbar\,\omega_0\, j - \frac{\chi^2}{4w}$$

and

$$H_{j+1,j}^{\pm} = H_{j,j+1}^{\pm} = \tfrac{1}{2}\left[\omega_0 \frac{\chi^2}{w}\right]^{\frac{1}{2}} \sqrt{j+1}.$$

Each of the eigenvectors of H^{+} (H^{-}) gives the coefficients c_j^{+} (c_j^{-}) in the expansion (4) corresponding to a symmetric (anti-symmetric) stationary state of the FED. The associated eigenvalue of H^{+} (H^{-}) is the energy of that stationary state. The expansion coefficients and energies of the two lowest-energy FED eigenstates are listed in Table II.

Table II: Exact FED Stationary States						
State	Energy	c_0	c_1	c_2	c_3	$c_{j>4}$
ψ^+	-1.76 x 10^{-22} Joules	.704	-.060	.005	.000	.000
ψ^-	1.25 x 10^{-22} Joules	.698	-.113	.009	-.001	.000

It is significant that the coefficients c_j are not the same for the two states in Table II. This means that unlike for $\chi = 0$, one cannot generate the anti-symmetric state by "anti-symmetrizing" the symmetric state. Another difference between the two states is that while the energies of both are lower than the $\chi = 0$ values $E^{\pm} = \pm 1.55\text{x } 10^{-22}$ Joules, the anti-symmetric state has been lowered .30x 10^{-22} Joules while the symmetric state has been lowered only .21x 10^{-22} Joules. These observations indicate that an accurate characterization of a superposition wavepacket requires knowledge of *all* the stationary states involved in the superposition. Often in studies of larger systems only the lowest-energy eigenstate is computed.

4. Local Mode Wavepackets

For $\chi = 0$ and for the $\chi \neq 0$ exact calculations, the two lowest-energy stationary states are used to construct a nonstationary wavepacket that concentrates the Amide-I excitation probability on site 1 at time $t = 0$. In each case, the site 1 excitation probability $P_1(t)$ oscillates periodically in time for t>0. Four quantities are used to characterize the local mode wavefunctions in order to make comparisons between the exact results, the linear limit, and the DST approximation. These quantities are: a) the expectation of the energy for the localized wavepacket $\psi_L(t)$, that is, $<\psi_L(t)|H_{FED}|\psi_L(t)>$, b) the initial site-1 excitation probability $P_1(0)$, c) the corresponding Einstein oscillator displacement $<Y_1(0)>$, and d) the frequency ω at which $P_1(t)$ oscillates. These four quantities are listed in Table III for the three cases of interest.

Table III: FED Local Modes				
Case	$< H_{FED} >$	$P_1(0)$	$<Y_1(0)>$	ω (1/psec)
Linear ($\chi = 0$)	0	1	0	2.94
DST ($\chi = 48.5$ pN)	-.24 x 10^{-22} Joules	1	-.99 pm	2.93
Exact ($\chi = 48.5$ pN)	-.25 x 10^{-22} Joules	.998	-1.05 pm	2.85

4.1. Effects of Nonzero χ

The local modes of the FED show three effects of the Amide-I/Einstein-oscillator interaction (the classical nonlinearity).

1. The average energy of a local mode wavepacket decreases for $\chi \neq 0$.
2. The stationary state wavefunctions deform so that the expectation of the coordinate Y_i is correlated with the probability of amide-I excitation $P_i(t)$.
3. The frequency at which the quantum probability oscillates between two sites decreases. In terms of quantum-mechanical data, this corresponds to a decrease in the splitting between the two lowest energy levels of the FED system.

These qualitative effects are found both in the exact quantum theory and in the DST approximation.

The wavepacket created from the exact stationary states in Table II was the simple linear combination

$$\psi_L(t) = \frac{1}{\sqrt{2}}(\psi^+(t) + \psi^-(t)).$$

This wavepacket gives $P_1(0) = .998$, the largest value possible for a superposition of the first two exact states. Theoretically, the two lowest-energy quantum states may be combined in any normalized superposition. A wavepacket for which the contribution from ψ^+ is larger than that from ψ^- will give a smaller maximum value for $P_1(t)$ but the wavepacket will have a lower average energy. For any superposition of ψ^+ and ψ^-, $P_1(t)$ oscillates at the frequency $\omega = 2.85$ psec^{-1}.

4.2. Errors in the DST Approximation

The DST approximation underestimates the effects of nonzero χ: the DST gives an average energy which is slightly higher than exact value, an Einstein oscillator displacement which is smaller than the exact value, and a probability transfer frequency which is greater than the exact value. In addition, the DST equation gives no indication that in the exact quantum theory a superposition of the two lowest-energy stationary states can give at most $P_1(t) = .998$.

It is not possible to make quantitative conjectures about the quantum dynamics of extended systems based on the results presented here. However, this information about the qualitative features of the FED wavefunctions and their relation to the DST approximation may be useful knowledge to those using such approximations to study more complicated systems.

Acknowledgements

It is a pleasure to thank David W. Brown and Herb Shore for helpful discussions and the President's Fellowship Program of the University of California for financial support.

References

[1] K. V. Reddy, D. F. Heller, and M. J. Berry, Highly vibrationally excited benzene: Overtone spectroscopy and intramolecular dynamics of C_6D_6, C_6D_6, and partially-hydrogenated or substituted benzenes, J. Chem. Phys. **76**, 2814-37 (1982).

[2] B. R. Henry, Local modes and their application to the analysis of polyatomic overtone spectra, J. Phys. Chem. **80** 2160-4 (1976).

[3] L. J. Bernstein, J. C. Eilbeck and A. C. Scott, The quantum theory of local modes in a coupled system of nonlinear oscillators, Nonlinearity **3**, 293-323 (1990).

[4] A. S. Davydov, The Theory of Contraction of Proteins under their Excitation, J. Theor. Biol. **38**, 559-69 (1973).

[5] P. L. Christiansen and A. C. Scott, eds, Introduction to Section I, *Davydov's Soliton Revisited*, Proceedings of the NATO-Midit Advanced Workshop on "Self-Trapping of Vibrational Energy in Protein", Plenum (1990) and references therein.

[6] Luca Bonci, Paolo Grigolini and David Vitali, Beyond the semi-classical approximation of the discrete nonlinear Schrödinger equation: Collapses and revivals as a sign of quantum fluctuations, Phys. Rev. A **42**, 4452-61 (1990).

[7] Andreas Köngeter and Max Wagner, Exotic exciton-phonon states and bottleneck for self-trapping, J. Chem. Phys. **92**, 4003-11 (1990).

[8] J. C. Eilbeck, P. S. Lomdahl and A. C. Scott, The Discrete Self-Trapping Equation, Physica **16D**, 318-38 (1985).

[9] David W. Brown, Katja Lindenberg and Bruce J. West, Phys. Rev. A **33**, 4104 (1986); David W. Brown, Bruce J. West and Katja Lindenberg, Phys. Rev. A **33**, 4110 (1986).

[10] A. A. Maier, Self-switching of light in an optical coupler, Sov. J. Quantum Electron. **14**, 101-4 (1983).

[11] V. M. Kenkre and D. K. Campbell, Self-trapping on a dimer: Time-dependent solutions of a discrete nonlinear Schrödinger equation, Phys. Rev. B **34**, 4959-61 (1986).

[12] Herbert B. Shore and Leonard M. Sander, Ground State of the Exciton-Phonon System, Phys. Rev. B **7**, 4537-4546 (1973).

KINKS IN DISORDERED CONJUGATED POLYMERS

F.Bronold and K.Fesser
Physikalisches Institut, Universität Bayreuth
D-8580 Bayreuth, Germany

Abstract: The influence of disorder on the structure of kinks in conjugated polymers is studied. Towards this end the quasi-classical Green's function equations are solved on the imaginary frequency axis and analytically continued to real frequencies via a Padé method. A substantial broadening of the kink on increasing disorder strength is found.

I. Introduction

One of the unsolved problems in the area of conjugated polymers is the detailed nature of the metal-insulator transition as found experimentally upon doping. Several attempts have been made which take into account the formation of nonlinear excitations such as kinks and polarons as more and more charges are introduced into the system. In most cases, however, a detailed microscopic description of the interaction of these excitations with the dopants has not been used. Only a qualitative change such as gap decreasing due to disorder or band broadening due to doping has been considered. (For a short review see Ref.1) Here we want to study in detail the change of electronic and lattice structure of kinks as the impurity concentration is increased. The interaction of a single kink with a single impurity within the t-matrix formalism has been studied before [2]. The generalization to the case of many impurities appears to be difficult. In our approach here we include the influence of disorder via a uniform background which is obtained through averaging over an impurity distribution. We expect to obtain thus some of the important aspects of the many impurity situatuion. On a general scale our approach might be viewed as an example how nonlinear structures are modified under the influence of external randomness starting from a microscopic description of the physical situation.

This paper is organized as follows: in the next Chapter we decsribe the microscopic model and give a short derivation of the coupled system of equations which we have to solve. Symmetries and asymptotics which are imposed for physical reasons will be discussed. In a following section we describe the numerical procedure which we have used in order to obtain a selfconsistent solution. Some remarks on the problem of a numerical analytic continuation are given. Finally we discuss the results for a kink in detail and close with some prospects on the polaron.

II. Model

As in most cases the relevant physics of an electronic system such as a conjugated polymer is governed by the states around the Fermi energy. Since this is clearly an approximation to the real situation one can nevertheless describe the universal features of a whole class of materials. In the case of conjugated polymers this approach leads to the widely used Su-Schrieffer-Heeger model [3] where only an effective coupling of one-dimensional electrons to the lattice is taken into account. The continuum description [4] within this model is a second step with respect to the universal features of these materials, the final Hamiltonian reads

$$H = \sum_s \int dx \; \psi_s^+(x) \; \{-i \; \sigma_3 \; \partial_x + \Delta(x) \; \sigma_1 \; \} \; \psi_s(x) + 1/2\lambda \int dx \; \Delta^2(x) \tag{1}$$

with the 2-spinor $\psi(x)$ describing left- and right-moving electrons, $\Delta(x)$ is the lattice order parameter (dimerization), and all quantities have been scaled such that the electron-phonon coupling constant λ appears in front of the lattice elastic energy. In this formalism the influence of bond impurities are represented by an additional term

$$H_{imp} = \sum_s \int dx \; \psi_s^+(x) \sum_a U \; \sigma_1 \; \delta(x-x_a) \; \psi_s(x) \tag{2}$$

with U the strength and x_a the random position of the impurities. The order parameter has to be determined selfconsistently as the minimum of the total energy.

Our approach to solve this problem uses the conventional Green's function technique in the quasiclassical approximation. The derivation of the corresponding equations of motion has been given elsewhere [5], here we indicate the essential steps only. Starting from the Dyson equation for the matrix Green's function G the corresponding equation for the impurity averaged function g = <G> with a selfenergy Σ can be derived. With the usual left-right trick this imhomogeneous equation is transformed into a homogeneous one with an additional normalization condition.

Using the Born approximation (weak scattering limit) for the selfenergy the final system of equations reads

$$\partial_x b_1(x,\omega) = 2i\,\Delta(x)\,b_4(x,\omega) - 2i/\tau\,b_4(x,\omega)\,b_5(x,\omega) \tag{3a}$$

$$\partial_x b_4(x,\omega) = 2\omega\,b_5(x,\omega) - 2i\,\Delta(x)\,b_1(x,\omega) + 2i/\tau\,b_1(x,\omega)\,b_5(x,\omega) \tag{3b}$$

$$\partial_x b_5(x,w) = -2\omega\,b_4(x,\omega) \tag{3c}$$

$$1 = b_1^2(x,\omega) + b_4^2(x,\omega) + b_5^2(x,\omega) \tag{3d}$$

$$\Delta(x) = -i\pi\,\lambda\,\sum_s \int_{C_s} d\omega/2\pi\,\exp(i\omega\varepsilon)\,b_5(x,\omega) \tag{3e}$$

$$N(x,\omega) = \mathrm{Im}\,2i\,b_1(x,\omega) \tag{3f}$$

with b_n the components of the Green's function g, (3d) the normalization condition, (3e) the selfconsistency equation, and (3f) the electronic density of states. $1/\tau = c\,U^2$ describes the influence of the impurities with c concentration. In the absence of impurites ($\tau=\infty$) one can give analytical expressions [6] for the homogeneous ground state (Δ=const.) as well as kink and polaron excitations with a spatially structured $\Delta(x)$.

One observes that in these solutions square root singularities around the gap determine the interesting functions. This suggests to go over to imaginary frequencies for a numerical solution. Decomposing the Green's function into imaginary and real part $b_n=R_n+iI_n$ and using fundamental symmetries of the spectral function we arrive at $I_5=I_1=R_4=0$ along the imaginary frequency axis so that we have only three coupled differential equations instead of six.

In additon we have $\Delta(x)= s\ \Delta(-x)$, $R_1(x)=R_1(-x)$, $I_4(x)= -s\ I_4(-x)$, $R_5(x)= s\ R_5(-x)$ with s=+1 for the ground state (and the polaron) and s=-1 for the kink. Also the relations $R_1(i\nu)=-R_1(-i\nu)$, $I_4(i\nu)=-I_4(-i\nu)$, and $R_5(i\nu)=R_5(-i\nu)$ along the imaginary frequency axis can be derived. As boundary conditions we have $R_5(x=0)=0$ from symmetry and $I_4^{kink}(x_{max}) = I_4^{hom}(x_{max})$ since far from the center of the kink the structure has relaxed to the homogeneous case.

II. Numerical Procedure

For a numerical treatment we confine ourselves to a finite system in both space x and frequency ν which leads to a fixed frequency cut-off ν_{max}. The scaling property of the equations (3) $\nu'=\nu/\Delta_0$, $x'=x\Delta_0$ together with $\Delta(x)=\Delta_0 f(x)$ combined with the asymptotic behavior $f(x\text{->}\infty) = 1$ leads in the selfconsistency equation (3e) to an effective coupling constant $\lambda(\eta)$, $\eta=1/\tau\Delta_0$,

$$\lambda(\eta) = [\ 2 \int_0^{\nu'_{cut}} d\nu'\ b_5^{hom}(i\nu';\eta)\ ]^{-1} \tag{4}$$

if one wants to preserve the property that the scaled quantitiy Δ is indeed the correct asymptotic value for x going to infinity. In consequence the physical space which we can cover by this procedure is enhanced by a factor $1/\Delta_0(\eta)>1$.

The coupled equations (3) are now solved by an iteration method: for a given function f(x) the differential equations (3a-c) are solved by a relaxation procedure. From the new Green's function a new order parameter f(x) is calculated and put back into (3a-c) until convergence is reached. For the determination of the density of states the Green's function b_1 has to be continued to real frequencies. It turned out that for the ground state as well as for the kink this can be achieved with sufficient accuracy through a Padé approximation (using the Thacher-Tuckey algorithm [7]). Unfortunately in order to solve the selfconsistency equation for the polaron this continuation has to be performed at every step of the iteration.

We have tested our method with the case of a homogeneous order parameter where the exact solution can be given analytically [5]. These results will be given elsewhere [8], the overall agreement is very good including the numerical analytic continuation.

IV. Dirty Kink

In Fig.1 we show the results for the spatial structure of the kink order parameter f(x) for different values of disorder strength η. A substantial broadening can be observed as the impurity concentration is increased.

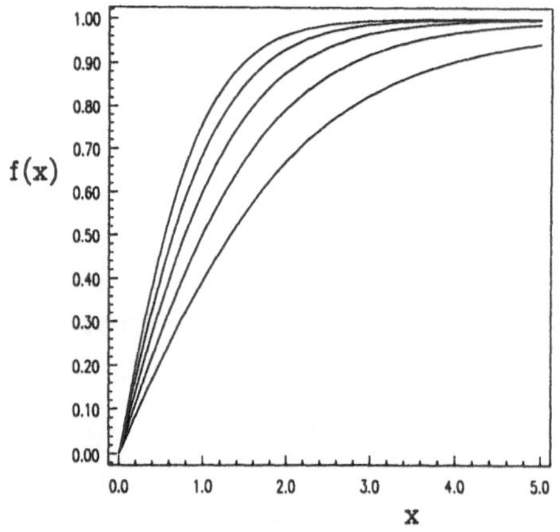

Fig.1: Spatial structure of the kink order parameter $f(x)=\Delta(x)/\Delta_0$ for different impurity concentrations: η=0.0 (upper curve), η=0.8 (lower curve), in steps of 0.2.

This broadening can be quantified by analyzing the behavior at the origin. Assuming a linear dependence $f(x) \sim x/\zeta$ (which is motivated by the exact result for the clean case) we can calculate the width ζ as function of impurity concentration. Keeping in mind

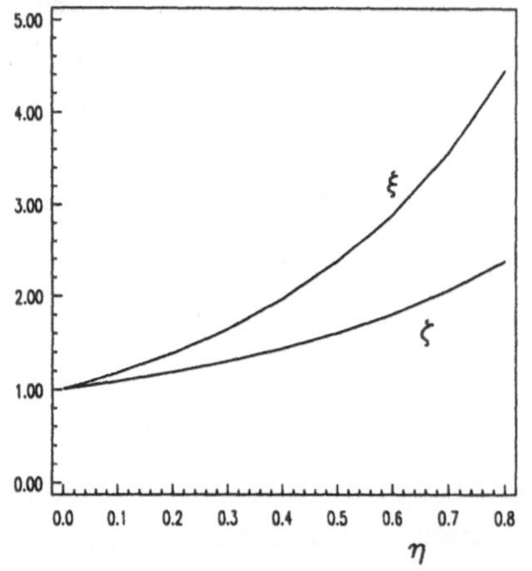

Fig.2: Kink width as function of impurity concentration (see text).

that we have used scaled variables the physical width ξ is given by $\xi = \zeta / \Delta_0(\eta)$. Both quantities are displayed in Fig.2. One can see that as η approaches the critical value $\eta=1$ the width of the kink appears to diverge. Unfortunately due to the finite lenght used in the solution of (3) a detailed investigation of the critical region is beyond the present approach.

This can also be seen clearly in Fig.3 where we show the asymptotic value $R_1(x_{max})$ as function of impurity concentration. As stated in the previous section this should approach the value of zero corresponding to the homogeneous solution. Only up to $\eta < 0.4$ we are able to reproduce this boundary value with sufficient accureteness.

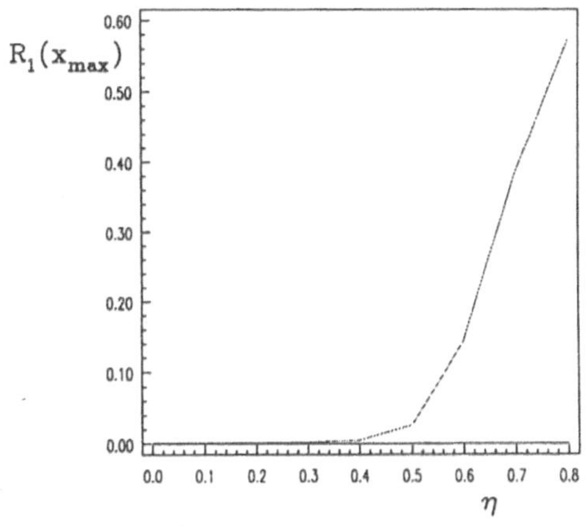

Fig.3: Asymptotic value $R_1(x_{max})$ of the kink as function of impurity concentration.

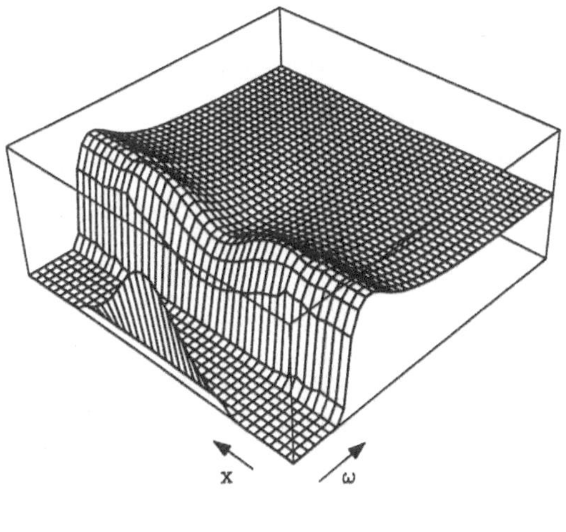

Fig.4: Electronic density of states for a kink as function of energy ω and spatial coordinate x (x=0 being the center of the kink) for impurity parameter $\eta=0.1$.

In Fig.4 we finally display the electronic density of states obtained from the Green's function via a numerical analytical continuation. The pole struture due to the kink at zero frequency is well reproduced, in the Figure the calculated values have been scaled down by a factor 10^5. This pole modifies the density of extended states beyond the gap for $|\omega|>\Delta_0$ as well. Far away from the kink ($|x|\gg 1$) the system relaxes into the ground state.

V. Summary

In conclusion we have shown how a nonlinear excitation such as the kink in conjugated polymers is influenced by the presence of randomly distributed impurities starting from a microscopic description of these impurities. We find that the spatial structure is substantially broadened, the electronic properties change drastically as the gap decreases upon doping.

It turns out to be more difficult to study the polaron with the method presented here since at each step of the iteration an analytic continuation is necessary. The reason for this is the more complicated pole structure of the polaron state. However, as far as the doping process in the whole class of conjugated polymers is concerned, the polaron is the more important object to study. We shall address this question in a future publication [8].

We close with a technical remark: we have studied various methods for the analytic continuation problem [9]. For the purpose of this paper the method described in Chap. 3 gave satisfactory results; the polaron problem, however, poses more serious constraints on the quality of the continuation procedure.

Acknowledgement: We thank J.Heym, W.Pesch and D.Rainer for discussions. This work was performed within Sonderforschungsbereich SFB213 (TOPOMAK), Bayreuth.

References

1. K.Fesser, Synth.Met. (1991) (in press); Y.Wada, preprint (1991)
2. Y.Wada, K.Fesser (unpublished)
3. W.P.Su, J.R.Schrieffer, A.J.Heeger, Phys.Rev. B 22, 2099 (1980)
4. H.Takayama, Y.R.Lin-Liu, K.Maki, Phys. Rev. B 21, 2388 (1980)
5. K.Fesser, J.Phys. C 21, 5361 (1988)
6. K.Iwano, Y.Wada, J.Phys.Soc.Jpn. 58, 602 (1989)
7. G.A.Baker,jr., P.R.Graves-Morris, Encyclopedia of Mathematics and its Applications 13, 11 (1981)
8. F.Bronold, K.Fesser, in preparation
9. F.Bronold, diploma thesis (1991)

A DISCRETE SELFTRAPPING EQUATION MODEL FOR SCHEIBE AGGREGATES

O. Bang and P.L. Christiansen
Laboratory of Applied Mathematical Physics
The Technical University of Denmark
Building 303, DK-2800 Lyngby, Denmark

Abstract. A discrete nonlinear model for the dynamics of Scheibe aggregates is proposed. The collapse of the collective excitations found by Möbius and Kuhn is described in the isotropic case as a shrinking ringwave which is eventually absorbed by an acceptor molecule.

1. Introduction

Recently, Huth et al. proposed a nonlinear continuum model for the energy transfer in Scheibe aggregates [1]. These are highly ordered molecular monolayers, which can be produced by Langmuir-Blodgett technique [2,3]. Oxycyanine dyes, e.g., are used as donor molecules and thiacyanine dyes as acceptor molecules. Even with a donor to acceptor ratio as low as 10^4, the aggregates exhibit highly efficient transfer of energy from impinging photons via excited host molecules to acceptor guests [4,5]. In Ref. [4] it is found that the coherent exciton picture provides an adequate description of the experimental results, which indicated a lifetime of the coherent exciton, t_{life}, before it is absorbed by an acceptor molecule, of about 10^{-10} s [5]. The exciton involves approximately 10^4 molecules. In Ref. [6] the isotropic continuum model proposed in [1] was used for a qualitative prediction of the lifetime. Here the dynamics of the ringwave solution to the cubic Schrödinger equation in two spatial dimensions [7] is essential.

In the present paper we introduce a discrete model of the Scheibe aggregate based on the discrete selftrapping equation (DST) [8]. The dynamics of the ringwave in this discrete case is investigated and results concerning absorption at the acceptor molecule are included in the isotropic case. Further ongoing work concerning the discrete model will be reported elsewhere [9,10].

2. Discretization of the continuum model

The continuum model proposed in [1] leads to the cubic Schrödinger equation for the wave function of the molecular excitation u(r,t)

$$iu_t + u_{rr} + r^{-1} u_r + 2|u|^2u = 0 \qquad (1)$$

in dimensionless variables [6] in the case of two spatial dimensions and circular symmetry. Here r is the radial coordinate and t is the time. The first conserved quantity becomes

$$I_1 = \int_0^\infty |u|^2 r dr = \alpha \ell / 2\pi \ , \qquad (2)$$

where α is the anharmonicity parameter and ℓ is the molecular spacing in the Scheibe aggregate. For realistic values of the physical parameters $I_1 = 5.55$ [6]. Under certain conditions, an initial circular ringwave was found in [7] to shrink and collapse at the centre of the ring in a finite time, giving rise to blow-up of the excitation amplitude at the centre. With N_0 (= 10^4) molecules inside the ringwave the initial radius becomes $r_0 = 50.9$ [6] (yielding an initial amplitude of the ringwave $u_0 = I_1/2r_0 = 0.0545$). Furthermore, if the initial radial velocity $r_0' = 0$, the theory predicts a collapse time $t_{collapse} = 809$ [6]. In the following, we use the scaling invariance of Eq. (1) $t \to \beta t$, $r \to \beta^{1/2} r$, $u \to \beta^{-1/2} u$, $I_1 \to I_1$ with the constant $\beta = 1024$ yielding $r_0 = 1.59$, $u_0 = 1.74$, and $t_{collapse} = 0.790$.

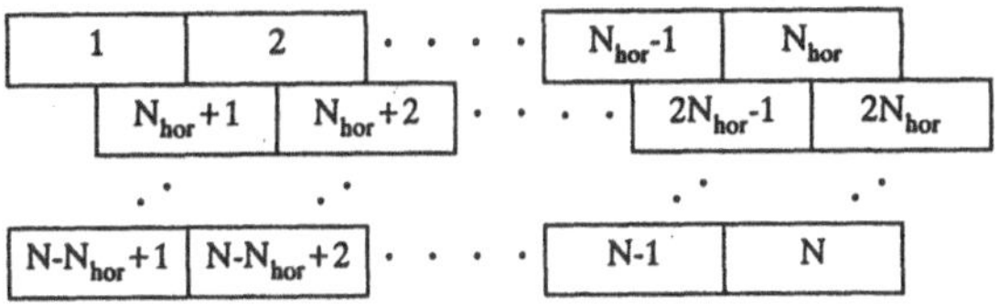

Figure 1. Position and numbering of the molecules in the brickstone work model of the Scheibe aggregate [5]. N is the total number of molecules and N_{hor} is the number of molecules in one horizontal row of the aggregate. Both N and N_{hor} are chosen to be odd.

Figure 1 shows Kuhn and Möbius' brickstone work model of the Scheibe aggregate [5]. Each donor and acceptor molecule is a dipole which is represented as a brickstone. Clearly, the molecular monolayer is anisotropic. However, assuming that the length is twice the width of each brickstone and taking only the dipole-dipole interactions between a molecule and its four nearest neighbours into account, the resulting model agrees with a direct discretization of Eq. (1). Thus replacing $u_{rr} + r^{-1}u_r = u_{xx} + u_{yy}$ by the central difference

$$\left(\sum_1^4 A_{\text{nearest neighbours}} - 4\,A\right) / \lambda^2$$

we get the discrete selftrapping equation (DST)

$$i\dot{\underline{A}} + i\,\mathrm{diag}(\underline{\alpha})\underline{A} + \gamma\,\mathrm{diag}(|\underline{A}|^2)\underline{A} + \epsilon \underline{\underline{M}}\underline{A} = 0 \ . \qquad (3)$$

Here $u(n_x\lambda, n_y\lambda, t) \to A_n(t)$, A_n denoting the excitation of molecule number n, placed at $(x,y) = (n_x\lambda, n_y\lambda)$, n_x and n_y being integers and λ the distance between neighbouring molecules. $\underline{A}$ is the column vector $(A_1, A_2, \cdots, A_N)$, $N = N_{hor} \times (2N_{hor} - 1)$ being the total number of molecules and N_{hor} the number of molecules in each row in Fig.1.

In the DST-model each dipole molecule is described as a nonlinear oscillator, the nonlinearity entering into Eq. (3) via the diagonal $N \times N$ matrix (zero elements not shown)

$$\gamma \, \mathrm{diag}(|\underline{A}|^2) = \gamma \begin{bmatrix} |A_1|^2 & & & \\ & |A_2|^2 & & \\ & & \ddots & \\ & & & |A_N|^2 \end{bmatrix} \tag{3a}$$

γ being the nonlinearity parameter ($\gamma = 2$ when Eq. (3) is a discretization of Eq. (1)). The nonlinear oscillators are coupled via the symmetrical $N \times N$ matrix (zero elements not shown)

$$\varepsilon \underline{\underline{M}} = \varepsilon \left[\begin{array}{cccccccccccccccccccc}
\multicolumn{5}{c}{N_{hor}} & \multicolumn{5}{c}{N_{hor}} & \multicolumn{5}{c}{N_{hor}} & \multicolumn{5}{c}{N_{hor}} \\
&&&&&1 \\
&&&&&1&1 \\
&&&&&&\ddots&\ddots \\
&&&&&&&1&1 \\
&&&&&&&&1&1 \\
1&1&&&&&&&&&1&1 \\
&1&\ddots&&&&&&&&&1&1 \\
&&\ddots&1&&&&&&&&&\ddots&\ddots \\
&&&1&1&&&&&&&&&1&1 \\
&&&&1&&&&&&&&&&1 \\
&&&&&1 \\
&&&&&1&1&&&&&&&&&1&1 \\
&&&&&&1&\ddots&&&&&&&&&1&1 \\
&&&&&&&\ddots&1&&&&&&&&&\ddots&\ddots \\
&&&&&&&&1&1&&&&&&&&&1&1 \\
&&&&&&&&&&&&&&&&&&&1 \\
&&&&&&&&&&&1 \\
&&&&&&&&&&&1&1 \\
&&&&&&&&&&&&1&\ddots \\
&&&&&&&&&&&&&\ddots&1 \\
&&&&&&&&&&&&&&1&1
\end{array} \right] , \tag{3b}$$

ε being the dispersion parameter ($\varepsilon = 1/\lambda^2$. Diagonal terms, -4, have been removed from Eq. (3b) by a gauge transformation). Losses, not included in Eq. (1), enter via the term $i \, \mathrm{diag}(\underline{\alpha})\underline{A}$, where

$$\mathrm{diag}(\underline{\alpha}) = \begin{bmatrix} \alpha_1 & & & \\ & \alpha_2 & & \\ & & \ddots & \\ & & & \alpha_N \end{bmatrix} . \tag{3c}$$

Here,

$$\alpha_i = \begin{cases} \alpha_{acc} & \text{for } i = (N+1)/2 \\ \alpha_{don} & \text{for } i \neq (N+1)/2 \quad , \end{cases} \tag{3d}$$

where α_{acc} is the loss coefficient for the oscillator placed at site n = (N+1)/2, modelling the absorption at the acceptor molecule (at the centre of the ringwave). Radiative losses, represented by damping at the donor molecules, are neglected (α_{don} = 0).

Initial data for the numerical solution of the DST equation, presented in the following section, are obtained by sampling the initial ringwave with radius $r_0 = \lambda\sqrt{N_0/\pi}$.

3. Numerical results

Figure 2 shows the time evolution of the ring wave in the DST model without losses corresponding to the continuum model of the actual Scheibe

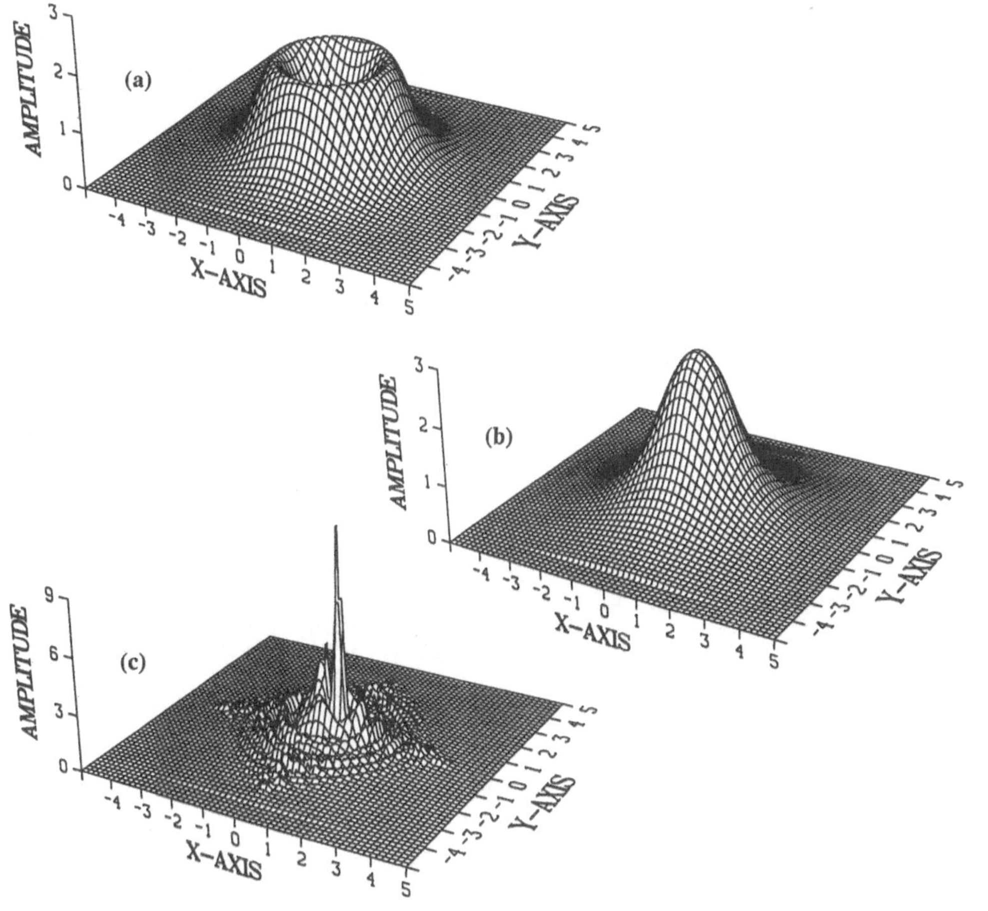

Figure 2. Evolution of ringwave in DST model. $\varepsilon = 314.4$, $\gamma = 2$, $r_0 = 1.59$, $I_1 = 5.55$, $N_0 = 10^4/4$, $\alpha_{acc} = 0$. t = (a) 0, (b) 0.6, (c) 0.75.

aggregate [6] with the particular scaling described in Section 2. For computational reasons, N_0 was reduced from 10^4 till $10^4/4$. It was checked that the time evolution does not depend critically on the choice of N_0. (Thus a further reduction of N_0 till $10^4/12$ did not produce any significant change in the computational results.)

Initially (Fig. 2a-b), the ringwave is seen to contract as predicted by the isotropic continuum model [7]. However, no matter how fine the grid may be, the amplitude of the shrinking ringwave in the centre area will eventually reach such a magnitude that the resolution of the grid becomes insufficient. As a consequence, the discrete model cannot reproduce the blow-up any further and dispersive radiation among the coupled oscillators results (Fig. 2c). In the scaled continuum model of the Scheibe aggregate the collapse time is 0.790 while the time needed for maximal excitation at the centre was found to be 0.59 in the corresponding DST-model. One reason for this difference is the fact that the requirement for the validity of the continuum perturbation theory [7], $I_1 \gg 4$, is barely fulfilled ($I_1 = 5.55$). In Figure 3, we compare the radius of the

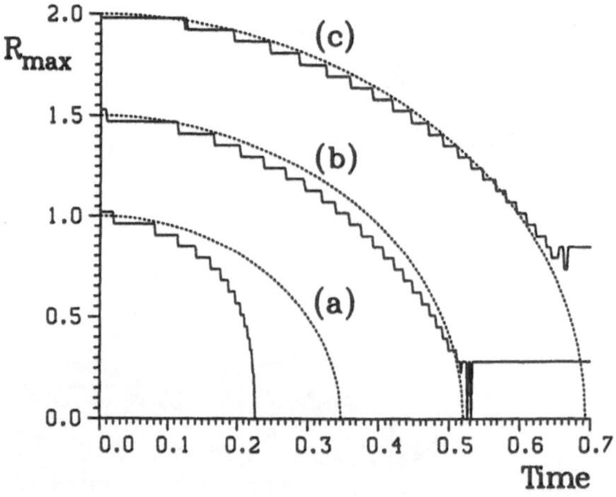

Figure 3. Ringwave radius, R_{max}, as a function of time. Dotted curves: continuum model. Full curves: DST model ($\varepsilon = 312.5$, $\gamma = 2$, $u_0 = 2.5$, $\alpha_{acc} = 0$). (a) $I_1 = 5.0$, (b) $I_1 = 7.5$, (c) $I_1 = 10.0$.

ringwave, as a function of time, in the continuum model and the DST-model for different values of I_1. The larger I_1, the better agreement between the two models and the larger the radius at which the ringwave begins to disperse.

In the continuum model [6] the absorption by the acceptor molecule was neglected. In the present DST-model where the discreteness prevents completion of the collapse we now add loss at the acceptor site. Figure 4 shows the amplitude at this site as a function of time in the lossfree

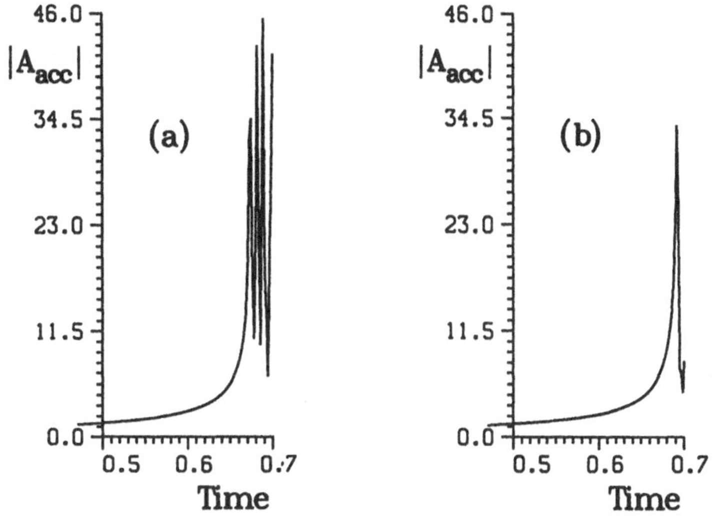

Figure 4. Centre amplitude, $|A_{acc}|$, as a function of time. DST-model (ε = 314.4, γ = 2, u_0 = 1.72, I_1 = 5.5). (a) α_{acc} = 0, (b) $\alpha_{acc} = 10^2$.

case (a) and for (b) $\alpha_{acc} = 10^2$. The introduction of attenuation is seen to compete with the dispersion due to the discreteness of the model, and wins the competition, eventually. In this manner, a substantial part of the ringwave excitation is transmitted to the acceptor molecule. A delay in the shrinking of the ringwave due to the acceptor loss is also observed.

Conclusion

In the continuum model of the Scheibe aggregate the coherent excitations collapse at the centre of the ringwave in finite time. The molecular structure of the aggregate leads to a discrete model in which the collapse cannot be completed because of dispersion. Addition of absorption at the acceptor molecule collects the excitation at this site. A detailed study of the competition between dispersion and dissipation will be presented elsewhere.

Acknowledgement

G. Vitiello, S. Pagano, J.C. Eilbeck, H. Feddersen, and M.P. Soerensen are acknowledged for helpful discussions.

References

[1] G.C. Huth, F. Gutmann, and G. Vitiello, Phys. Lett. A 140 (1989) 339.

[2] L.M. Blinov, Russ. Chem. 52 (1983) 173.

[3] M. Sugi, J. Mol. Electron. 1 (1985) 3.

[4] D. Möbius and H. Kuhn, Israel J. Chem. 18 (1979) 375.

[5] D. Möbius and H. Kuhn, J. Appl. Phys. 64 (1988) 5138.

[6] P.L. Christiansen, S. Pagano, and G. Vitiello, Phys. Lett. A 154 (1991) 381.

[7] P.S. Lomdahl, O.H. Olsen, and P.L. Christiansen, Phys. Lett. A 78 (1980) 125.

[8] J.C. Eilbeck, P.S. Lomdahl, and A.C. Scott, Physica 16 D (1985) 318.

[9] P.L. Christiansen, O. Bang, S. Pagano, and G. Vitiello, "The Lifetime of Coherent Excitations in Continuous and Discrete Models of Scheibe Aggregates", Nanobiology (to appear).

[10] O. Bang and P.L. Christiansen, "A Discrete Non-Linear Model of Collective Excitations of Langmuir-Blodgett Scheibe Aggregates" (to appear).

COMPUTER SIMULATION OF CARDIAC ARRHYTHMIAS AND OF DEFIBRILLATING ELECTRIC SHOCKS. EFFECTS OF ANTIARRHYTHMIC DRUGS

P. Auger[1], A. Coulombe[2], P. Dumée[3], M-C Govaere[3],
J-M. Chesnais[2], A. Bardou[3].
1. Laboratoire d'Ecologie, B.P. 138, Faculté des Sciences, 21000 Dijon, France.
2. Laboratoire de Physiologie comparée, Université d'Orsay, 91405 Orsay, France.
3. INSERM U256, Hôpital Broussais, 96 rue Didot, 75674 Paris Cédex 14, France.

INTRODUCTION

To describe the heart tissue, i.e. a set of a large number of coupled cells, many models use a network of electrically coupled elements corresponding to cells with resistances and capacitances. Sets of coupled differential equations governing the time evolution of trans-membrane potentials and of internal parameters or excitabilities are used. Bidimensional models have been developped and we refer to [3], [8]. These computer simulations calculate time by time and for each element of the network the trans-membrane potential and they allow to describe the propagation of the depolarizing wave through the tissue cell by cell. Nevertheless, these models involve a lot of variables and their simulations consume a lot of computer times. Few tridimensional models of this type have been developped such as [9].

In order to limit the consumption of computer times, several authors use simpler models. The heart tissue is represented as a network of elements that can depolarize the nearest excitable elements. For instance, in such a time discrete bidimensional model, any newly depolarized cell can depolarize at the following time interval any excitable cell in a small neighborhood, either the four nearest cells or the eigth nearest cells. In these models, the trans-membrane potential is not calculated continuously with time from chosen differential equations. A simple law of propagation is used and computer simulations are realized. We refer to [6-7]. In our bidimensional model of the ventricle, we use a particular law of wave propagation, the Huygens' construction method. This model has allowed us to simulate different reentry mechanisms inducing self-sustained waves and we refer to [1-2].

1. MECHANISMS INDUCING SELF-SUSTAINED CONDUCTION TROUBLES

1.1 Presentation of the model

A square surface element of the ventricle is represented by a network of 2500 points. Each point (i,j) of the surface element, where i and j $\in$ [1, 50], corresponds to a group of few cardiac cells which can be found in a discrete set of cellular states. To each point (i,j) is associated a state matrix taking integer values S(i,j) and varying with the cellular states. Cells can be excitable (S(i,j)=0) or depolarized (S(i,j)≠0). It is a time discrete model and we calculate at equal time intervals Dt the states of the different groups of cells. The refractory period can differ from one point (i,j) to another (k,l) but in general it is fixed at the beginning of each simulation, say R(i,j) at point (i,j). When cells are depolarized, the state matrix varies suddenly from 0 to R(i,j). Then, at each time interval Dt, the state matrix decreases to give R(i,j)-1 after time Dt, R(i,j)-2 after 2Dt, . . . , 1 after {R(i,j)-1}Dt, and finally becomes again equal to 0 after R(i,j)Dt. In this situation, the corresponding cells are again excitable and can be depolarized another time. There are R(i,j)+1 possible state transitions 0 → R → R-1 → R-2 → . . . → 1 → 0, and thus the refractory period is equal to the time necessary for them, i.e. {R(i,j)+1}Dt.

The propagation law is the Huygens' construction method. In this way, it is assumed that during the time interval Dt, each newly depolarized group of cardiac cells at a point (i,j) is able to depolarize any excitable cardiac cell which can be found in a small neighborhood defined by a circle centered on this point (i,j) with a radius R(i,j) = c(i,j).Dt. This radius R(i,j) corresponds to the distance covered by the wave from this point during the time interval Dt. Computer simulations are presented as sequences of electrical mappings giving the state of each point (i,j) of the surface element at consecutive time intervals, t=0, Dt, 2Dt, ... , NDt, where N+1 is the number of electrical mappings. We use different grey intensities in order to visualize states of cardiac cells.

1.2. Unidirectional blocks

Unidirectional blocks occur when for a given axis of propagation, the wave can only propagate in one direction. In our simulations, a unidirectional block is represented by a segment of length l located at the middle of the surface element. If the wave moves from the left side, it cannot pass through it. Reversely, a wave coming from the right side can pass through it. The conduction is allowed in one direction and is forbidden in the opposite direction. The computer simulation presented on figure 1 shows that the wave initialized on the left side turns around the unidirectional block corresponding to a length of 32 points and then reenters. This is the triggering process of a self-sustained circus motion. These simulated patterns are in good agreement with experimental ones obtained by epicardial and endocardial electrical mappings realized on dog hearts during ventricular tachycardia and we refer to [4]. Also, it can be noticed that these patterns are very similar to those obtained by computer simulations from [6] or still [9]. Such reentries and periodic rotating waves can also be obtained by considering an area of ectopic cells contiguous to an area of cells in refractory periods.

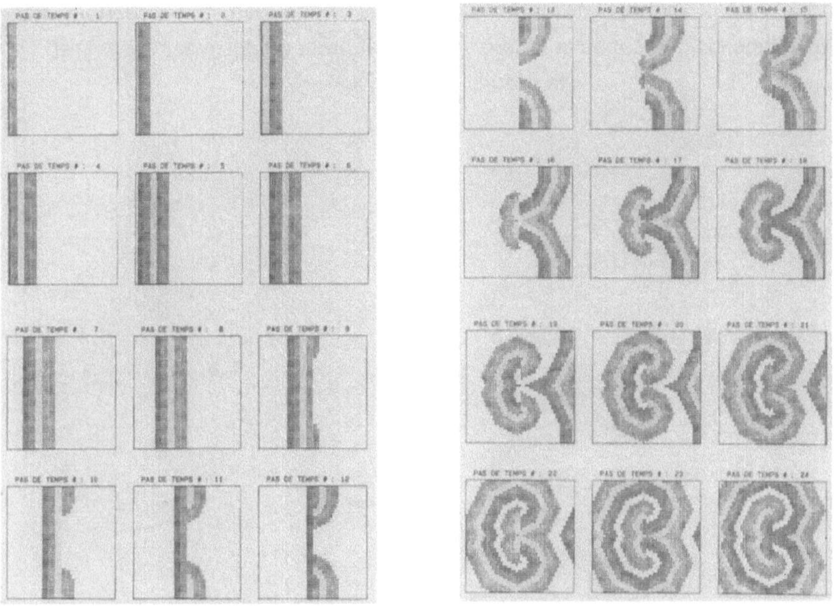

Figure 1: Unidirectional block inducing a self-sustained periodic reentry.

1.3. Role of anisotropic conduction on the initialization of reentries

Till now, we considered isotropic conduction conditions. As a matter of fact, it has been shown experimentaly that the conduction velocity was connected to the relative orientation of the axis of propagation with respect to the heart fibres direction. The conduction velocity is about three times larger for an axis of propagation parallel to the heart fibres direction than for a perpendicular one.

In order to take into account the anisotropy of conduction, now we shall consider an homogeneous surface element of the ventricle composed of cardiac fibres with an horizontal direction. Consequently, we have modified the Huygens construction principle so that the wave front position is now given by the envelope of ellipses instead of circles with long axis three times larger than the short axis corresponding to the heart fibres direction, i.e. the horizontal axis. Typical patterns are presented on figure 2. In previous works, we have shown that the anisitropy can play an important role in the triggering of reentries. Particularly, if the unidirectional block is perpendicular to the heart fiber axis, the threshold size of the unidirectional block is decreased in a ratio three. For further details, we refer to [1-2].

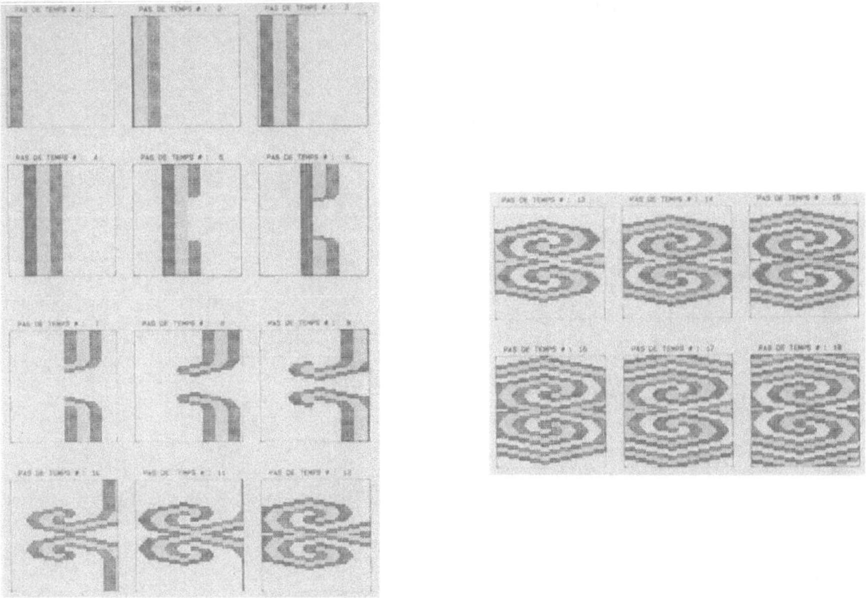

Figure 2: Unidirectional block in anisotropic conduction conditions showing that a small block (perpendicular to heart fibres) can induce a reentry.

2. ELECTRICAL DEFIBRILLATING SHOCKS

This section presents simulations of electrical defibrillating shocks. First of all, a self-sustained conduction trouble is initialized. In a second step, we simulate an electric shock delivered to the surface element of the ventricle. In order to simulate various intensities of the electric shock, the average percentage P of excitable cardiac cells which are depolarized by it is chosen differently from one sequence to another.

2.1. Low percentage of depolarized cells

At time 6, an electric shock is delivered to the ventricle surface element with a percentage P = 2%. The following time interval 7 displays the effect of the shock. Figure 3 (a) shows that the percentage of depolarized cells is too low in order to stop the reentry. The pattern of the electrical mapping is similar after the shock than before it. Such a low percentage of depolarized cells is not efficient. Now, let us increase the energy of the electric shock and let us consider a high energy shock with a large value of P.

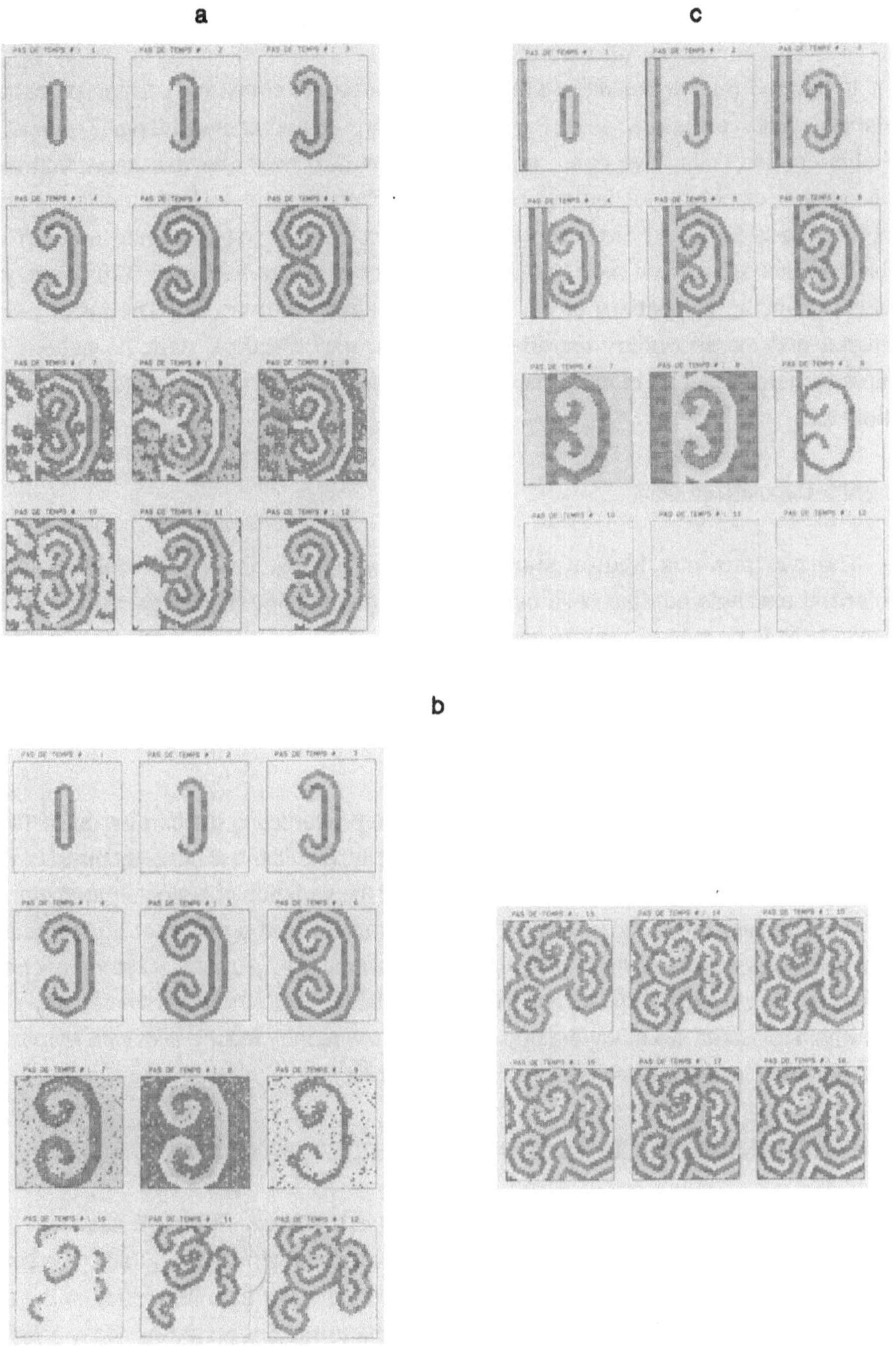

Figure 3 : (a) Defibrillating electric shock depolarizing 2% of excitable cells. (b) High percentage P = 90% defibrillating electric shock. (c) 100% defibrillating shock.

2.2. High percentage of depolarized cells

In figure 3 (b), we present a simulation of an electric shock with a high percentage of excitable depolarized cardiac cells, P = 90%. One can see that at time interval 7, almost all the excitable cardiac cells have been depolarized. The next time intervals show that the very small number of cells that has not been depolarized is able to initialize again multiple reentries. These secondary reentries are sources of several self-sustained rotating waves. Under these conditions, one sees that this electric shock is also inefficient. But contrary to the previous example, the pattern of the electrical mapping is changed. The initial pattern is destroyed and the secondary reentries generate a new electrical periodic pattern. These multiple reentries will lead to an uncoordinated contraction of the heart which can be related to fibrillation.

2.3. 100% Depolarized cells

The two previous figures show that at the time of the shock delivery, the non depolarized excitable cardiac cells can initialize again self-sustained conduction troubles. Consequently, to be sure to stop the reentry, it is necessary to deliver an electric shock which depolarizes 100% of the excitable cells such as on Figure 3 (c).

3. EFFECTS OF ANTIARRHYTHMIC DRUGS

Drugs act at the cellular level by modifying the properties of the cardiac cells. They can make vary the refractory period, the conduction velocity, or still the excitability threshold. In the next computer simulations, we consider the effect of the variation of a single parameter on the reentry mechanism. As a consequence, we do not simulate the action of a particular drug which can have complex actions on several parameters. But, it is rather a computer experiment allowing to study the action of the variation of a single parameter while all the other remain constant. We study the dependence of the reentry mechanism with respect to the refractory period, (either by decreasing or increasing it).

3.1. Effect of a decrease in refractory period on the reentry mechanism

In this simulation, firstly a reentry mechanism is triggered. Then, it is assumed that the refractory period is decreasing during the sequence of about a ratio 1/3. Firstly, one sees that the spatial size of the reentrant area decreases. Secondly, the time frequency of the reentry increases. Under these conditions, a decrease in the refractory period allows the reentry to develop in smaller space areas at a higher rhytm. In this way, a decrease in the refractory period has an arrhythmogenic efect and favors the reentry mechanism.

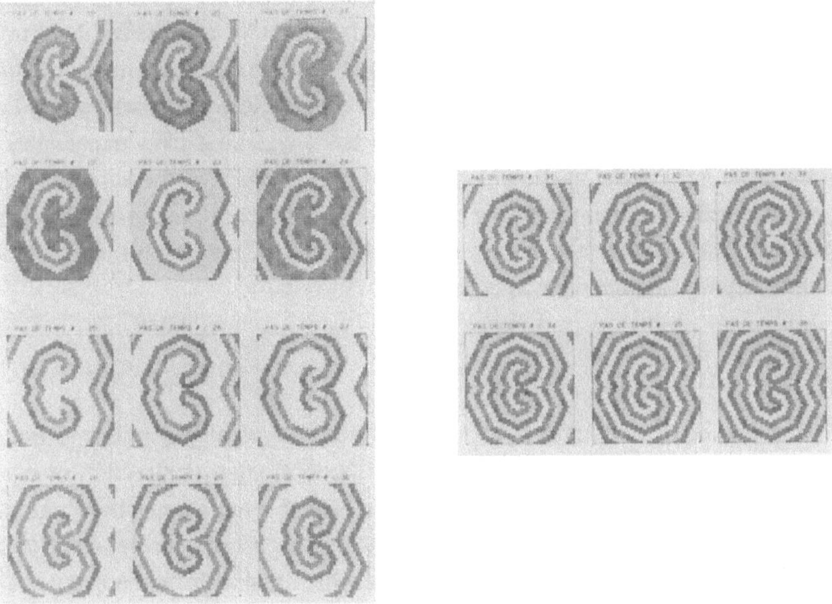

Figure 4: Effect of a decrease in the refractory period on a periodic reentry.

3.2. Effect of an increase in the refractory period on the reentry mechanism

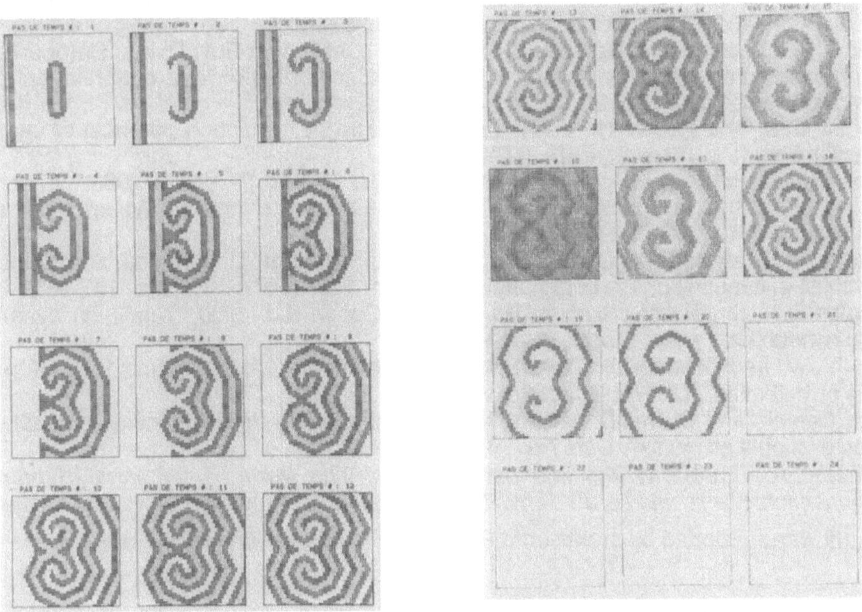

Figure 5: Effect of an increase in the refractory period on a periodic reentry.

In the next simulation of figure 5, we assume the inverse process of figure 4, i.e. an increase in the refractory period. In this sequence, a reentry is initialized by a unidirectional block. This increase in the refractory period can result of the action of an antiarrhythmic drug of class III. Figure 5 shows that the wave front of the periodic wave is now going to meet cells with longer refractory periods. While before, the tissue was becoming immediately reexcitable when the wave front of the next reentering wave was arriving, on the contrary, now the wave front meets cardiac cells in refractory period and thus cannot propagate anymore. This is a desynchronization between the travelling wave periodic motion and the periodicity of the depolarization-repolarization process.

CONCLUSION

Our computer simulations allow us to study several mechanisms inducing self-sustained conduction troubles leading to an uncoordinated contraction of cardiac fibres related to ventricular tachycardia and fibrillation. These simulations show that an important parameter is the wave length which is defined by the product of the conduction velocity by the refractory period duration. Grossly, any process which makes increase this wave lenght has an antiarrhythmic effect. On the contrary, any process which makes decrease the wave lenght is going to favor reentries and arrhythmias.

REFERENCES

[1] Auger P., Bardou A., Coulombe A. and Degonde J. 1988 b. "Computer simulation of ventricular fibrillation". *Mathl Comput. Modelling.* **11** , 813-822.
[2] Auger P., Coulombe A., Govaere M-C., Chesnais J-M., Von Euw D. and Bardou A. 1989. "Computer simulations of mechanisms of ventricular fibrillation and defibrillation". *Innov. Techn. Biol. Med.* **10** , 299-312.
[3] Beeler G.W. and Reuter H. 1977. "Reconstruction of the action potential of ventricular myocardial fibres". *J. Physiol.* , **268** , 177-210.
[4] Downar E., Harris L., Mickleborough L., Shaikh N. and Parson I. 1988. "Endocardial mapping of ventricular tachycardia in the intact human ventricle: Evidence for reentry mechanisms". *JACC*, **11**, 783-791.
[5] Janse M.J. and Kleber A.G. 1981. "Electrophysiological changes and ventricular arrhythmias in the early phase of regional myocardial ischemia". *Circ. Res.*, **49** , 1069-1081.
[6] Kaplan D.T., Smith J.M., Saxberg Bo E.H. and Cohen R.J. 1988. "Nonlinear dynamics in cardiac conduction". *Math. Biosci.*, **90** ,19-48.
[7] Smith J.M. and Cohen R.J. 1984. "Simple finite element model accounts for wide range of cardiac arrhythmias". *US Nat. Acad. Sci. Pro.*, **81** , 233-237.
[8] Van Capelle and Durrer D. 1980. "Computer simulation of arrhythmias in a network of coupled excitable elements". *Circ. Res.*, **47** , 454-466.
[9] Winfree A.T. 1987. *When time breaks down. The three dimensional dynamics of electrochemical waves and cardiac arrhythmias.* Princeton : Princeton University Press.

This work was supported by grants from Région de Bourgogne and from CNAMTS.

PART III

LATTICE EXCITATIONS AND LOCALISED MODES

NUMERICAL STUDIES OF SOLITONS ON LATTICES

J. C. Eilbeck
Department of Mathematics, Heriot-Watt University
Edinburgh EH14 4AS, UK

Abstract

We use path-following methods and spectral collocation methods to study families of solitary wave solutions of lattice equations. These techniques are applied to a number of 1-D and 2-D lattices, including an electrical lattice introduced by Remoissenet and co-workers, and a 2-D lattice suggested by Zakharov, which in a particular continuum limit reduces to the Kadomtsev-Petviashvili equation.

1 Introduction

We consider here the study of solitary waves on lattices, as a special case of the more general problem of energy transport in lattice models. For example, consider the atomic lattice with Lagrangian

$$L = \sum_n \{\frac{1}{2}\dot{\alpha}_n^2 - V(\alpha_{n+1} - \alpha_n)\} \tag{1}$$

where α_n is the displacement of the nth particle from its equilibrium position, and $V(\alpha_{n+1} - \alpha_n)$ is some interaction potential. If the relative displacement of the nth bond is defined to be $u_n = \alpha_{n+1} - \alpha_n$, then the equation of motion becomes

$$\frac{d^2}{dt^2}u_n = V'(u_{n+1}) - 2V'(u_n) + V'(u_{n-1}). \tag{2}$$

In general no analytic solutions of this or similar lattice equations are known, except for some special cases such as the Toda lattice [1] or the Ablowitz-Ladik lattice [2]. If we look for solitary wave solutions of (2), i.e. solutions of the form $u_n(t) = u(n - ct) \equiv u(z)$, (2) becomes

$$c^2\frac{d^2u(z)}{dz^2} = F(z+1) - 2F(z) + F(z-1) \tag{3}$$

with $F(z) = V'\{u(z)\}$. Although we cannot solve this equation analytically, except in some special cases, it is possible to solve it numerically by a variety of methods. One technique which turns out to be efficient and accurate is the spectral collocation method [3]. If we use this together with path-following methods, we can generate a whole family of solutions to (3) as one of the parameters, such as the wave speed c, varies. A general survey of spectral methods can be found in [4]: for an introduction to continuation methods see

[5]. One important point to note is that (3) has a continuum of periodic solutions as well as a solitary wave solution, and that it is necessary to pick out the solitary wave by imposing an extra integral condition which ensures that $u(z) \to 0$ as $|z| \to \infty$ [3, 6].

Finding a solitary wave solution to (3) tells us nothing about the stability of such a pulse as a solution of the full time-dependent problem (2), nor whether it possesses approximate soliton properties on collision with other waves (we do not expect to find *exact* soliton properties except for some special cases as mentioned earlier). To investigate this question, we need to integrate (2) numerically. Conventionally this has been done using Runge-Kutta methods. Recently Duncan et al. [7] have developed symplectic solvers for lattice equations which conserve the Hamiltonian of the system to a high degree of accuracy. Their results should be read in conjunction with those described below.

Fig. 1 shows a numerical integration of (3) with two solitary waves as initial conditions, prepared by J A Wattis.

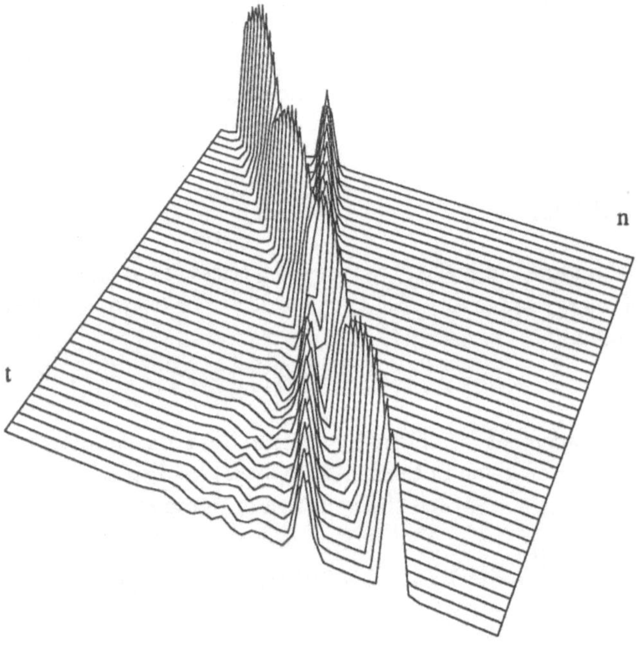

Figure 1: Collision of two solitary waves in the $V(u) = \frac{1}{2}(u^2 + u^4)$ lattice

The solitary waves emerge from the collision region with almost the same energy as before, with only a small oscillating tail left behind. For the rest of this paper we shall use the words "solitons" and "solitary waves" in a loose and interchangeable manner.

2 Some Applications

In [3] we considered the Toda lattice as a numerical test, and a lattice with $V(u) = \frac{1}{2}(u^2 + u^4)$. A more complicated example we consider below is the "electrical" lattice model worked on by Remoissenet and Michaux [8]. This is an electrical transmission line, which in the absence of loss terms has the equation (in dimensionless coordinates)

$$\frac{d^2}{dt^2}Q(V_n) = V_{n+1} - 2V_n + V_{n-1}, \tag{4}$$

where the charge Q is a nonlinear function of the voltage V.

The original motivation for our interest in lattices was a study of Davydov solitons on protein molecules [9]. A typical lattice equation from the semi-classical theory for such models is

$$\begin{aligned} i\hbar\frac{d}{dt}a_n &= E_0a_n - J(a_{n+1} - a_{n-1}) + \chi(\beta_{n+1} - \beta_{n-1})a_n \\ M\frac{d^2}{dt^2}\beta_n &= w(\beta_{n+1} - 2\beta_n + \beta_{n-1}) + \chi(|a_n|^2 - |a_{n-1}|^2) \end{aligned} \tag{5}$$

Here $|a_n(t)|^2$ is the probability of finding a quantum of bond energy at site n, $\beta_n(t)$ is the longitudinal displacement of the nth amino acid in the protein, $E/\hbar$, J, χ, M and w are real physical constants. We have not yet developed the code to treat coupled systems of equations of this complexity, but this presents no problems in principle, except as discussed below. A simpler approximation to this system which has been studied by various authors is the so-called Discrete Self-Trapping (DST) equation [10], which with nearest neighbour couplings becomes a discrete Nonlinear Schrödinger (DNLS) equation

$$i\frac{d}{dt}A_n + \gamma|A_n|^2A_n + \epsilon(A_{n+1} + A_{n-1}) = 0 \tag{6}$$

Here $A_n(t)$ is complex, and γ and ϵ are real parameters (not necessarily small). Finding travelling waves for this system is more difficult that the normal lattice equations, since in general a travelling wave will be a travelling wave *envelope* modulating a carrier wave travelling at a different velocity. This problem is treated by Feddersen elsewhere [11]. Solving the Davydov equations will be even more involved.

A more straightforward extension of the basic method described in the Introduction is to treat simple 2D problems. For example, we can generalise (2) to a 2D square lattice

$$\frac{d^2}{dt^2}u_{n,m} = V'(u_{n+1,m}) + V'(u_{n-1,m}) + V'(u_{n,m+1}) + V'(u_{n,m-1}) - 4V'(u_n). \tag{7}$$

Looking for a travelling wave solution, with wave front at an angle θ to the m−axis, we use $u_{n,m}(t) = u(n\cos\theta + m\sin\theta - ct) \equiv u(z)$ to get the equation corresponding to (3) (c.f. [12])

$$c^2\frac{d^2u(z)}{dz^2} = F(z+\cos\theta) + F(z-\cos\theta) + F(z+\sin\theta) + F(z-\sin\theta) - 4F(z) \tag{8}$$

This can be solved with the same techniques as above. It is straightforward to show that $u(z)$ satisfies the same integral constraint as in the 1-D case, i.e.

$$c^2\int_{-\infty}^{\infty} u(z)\,dz = \int_{-\infty}^{\infty} F(z)\,dz \tag{9}$$

Another, more anisotropic 2D lattice, is the following, first suggested by Zakharov [13]

$$\frac{d^2}{dt^2}v_{n,m} = v_{n+1,m} - 2v_{n,m} + v_{n-1,m} + \epsilon^2(v_{n,m+1} - 2v_{n,m} + v_{n,m-1})$$
$$-a\epsilon^2(v_{n+1,m}^2 - 2v_{n,m}^2 + v_{n-1,m}^2) \tag{10}$$

Here $\epsilon \ll 1$ is a small parameter and a is an $O(1)$ parameter. The lattice has a weak non-linearity along the n-axis and has weak linear coupling in the m direction. It can be shown that in a particular continuum limit this lattice becomes the Kadomtsev-Petviashvili (KP) equation.

$$(24v_\tau - 24avv_z + v_{zzz})_z + 12v_{ww} = 0, \tag{11}$$

3 The "electrical" lattice

The simplest version of the model (4) is to take $Q(V) = V - aV^2$. Since a can be taken out of the calculation by a rescaling of V, we take $a = 1$. The solution of the equation corresponding to (3) proceeds in a similar way to the cases described in [3], and Fig. 2 shows the results of the calculation. In this figure, the solid line shows the height of the

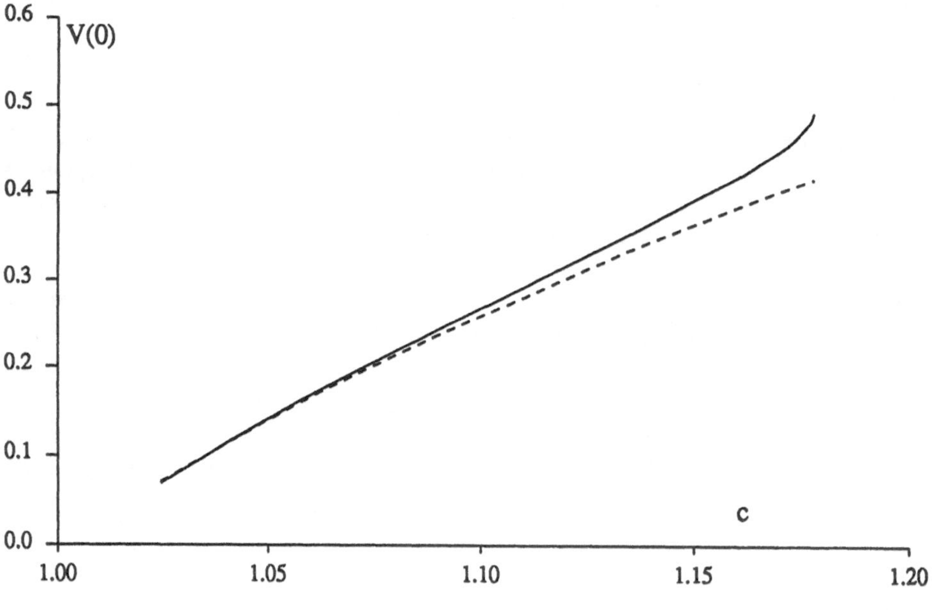

Figure 2: $V(0)$ v. c for the quadratic electrical lattice.

soliton as a function of c from the numerical calculation, and the dashed line shows the continuum approximation to the solution, which takes the form $V(0) = \frac{3}{2}(1 - 1/c^2)$ for the given $Q(V)$. For small amplitudes the agreement between the two is seen to be good. When the velocity is close to 1, the numerical calculation fails, because the soliton width is of the same order as that of the periodic boundary conditions (this could easily by cured by working on a larger interval). However the program also failed near the upper end of the curve, where $V(0)$ gets close to $\frac{1}{2}$. In this region the number of Fourier modes required to ensure accuracy grows very large, and eventually for large enough c, above $c \approx 1.172$, no solution can be found. If the solutions are plotted out for various values of c, the cause of the problem becomes obvious. Fig. 3 shows such a plot, for solitons corresponding to $c \approx 1.024$, 1.068, and 1.172 respectively.

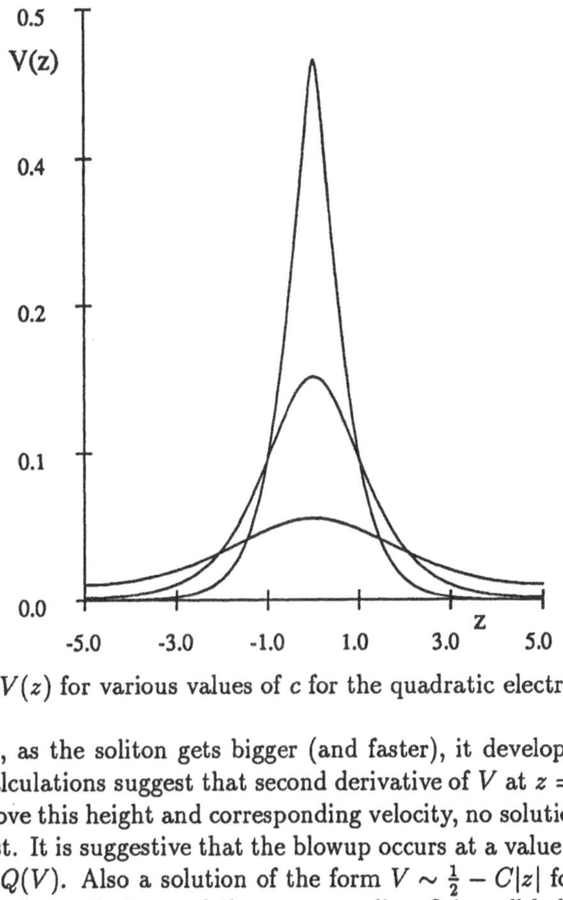

Figure 3: $V(z)$ for various values of c for the quadratic electrical lattice.

It is clear that, as the soliton gets bigger (and faster), it develops a sharp peak. In other words the calculations suggest that second derivative of V at $z = 0$ is blowing up as $V(0) \to \frac{1}{2}$, and above this height and corresponding velocity, no solutions of the travelling wave equation exist. It is suggestive that the blowup occurs at a value of V corresponding to a maximum of $Q(V)$. Also a solution of the form $V \sim \frac{1}{2} - C|z|$ for some constant C has the right type of singularity, and the corresponding Q is well-behaved at this point. However, we have not yet succeeded in proving any firm results along these lines.

Although of mathematical interest, this behaviour may not be physically relevant, since the approximation $Q(V) \approx V - aV^2$ is invalid at these values of V. We repeated the calculation with a better approximation, $Q(V) = V - 0.18V^2 + 0.021V^3 - 0.0021V^4$ (constants supplied by M. Remoissenet). The results were qualitatively the same as those shown in Figs. 2 and 3, with blowup near $c \approx 1.216, V(0) \approx 3.98$, close to the value of V (≈ 4.11) at which $Q(V)$ has a maximum for these parameters. In a final effort to avoid blowup, Remoissenet supplied an even more accurate fit to Q, $Q(V) = V - aV^2 + bV^3 - cV^4 + dV^5 - eV^6$, with $a = 0.19086, b = 0.0223199, c = 0.00166791, d =$

$0.749668 \times 10^{-4}, e = 0.159795 \times 10^{-5}$. When the program was rerun with this new $Q(V)$ we again recorded a blowup phenomena similar to Figs. 2 and 3, this time at $c \approx 1.433, V(0) \approx 9.265$, again close to the value of V (≈ 9.86) at which $Q(V)$ has a maximum. However, these values of the scaled variable V lie outside the physical range of the electrical components involved in the network.

4 2-D results

4.1 The $V(u) = \frac{1}{2}(u^2 + u^4)$ lattice

We can solve (7) with a simple extension of the methods used in 1D (of course the resulting advance-delay equation (8) is still one-dimensional, but there are now two delay constants and two advance constants). One interesting point revealed by (8) is that when $\theta = \pi/4$, the equation reduces to the usual 1-D model except that z is scaled by $\sqrt{2}$. This means that any solitary wave solution of the 1-D equation will also propagate at $\pi/4$ to the axes with the same velocity and height but with a width reduced by a factor of $\sqrt{2}$.

For fixed θ, the curves of $u(0)$ v. c look very like the 1-D case, c.f. Fig. 2 in [3]. For values of θ between 0 and $\pi/4$, the height of the solitary wave is slightly greater than for one travelling along an axis with the same velocity. Fig. 4 shows a graph of $u(0)$ v. θ for solitary waves with fixed $c = 2$.

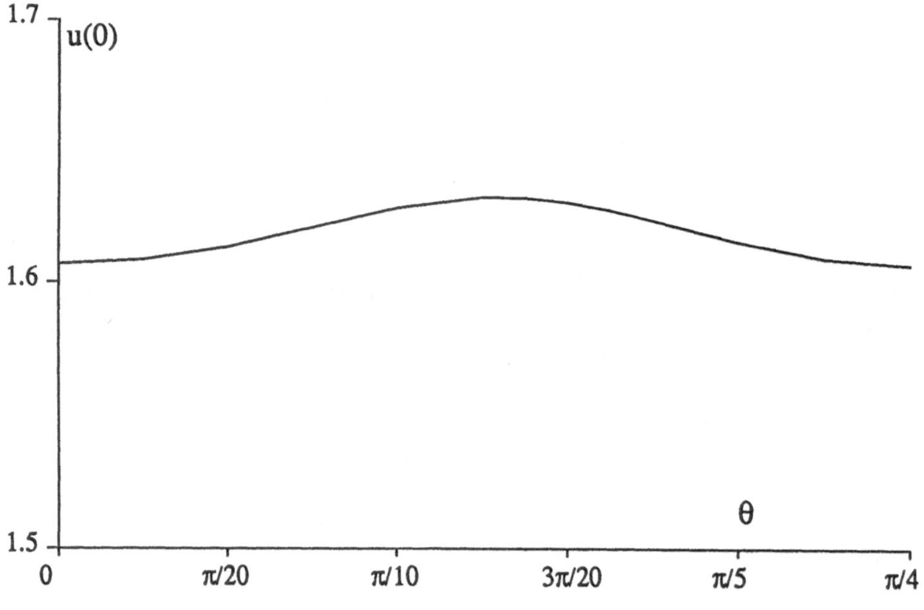

Figure 4: $u(0)$ v. θ for the $V(u) = \frac{1}{2}(u^2 + u^4)$ lattice, $c = 2$.

This graph is symmetric around $\theta = \pi/4$ and periodic with period $\pi/2$. The graph is not quite symmetric around $\pi/8$: a study of a continuum expansion shows the same qualitative features. Another interpretation is that solitary waves travelling at angles between 0 and $\pi/4$ with fixed *height* will travel slightly slower than those travelling along an axis or a diagonal.

4.2 KP lattice

For the problem (10), the 1-D technique again carries over simply to the 2-D case. This time there is no symmetry except the trivial one for rotations through π. In the limit $\theta \to \pi/2$, where θ is the direction of propagation relative to the "nonlinear" n-axis, we expect no solutions, since the equation reduces to a linear 1-D problem. For $\theta = 0$ the equation reduces to a 1-D problem with $V(v) = v^2 + kv^3$.

A continuum calculation suggest that $c \to \cos\theta$ as the pulse amplitude $\to 0$. Fig 5, taken from [13], shows the results of calculations giving a plot of pulse height against c for various angles of propagation to the n-axis. The solid line shows the numerical result,

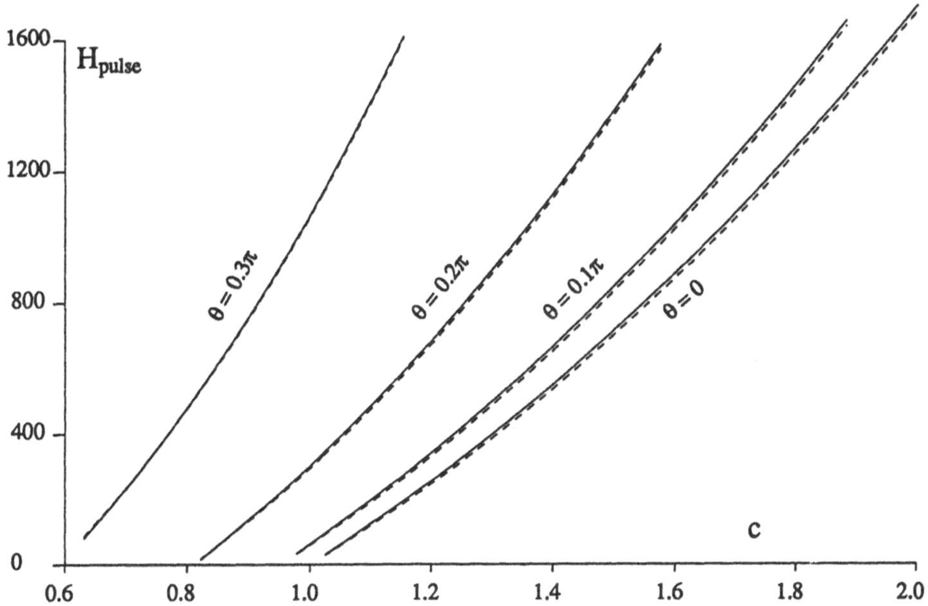

Figure 5: H_{pulse} v. c for the KP lattice

whereas the dashed line shows a heuristic formula $v(0) \approx (c^2 - \cos^2\theta)Q(0)/a\epsilon^2\cos^2\theta$, where $Q(w)$ is a universal function with $Q(0) \approx 1.3977$. This fit to the pulse height, derived from a modified asymptotic formula, is surprising good, although the corresponding pulse shapes are not accurate, except in the limit $c \to \infty$.

Acknowledgements

I am grateful to the NATO Special Programme Panel on Chaos, Order and Patterns for support for a collaborative programme, to the SERC for research funding under the Nonlinear System Initiative, and to the EC for funding under the Science programme SCI-0229-C89-100079/JU1.

References

[1] M Toda. *Theory of Nonlinear Lattices.* Springer, Berlin, 1981.

[2] M J Ablowitz and J F Ladik. A nonlinear difference scheme and inverse scattering. *Stud. Appl. Math.*, 55:213–229, 1976.

[3] J C Eilbeck and R Flesch. Calculation of families of solitary waves on discrete lattices. *Phys. Lett. A*, 149:200–202, 1990.

[4] C Canuto, M Y Hussaini, A Quarteroni, and T A Zang. *Spectral Methods in Fluid Mechanics.* Springer-Verlag, Berlin, 1988.

[5] R Seydel. *From Equilibrium to Chaos - Practical Bifurcation and Stability Analysis.* Elsevier, London, 1988.

[6] D Hochstrasser, F G Mertens, and H Büttner. An iterative method for the calculation of narrow solitary excitations on atomic chains. *Physica*, D35:259–266, 1989.

[7] D B Duncan, C H Walshaw, and J A D Wattis. A symplectic solver for lattice equations. (these proceedings), 1991.

[8] M Remoissenet and B Michaux. Electrical transmission lines and soliton propagation in physical systems. In G. Maugin, editor, *Continuum Models and Discrete Systems, Proceedings of the CMDS6 Conference, Dijon 1989*, London, 1990. Longman.

[9] P L Christiansen and A C Scott. *Davydov's Soliton Revisited.* Plenum, New York, 1990.

[10] J C Eilbeck, P S Lomdahl, and A C Scott. The discrete self-trapping equation. *Physica D: Nonlinear Phenomena*, 16:318–338, 1985.

[11] H Feddersen. Solitary wave solutions to the discrete nonlinear Schrödinger equation. (these proceedings), 1991.

[12] O A Druzhinin and L A Ostrovskii. Solitons in discrete lattices. Preprint, to be published in Phys. Lett. A, 1991.

[13] D B Duncan, J C Eilbeck, C H Walshaw, and V E Zakharov. Solitary waves on a strongly anisotropic KP lattice. Submitted to Phys. Lett. A., 1991.

A SYMPLECTIC SOLVER FOR LATTICE EQUATIONS

D.B. Duncan, C.H. Walshaw and J.A.D. Wattis
Department of Mathematics, Heriot-Watt University
Edinburgh EH14 4AS, Scotland

Abstract

We describe an Ordinary Differential Equation solver for lattice dynamics equations in Hamiltonian form, which is more accurate, more efficient and easier to programme than the commonly used Runge-Kutta methods. An important feature of the solver is that it preserves the symplectic nature of the differential equations. We illustrate the application of scheme in a variety of examples of one and two space dimensional lattices, including the Toda lattice and a discrete version of the K.P. equation. We also show some comparisons with standard Runge-Kutta methods.

1 Introduction

We describe an Ordinary Differential Equation solver for lattice equations which can be written in the form

$$\frac{d^2}{dt^2}q_j = -\frac{\partial}{\partial q_j}G(\underline{q}) \tag{1}$$

for integers j, where q_j is the displacement of the jth lattice point from its equilibrium position and $\underline{q}$ is a vector containing the q_j's. For example, the choice

$$G(\underline{q}) = \sum_j V(q_j - q_{j-1}) \quad \Rightarrow \quad \frac{d^2}{dt^2}q_j = V'(q_{j+1} - q_j) - V'(q_j - q_{j-1}) \tag{2}$$

for integers j, where $V(\Delta q)$ is the potential. The well-known Toda lattice [9] is obtained with potential $V(s) = a\exp(-bs)/b + as$ where a, b are constants and $a, b > 0$. It is also common to see (2) written as

$$\frac{d^2}{dt^2}s_j = V'(s_{j+1}) - 2V'(s_j) + V'(s_{j-1}) \tag{3}$$

for integers j, where $s_j = q_j - q_{j-1}$ is the deviation of the bond length from its equilibrium. The extension to lattices in more than one space dimension is straightforward.

The rest of the paper is organised as follows. In section 2 we describe the symplectic Ordinary Differential Equation solver and compare its structure with the commonly used explicit classical Runge-Kutta method. In section 3 we describe the results of a series of experiments to evaluate the accuracy, Hamiltonian conservation and efficiency of the symplectic solvers applied to lattice equations in one and two space dimensions.

2 Symplectic ODE Solver

In recent years, much effort has been devoted to the development of solvers which preserve the symplectic properties of Hamiltonian systems of Ordinary Differential Equations. Surveys of this work can be found in [8] and [2]. Briefly, the main arguments in support of these methods are: the symplectic property determines much of the Hamiltonian dynamics and so should be preserved by the approximate solver; symplectic methods produce approximate solutions which stay close to the exact solution manifold over very long time periods; in general, although they do not exactly conserve the Hamiltonian and other quantities, they approximately conserve them to high precision with little change over long time periods (see [6] for example). The main reason that they are not in common use (apart from the fact that they are not yet widely known) is that the easily obtainable symplectic schemes are fully implicit Runge-Kutta methods, which are very computationally expensive since they involve the solution of systems of nonlinear algebraic equations at each time step. However, the special form of the lattice equations we are considering allows the use of a very cheap symplectic scheme which does not suffer this drawback.

We adapt a fourth order accurate, fixed time step, symplectic algorithm described in [1, 10] for lattice equations. We concentrate on one dimensional lattices, and the extension to two dimensional lattices is obvious. This scheme is designed for ODEs with a separable Hamiltonian form i.e.

$$\dot{p}_j = -\frac{\partial}{\partial q_j} H(\underline{p}, \underline{q}) \quad , \quad \dot{q}_j = \frac{\partial}{\partial p_j} H(\underline{p}, \underline{q}) \quad \text{with} \qquad H(\underline{p}, \underline{q}) = F(\underline{p}) + G(\underline{q}) \; . \tag{4}$$

The lattice equation (1) can be written in a particularly simple, separable form as

$$\dot{p}_j = -\frac{\partial}{\partial q_j} G(\underline{q}) \quad , \quad \dot{q}_j = p_j \quad \text{with} \qquad H(\underline{p}, \underline{q}) = G(\underline{q}) + \frac{1}{2} \sum_j (p_j)^2 \; . \tag{5}$$

The symplectic scheme to advance the solution from time $n\tau$ to $(n+1)\tau$ using a fixed time step size τ is written in pseudo-code as

(a) For all j do $q_j := q_j + b_1 p_j$.

(b) For $s = 2, \ldots, 4$ do

   (b.1) For all j do $p_j := p_j + a_s A_j(\underline{q})$.

   (b.2) For all j do $q_j := q_j + b_s p_j$.

where

$$A_j(\underline{q}) = -\frac{\partial}{\partial q_j} G(\underline{q})$$

and the constants b_s, a_s are given by

$$a_2 = a_4 = \tau(2 - 2^{1/3})^{-1}, \quad a_3 = \tau(1 - 2^{2/3})^{-1},$$

$$b_1 = b_4 = \tau(2 + 2^{1/3} + 2^{-1/3})/6, \quad b_2 = b_3 = \tau(1 - 2^{1/3} - 2^{-1/3})/6 \; .$$

At the start of step (a), the vectors $\underline{p}, \underline{q}$ contain the values of approximate solution at time $n\tau$ and at the end of step (b), they contain the approximate solution at time $(n+1)\tau$.

The symbol := indicates that we evaluate the right hand side and overwrite the storage locations on the left with the result. It is important to complete step (b.1) before starting step (b.2).

The computational cost of the algorithm is dominated by the three $\underline{A}(\underline{q})$ evaluations required for each time step and the main storage requirement is space for two vectors $\underline{p}, \underline{q}$. In comparison, the Classical 4th order Runge-Kutta method, which is often used in the solution of lattice equations, uses four evaluations of vector function $\underline{A}(\underline{q})$ and a large number of other vector operations each time step. If we use a simple coding of the Runge-Kutta method, then the storage required is 5 pairs of vectors the same size as $\underline{p}, \underline{q}$. With more complicated coding and at the expense of vectorisation, the storage can be cut to 1 pair of vectors. The symplectic scheme requires fewer auxiliary calculations also and the net result is that it is cheaper than the Runge-Kutta on all counts. This is confirmed in the experiments reported in the next section which showed it to take about 0.75 of the time required by the Runge-Kutta for each time step.

3 Numerical Results

In this section we describe the results of computations designed to investigate the stability of travelling wave solutions on one and two space dimensional lattices. In most cases we use initial data obtained by the approximation method described in [4, 5] which uses path following and spectral methods. We test the solver on finite lattices with periodic boundary conditions, so that the solitary wave is in effect travelling around a ring in the one space dimension case and over a torus in two.

We first consider lattice equations in one space dimension which can be written as (2) and then rewritten in first order Hamiltonian form as

$$\dot{p}_j = V'(q_{j+1} - q_j) - V'(q_j - q_{j-1}) \quad , \quad \dot{q}_j = p_j \,. \tag{6}$$

We consider various forms of potential energy function $V(\cdot)$, including the Toda and electrical potentials and the quartic polynomial potential

$$V_4(s) = \tfrac{1}{2}s^2 + \tfrac{1}{4}as^4 \,. \tag{7}$$

The change of variables,

$$s_j = q_{j+1} - q_j \,, \quad r_j = p_{j+1} - p_j \tag{8}$$

allows us to rewrite (6) again to obtain,

$$\dot{r}_j = V'(s_{j+1}) - 2V'(s_j) + V'(s_{j-1}) \quad , \quad \dot{s}_j = r_j \,. \tag{9}$$

The direct application of the symplectic scheme to lattice equations in the non-Hamiltonian form above is exactly equivalent to applying the scheme to the Hamiltonian form (6) and computing s_j from $s_j = q_j - q_{j-1}$, thus preserving the useful features of the scheme.

In all the examples tested, the Hamiltonian is conserved with considerable accuracy. The quartic potential results with $\tau = 0.01$ shown in figure 1 show an apparently constant upper bound on the error in the Hamiltonian up to (and beyond) 4 million time steps.

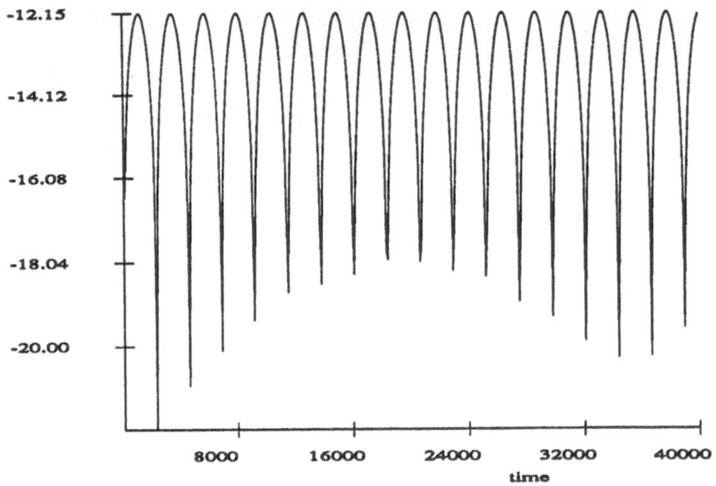

Figure 1: Graph of $\log_e(\frac{|H(t)-H(0)|}{H(0)})$ vs time for the symplectic solver on the quartic potential showing the large time control of $H(t)$. Parameters $\tau = 0.01$, 50 lattice points, $c \approx 3.8$.

For a given potential, the relative size of the error in the Hamiltonian is dependent on the time step size used and the height (and hence the speed) of the initial pulse. One would expect the relative error in the Hamiltonian to vary as $O(\tau^4)$ for small enough τ in this fourth order accurate scheme, and experiments on the quartic potential with different values of τ confirm this. For example, successively halving τ from $\tau = 0.04$ shows the maximum error decreasing by factors 15.5, 19.1, 17.6 etc. which is close to the predicted value 16.

We find apparent long time instabilities in the pulse solutions of the quartic potential problem when the initial data is perturbed slightly (actually by a mismatch between the number of lattice points used in the path following code and in the ODE solver). In this case, the pulse propagates as expected initially, but ripples develop after some time, at the expense of the pulse's height and speed. As the wave travels around the ring it interacts with these small linear waves, accelerating its decay. This behaviour is illustrated in figure 2, where the wave is travelling around a lattice of 50 nodes. We should remark that initially the wave has travelled around the ring slightly *more* than three times between each plot, and at later times slightly *less* than three times and has not reversed its direction (as it appears). These same examples have been calculated using the Classical fourth order Runge-Kutta scheme where ripples also develop and a similar picture is seen. However the Runge-Kutta scheme damps the solution causing a steady decrease in the Hamiltonian, at first more severely, but later less so, once the soliton has been mostly eroded. See figure 3 for a more quantitative description.

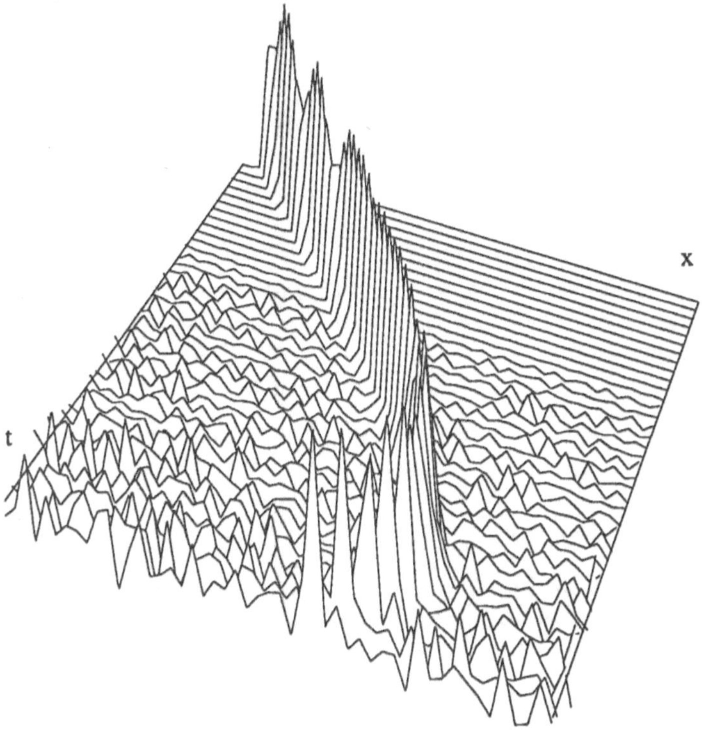

Figure 2: Solution profiles on quartic lattice at regular time intervals with about three revolutions between each plot. See text for details.

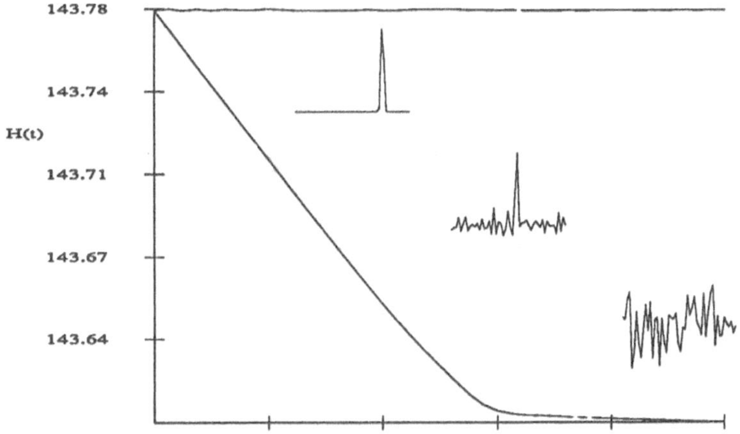

Figure 3: The main graph shows the behaviour of the Hamiltonian for the Symplectic scheme (upper line) and for the classical Runge-Kutta scheme (lower). The insets show the solution profile at times roughly corresponding to their position on the main graph.

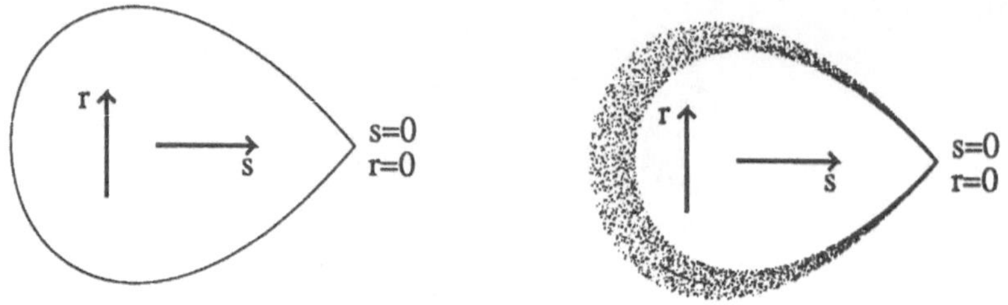

Figure 4: Phase plane portraits of a solitary wave in the lattice equation (9) with the Toda potential using the symplectic algorithm (left) and the Classical Runge-Kutta (right). 400000 time steps shown.

A slightly different example is given by the problem of electrical transmission lines [7, 5]. The simplest form of the equation of motion is

$$\frac{d^2}{dt^2}(w_n - aw_n^2) = w_{n+1} - 2w_n + w_{n-1}. \tag{10}$$

where $a > 0$. To put this into the form of equation 9, the substitution $u_n = w_n - aw_n^2$ is used. This is not a 1-1 transformation, so we have to take care in inverting it. Solving the quadratic and taking the negative branch, we obtain $V'(u) = (1 - \sqrt{1-4au})/2a$ giving,

$$V(u) = \frac{(1-4au)^{\frac{3}{2}}}{12a^2} + \frac{u}{2a} - \frac{1}{12a^2} \tag{11}$$

and we can solve this as before. The very sharp peak or cusp which appears at larger speeds (see [5]) makes determining the initial data harder, but once it is found the symplectic algorithm propagates this unusual solitary wave with little distortion and at the correct speed. The Hamiltonian was conserved to 9 significant figures in these tests.

Another clear indication of the advantages of the symplectic scheme can be seen in the phase plane portraits shown in figure 4. These show a solitary wave on the Toda lattice of 32 points, sampled at the same mesh point every 10 time steps for 400,000 steps and computed using the symplectic and Runge-Kutta methods with identical step sizes $\tau = 1/8$. The symplectic results are very sharp, but the Runge-Kutta results show the solution decaying in amplitude.

Finally we turn to a two dimensional example. We examine the KP lattice described in [3]

$$\begin{aligned}\frac{d^2}{dt^2}s_{i,j} &= (s_{i+1,j} - 2s_{i,j} + s_{i-1,j}) + \epsilon^2(s_{i,j+1} - 2s_{i,j} + s_{i,j-1}) \\ &+ \frac{1}{4}\epsilon^2(s_{i+1,j}^2 - 2s_{i,j}^2 + s_{i-1,j}^2).\end{aligned} \tag{12}$$

which can be solved using the symplectic method, modified to cover all the points in this two dimensional grid. We again show phase portraits of solitary waves in figure 5. We use a 150×200 mesh with appropriate periodic boundary conditions, time step size $\tau = 1/16$ and waves at angles 0 and $\tan^{-1}(3/4)$ to the i-axis at speeds close to 1. The phase portraits are again sharp over many timesteps, apart from the inner one (after magnification) which was given slightly perturbed initial data.

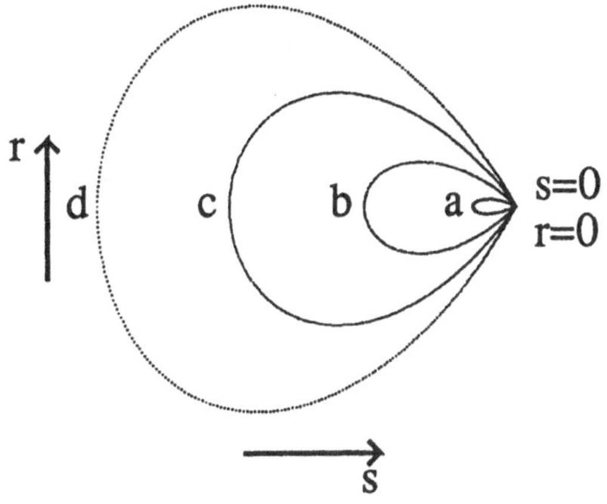

Figure 5: Phase plane diagram for the discrete KP equation on a two dimensional lattice showing (a) $\theta = 0, c = 1.053$ (b) $\theta = 0, c = 1.175$ (c) $\theta = \tan^{-1} 3/4, c = 1.054$ (d) $\theta = \tan^{-1} 3/4, c = 1.155$. (Note that $r = \dot{s}$)

4 Conclusion

The symplectic ordinary differential equation solver we describe is a useful tool for investigations of solitary waves on discrete lattices. It is simple, efficient and accurate.

Acknowledgements

We are grateful to the NATO Special Programme Panel on Chaos, Order and Patterns for support for a collaborative programme, to the SERC for research funding under the Nonlinear System Initiative, and to the EC for funding under the Science programme SCI-0229-C8C-100079/JU1.

References

[1] J Candy & W Rozmus, *A symplectic integration algorithm for separable Hamiltonian functions*, J. Comp. Phys., **92**, 230–256 (1991).

[2] P J Channell & J C Scovel, *Symplectic integration of Hamiltonian systems*, Nonlinearity, **3**, 231–259 (1990).

[3] D B Duncan, J C Eilbeck, C H Walshaw and V E Zakharov, *Solitary waves on a strongly anisotropic KP lattice*, submitted to Phys. Lett. A, (1991).

[4] J C Eilbeck & R Flesch, *Calculation of families of Solitary Waves on discrete lattices*, Phys. Lett. A, **149**, 200–202 (1990).

[5] J C Eilbeck, *Numerical studies of solitons on lattices*, these proceedings, (1991).

[6] D I Pullin & P G Saffman, *Long-time symplectic integration: the example of four vortex motion* Proc. R. Soc. Lond. A, **432**, 481–494 (1991).

[7] M Remoissenet & B Michaux *Electrical Transmission Lines and Soliton Propagation in Physical Systems*, Proceedings of the CMDS6 Conference, Dijon 1989, 313–324. Ed G Maugin. Longman (1990).

[8] J M Sanz-Serna, *Runge-Kutta schemes for Hamiltonian systems*, BIT, **28**, 877–883 (1988).

[9] M Toda, Theory of nonlinear lattices, 2nd Edition. Springer, Berlin (1988).

[10] Qin Meng-Zhao, Wang Dao-Liu & Zhang Mei-Qing, *Explicit symplectic difference schemes for separable Hamiltonian systems*, to appear in J. Comp. Math.

Solitary Wave Solutions to the Discrete Nonlinear Schrödinger Equation

H. Feddersen
Department of Mathematics
Heriot-Watt University
Edinburgh EH14 4AS
Scotland, UK

Abstract

The existence of various solitary wave solutions to the (nonintegrable) discrete nonlinear Schrödinger equation is demonstrated numerically.

1 Introduction

The discrete nonlinear Schrödinger (NLS) equation appears in numerous applications of nonlinear dynamics [1, 2, 3, 4]. In these applications the nonintegrable discrete NLS equation is well approximated by the continuous cubic NLS equation which has well known soliton solutions [5]. A numerical time integration of the discrete NLS equation with the soliton solution to the continuum approximation used as initial condition suggests that a stable solitary wave may exist [6], although the resulting solution is not a perfect solitary wave.

The purpose of this paper is to find families of solitary wave solutions to the discrete NLS equation numerically. This is done by using a very efficient spectral collocation method coupled with path–following and bifurcation techniques [7, 8, 9, 10, 11]. This method allows us not only to find the expected solitary wave corresponding to the standard NLS soliton, we also find "dark" and multiple solitary waves as well as periodic travelling waves.

The discrete NLS equation we will be concerned with is

$$i\frac{dA_j}{dt} + \gamma|A_j|^2 A_j + A_{j+1} + A_{j-1} = 0 \tag{1}$$

with periodic boundary conditions $A_{j+L} = A_j$, where L is the number of lattice points. Hence, all the solutions we find are periodic with period L. For large L we can expect to find good approximations to solitary waves which have infinite period.

Eq. (1) has two constants of motion [12], the Hamiltonian

$$H = -\sum_{j=1}^{L}(\frac{\gamma}{2}|A_j|^4 + A_{j+1}A_j^* + A_{j+1}^*A_j) \tag{2}$$

with the canonical variables A_j and iA_j^*, where A_j^* is the complex conjugate to A_j, and the norm

$$N = \sum_{j=1}^{L} |A_j|^2. \tag{3}$$

Hence, Eq. (1) is integrable when $L = 2$, a nonlinear dimer [13], but nonintegrable for $L > 2$ [14].

Throughout we will normalise Eq. (1) such that $N = 1$. Note that this is equivalent to normalising the parameterless discrete NLS equation such that $N = \gamma$ which is sometimes useful in the numerical calculations.

The continuous NLS equation

$$i\frac{\partial u}{\partial t} + |u|^2 u + \frac{\partial^2 u}{\partial^2 x} = 0 \tag{4}$$

reduces to Eq. (1), with $\gamma = h^2$, under the finite difference discretisation $\partial^2 u/\partial^2 x \rightarrow (u_{j+1} - 2u_j + u_{j-1})/h^2$ followed by the gauge transformation $u_j = A_j \exp(-2it/h^2)$ and the scaling of time $t \rightarrow h^2 t$. Here h is the distance between adjacent lattice points. Thus, the nonlinearity $\gamma = h^2$ should be small in order that Eq. (1) be a good approximation to the NLS equation (4).

As a finite difference approximation to the NLS equation, the equation

$$i\frac{dA_j}{dt} + \gamma|A_j|^2 \frac{A_{j+1} + A_{j-1}}{2} + A_{j+1} + A_{j-1} = 0 \tag{5}$$

which also reduces to the NLS equation in the continuum limit is a better approximation in the sense that it is completely integrable with soliton solutions which have been found by using the inverse scattering method [15, 16]. However, here we will study Eq. (1) in its own right as a model for several discrete physical systems.

2 Numerical Method

We are interested in travelling wave solutions to Eq. (1), i.e., guided by the form of the soliton solution to the NLS equation, we will seek solutions of the form

$$A_j(t) = A(j - ct)e^{i(kj-\omega t)} = A(z)e^{i(kj-\omega t)} \tag{6}$$

where c/h is the velocity of the envelope of the travelling wave. The periodic boundary conditions $A(z + L) = A(z)$ requires k to be of the form

$$k = \frac{2\pi m}{L} \tag{7}$$

where m is an integer. With the ansatz (6) inserted into (1) we find that $A(z)$ must satisfy the complex nonlinear differential–delay equation

$$-icA'(z) + (kc + \omega)A(z) + \gamma|A(z)|^2 A(z) + e^{ik}A(z+1) + e^{-ik}A(z-1) = 0. \tag{8}$$

The solutions to Eq. (8) are approximated by the finite series

$$A(z) \approx \sum_{p=0}^{n-1} a_p \cos\frac{2p\pi z}{L} + i\sum_{p=1}^{n} b_p \sin\frac{2p\pi z}{L}, \tag{9}$$

where a_p and b_p are real coefficients which are determined by requiring that Eq. (8) be satisfied in n collocation points with the approximate solutions (9) inserted [10]. Thus Eq. (8) is reduced to a set of $2n$ real, nonlinear, algebraic equations with $2n$ unknowns, which can be solved numerically by path–following methods which are based on an Euler predictor/Newton–Raphson iteration scheme [7, 11]. It is also possible to detect bifurcation points and path–follow bifurcating branches as is demonstrated in the next sections.

3 Stationary Solutions

The numerical procedure described in the previous section requires a suitable starting guess. This can for example be a simple analytical solution or a solution to the continuum approximation to Eq. (1). It turns out that the constant solutions of Eq. (8), $A(z) = \Phi$, are very useful as the interesting solitary wave solutions bifurcate from these solutions. The constant solutions of Eq. (8) will be referred to as *stationary* solutions [12] as they correspond to the solutions to Eq. (1)

$$A_j(t) = \Phi_j e^{-i\Omega t} \tag{10}$$

with $\Phi_j = \Phi \exp(ikj)$ and $\Omega = kc + \omega$.

By inserting $A(z) = \Phi$ into Eq. (8) and using the normalisation $N = 1$ we obtain the following relation between the parameters for the stationary solutions:

$$\omega = -\frac{\gamma}{L} - kc - 2\cos k. \tag{11}$$

In order to find possible bifurcation points we perturb the stationary solutions by a periodic function $a(z)$, $|a(z)| \ll |\Phi|$,

$$A(z) = \Phi + a(z). \tag{12}$$

If $a(z)$ is expanded as $a(z) = \sum_p x_p \exp(2i\pi pz/L) + i\sum_p y_p \exp(2i\pi pz/L)$ and if terms of order $\mathcal{O}(|a|^2)$ are neglected when (12) is inserted into (8) we find for $\Phi^2 = 1/L$, after some algebra, that the following matrix equation must be satisfied:

$$\begin{bmatrix} \gamma/L + \alpha & i\beta \\ -i\beta & \alpha \end{bmatrix} \begin{bmatrix} x_p \\ y_p \end{bmatrix} = \begin{bmatrix} 0 \\ 0 \end{bmatrix}, \tag{13}$$

where

$$\alpha = \cos\frac{2\pi m}{L}(\cos\frac{2\pi p}{L} - 1), \tag{14}$$

$$\beta = \frac{\pi pc}{L} - \sin\frac{2\pi m}{L}\sin\frac{2\pi p}{L}. \tag{15}$$

Bifurcation points occur where the determinant is 0, i.e. when the nonlinearity γ is

$$\gamma = L(\frac{\beta^2}{\alpha} - \alpha). \tag{16}$$

Near the bifurcation points the bifurcating solutions are approximately

$$A^{(p)}(z) \approx \Phi + \phi e^{2i\pi pz/L} \tag{17}$$

for $|\phi| \ll |\Phi|$. Provided the size of the lattice, L is sufficiently large $A^{(p)}(z)$ will evolve into a p–solitary wave solution as γ is increased. This is demonstrated numerically in the following section.

4 Solitary Waves

The path-following method allows us to find a whole family of solutions as one parameter is varied. In the following the varying parameter will be ω while c will be fixed. Fig. 1 shows paths of $p = 1$ solitary waves for two different velocities. They both bifurcate from stationary solutions which are represented by the straight lines according to (11) in the figure. The numerical solutions are obtained for $n = 25$ modes in the expansions (9).

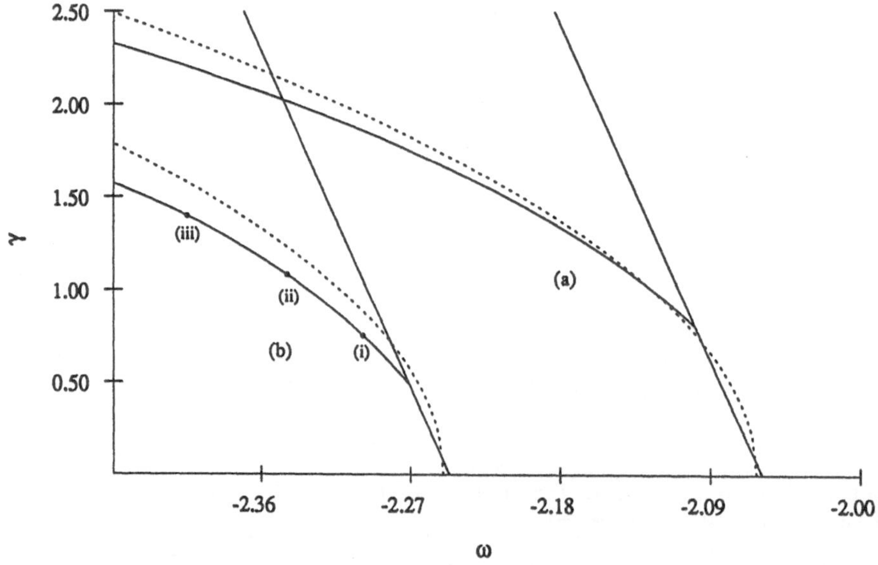

Figure 1: Solitary wave solutions bifurcating from stationary solutions for a lattice of size $L = 20$. (a) $c = 0.5$, (b) $c = 1$. The waveforms corresponding to the three points (i)–(iii) are shown in Fig. 2. The dashed lines show the relation between γ and ω for the soliton solutions to the corresponding continuous NLS equations.

The dashed lines show the paths for the soliton solutions to the corresponding continuous (parameterless) NLS equations. These solutions are given by [17]

$$A(z,t) = Q\operatorname{sech}\left(\frac{Qz}{\sqrt{2}}\right)e^{i(cz/2-\omega t)} \tag{18}$$

where $z = x - ct$ and $Q^2 = -2(\omega + 2 + c^2/4) > 0$. The relation between γ and ω is found as

$$\gamma = \int_{-\infty}^{+\infty} |A(z,t)|^2 dz = 4\sqrt{-(\omega + 2 + c^2/4)}. \tag{19}$$

Since all solitary waves in the discrete NLS equation (1) appear to bifurcate from stationary solutions we will consider the occurrences of the bifurcation points. The nonlinearity γ will everywhere be assumed positive. For $p = 1$, γ (as given in (16)) is positive only for the integer m nearest $(L/2\pi)\text{Arcsin}(c/2)$ when $k < \pi/2$. For $c > 2$ and $k < \pi/2$ Eq. (16) yields a negative value of γ. Hence, there are no ("bright") 1–solitary waves for $c > 2$, i.e. the maximum speed of the (bright) 1–solitary wave is $2/h$ [18]. This result also holds for the integrable, discrete NLS equation (5).

For $\pi/2 < k < 3\pi/2$ we find a number of 1–solitary wave solutions to Eq. (1), but they are all "dark" solitary waves. An example is shown in Fig. 4. Corresponding dark soliton solutions to the integrable, discrete NLS equation have not been found.

Consider the $c = 1$ solitary wave solution (Fig. 1) to the discrete NLS equation. The waveform for different values of γ is shown in Fig. 2 while Fig. 3 shows the result of a numerical time integration of Eq. (1) where the initial condition is taken as the "middle" one of the computed waves in Fig. 2. This time integration shows that the collocation method gives a very accurate approximation to a solitary wave and that the actual solitary wave is stable.

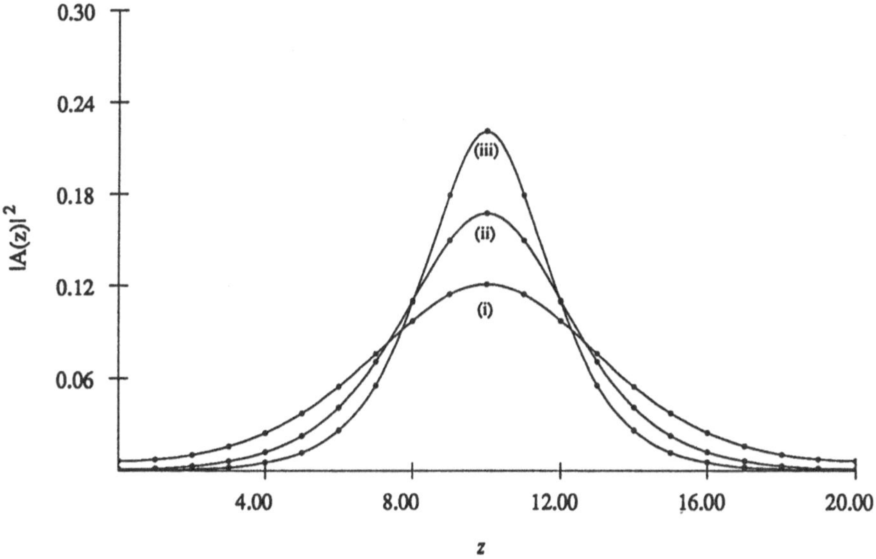

Figure 2: Waveforms of the solutions corresponding to the points (i)–(iii) in Fig. 1.

To get an indication of the rate of convergence of the solitary wave solution as the number of collocation points and modes, n in the expansion (9) is increased consider Eq. (8). This equation is only satisfied if the numerical solution (9) is exact. If it is not exact then the 0 on the right hand side of Eq. (8) will be replaced by a function $r(z)$ which is 0 in the collocation points. Fig. 5 shows how the numerical solution converges when n is increased. Here the error of the numerical solution is defined as $\max(|\mathrm{Re}[r(z)]|, |\mathrm{Im}[r(z)]|)$. The graph clearly suggests superalgebraic convergence [11].

Eq. (1) is not completely integrable. One implication of this is that the solitary wave paths stop at some point as γ is increased so solitary waves only exist for sufficiently small values of γ. For the two examples in Fig. 1 numerical calculations show that the paths stop at $\gamma \approx 2.4$ for $c = 0.5$ and at $\gamma \approx 1.8$ for $c = 1.0$.

Finally, Fig. 6 shows an example a double–solitary wave solution to the discrete NLS equation (1). This solution has been found using the procedure described above, i.e. it bifurcates from a stationary solution. A numerical time integration of Eq. (1) shows that this solution is stable.

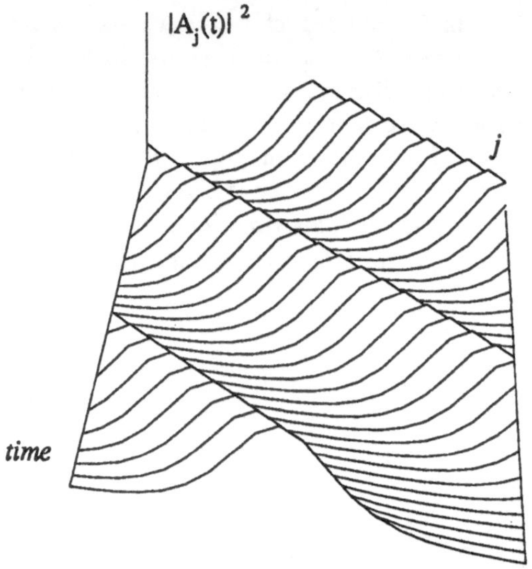

Figure 3: Numerical integration of the discrete NLS equation with the wave (ii) as initial condition showing a perfect solitary wave propagating with speed $c = 1$.

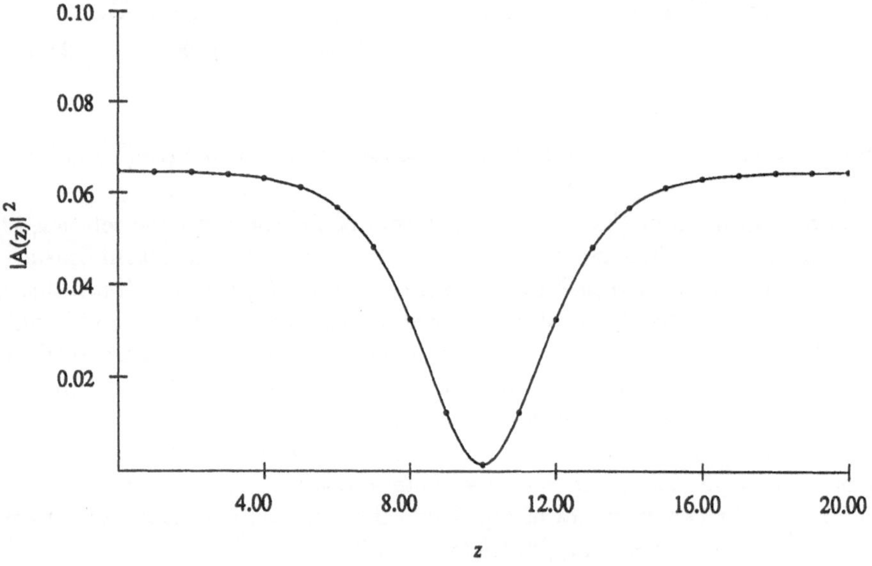

Figure 4: "Dark" solitary wave. $L = 20, c = 1, m = 9, \omega = -1.21, \gamma = 4.9$.

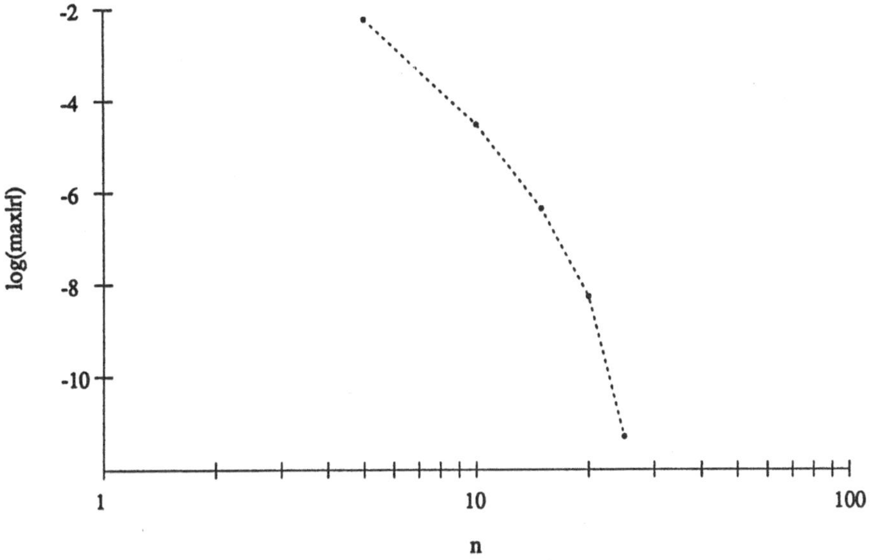

Figure 5: The error of the numerical solution as a function of the number of modes suggesting superalgebraic convergence.

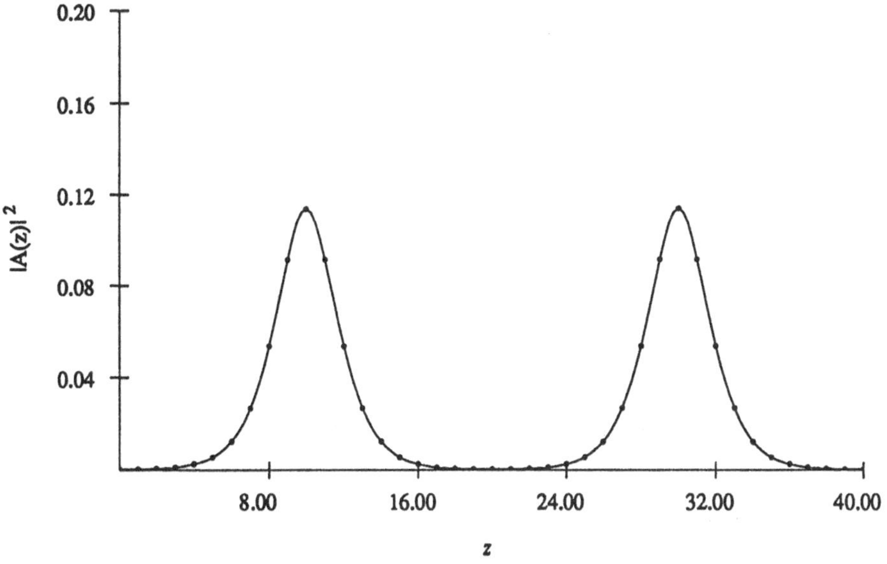

Figure 6: Double solitary wave. $L = 40, c = 1, m = 3, \omega = -2.42, \gamma = 2.9$.

Acknowledgements

I would like to thank J C Eilbeck for helpful suggestions and R Flesch for his contribution to the numerical path-following code. Also, I would like to acknowledge support for this research from the NATO Special Programme Panel on Chaos, Order and Patterns, the SERC for a program of research under the Nonlinear System Initiative, the EC for funding under the Science programme SCI-0229-C89-100079/JU1, and Simon Spies Fonden.

References

[1] A S Davydov and N I Kislukha. Solitary excitons in one-dimensional molecular chains. *Phys. Stat. Sol. (b)*, 59:465–470, 1973.

[2] P Baňacký and A Zajac. Theory of particle transfer dynamics in solvated molecular complexes: analytic solutions of the discrete time-dependent nonlinear Schrödinger equation. I. conservative system. *Chem. Phys.*, 123:267–276, 1988.

[3] D N Christodoulides and R I Joseph. Discrete self-focussing in nonlinear arrays of coupled waveguides. *Optics Letters*, 13:794–796, 1988.

[4] H-L Wu and V M Kenkre. Generalized master equations from the nonlinear Schrödinger equation and propagation in an infinite chain. *Phys. Rev. B*, 39:2664–2669, 1989.

[5] A C Scott, F Y F Chu, and D W McLaughlin. The soliton: a new concept in applied science. *Proc. IEEE*, 61:1443–1483, 1973.

[6] J C Eilbeck. Numerical simulations of the dynamics of polypeptide chains and proteins. In Chikao Kawabata and A R Bishop, editors, *Computer Analysis for Life Science - Progress and Challenges in Biological and Synthetic Polymer Research*, pages 12–21, Tokyo, 1986. Ohmsha.

[7] J C Eilbeck. Numerical studies of solitons on lattices. (These proceedings), 1991.

[8] H B Keller. Numerical solution of bifurcation and nonlinear eigenvalue problems. In P H Rabinowitz, editor, *Applications of Bifurcation Theory*, New York, 1977. Academic.

[9] D W Decker and H B Keller. Path following near bifucation. *Comm. Pure Appl. Math.*, 34:149–175, 1981.

[10] J C Eilbeck. The pseudo-spectral method and path following in reaction-diffusion bifurcation studies. *SIAM J. Sci. Statist. Comput.*, 7:599–610, 1986.

[11] J C Eilbeck and R Flesch. Calculation of families of solitary waves on discrete lattices. *Phys. Lett. A*, 149:200–202, 1990.

[12] J C Eilbeck, P S Lomdahl, and A C Scott. The discrete self-trapping equation. *Physica D: Nonlinear Phenomena*, 16:318–338, 1985.

[13] V M Kenkre. The discrete nonlinear Schrödinger equation: nonadiabatic effects, finite temperature consequences, and experimental manifestations. In P L Christiansen and A C Scott, editors, *Davydov's soliton revisited*, London, 1990. Plenum. (To be published).

[14] L Cruzeiro-Hansson, H Feddersen, R Flesch, P L Christiansen, M Salerno, and A C Scott. Classical and quantum analysis of chaos in the discrete self-trapping equation. *Phys. Rev. B*, 42:522–526, 1990.

[15] M J Ablowitz and J F Ladik. A nonlinear difference scheme and inverse scattering. *Stud. Appl. Math.*, 55:213–229, 1976.

[16] M J Ablowitz and J F Ladik. Nonlinear differential–difference equations and Fourier analysis. *J. Math. Phys.*, 17:1011–1018, 1976.

[17] P G Drazin. *Solitons*. Cambridge Univ. Press, Cambridge, 1983.

[18] A V Zolotariuk and A V Savin. Solitons in molecular chains with intramolecular nonlinear interactions. *Physica D*, 46:295–314, 1990.

ASYMPTOTIC BI-SOLITON IN DIATOMIC CHAINS

Jérôme LEON
Département de Physique Mathématique, Université Montpellier II, 34095 MONTPELLIER cdx05 FRANCE

Abstract

The light scattering in a diatomic chain of nonlinearily coupled oscillators is studied on the basis of classical Hamiltonian equation of motion in the continuum limit. The basic process is a *localized Brillouin scattering* and we prove that the nonlinear interaction of the light-wave with the phonon-wave results in a strong localization and a mutual trapping of the acoustic wave and the reflected light wave. This is shown to corresponds to the exchange of a given *acoustic particle* whose energy and momentum depend only on the elastic parameters of the chain. We conclude that the nonlinear coupling induces the existence of a new energy level which value does not depend on the initial condition or any other external constraint or parameter. The asymptotic state consists in a sonic wave front followed by two localized structures which eventually coalease onto the wave front.

This report is a shortened version of [1] completed with a more detailed discussion of the asymptotic behavior of the generic solution and some figures.

We study the nonlinear effects in one-dimensional sytems of coupled oscillators, which is an essential problem in physics and biophysics (hydrogen-bond systems) [2]. Many phenomena (proton conductivity in ice [3], anomalous infrared absorption in crystalline acetanilide (ACN) [4], lossless energy transport along alpha-helical proteins [5]) are suspected to be related to the nonlinear nature of the interaction of two different types (HF vs BF) of vibrations along quasi one-dimensional chains of atoms or molecules.

We consider here the classical model of a diatomic chain of nonlinearily coupled oscillators represented as the following diatomic chain (the vertical axis represent a reference equilibrium frame), with the three elastic parameters $\omega_0, S/M$ and s/m:

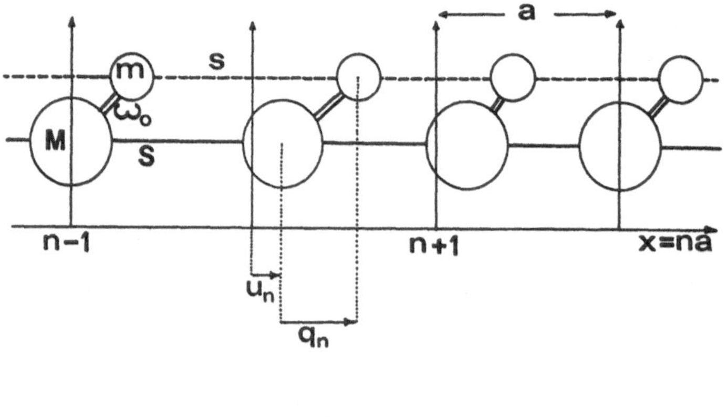

The oscillators u_n (low frequency) represent the motion of the molecule in the crystal (acoustic phonons) and they are associated with the hamiltonian:

$$H_L = \frac{1}{2}\sum_n M\dot{u}_n^2 + S(u_{n+1} - u_n)^2 + S(u_n - u_{n-1})^2 \tag{1}$$

where M is the mass and S the spring constant accounting for the hydrogen bond. The index L stands for *Lattice.*

The oscillators q_n (high frequency) represent the motion of one atom (or group of atoms) of the molecule (optical mode) and they are associated with the hamiltonian:

$$H_V = \frac{1}{2}\sum_n m[\dot{q}_n^2 + \omega_0^2 q_n^2] + s(q_{n+1} - q_n)^2 + s(q_n - q_{n-1})^2 \tag{2}$$

where ω_0 is the frequency of the isolated oscillator and where the spring constant s allows for the propagation of the vibration q_n along the chain. Here the index V stands for *Vibration.*

Both oscillators are nonlinearily coupled through the following interaction hamiltonian [5, 6, 7]

$$H_I = \frac{1}{2}\sum_n \frac{C_1}{2}(u_{n+1} - u_{n-1})q_n^2 + C_2[(u_{n+1} - u_n)q_{n+1} + (u_n - u_{n-1})q_{n-1}]q_n \tag{3}$$

where C_1 and C_2 are the interaction constants.

The equations of motion result in the following coupled system

$$M\ddot{u}_n = -\frac{\partial H}{\partial u_n}, \quad m\ddot{q}_n = -\frac{\partial H}{\partial q_n}, \quad H = H_L + H_V + H_I. \tag{4}$$

The masses M and m, together with the spring constants S and s, are in quite different scales which allows us to define a scaling parameter ϵ as (see also (12) below):

$$\epsilon^2 = \frac{S}{s}\frac{m}{M}. \tag{5}$$

Now we can precisely state the problem we are concerned with: describe the nonlinear dynamics of the scattering of light (the incident radiation is represented by a forced oscillation of q_n at, say, $n = +\infty$) at a frequency close to ω_0, in the case when $\epsilon \ll 1$.

The process which we consider is the Brillouin back-scattering of an incident wave (ω, k) according to the following selection rules:

$$\omega = \Omega + \omega_r, \qquad k = K + k_r, \tag{6}$$

where (Ω, K) is the sound wave (with dispersion law $\Omega = vK$ for small wave numbers, consistently with the continuum limit later adopted) and (ω_r, k_r) is the back-scattered wave (therefore $k_r \simeq -k$, and hence $K \simeq 2k$).

We will prove that the reflected wave is localized, and hence the back-scattered wave is not directly observable, bounded to the acoustic wave, and that the three waves do obey (6) according to the following actual distribution:

$$\begin{array}{cccc} \omega & = \Omega^{(0)} - \Omega_B & + & \omega_r^{(0)} + \Omega_B \\ k & = K^{(0)} - K_B & + & k_r^{(0)} + K_B \\ \textit{Incident} & \textit{Acoustic} & & \textit{Reflected} \end{array} \tag{7}$$

where we have defined

$$\Omega^{(0)} = vK^{(0)}, \quad K^{(0)} = 2k, \quad \omega_r^{(0)} = \omega - \Omega^{(0)}, \quad k_r^{(0)} = -k, \tag{8}$$

$$\Omega_B = 2\epsilon^2\omega, \qquad K_B = \Omega_B/v. \tag{9}$$

The word *incident* indicates a wave propagating from right to left (the input zone is $n = +\infty$), and ω is the *free* parameter.

As a consequence, the nonlinear coupling of the vibration q_n with the acoustic phonon u_n results in the exchange of the energy Ω_B and momentum K_B (obeying the dispersion law of the acoustic wave), which then can be thought of as representing a *binding acoustic particle.* The shift in frequency ($\Omega_B \simeq 2\epsilon^2\omega_0$) depends *only* on the elastic constants of the chain and *not* on the coupling constants or any other adjustable parameter as the energy of the acoustic soliton.

Starting with the equation of motion (4), we first go to the continuum limit and obtain the following system of coupled wave equations:

$$q_{tt} - c^2 q_{xx} + \omega_0^2 q = -\alpha u_x q, \tag{10.a}$$

$$u_{tt} - v^2 u_{xx} = \beta q_x q, \tag{10.b}$$

where we have defined

$$c^2 = a^2\frac{s}{m}, \; v^2 = a^2\frac{S}{M}, \; \alpha = (C_1 + 2C_2)\frac{a}{m}, \; \beta = (C_1 + C_2)\frac{a}{M}, \tag{11}$$

and we have now

$$\epsilon = \frac{v}{c} \Rightarrow \Omega_B = 2(\frac{v}{c})^2\omega. \tag{12}$$

The above system is identical to the equation discussed in [7] where it is obtained without the approximation of a nearest-neighbour interaction.

The usual approach of a coupled system like (9-10) is to assume a quasi stationary acoustic wave $u(x,t)$ so that $u_x \propto q^2$. Then the slowly varying envelope of the wave $q(x,t)$ evolves according to the nonlinear Schrödinger equation (NLS). As a consequence, if a soliton solution to NLS is assumed (first problem: the mechanism of the *soliton creation* is not described) it has a non-zero phase and the resulting phase of the wave $q(x,t)$ is shifted from its value $\Omega^{(0)}$ of an amount depending on the soliton energy (second problem: the soliton energy is a *free* parameter).

Here we perform a multiscal analysis [8] on the system (9-10) by looking for small amplitude, slowly varying envelope solutions under the form:

$$q(x,t) = \epsilon a_1(\xi,\tau)\exp[i\phi_1] + \epsilon a_2(\xi,\tau)\exp[i\phi_2] + \text{c.c.}+$$

$$+\sum_{n=2}^{\infty}\epsilon^n a_1^{(n)}(\xi,\tau)\exp[in\phi_1] + \epsilon^n a_2^{(n)}(\xi,\tau)\exp[in\phi_2] + \text{c.c.} \tag{13.a}$$

$$u(x,t) = \epsilon\Psi_1(\xi,\tau)\exp[i\phi] + \epsilon^2\Psi_0(\xi,\tau)+$$

$$+\sum_{n=2}^{\infty}\epsilon^n\Psi_n(\xi,\tau)\exp[in\phi] + \text{c.c.} \tag{13.b}$$

$$\phi_1 = \omega t + kx, \; \phi_2 = \omega_r t + k_r x, \; \phi = \Omega t + Kx$$

$$\xi = \epsilon x, \ \tau = \epsilon^2 t. \tag{13.c}$$

Note that the dispertion relations for the wave $q(x,t)$ read

$$\omega^2 = \omega_0^2 + c^2\, k^2, \tag{14}$$

$$\omega_r^2 = \omega_0^2 + c^2\, k_r^2. \tag{15}$$

The above expressions (13) are inserted in (10) and the leading orders in ϵ give, after averaging over fast oscillations (which means practically: look at the coefficients of $\exp[i\phi]$):

$$\begin{gathered} \Psi_\tau - c\Psi_\xi = \gamma a_1 a_2^*, \\ a_{1,\xi} = \Psi a_2, \qquad a_{2,\xi} = \Psi^* a_1. \end{gathered} \tag{16}$$

We have set hereabove

$$\Psi = \frac{\alpha}{c^2}\Psi_1, \quad \gamma = \frac{\alpha\beta'}{2c^3}, \quad \beta = \epsilon\beta', \tag{17}$$

(the last definition is consistent with (11) and (5)), and we have used the selection rules (6) together with (14), (15) and $\Omega = vK \equiv \epsilon c K$ from the very definition of ϵ.

The relation (14) determines k from the input ω. The relation (15) together with (6) (and $\Omega = vK$) can be solved and we obtain

$$K = [2k - 2\epsilon^2 \frac{\omega}{v}][1 - 2k\epsilon^2]^{-1}, \tag{18}$$

which implies the relations (7),(8) and (9) for small k, at order $k\epsilon^2$.

Since we are dealing with the problem of the absorption of an incident (electromagnetic) wave (ω, k) at, say $x = +\infty$, with given amplitude A, we complete the system (14) with the following boundary values

$$a_1(\xi,\tau) \to A, \ (\xi \to +\infty), \qquad a_2(\xi,\tau) \to 0, \ (\xi \to -\infty), \tag{19}$$

and an initial condition $\Psi(\xi, 0)$ in $L^2(\Re)$. It is worth remarking that, due to (19), the system (16) *implies* the property

$$\Psi(\xi,0) \to 0, (\xi \to \pm\infty) \Rightarrow a_2(\xi,0) \to 0, (\xi \to +\infty). \tag{20}$$

The essential point is now that the system (16) with boundary values (19) is *integrable* in the sense of the spectral transform theory [9] (it has actually a structure similar to that of the well known "self- induced- transparency" equation governing the light pulse propagation in a two level system [10]). In the spectral transform theory, the system (16) is said to have a *singular dispersion law* [11] and we have developed the general formalism for such classes of nonlinear coupled evolutions on the basis of the $\bar{\partial}$ problem formulation of the spectral transform [12] (the most general class of integrable evolutions of type (16) is displayed in [13]).

We have proved in [2] that, for any initially localized initial acoustic wave $\Psi(\xi, 0)$, the general solution of (16) has the following *universal* asymptotic form for $\tau \to +\infty$: it vanishes for $\xi + c\tau < 0$, and, for $\xi + c\tau > 0$ it obeys

$$a_1 \to \ A[1 + |\eta(\xi,\tau)|^2][1 - |\eta(\xi,\tau)|^2]^{-1}, \tag{21}$$

$$a_2 \to 2iA\eta(\xi,\tau)[1-|\eta(\xi,\tau)|^2]^{-1}, \tag{22}$$

$$\Psi \to -4i\sqrt{\frac{\Delta\tau}{\xi+c\tau}}\,\bar{\eta}(\xi,\tau)[1-|\eta(\xi,\tau)|^2]^{-1}, \tag{23}$$

where the function $\eta(\xi,\tau)$ is given by

$$\eta(\xi,\tau) = 2i\rho\frac{\sqrt{\pi y}}{\Delta\tau}[\sum_{n=1}^{N-1} A_n y^{-n} + \mathcal{O}(y^{-N})]\ \exp[y/2],$$

$$A_0 = 1,\quad A_1 = -27/16,\quad A_2 = 8385/2^9,\quad A_3 = -2589825/2^{13},$$

$$y(\xi,\tau) = 8\sqrt{\Delta\tau(\xi+c\tau)},\quad \Delta = |A|^2\gamma/4. \tag{24}$$

It is then an easy task to go back to the physical solution $u(x,t), q(x,t)$ of the system (10) by using successively (17), (13) and (12). So we have at first order in ϵ and for large t and for $x > -vt$:

$$q(x,t) = \epsilon A\exp[i(\omega t + kx)]\frac{1+|\eta|^2}{1-|\eta|^2}+$$

$$+2i\epsilon A\exp[i(\omega_r^{(0)}+\Omega_B)t + i(k_r^{(0)}+K_B)x]\frac{\eta}{1-|\eta|^2} + \text{c.c.} \tag{25}$$

$$u(x,t) = -4i\epsilon\frac{c^2}{\alpha}\sqrt{\frac{\Delta\tau}{\xi+c\tau}}\exp[i(\Omega^{(0)}-\Omega_B)t + i(K^{(0)}-K_B)x]\frac{\bar{\eta}}{1-|\eta|^2} + \text{c.c.} \tag{26}$$

while for $x < -vt$

$$q(x,t) = \epsilon A\exp[i(\omega t + kx)],\quad u(x,t) = 0. \tag{27}$$

The above behaviour of the solution allows us to deduce the following properties of the system of coupled wave equations (10):

1- The frequencies and wave numbers of the waves q and u do obey the relation (7): the scattered part of the electromagnetic wave q and the acoustic wave u have exchanged the *acoustic particle* (Ω_B, K_B).

2- Although these asymptotic behaviours present singularities (for $|\eta|^2 = 1$), the solution (q,u) itself is never singular (the saddle point expansion becomes exact only for t strictly infinite).

3- The localized wave $u(x,t)$ has been *created* out of the continuous spectrum and it has the structure of a sonic wave front followed by two supersonic localized coherent structures which eventually reach the wave front. The scattered light wave is bounded to these localized structures.

We have drawn in the figures 1 to 3 the acoustic wave $u(x,t)$ in the rest frame of sound wave (moving at velocity $-v$) for the following choices of the parameters:

$$v = 3800ms^{-1},\quad c = 56000ms^{-1},\quad K = 100cm^{-1}$$

$$\gamma|A|^2 = 0.1\ S.I.,\quad \alpha c^{-2} = 1\ S.I.\quad \rho = 0.01$$

at different times (from 10 to 10^4).

These parameters have been chosen for representing the crystal of acetanilide. These values imply that the nonlinear binding mode has a frequency of $15cm^{-1}$. Actually the value of c (velocity of the propagation of the Amide-I vibration) has been inferred from the value of Ω_B. The good point is that this value of c lies in the range of acceptable values.

For the fig. 1 we have taken the saddle point expansion at order 1 (that is $N = 1$ in (24)), for the fig. 2 at order 2 and 3 for the fig. 3. Then it is clear that indeed there is a bisoliton (or bicaviton) stucture, which appears better at order three.

Finally the proof that the expansion converges as $t \to \infty$ can be inferred from the figure 4 where we have drawn (solid lines) the curves

$$f_N(y) = [\sum_{n=1}^{N} A_n y^{-n}]^2\, y \tag{28}$$

(see (24)) for different N. The dashed line is the curve

$$(\frac{\gamma|A|^2}{8\rho\sqrt{\pi}})^2 \tau^2 e^{-y}. \tag{29}$$

The intersections of (28) with (29) give the positions of the singularities of $u(x,t)$.

It is clear that all curves (28) have for $N \geq 1$ two asymptotes, and therefore that at any order *except the order* 0, there exists a time t_m such that, for $t > t_m$ the solution is as close as we want to the solution at next order. This proves both the convergence of the series and the fact that the asymptotic solution consists *always* in **two** nonlinear coherent structures, see also [14].

References

[1] J. LEON, Phys. Lett. **A152**, 178 (1991).

[2] M. BARTHES and J.LEON, (Editors), "Nonlinear Coherent Structures", Lecture Notes in Physics **353**, Springer Verlag (1990).

[3] M.PEYRARD, S. PNEVMATICOS, N. FLYTZANIS, Phys. Rev. **A 36**, 903 (1987), D. HOCHTRASSER, H. BUTTNER, H. DESFONTAINES, M. PEYRARD, Phys. Rev. **A 38**, 5332 (1988), S. PNEVMATICOS, Phys. Rev. Lett. **60**, 1534 (1988).

[4] G. CARERI, U. BUONTEMPO, F. GALLUZI, A.C. SCOTT, E. GRATTON, E. SHYAMSUNDER, Phys. Rev. **B 30**, 4689 (1984).

[5] A.S. DAVYDOV, Phys. Scripta **20**, 387 (1979).

[6] J.C. EILBECK, P.S. LOMDAHL, A.C. SCOTT, Phys. Rev. **B 30**, 4703 (1984).

[7] S. TAKENO, Prog. Theor. Phys. Lett., **73**,853 (1985).

[8] the technique of multiscale analysis for evolution equations has been studied in details in: F. CALOGERO and V. ECKAUS, Inverse Problems **3**, 229 (1987).

[9] R.K. DODD, J.C. EILBECK, J.D. GIBBON, H.C. MORRIS, "Solitons and Nonlinear Wave Equations", in Academic Press (1982). F.CALOGERO, A.DEGASPERIS, "Spectral Transform and Solitons", North Holland (1982). M.J.ABLOWITZ, H.SEGUR, "Solitons and the Inverse Scattering Transform", SIAM, Philadelphia (1981).

[10] G.L.LAMB Jr,Phys. Rev. **A 9**, 422 (1974) and **A 12** 2052 (1975).

[11] D.J.KAUP, A.C.NEWELL, Adv. Math. **31**, 67 (1979).

[12] J.LEON, Phys. Lett. **A 123**, 65 (1987) and J. Math. Phys. **29**, 2012 (1988).

[13] J.LEON, Phys. Lett. **A 144**, 444 (1990).

[14] J.LEON, Phys. Rev. Lett. **66**, 1587 (1991).

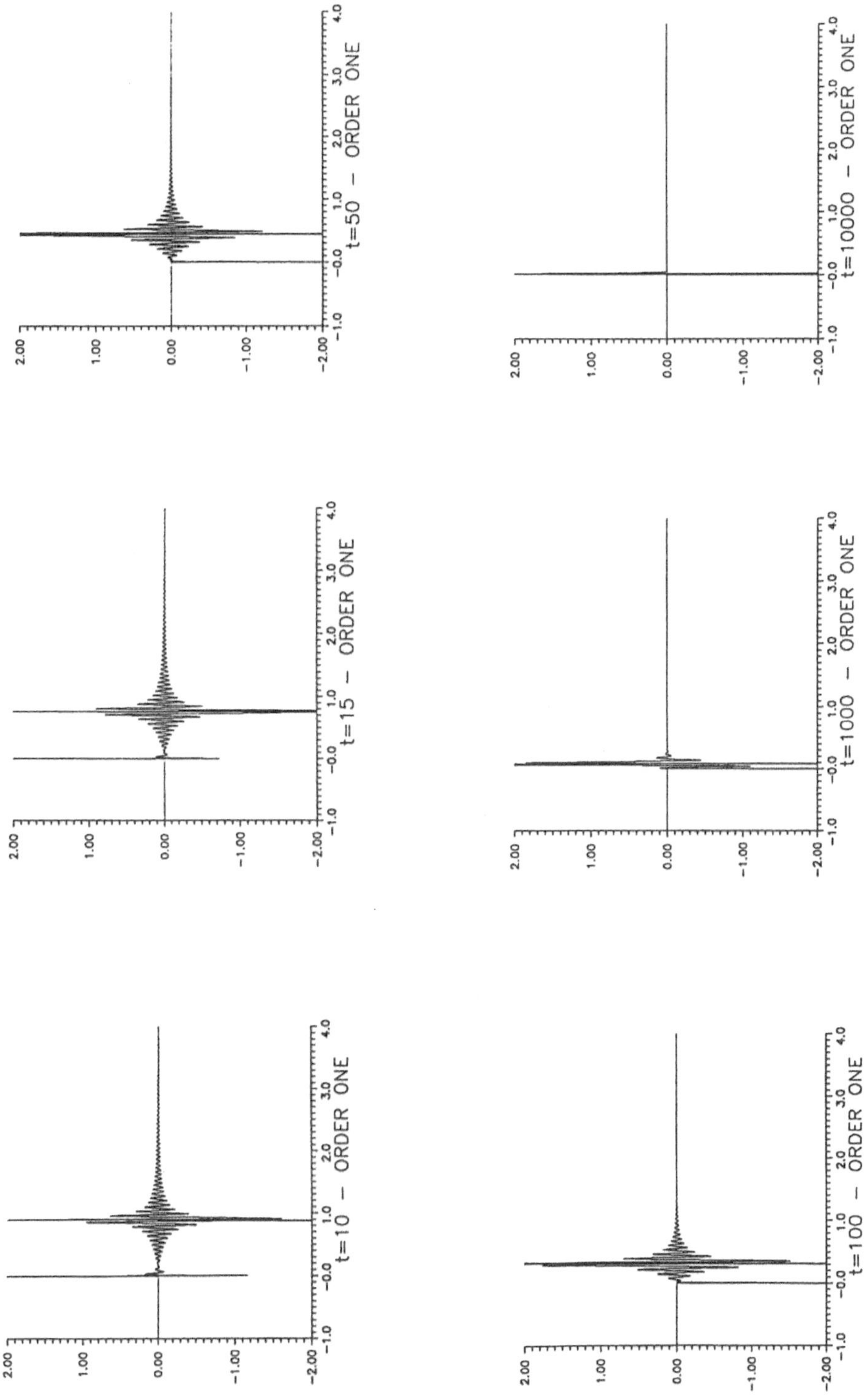
t=50 – ORDER ONE
t=15 – ORDER ONE
t=10 – ORDER ONE
t=10000 – ORDER ONE
t=1000 – ORDER ONE
t=100 – ORDER ONE
2.00
1.00
0.00
-1.00
-2.00
-1.0
-0.0
1.0
2.0
3.0
4.0

FIG. 1

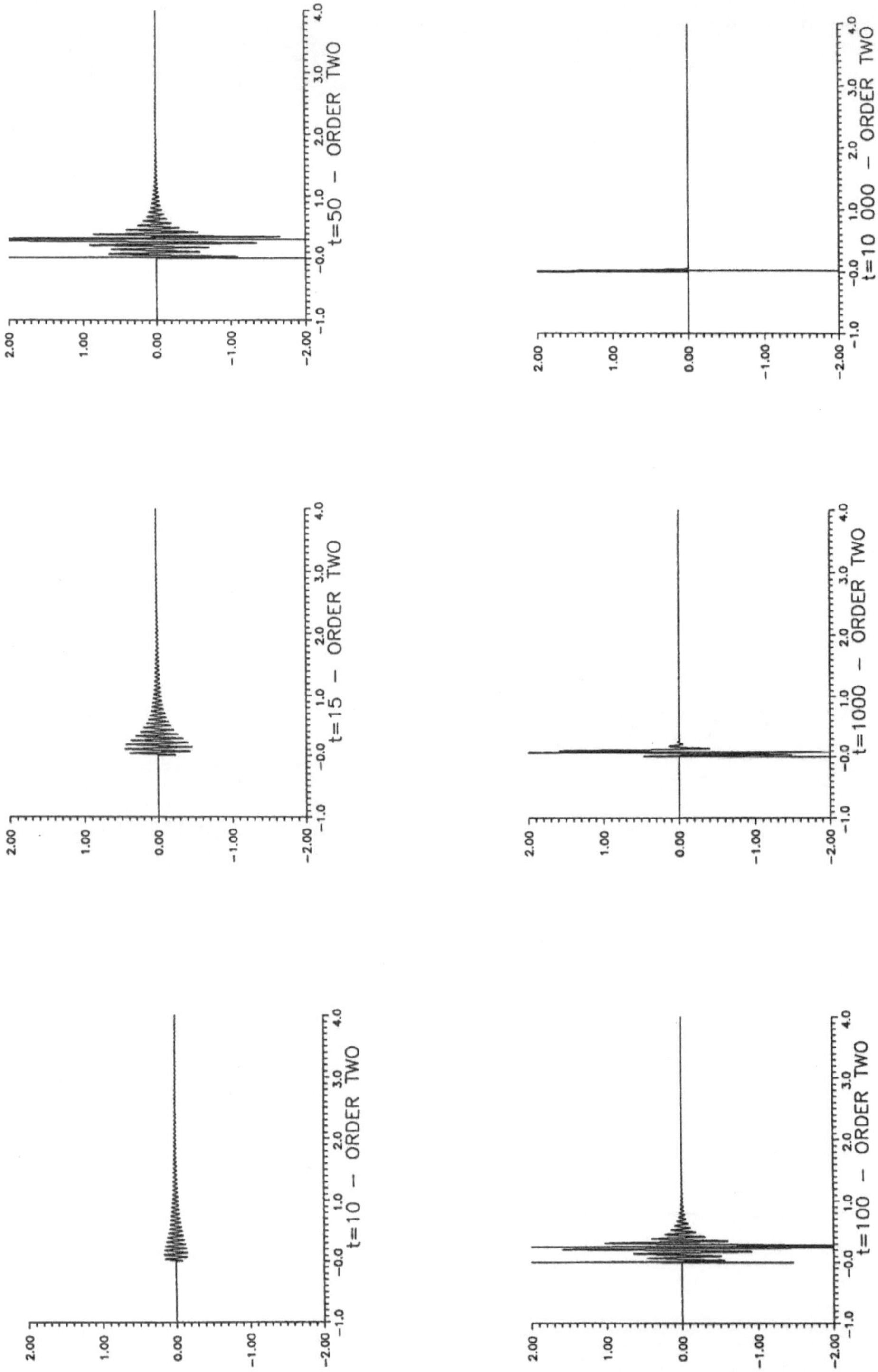
t=50 - ORDER TWO
t=10 000 - ORDER TWO
t=15 - ORDER TWO
t=1000 - ORDER TWO
t=10 - ORDER TWO
t=100 - ORDER TWO

FIG. 2

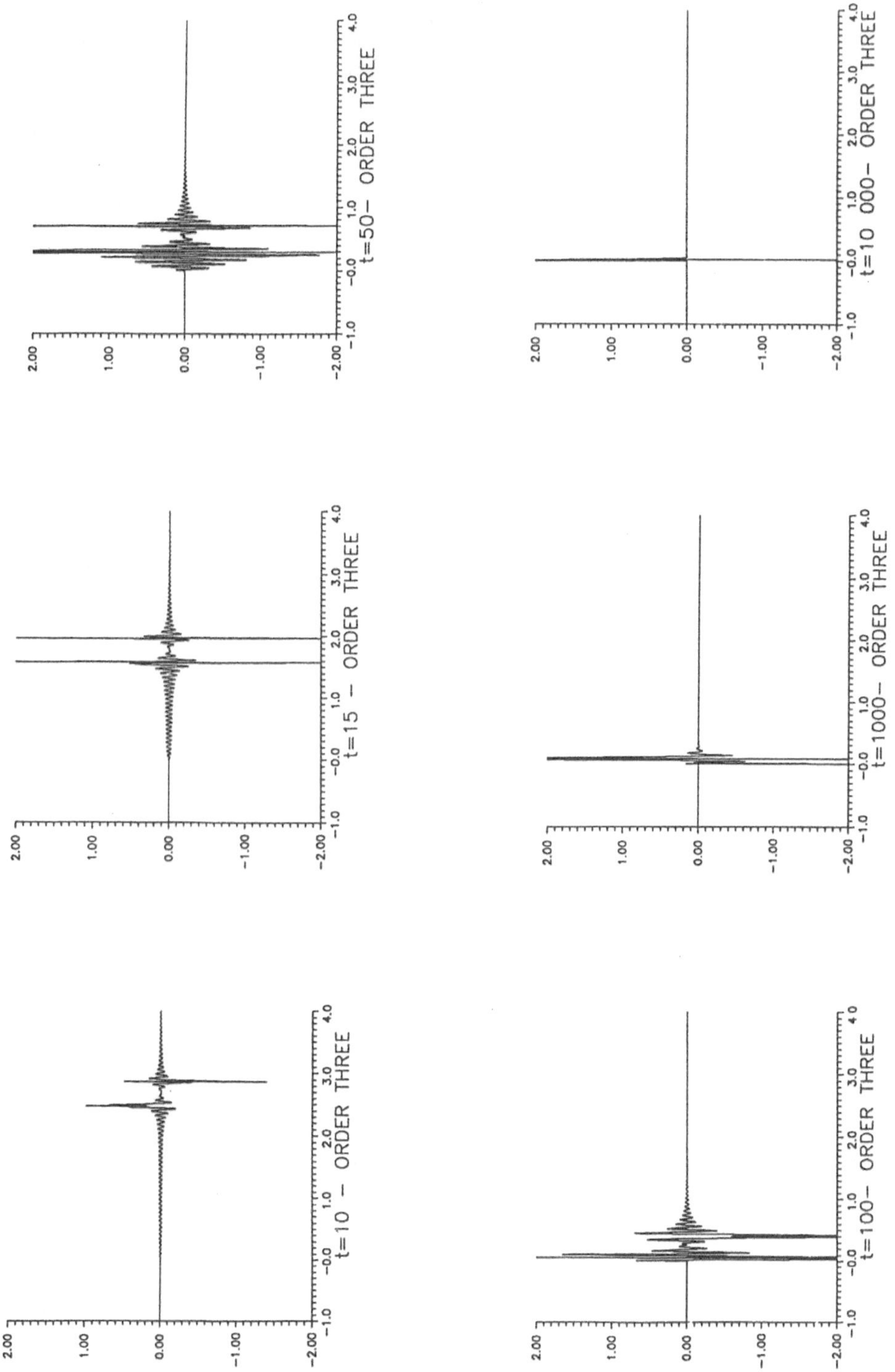
t=50- ORDER THREE
t=15 - ORDER THREE
t=10 - ORDER THREE
t=10 000- ORDER THREE
t=1000- ORDER THREE
t=100- ORDER THREE

FIG. 3

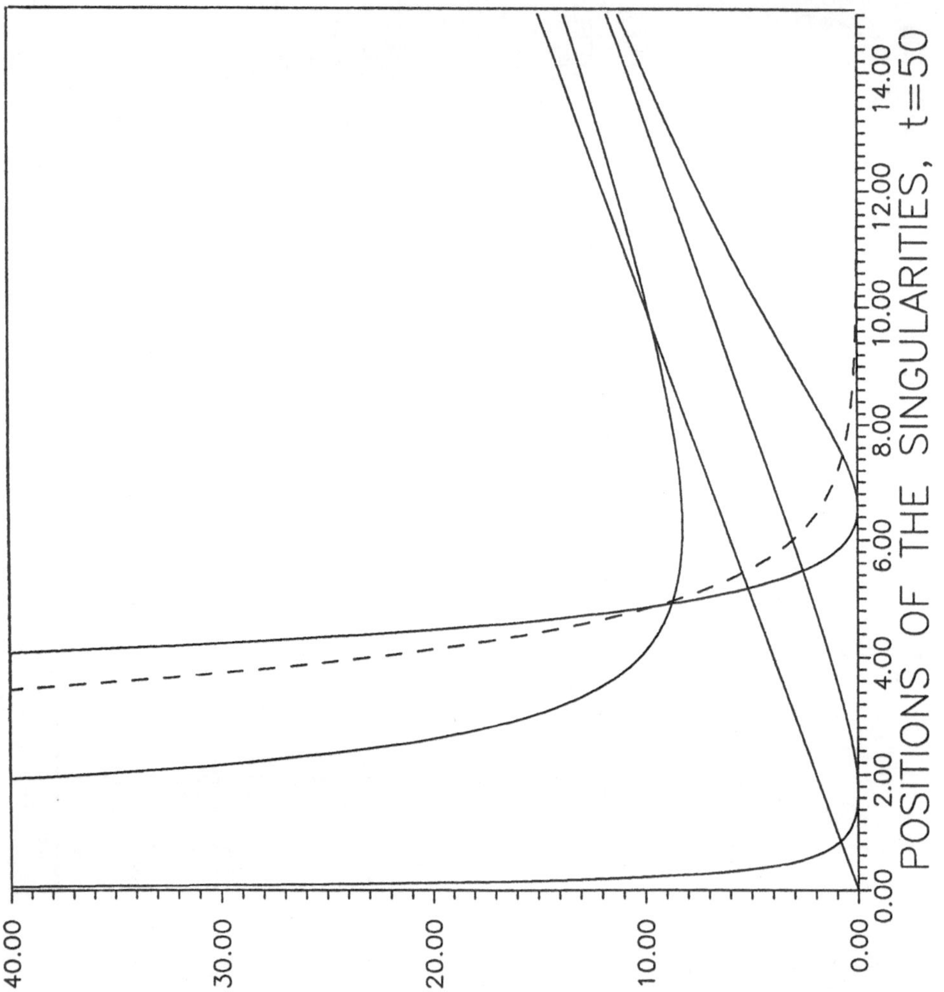
POSITIONS OF THE SINGULARITIES, t=50
0.00
2.00
4.00
6.00
8.00
10.00
12.00
14.00
0.00
10.00
20.00
30.00
40.00

FIG 4

NONLINEAR DYNAMICS OF LOCALIZED STRUCTURES AND PROTON TRANSFER IN A HYDROGEN-BONDED CHAIN MODEL INCLUDING DIPOLE INTERACTIONS

I. CHOCHLIOUROS and J. POUGET
Laboratoire de Modélisation en Mécanique (associé au CNRS)
Université Pierre et Marie Curie
4 Place Jussieu, 75252 PARIS Cédex 05
FRANCE

Abstract : The transport of energy in H-bonded chains is really an extremely important problem, because of its close connection with basic phenomena in biological systems. We consider a lattice model which is made of two one-dimensional harmonically coupled sublattices corresponding to the oxygens and protons, the two sublattices being coupled. The study becomes more interesting when we introduce the **dipole-dipole interactions.** As a microscopic dipole is created by the proton motion, it may affect the response of the nonlinear excitations propagating along the chain. We are looking for a solution for which the motion of oxygen ions can be neglected. A Φ^6 **equation** is found , which admits nonlinear excitations of **solitary wave type.** We distinguish different classes of solutions for the description of the proton motion. Analytical expressions and the necessary conditions for the existence of these types of solutions are given. The introduction of the dipole interaction produces an influence on the electric field of the system which means that the proton motion is also affected and this makes the **proton conductivity** much easier. Numerical simulations are presented for special cases. Finally, possible further extensions of the work are discussed.

1.- INTRODUCTION.

Among the whole variety of condensed-matter systems where the soliton concept can be used are **quasi-one-dimensional molecular systems** in biology [1]. The transport of protons in H-bonded chains is a very interesting problem since it can explain fundamental properties of life [2]. It can be also used in order to explain **proton mobility and electric conductivity** in ice [5,9,10].
The aim of this work is to study the influences of the dipole-dipole interactions on protonic conductivity in H-bonded chains. At the same time an analytical study of the nonlinear dynamics of the proton motion is also proposed.
In order to reach our goal we construct a one-dimensional lattice model based on the **Antonchenko-Davydov-Zolotaryuk model** [3], for H-bonded chains. From previous scientific works [4,6,7,8], there is already a satisfactory number of results both analytical and numerical, which contribute to the validity of the ADZ model. This model was really a successful attempt to approximate and explain mechanisms which occur in

the atomic scale study. We also introduce the interatomic potentials involved in the model and we include -the important point of our work- the dipole interactions due to the proton motions. The existence of electric dipoles along the chain may affect the response of the system and the proton conductivity becomes much easier. Since the discrete system is not manageable, we are faced with the continuum approximation of the microscopic model.

After some calculations we find a Φ^6 equation. This means that now we have the possibility of many localized solutions, according to several selections and conditions. The protonic conductivity is caused by ionic and orientational defects as [1] explained in previous works. What is important to note here is also the possibility of many further extensions of the work.

2.- CONSTRUCTION OF THE MODEL.

The model consists of two one-dimensional interacting sublattices. These are the proton sublattice and the heavy-ion sublattice as shown in Fig.1. Protons and "oxygens" are harmonically coupled.

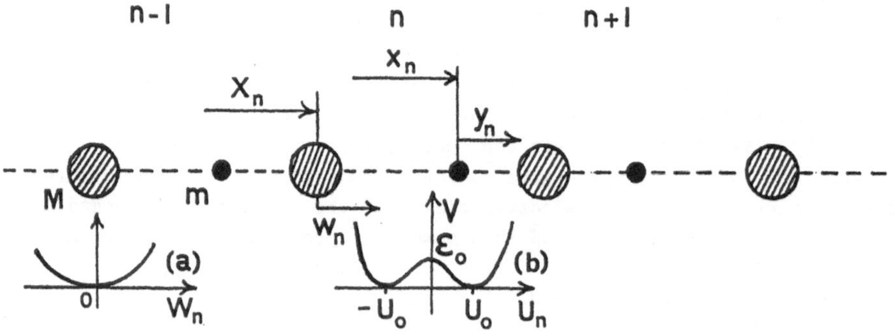

Fig. 1 : One-dimensional lattice model for a hydrogen-bonded diatomic chain. (a) harmonic potential for heavy ions and (b) double-well potential for protons.

Each proton lies between a pair of heavy ions usually called as "oxygens". There are covalent and hydrogen bonds connecting proton with the two neighbouring oxygens. When a proton moves from a position closer to the one of the oxygens to a position closer to the next one, then the two bonds exchange their positions. The double-well acting on protons and allowing this jump can be approximated [11] by the expression

$$U(y_n) = \epsilon_0 \left(1 - y_n^2/y_0^2\right)^2 . \qquad (1)$$

A schematical representation is given in Fig.1. In Eq.(1), y_n denotes the displacement of the n-th proton with respect to the center of the oxygen pair in which it is located, ϵ_0 is the potential barrier and $2y_0$ is the distance between the two minima of the double-well potential. The coupling between oxygens and protons can provide a mechanism [6] which changes the potential barrier that protons have to overcome to jump from one molecule to the other and makes their motion easier.

3.- EQUATIONS OF THE MODEL.

The total Hamiltonian of the system is

$$H_{tot} = H_p + H_o + H_{int} + H_{dd} \ . \tag{2}$$

The proton part of the Hamiltonian is

$$H_p = \sum_n \left[\frac{1}{2}m\dot{y}_n^2 + U(y_n) + \frac{1}{2}m\omega_1^2(y_{n+1} - y_n)^2\right] \ . \tag{3}$$

The first term denotes the kinetic energy of each proton, the second is due to the double-well potential while the last term represents the harmonic coupling with the characteristic frequency ω_1 between neighbouring protons (m is the proton mass). The oxygen part of the Hamiltonian can be written as

$$H_o = \sum_n \left[\frac{1}{2}M\dot{w}_n^2 + \frac{1}{2}M\Omega_0^2 w_n^2 + \frac{1}{2}M\Omega_1^2(w_{n+1} - w_n)^2\right] \ . \tag{4}$$

Here we consider only the relative displacement w_n of an oxygen pair, because a possible variation of the O-O distance can modulate the double-well undergone by the protons. The first term of H_o denotes the kinetic energy of oxygens, the second term denotes the coupling between oxygens of the same cell while the third term describes the harmonic coupling between neighbouring oxygen pairs and introduces in the model the dispersion of an optical mode (Ω_0 and Ω_1 are characteristic frequencies of the optical mode, M is the oxygen mass).
The Hamiltonian H_{int} is derived from the **dynamic interaction** between the two sublattices and describes how the double-well is modulated by the variation of the O-O distance. It is written as in ADZ model [6]

$$H_{int} = \sum_n \delta w_n \left(y_n^2 - y_0^2\right) \ . \tag{5}$$

Its physical meaning is the **lowering of the potential barrier** due to the oxygen displacements. (δ measures the strength of the coupling and determines the amplitude of the distortion in the oxygen sublattice).
The Hamiltonian H_{dd} is derived from the dipole interactions. The existence of electric dipoles in the chain, because of the electric charges, leads us to account for the **mutual interaction betweeen the dipoles**. In the present case, we assume that the distance r between neighbouring dipoles does not depend on the lattice displacement and all vectors of the dipole moment P_n are in the same direction (all are alligned). We find that

$$H_{dd} = \bar{\beta} \sum_n P_n P_{n+1} \ . \tag{6}$$

($\bar{\beta}$ is a constant which may account for the environment of the chain). The dipole moment induced by the proton motion must be zero when proton is at either position of the oxygens or when it is at the middle of the distance joining the oxygen pair where the interactions are opposite. The law which the dipole has to agree with is presented in Fig.2.

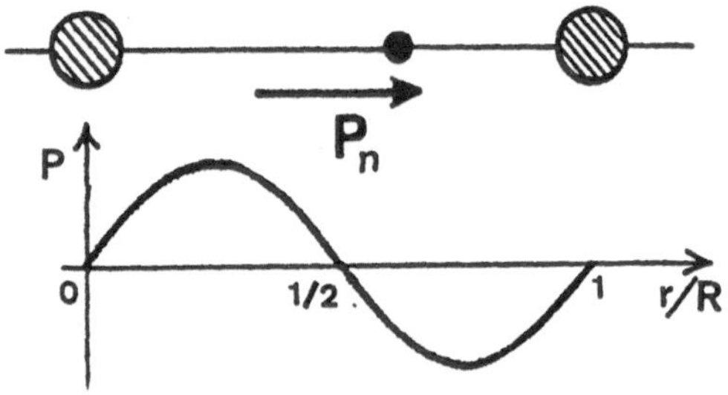

Fig. 2 : Dipole as a function of the proton position with respect to the neighbouring heavy ion positions.

The simplest form for the dipole moment P to approximate the structure of the curve in Fig.2 can be given [12] by a polynomial of the third degree

$$P_n = \bar{a}(x_n - X_n)(x_n - X_{n+1})\left[x_n - \frac{1}{2}(X_n + X_{n+1})\right]. \tag{7}$$

(Where $\bar{a}$ is a constant). The absolute positions x_n and X_n are shown in Fig.1 and we have

$$x_n = n\ell_0 + \frac{1}{2}\ell_0 + y_n \ , \quad X_n = n\ell_0 + w_n \ , \quad X_{n+1} = (n+1)\ell_0 + w_{n+1} \ . \tag{8}$$

We note that ℓ_0 is the lattice spacing. The equations of motion of the system form a set of coupled nonlinear differential-difference equations which derived from the Hamiltonian (1) and takes on the form

$$\frac{d^2 y_n}{dt^2} = \omega_1^2(y_{n+1} - 2y_n + y_{n-1}) + (4\epsilon_0/my_0^2)(1 - y_n^2/y_0^2)2y_n - (2\delta/m)w_n y_n$$
$$- \bar{\beta}\frac{\partial P_n}{\partial y_n}(P_{n+1} + P_{n-1}) \ , \tag{9a}$$

$$\frac{d^2 w_n}{dt^2} = \Omega_1^2(w_{n+1} - 2w_n + w_{n-1}) - \Omega_0^2 w_n - (\delta/M)(y_n^2 - y_0^2)$$
$$- \bar{\beta}\frac{\partial P_n}{\partial w_n}(P_{n+1} + P_{n-1}) - \bar{\beta}\frac{\partial P_{n-1}}{\partial w_n}(P_n + P_{n-2}) \ . \tag{9b}$$

We study that case in which the heavy-ion sublattice can be considered as "frozen". The motion of the oxygens is not remarkable as the motion of the protons, which is very important. This approximation is related to the inertia of the oxygen sublattice which cannot follow the fast proton motion especially for large velocities and suggests that a solution could involve only the proton displacement while oxygens stay at rest and do not participate in the motion. We can assume that w is very small compared to y. In the atomic lattice y is of the order of few Å and for this reason we can suppose that w is almost zero.

Now we consider only the Hamiltonians H_p and H_{dd}. We introduce the units E_0 for energy, t_0 for time and ℓ_0 for length. We find the derived units $m_0=E_0 t_0^2/\ell_0^2$ for mass and $f_0=\epsilon_0/\ell_0$ for force. We introduce the dimensionless parameters $\tilde{m}=m/m_0$, $\epsilon=\epsilon_0/E_0$, $\chi_1=-\bar{\beta}\,\bar{a}^{-2}\ell_0^6/E_0\tilde{m}$. We also consider $u_n=y_n/\ell_0$, $\tilde{H}=H/m\ell_0^2$. The expression for the energy of the system becomes

$$\tilde{H} = \sum_n \left[\frac{1}{2}\dot{u}_n^2 + \frac{1}{2}\omega_1^2(u_{n+1}-u_n)^2 + \frac{1}{4}G_0 u_0^2\left(1 - u_n^2/u_0^2\right) - \chi P_n P_{n+1}\right], \tag{10}$$

where the electric dipole is merely given by

$$P_n = u_n\left(u_n^2 - 1/4\right). \tag{11}$$

The equation of the discrete system describing the proton motion reduces to

$$\begin{aligned}\ddot{u}_n &= \omega_1^2(u_{n+1} - 2u_n + u_{n-1}) + G_0 u_n\left(1 - u_n^2/u_0^2\right)\\ &+ \chi\left(3u_n^2 - 1/4\right)\left[u_{n-1}\left(u_{n-1}^2 - 1/4\right) + u_{n+1}\left(u_{n+1}^2 - 1/4\right)\right].\end{aligned} \tag{12}$$

Where we have previously set $G_0=4\epsilon_0/\ell_0^2 u_0^2$ and $\chi=\chi_1 E_0/m\ell_0^2$.

4.- CONTINUUM APPROXIMATION OF THE DISCRETE SYSTEM.

Further step to the somewhat rough simplification consists of considering the continuum approximation. The equation of the proton motion -if we consider terms up to the second order- can be written as follows

$$u_{tt} - \left[1 + \chi(3u^2 - 1/4)^2\right]u_{xx} - \alpha_1 u + \beta_1 u^3 - \gamma_1 u^5 = 0, \tag{13}$$

where we have set

$$\alpha_1 = G_0 + \chi/8, \qquad \beta_1 = G_0/u_0^2 + 2\chi, \qquad \gamma_1 = 6\chi. \tag{14}$$

We can notice that all three parameters α_1, β_1, γ_1 depend on χ, which χ is also present in the factor accompanied u_{xx}. This last remark makes further investigation extremely difficult. In order to simplify the procedure we propose the following hypothesis : we set $1 + \chi(3u^2-1/4) = \tilde{C}_0$ as a function of u. This function has two minima. For certain value of χ we can find the mean value $\langle\tilde{C}_0\rangle$ and then we set $\langle\tilde{C}_0\rangle$ as the coefficient of u. This rough approximation can be done when there is no great difference between the greatest and the lowest values of the function. This occurs if the coefficient χ is considered small enough such that $|\chi|\ll 1$. We now set $\langle\tilde{C}_0\rangle=\omega^2$ and the equation for study becomes

$$u_{tt} - \omega^2 u_{xx} - \alpha_1 u + \beta_1 u^3 - \gamma_1 u^5 = 0 \,. \tag{15}$$

We use the change of variables $u = \alpha U$, $t = \gamma T$, $x = \beta X$. Then we select $\alpha = \sqrt{\beta_1/\gamma_1}$, $\gamma = \sqrt{\gamma_1}/\beta_1$, $\beta = \omega\sqrt{\gamma_1}/\beta_1$ (it must be $\gamma_1>0$, $\beta_1>0$). We also set $A=\alpha_1\gamma^2$. The equation for study takes on the following form

$$U_{TT} - U_{XX} = AU - U^3 + U^5 \,. \tag{16}$$

We are looking for localized solutions with constant profile, moving at a characteristic velocity υ, that is for solutions $U=U(\xi)=U(X-\upsilon T)$. We do not dwell on the algebraic manipulations for finding out the different classes of solutions.

5.- DIFFERENT TYPES OF LOCALIZED SOLUTIONS.

The equation presents a symmetry, so when U is a solution then -U is also a solution. We can set $U_1 = U(\xi\to+\infty)$ and $U_2 = U(\xi\to\infty)$. We distinguish different types of solutions as we try to approach U_2 beginning by U_1, or the inverse.
For the solution of the type I (pulse), we have $U_1 = U_2 = 0$. We obtain the expression

$$U = \pm \frac{U_m}{[1 + P\sinh^2(\Omega\xi/2)]^{1/2}} \,. \tag{17}$$

Where it is $U_m^2 = \pm 4A/\left(\sqrt{1 - A/A_0} \pm 1\right)$ and $P = 2\sqrt{1 - A/A_0}/\left(\sqrt{1 - A/A_0} \pm 1\right)$. The sign (+) corresponds to supersonic waves ($|\upsilon|>1$) and for this case we have also the condition $0<A<(3/16)$. The sign (-) corresponds to subsonic waves ($|\upsilon|<1$) and for this case it must be necessarily considered $A<0$. In both cases we have $A_0=(3/16)$ and $\Omega^2 = 4A/(\upsilon^2-1)$.
The solution of the type III presents a kink. In this case we have two opposite non zero values ($U_1=U_0$, $U_2=-U_0$). The expression becomes

$$U = \frac{\pm U_0 \tanh z}{[1 + P(1-\tanh^2 z)]^{1/2}} \,. \tag{18}$$

Where it is $z=(1/2)\Omega\xi$, $P= U_0^2/\left(2U_0^2 - 3/2\right)$, $\Omega^2 = 4U_0^2\left(U_0^2 - 1/2\right)/(\upsilon^2 - 1)$. U_0 is defined by $\left(U_0^{\pm}\right)^2 = (1\pm\sqrt{1 - 4A})/2$, where the sign (+) corresponds to supersonic waves (it must be additionally $A<(3/16)$ and $A\neq 0$), and the sign (-) corresponds to subsonic waves (for the definition it must be additionally set $A<(1/4)$ with $A\neq 0$, $A\neq(15/64)$).

Finally, the solution of the type IV corresponds to a case in which it is $U_1 = U_2 \neq 0$. It is found

$$U = \pm \frac{U_0}{\left[P(\tanh^2 z - 1) + \tanh^2 z\right]^{1/2}} . \tag{19}$$

Where it is $P = U_0^2/\left(2U_0^2 - 3/2\right)$, $\Omega^2 = 4U_0^2\left[U_0^2 - 1/2\right]/(v^2 - 1)$. For the constant value U_0 we have $\left(U_0^{\pm}\right)^2 = (1 \pm \sqrt{1 - 4A})/2$. The sign (+) corresponds to supersonic waves which can be defined when we also consider $(3/16) < A < (1/4)$. The sign (-) corresponds to subsonic waves. (Now we have to consider $A < (1/4)$ with $A \neq 0$, $A \neq (15/64)$). Finally, a particular case (type II) occurs for which $A = (3/16)$ and the solution represents a kink describing a transition from the state $U_1 \neq 0$ to the state $U_2 = 0$.

6. NUMERICAL RESULTS.

We present numerical simulations corresponding to the solutions of the types I and III. We examine the evolution of a localized solution in time and we consider a certain number of particles in each case. The numerical simulations are performed directly by means of the set of discrete equations (12).
The kink-like solution of the type III corresponds to a proton displacement from the state $-U_0$ to the state $+U_0$ which are wells of the Φ^6 potential. This solution is remarkably **stable** as it is shown in Fig.3a. On the contrary, the pulse solution of the type I shown in Fig.3b presents an **instability**. After a short lapse of time, the pulse splits into two pulses travelling in opposite directions accompanied with rather large perturbations behind them. This makes difficult further investigations.

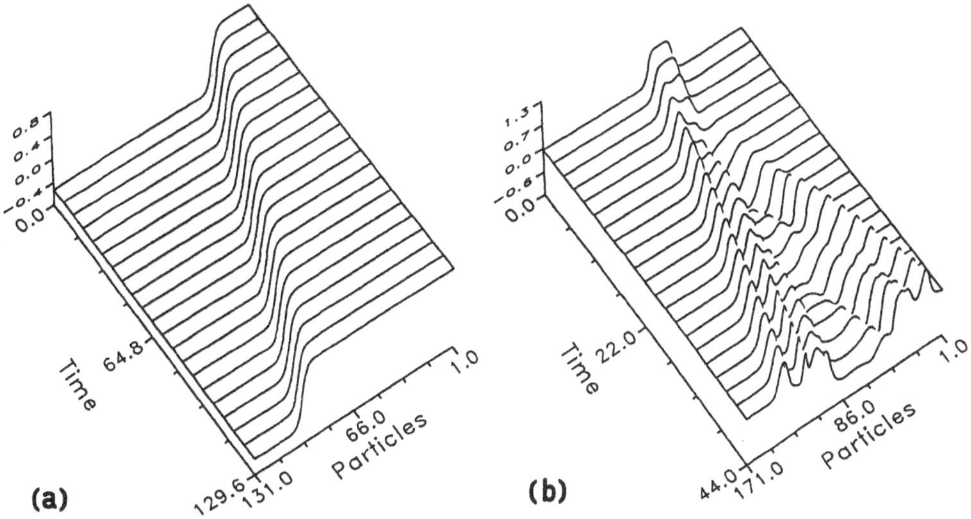

Fig. 3 : Numerical simulations of the proton motion on the lattice, (a) stable kink-like solution (type III) and (b) unstable pulse solution (type I).

7.CONCLUSIONS.

We have studied the influence of the dipole-dipole interaction on the proton motion. This kind of interaction produces an influence on the electric field of the system which affects to the proton motion and proton conductivity becomes much easier. The description of ionic and orientational defects associated with the protonic conductivity remains the same as in Φ^4 case, previously studied by other scientists [6]. It is necessary to remark that in our case we obtain many possibilities for the solution according to the values of the parameters, since the resulting equation possesses stable, unstable and metastable steady states. Another important point of the whole study is that we have to return to the problem of the continuum approximation and examine in detail all the possible cases. The model can be extended by the introduction of an external electric field applied on protons and also the introduction of damping. We can additionally consider a rotational motion of the dipoles, something which is closer to the real system, especially if we deal with nonlinear atomic chain. We can also examine the influence of second nearest-neighbouring interactions and more. At length, discreteness effects can be a source of very fascinating phenomena occuring at the microscopic scale.

REFERENCES.

[1] A. S. DAVYDOV, *Biology and Quantum Mechanics* (Pergamon, New York, 1982).

[2] J. F. NAGLE and H. J. MOROWITZ, *Proc. Nat. Acad. Sci. USA* 75, 298 (1978).

[3] V. Ya. ANTONCHENKO, A. S. DAVYDOV and A. V. ZOLOTARYUK, *Phys. Status Solidi B* 115, 631 (1983).

[4] E. W. LAEDKE, K. H. SPATSCHEK, M. WILKENS Jr. and A. V. ZOLOTARYUK, *Phys. Rev. A* 32, 1161 (1985).

[5] J. D. BERNAL and R. H. FOWLER, *J. Chem. Phys.* 1, 515 (1933).

[6] M. PEYRARD, St. PNEVMATIKOS and N. FLYTZANIS, *Phys. Rev. A* 36, 903 (1987).

[7] D. HOCHSTRASSER, H. BUTTNER, H. DESFONTAINES and M. PEYRARD, *Phys. Rev. A* 38, 5332 (1988).

[8] H. WEBERPALS and K. H. SPATSCHEK, *Phys. Rev. A* 36, 2946 (1981).

[9] L. ONSAGER, in *Physics and Chemistry of Ice* edited by E. Whalley, S. J. Jones and L. W. Crold (Royal Society of Canada, Ottawa, 1973) pp.7-12.

[10] N. BJERRUM, *Science* 115, 385 (1952).

[11] R. JANOSCHEK, in *The Hydrogen Bond: Recent Developments in Theory and Experiments*, edited by P. Schuster, G. Zundel and C. Sandforty (North Holland, Amsterdam, 1976), p. 165.

[12] E. WHALLEY, *Journal of Glaciology*, Vol. 21, 13 (1978).

RESONANT STATES IN THE PROPAGATION OF WAVES IN A PERIODIC , NON-LINEAR MEDIUM

J. Coste , J. Peyraud
Laboratoire de Physique de la Matière Condensée - URA 190
Université de Nice-Sophia Antipolis, Parc Valrose, 06034 Nice Cedex (France)

A - CLASSICAL WAVES

Let us first consider a transparent linear medium where the refractive index is periodically modulated:

$$n^2 = k_0^2\left[1 + \varepsilon v(x)\right] \qquad \text{(in reduced units)}$$

where $v(x + l_o) = v(x)$ (l_o : modulation period). There exist unstable solutions of the wave equation in the gaps surrounding Bragg resonances. These resonances are defined by $\varphi = k_o a = n\pi$, φ being the phase shift of the wave over l_o.
In the case considered in this first part, the modulation is harmonic ($v(x) = 2\cos 2x$), and the gap associated with first Bragg resonance is defined by : $k_0^2 \in \left[k_-^2 , k_+^2\right]$, where $k_\pm^2 = 1 \pm \varepsilon$

Let us now add a non linearity (due to Kerr effect). Then : $n^2 = k_o^2 + 2\varepsilon \cos 2x + |\phi|^2$ ϕ being the wave amplitude. We first look for stationary (monochromatic) solutions of the wave equation $(\partial_x^2 - n^2 \partial_\tau^2)\, \phi = 0$, that is of the form $\phi = \phi(x)\, e^{i\tau}$ $(\tau = \omega_0 t)$.

The problem is conveniently studied (3,4) with the help of the Poincaré map for variables $X_j = \phi(j\pi)$, $Y_j = (d/dx)\, \phi(j\pi)$ (the modulation period l_o is here equal to π).

It appears that the mapping

$$\begin{pmatrix} X_{j+1} \\ Y_{j+1} \end{pmatrix} = \begin{pmatrix} f(X_j, Y_j) \\ g(X_j, Y_j) \end{pmatrix}$$

is non integrable. The fondamental bifurcations of this dynamical system are the Arnold strong resonances : $\varphi = 2\pi/n$ $(n = 1,2,3,4)$, where φ is the rotation number around the origin (which is also the wave phase shift over one period). We note that two of these values $(n = 1,2)$ coïncide with Bragg resonances.

n = 1 resonance ($\varphi = 2\pi$: 2nd Bragg order). We obtain the following bifurcation diagram :

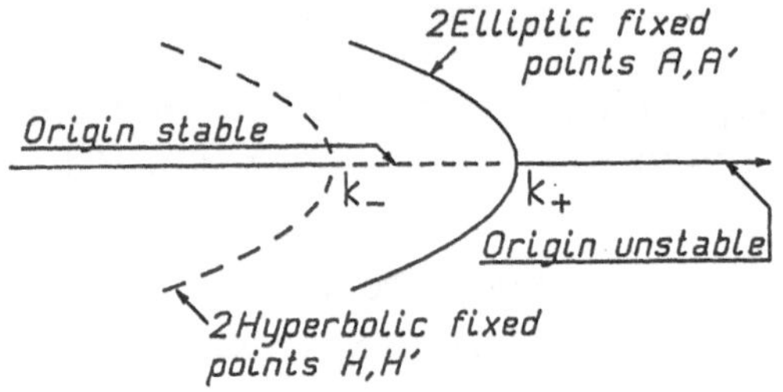

Fig.1 Bifurcation diagram for n=1 resonance

Neighborhood of k_+ : $k_o^2 = k_+^2 - \eta$. The origin is unstable inside the gap, and the phase portrait has the following form :

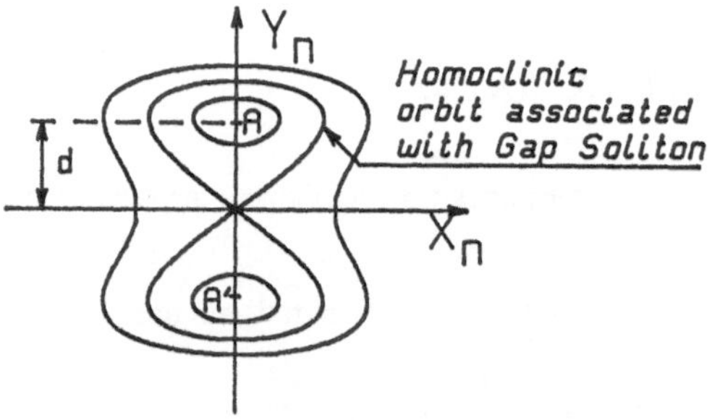

Fig.2 Phase portrait inside the gap

The two homoclinic orbits are associated with the celebrated gap solitons(1,2,3,4), which are immobile localized structures. The distance d of elliptic fixed points to the origin, which is a measure of the soliton amplitude, goes to zero when $\eta \rightarrow 0$. The non integrability of the mapping manifests itself by the stochastic behavior of the orbits near the origin. As a consequence, one-peak solitonic solutions are not allowed in large systems. Indeed the orbit points always escape from the origin after a finite number of iterations. Therefore the solutions are multi-peaks, with random inter-peak distances l_i, the average $l = \langle l_i \rangle$ diverging when $\eta \rightarrow 0$ like $\eta^{-\nu}$. The critical exponent is found to be $\nu \approx 1.4$.

When $\eta \rightarrow 0$ the rotation number around A or A' goes to zero and the mapping becomes integrable (or close to an integrable one). Then it can be shown that the soliton amplitude A obeys an ODE of the form :

$$\frac{d^2A}{dy^2} - A + |A|^2 A = 0 \tag{1}$$

y being a large-scale spatial variable. Eq. (1) admits a solution of the NLS type.

Neighborhood of k_- : $k_o^2 = k_-^2 - \eta$. The origin is now stable (we are outside the gap) and the phase portrait has the following form :

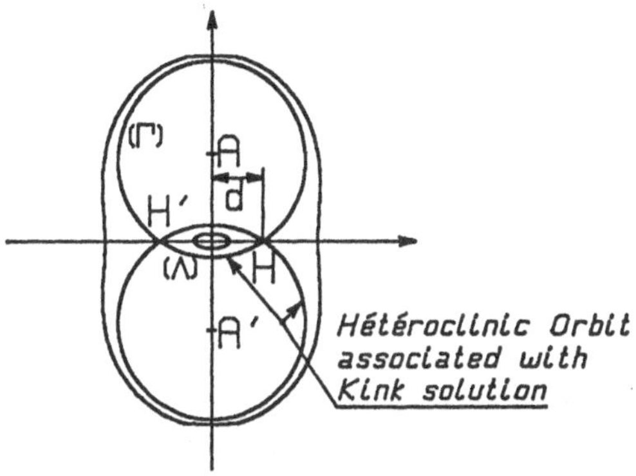

Fig.3 Phase portrait for $k_o^2 = k_-^2 - \eta$

It appears, in addition to elliptic points A, A'(which are now at finite distance from the origin), two hyperbolic points H, H' whose distance d to origin goes to zero as $\eta \to 0$. The finite amplitude orbit (Γ) is strongly chaotic. On the contrary the small orbit (Λ) is "nearly integrable" and is associated with a kink-like solution.

n = 2 resonance (1st Bragg order). The phenomena are exactly the same, except that the sign of wave function changes after each period. We therefore call "alternate" the solitons and kinks obtained in this case.

n = 4 resonance ($\varphi = \pi/2$). This resonance occurs exactly in the middle of the transparence band. The bifurcation of the mapping around the origin (which is elliptic) generates a set of 4 fixed elliptic points and 4 hyperbolic points, and the union of heteroclinic orbits is a set of two entangled ellipses.

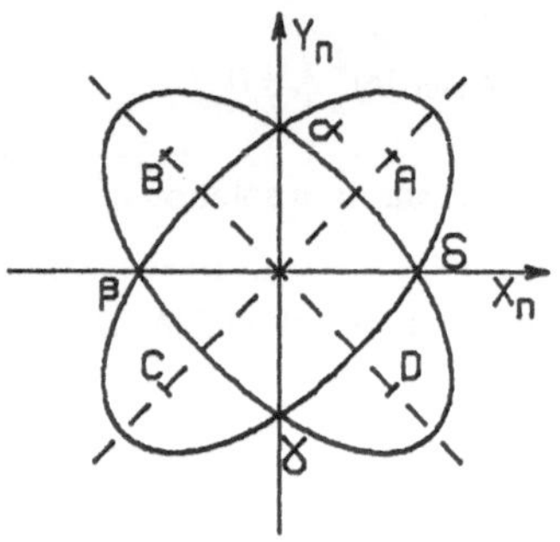

Fig 4 Phase portrait for n=4 . The "square structure".

We have called this configuration a " square".

n = 3 resonance ($\varphi = 2\pi/3$)

Here the bifurcation of the origin does not generates a strong resonance, because of the symmetry of the model (the non linearity is cubic). It would if some physical effect could break this symmetry, introducing a quadratic non linearity. In the present case this resonance can occur around an another elliptic fixed point of the mapping. We show on Fig.(5) this bifurcation around one of the fixed points of period-4 cycle considered above .

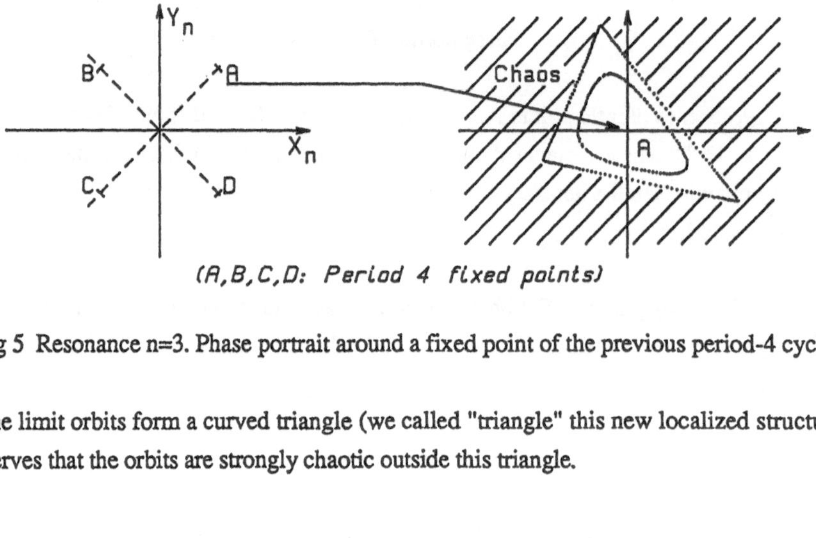

Fig 5 Resonance n=3. Phase portrait around a fixed point of the previous period-4 cycle.

The limit orbits form a curved triangle (we called "triangle" this new localized structure), and one observes that the orbits are strongly chaotic outside this triangle.

Unstationnary regimes

Considering now the propagation of a narrow wave packet around frequency ω_0, we have shown[5] that there exist solitonic solutions near the bifurcation points ($k_0^2 = k_+^2 - \eta$) which (for resonances n = 1, 2) take the form :

$\phi(x,t) \sim A \sin x\, e^{i\tau} + c.c.$

where A obeys the "slow" wave equation :

$$(i\partial_{\tau'} + \partial_y{}^2)A - A + |A|^2 A = 0 \qquad (2)$$

where y, τ' are an appropriate large scale variables.

Eq. (2) is isomorphic to NLS equation through gauge transformation $A \rightarrow A\, e^{-i\tau'}$. Rather unexpectedly the gap solitons are of the NLS type (near the bifurcation point). Their amplitude a is small and they propagate at low velocity v. Quantities a, v and η are related through the relation :

$$\alpha a^2 + \beta v^2 = \gamma\eta \qquad (3)$$

α, β, γ being numerical coefficients.

Concerning the square and triangle structures, a numerical study of the time-dependent problem have been made(6) showing that they are unstable, leading to chaotic regimes.

B - RESONANT POLARONIC STATES ON A 1-D PERIODIC CHAIN(7)

We consider the interaction of a periodic chain of atoms with one unbounded electron. The atoms are considered as classical scatterers (we neglect their quantum fluctuations in the limit of very large mass). The interaction potential of the electron with the nth atom at position x_n is $- V(x - x_n)$,

and the elastic energy of the atomic chain is : $\dfrac{\kappa_0^2}{2} \sum_n (x_{n+1} - x_n)^2$.

The interaction couples the x_n's and the electronic wave function $\phi(x,t)$, and this coupling may be described by the Ehrenfest equations obeyed by the atoms variables. In the classical limit, and neglecting the inertia of the atoms, one obtains :

$$\kappa_0^2 (u_n - u_{n-1}) = -\langle V'(x - x_n)\rangle \qquad (4)$$

where $u_n = x_{n+1} - x_n$ and < > means the quantum average with respect to ϕ.

The functional dependence of the x_n's on ϕ makes the Schrödinger equation

$$i\hbar\, \partial_t \phi = -\left\{ \frac{\hbar^2}{2m} \partial_x{}^2 + \sum_n V(x - x_n) \right\} \phi \qquad (5)$$

non linear. Solving Eqs.(4,5) amounts to study the propagation of the electronic wave on the nearly periodic atomic chain. The role of non linear Kerr effect in the previous problem is played here by the lattice deformation induced by the electron-lattice interaction.

In the simplest version of the model, V is taken δ - like :

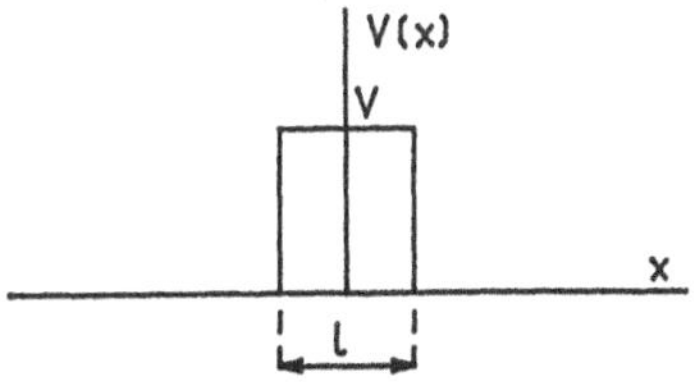

Fig 6 Shape of the interaction potential

with $V \to \infty$ and $l \to 0$ with $Vl \to \varepsilon$ finite. The Poincaré map connecting the fields over one period is for a stationary solution, $(\phi = \phi(x)\, e^{(i/\hbar)\, Et})$:

$$\begin{pmatrix} X_{n+1} \\ Y_{n+1} \end{pmatrix} = \frac{1}{\cos\mu} \begin{pmatrix} \cos\theta_n & \sin(\theta_n - \mu) \\ -\sin(\theta_n + \mu) & \cos\theta_n \end{pmatrix} \begin{pmatrix} X_n \\ Y_n \end{pmatrix}$$

$$\theta_{n+1} = \theta_n - \lambda\,(X_{n+1} Y^*_{n+1} + \text{c.c.})$$

where $\quad E = \dfrac{\hbar^2 k^2}{2m}\,, \quad \theta_n = k\, u_n + \mu\,, \quad \operatorname{tg}\mu = \dfrac{\varepsilon}{2k}$

$$X_n = \phi_n\,, \quad Y_n = \frac{\cos\mu}{k}\,\frac{d\phi_n}{dx}\,, \quad \lambda = \frac{2k^3 \sin\mu}{\cos^2\mu}$$

This mapping is conservative in (X, Y, θ) space, and it defines a non integrable dynamical system, whose qualitative properties are remarkably similar to those of the previous classical one. In particular it yields the same localized structures near the Arnold resonances, and these structures look like resonant polaronic states. However there is an additional constraint : the wave function must be normalized.

We shall briefly consider resonances $\theta = \pi/2$ and $\{\theta = 0, \pi\}$.

$\theta = \pi/2$ resonance. The two first equations of the mapping reduce to :

$$\begin{pmatrix} X_{n+1} \\ Y_{n+1} \end{pmatrix} = \begin{pmatrix} 0 & 1 \\ -1 & 0 \end{pmatrix} \begin{pmatrix} X_n \\ Y_n \end{pmatrix}$$

showing that $X_{n+1} Y_{n+1} = - X_n Y_n$. Therefore the u_n s are of the form $(-1)^n \alpha$ (α slow variable).

Again the system becomes integrable near the bifurcation point (the mapping reducing to an integrable set of 3 ODE). And again the limit orbits produce the set of two entangled ellipses, as in the classical problem. It must be noted that these solutions cannot be normalized in an infinite system. Indeed we find that

$$X_n \sim \left[1 - \tanh^2(\nu n)\right]^{1/2}, \quad Y_n \sim \tanh(\nu n)\left[1 + \tan h^2(\nu n)\right]^{-1/2}$$

where $\nu \approx k\, l_0 - (\pi/2)$. ($l_0$: lattice period). The graphs of Xn and Yn are sketched on Fig.(7).

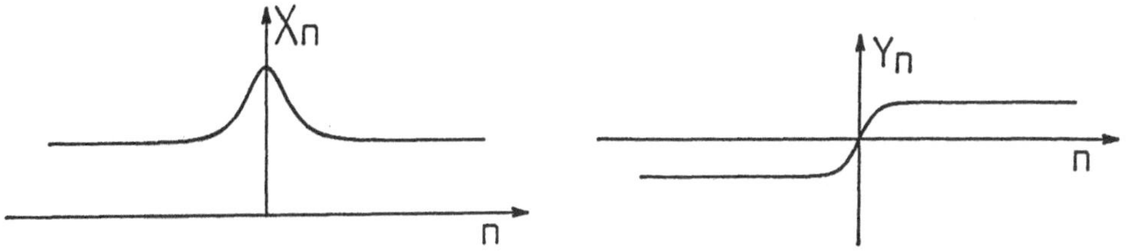

Fig 7 Graphs of Xn and Yn .

It is interesting to note that this resonant state corresponds to those found in acetylenic polymers. The Fermi wave number for these half filled band systems corresponds to $\theta_F = k_F l_o = \pi/2$ and one finds a ground states resulting from Peierls condensation (see Ref.(8)). This state is doubly degenerate (states A,B) and their exists a "solitonic structure" connecting A and B along the chain. Our "square structure" has clearly the same form. And indeed we are treating here a one electron version of the polymer problem, the electron wave number being equal to k_F.

Resonances $\theta = 0, \pi$

$\theta = \pi$ yields an "alternate soliton"
$\theta = 0$ yields an "non alternate soliton"

Let us consider the case $\theta = \pi$.

Putting $(\xi_n, \eta_n) = (-1)^n (X_n, Y_n)$, we obtain for $\theta = \pi$

$$\begin{pmatrix} \xi_{n+1} \\ \eta_{n+1} \end{pmatrix} = \frac{1}{\cos\mu} \begin{pmatrix} 1 & 0 \\ 2\sin\mu & 1 \end{pmatrix} \begin{pmatrix} \xi_n \\ \eta_n \end{pmatrix}$$

and the eigen values $s_{\alpha,\beta}$ of the linearized map around $\theta = \pi$, obey equation :

$$s^2 + 2\frac{\cos\theta}{\cos\mu} s + 1 = 0$$

This permits to define a gap, or forbidden band by $k_0 l \in [\pi - \mu, \pi + \mu]$.

The mapping becomes integrable near the bifurcation point giving a solitonic solution, and we are also able to treat the unstationary problem. We then obtain propagative solitons obeying equation

$$\left(\frac{2\,im}{\hbar}\partial_t + \partial_x{}^2\right)\eta - 2\,\mathrm{tg}\,\mu\left[\delta - \frac{\lambda}{2\sin\mu}|\eta|^2\right]\eta = 0 \qquad (6)$$

when $\delta = k\,l_0 - (\pi + \mu)$.

Eq. (6) has solitonic solutions of the form :

$$\eta \sim a\,\mathrm{sech}\,[a\,(x - wt)]\, e^{\left(i\pi \frac{w}{w_0}(x - wt)\right)}$$

where $w_0 = \hbar\,\pi / l_0$ and :

$$a^2 + \left(\frac{\pi w}{w_0}\right)^2 = 2\,\delta\,\mathrm{tg}\,\mu \qquad (7)$$

Finally we normalize ϕ, which implies : $a = \frac{\pi^3}{2}\,\mathrm{tg}\,\mu\,\left(\frac{m w_0^2}{M c_s^2}\right)$

(M atomic mass, c_s phonon velocity at $k = 0$).

Then the energy of the soliton can be written, with the help of relation (7) as :

$$E_s = E_0 + \frac{1}{2} m^* w^2 \text{ , with } m^* = -\frac{2\pi^2}{\varepsilon\, l_0}\, m_e \qquad (m_e \text{ electron mass}).$$

It is worth remarking that the theory can be extended to the case of potential peaks with finite width and amplitude. Then m^* may be of the same order of magnitude as m_e.

In the case of the $\theta = 0$ resonance, all goes along similar lines. The only differences are that the soliton is of the non alternate type, and $m^* \approx m_e$.

References

1. D.L. Mills, S.E. Trullinger, Phys. Rev. B, **36**, 947 (1987)
2. W. Chen, D.L. Mills, Phys. Rev. Lett., **58**, 160 (1987)
3. J. Coste, J. Peyraud, Phys. Rev. B, **39**, 13086 (1989)
4. J. Coste, J. Peyraud, Phys. Rev. B, **39**, 13096 (1989)
5. J. Coste, J. Peyraud, (1991) to be published
6 J. Peyraud, J. Coste , Phys. Rev B, **40**, 12201, 1989
7. J. Coste, J. Peyraud, (1991) to be published
8. A.J Heeger, S. Kivelson, J.R. Schrieffer, W.P. Su, Rev. Mod. Phys., 60, 3, 781, 1988

GAP SOLITONS IN 1D ASYMMETRIC PHYSICAL SYSTEMS

J.M. Bilbault, C. Tatuam Kamga and M. Remoissenet
Laboratoire O.S.C., Université de Bourgogne
6,Bd Gabriel,21100 Dijon, France

Abstract. We present a general approach for studying the nonlinear transmittance and gap solitons characteristics of asymmetric and one dimensional (1 D) systems in the low amplitude or Nonlinear Schrödinger limit. Included in this approach are some novel results on naturally asymmetric systems and systems where the symmetry is broken by an external constant force.

I. Introduction

Transmissivity near the gaps of a nonlinear system[1] of finite length exhibits bistability and can approach unity once the amplitude of the incoming sinusoidal wave is greater than a certain threshold which is frequency dependent and decreases with the length of the system. In the transmitting state, one has a nonlinear standing wave called[2,3] a "gap soliton ". Recent literature[1-3] has focused on symmetric systems, i.e. where the nonlinear potentiel (substrate or interaction potentiel) is symmetric: it only contains even powers of the characteristic field or of its gradient.

An interesting way to complete and extend our knowledge of the nonlinear response of finite systems with gaps is to analyse systems which are naturally asymmetric or systems where the symmetry is broken by the presence of an external force[4]. We present here a general approach for studying the nonlinear transmissivity and gap soliton characteristics of asymmetric 1D systems in the N.L.S. limit. We illustrate this methodology by application to the perturbed Sine-Gordon system.

II . General problem

In the one-dimensional arrangement illustrated in fig. 1, an incoming wave plane wave[1] of frequency ω, amplitude ϕ_0 and wave number $k_\ell = \omega / c_\ell$ in a linear medium propagates along the x direction and strikes at $x = 0$ a nonlinear medium of length L. The complex quantity R is the amplitude of the reflected wave measured with respect to ϕ_0 and similarly T is the amplitude of the transmitted wave at $x = L$ in the linear medium (3) expressed as a fraction of that of the incident wave.

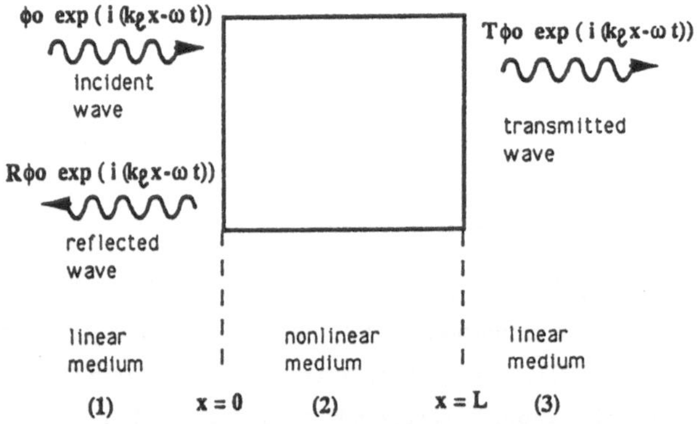

figure 1 : the incoming field $\phi(x,t) = \phi_0 \exp(i(k_\ell x - \omega t))$ strikes a nonlinear film (2) of length L. The transmitted wave emerges in the linear medium (3).

Inside the nonlinear medium we assume that the field $\Phi_n(t)$ obeys a generalized Klein-Gordon[4] (K G) lattice model equation :

$$\frac{\partial^2 \Phi_n}{\partial t^2} = \frac{c_0^2}{a^2} (\Phi_{n+1} + \Phi_{n-1} - 2\Phi_n) - \omega_0^2 \frac{dV(\Phi_n)}{d\Phi_n} . \tag{2.1}$$

Here n is the site number, $V(\Phi_n)$ is a nonlinear substrate potential ,the constants c_0 and ω_0 are the characteristic velocity and frequency of the system and a is the lattice parameter. In the low amplitude limit, we look for nonlinear collective oscillations in the bottom of the potential wells. For this purpose, assuming $\Phi_n = \varepsilon\phi_n + \Phi_0$ in eq. (2.1), where $\varepsilon << 1$ and Φ_0 is the ground state or potential minimum around which the oscillations will occur, and keeping terms to order ε^2, one gets:

$$\frac{\partial^2 \phi_n}{\partial t^2} = K (\phi_{n+1} + \phi_{n-1} - 2\phi_n) - (\omega'_0)^2 (\phi_n + \varepsilon\alpha\phi_n^2 + \varepsilon^2\beta\phi_n^3) , \tag{2.2}$$

where $K = c_o^2 / a^2$ and the coefficients ω'_0, α and β are determined by the shape of the potential. It is interesting to note here that $\Phi_o = 0$ and the second order term vanishes ($\alpha = 0$) when the potential wells are symmetric, as it is the case for the classical Sine-Gordon system where $dV(\Phi_n)/d\Phi_n = \sin \Phi_n$. We have $\Phi_o \neq 0$ when the potential wells are asymmetric, which is the case in SG system perturbed[4] by an external force $\mathcal{F}$.

Let us now consider oscillating solutions of the form:

$$\phi_n(t) = F_1 e^{i\theta_n} + c.c + \varepsilon[F_o + F_2 e^{2i\theta_n} + c.c] \tag{2.3}$$

where F_1, F_2 and F_3 are, respectively, the slowly varying amplitudes of the first harmonic, the dc and and second harmonic terms. These last two terms are introduced to take into account of the asymmetry of the potential, however we neglect higher harmonics. The phase defined as $\theta_n = kna - \omega t$ varies rapidly. Inserting (2.3) in (2.2) , equating dc, first and second harmonic terms and keeping terms to order ε^2 , we can relate F_0 and F_2 to F_1, and get :

$$\omega^2 = (\omega'_o)^2 + 4K \sin^2 \frac{ka}{2} + \varepsilon^2 (\omega'_o)^2 \left[-4\alpha^2 + \frac{2\alpha^2}{3 + \frac{16K}{(\omega'_o)^2} \sin^4 \frac{ka}{2}} + 3\beta\right] |F_1|^2 \tag{2.4}$$

Expanding now this general nonlinear dispersion relation eq. (2.4) in Taylor's serie about the carrier frequency ω_p and wave vector k_p yields[5]:

$$\omega - \omega_p = \left(\frac{\partial \omega}{\partial k}\right)(k-k_p) + \frac{1}{2}\left(\frac{\partial^2 \omega}{\partial k^2}\right)(k-k_p)^2 + \left(\frac{\partial \omega}{\partial |F_1|^2}\right)|F_1|^2 \tag{2.5}$$

Setting $\Omega = \omega - \omega_p$ and $K = k - k_p$,with $\Omega << \omega_p$ and $K << k_p$, eq. (2.5) represents the nonlinear dispersion relation $\Omega = f(K, |F_1|^2)$ of the wave envelope. In (2.5), the derivatives represent respectively the group velocity V_g, the group velocity dispersion P and the nonlinearity:

$$\left(\frac{\partial \omega}{\partial k}\right) = V_g, \qquad \left(\frac{\partial^2 \omega}{\partial k^2}\right) = \left(\frac{\partial V_g}{\partial k}\right) = 2P, \qquad Q = -\left(\frac{\partial \omega}{\partial |F_1|^2}\right) \tag{2.6}$$

Substituting now the derivative operators in eq. (2.5) by the coefficients defined in eq. (2.6) yields the Nonlinear Schrödinger Equation (N.L.S.):

$$i\,[F_{1t} + V_g F_{1x}] + P F_{1xx} + Q|F_1|^2 F_1 = 0 \tag{2.7}$$

We now consider a particular case, in order to illustrate how one can determine the transmittance and the envelope behaviour of a given system.

III. An example: the perturbed Sine-Gordon system.

We consider the specific case[4] of a perturbed SG system. In this case, the Sine-Gordon potential is:

$$V(\Phi_n) = 1 - \cos \Phi_n + \frac{\mathcal{F}}{\omega_o^2} \Phi_n \qquad (3.1)$$

Its minimum occurs at $\Phi_o = -\sin^{-1}(\mathcal{F}/\omega_o^2)$, while the coefficients of eq. (2.2) become[4] $(\omega'_o)^2 = \omega_o^2 \cos \Phi_o$, $\alpha = -\frac{1}{2} \tan \Phi_o$ and $\beta = -\frac{1}{6}$. The force $\mathcal{F}$ lowers the linear dispersion curve with respect to the unperturbed case ($\mathcal{F} = 0$), as shown on fig.2. This is enforced by the nonlinearity Q in N.L.S. eq. (2.7), which is positive in this case:

$$Q = \frac{\omega_0^2 \cos(\Phi_0)}{2\omega_p} \left[\tan^2(\Phi_0) - \frac{\frac{1}{2} \tan^2(\Phi_0)}{3 + \frac{16K}{\omega_0^2 \cos(\Phi_0)} \sin^4 \frac{k_p a}{2}} + \frac{1}{2} \right], \qquad (3.2)$$

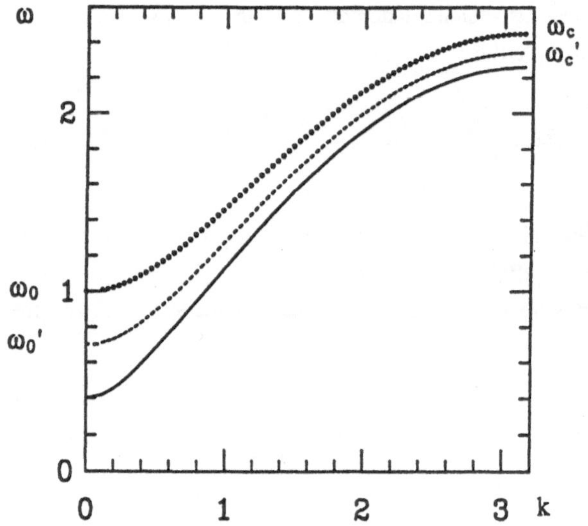

fig.2 Linear and nonlinear dispersion curves for a perturbed SG system. The external force lowers the linear curve (dashed line) with respect to the unperturbed case (dotted line), while the nonlinearity with an arbitrary amplitude enforces this lowering (solid line).

The interesting situations occur then near the gap edges of the linear dispersion curves, because the nonlinearity will change the transmission behaviour. We have:

- either $k_p = 0$, $\omega_p = \omega_0'$, $V_g = 0$ and $P = + c_0^2/2\omega_p$ near the lower gap
- or $k_p = \pi/a$, $\omega_p = \omega_c'$, $V_g = 0$ and $P = - c_0^2/2\omega_p$ near the upper gap.

In both cases,the angular frequency of the incoming wave is repered by the small detuning $\Omega = \omega - \omega_p$. We seek now envelope functions in the nonlinear medium of the form[6]:

$$F_1(x,t) = \phi_0 \sqrt{I(x)}\, \exp(i\theta(x)) \exp(-i\Omega t) \tag{3.3}$$

where the squared envelope function I(x) and the phase θ(x) are real. Putting the form (2.9) in N.L.S. eq. (2.7) gives after some calculations[6]:

$$\left(\frac{dI}{dx}\right)^2 = \mathcal{P}(I) \tag{3.4}$$

with:

$$\mathcal{P}(I) = -\left(2\frac{Q}{P}\phi_0^2 I^3 + 4\frac{\Omega}{P} I^2 - 4IB + 4A^2\right). \tag{3.5}$$

The constants of integration A and B are determined by the boundary conditions at the interfaces (see fig.1), namely the field and its first spatial derivative are continuous at x=0 and x=L. Moreover, these boundary conditions show that $I(x=L) = I_L$ is always a root of $\mathcal{P}(I)$, while limiting ourselves to low amplitude ϕ_0 and small detuning Ω, the two other roots of $\mathcal{P}(I)$ are real and positive. Finally, $\mathcal{P}(I)$ becomes:

$$\mathcal{P}(I) = -2\frac{Q}{P}\phi_0^2 (I - I_L)(I - I_+)(I - I_-) \tag{3.6}$$

where I_+ and I_- are given by:

$$I_+ = -\left(\frac{I_L}{2} + \frac{\Omega}{Q\phi_0^2}\right) + \sqrt{\left(\frac{I_L}{2} + \frac{\Omega}{Q\phi_0^2}\right)^2 + \frac{2I_L k_\ell^2 P}{Q\phi_0^2}} \tag{3.7.a}$$

$$I_- = -\left(\frac{I_L}{2} + \frac{\Omega}{Q\phi_0^2}\right) - \sqrt{\left(\frac{I_L}{2} + \frac{\Omega}{Q\phi_0^2}\right)^2 + \frac{2I_L k_\ell^2 P}{Q\phi_0^2}} \tag{3.7.b}$$

Integrating eq. (2.10) according to the ordering between I(x), I_L, I_+ and I_- and the sign of P, Q and Ω, leads to a Jacobi Elliptic[7] function expression with the parameter I_L. Using once more[6] the boundary conditions gives the following condition:

$$\left[\left(\frac{dI}{dx}\right)_{x=0}\right]^2 + 4k_\ell^2 (I_L + I(x=0))^2 - 16\, k_\ell^2\, I(x=0) = 0\,, \tag{3.8}$$

which permits to keep by numerical calculations the suitable values of I_L. Once I_L is determined, one can get I(x). Then, from eqs. (2.3), (3.3) and continuity at x = L, we get the transmissivity coefficient $|T|^2 = I_L$.

IV. Results for the perturbed Sine-Gordon system.

Using the approach presented in the previous sections, we calculate numerically the transmissivity and the square of the envelope function. We consider successively the lower gap and the upper gap of the dispersion curve (see fig.2), where the results are quite different.

A. Lower gap: we consider a system where L = 60 a, the velocities (defined in section II) are $c_0 = c_\ell = 4.5$ and: $\Omega = \omega - \omega_0' = -0.01$.

Our results, represented on fig.3.a , agree with the previous works[1,6] obtained for SG systems, i.e. the system presents bistabilities and hysteresis cycles. The external force lowers the threshold values, as seen on fig.3.a. This can be easily understood because Q, given by eq. (3.2) is a growing function of Φ_0 and $\mathcal{F}$.

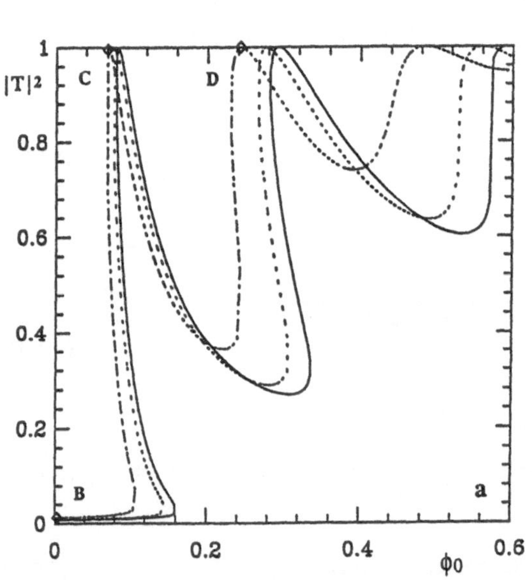

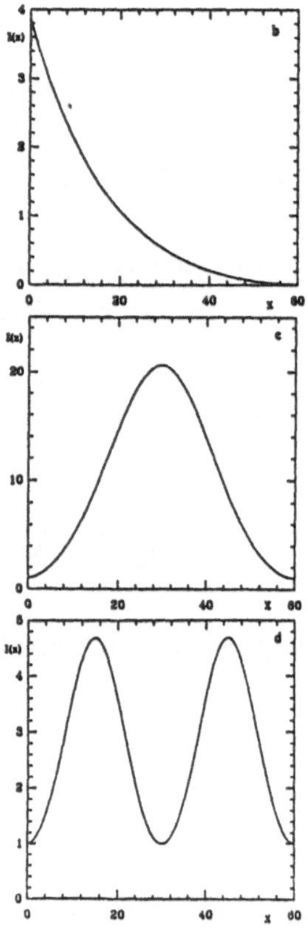

Fig.3.a Transmissivity $|T|^2$ versus the amplitude ϕ_0 of the incoming wave, when the frequency ω lies just below the lower gap edge (Ω=−0.01) for the unperturbed S-G system (solid line) and for the perturbed S-G system with a force $\mathcal{F}$ =0.39 (dashed line) or $\mathcal{F}$=0.72 (dotted line).

Fig.3.b,c and d: For $\mathcal{F}$= 0.72, the shape of I(x) at points B, C and D of fig.3.a is represented versus the coordinate x.

When ϕ_0 is weak, i.e. in the linear limit, the wave envelope is evanescent (fig.3.b). When ϕ_0 increases, the system reaches a certain threshold, which depends on the value of $\mathcal{F}$.

Then the system switches to a transmitting state and the envelope is that of a nonlinear standing wave (called a gap soliton). In this case, one has P.Q > 0. Then if one further increases ϕ_0, the transmissivity reaches successively several maxima, which correspond to different resonant modes (standing waves of fig.3.c and d) described by Jacobi Elliptic functions. For one resonant mode (fig.3.c), the maximum is at L/2 = 30.

B. Upper gap: the nonlinear system length is still L = 60 a , but now $c_0 = 2.$, $k\ell = 0.015$ and $\Omega = \omega - \omega_c' = -0.003$. Our results, represented on fig.4.a, still show that the threshold decreases with the external force. This can be explained by the fact that Q increases with $\mathcal{F}$. The squared envelope function, represented on fig.4.b, is characteristic of a standing wave behaviour: it now corresponds to P.Q < 0.By contrast to the previous case, for one resonant mode (standing wave on fig.4.c) at x = L/2 = 30 one has now a minimum. When ϕ_0 is further increased, the envelope finally becomes evanescent (fig.4.d); this can be explained by considering the dispersion curve (see fig. 2) and remarking that the nonlinearity tends to lower the curves. Then, ω lies inside the gap.

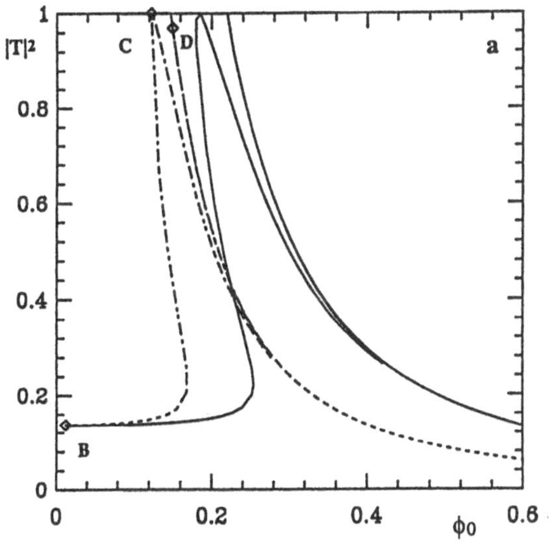

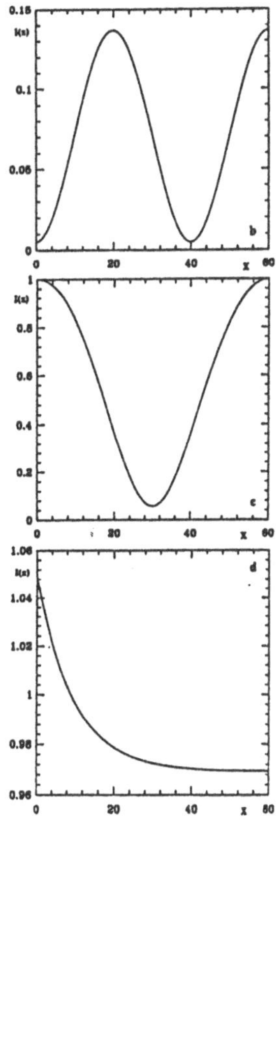

Fig.4.a Transmissivity $|T|^2$ versus the amplitude ϕ_0 of the incoming wave when the frequency ω lies just below the upper gap edge ($\Omega = -0.003$) for the unperturbed S-G system (solid line) and for the perturbed S-G system with a force $\mathcal{F}$ =0.72 (dotted line).
Fig.4.b,c and d: For $\mathcal{F}$= 0.72, the shape of I(x) at points B,C and D of fig.4.a is represented versus x.

The method presented here allows to investigate asymmetric systems. If the asymmetry results from a symmetry breaking, as for the perturbed S-G system, the external force allows to control the bistability or nonlinear switching. Note that our approach can also be used for a natural asymmetric system[4] like "ϕ^4 ".

References

[1] D.Barday and M.Remoissenet,Phys.Rev.B **41**, 10387 (1990).

[2] D.L.Mills and S.E.Trullinger,Phys.Rev.B **36**,947 (1987).

[3] C.Martijn de Stercke and J.E.Sipe,Phys.Rev.A **38**, 5149 (1988)

[4] M.Remoissenet,Phys.Rev.B **33**, 2386 (1986).

[5] Whitham , Linear and nonlinear waves (Academic Press, New York, 1972)

[6] W.Chen and D.L.Mills,Phys.Rev.B **35**, 524 (1987).

[7] P.F.Byrd and M.D.Friedman,Handbook of Elliptic integrals(Springer, Berlin, 1954)

EVIDENCE OF ENERGY DIFFUSION IN PURE ANHARMONIC DISORDERED CHAINS

R. Bourbonnais
HLRZ, KFA Jülich
Postfach 1913, D-5170 Jülich

R. Maynard
CRTBT, CNRS
B.P. 166X, 38042 Grenoble Cedex

Abstract: We present results of large scale simulations on vibrations of anharmonic disordered chain. We find that anharmonic effects tend to counter the localisation process and lead to diffusion of the energy on the lattice. For the anharmonic ordered case the energy is concentrated in large peaks as the global energy spread linearly in time.

Keywords: Vibrations, Non-Linear Effects, Localisation, Diffusion.

Introduction

Using a massively parallel computer (Connection Machine), we have studied a chain of atoms of mass m_i, interacting by harmonic and "quartic" anharmonic interactions described by the coefficients k_2 and k_4 in the hamiltonian which is written as:

$$H = \sum_i \frac{1}{2} m_i V_i^2 + \frac{k_2}{2} \sum_{i,j,(n.n.)} (U_i - U_j)^2 + \frac{k_4}{4} \sum_{i,j,(n.n.)} (U_i - U_j)^4 \qquad (1)$$

The $U_i(t)$ are the scalar amplitudes of vibration at time t and $V_i(t) = \dot{U}_i(t)$; the last two sums run over all pairs of nearest-neighbors. Disordered systems were simulated by having the masses m_i randomly distributed. In a perfect chain, $m_i = m$ and the hamiltonian (1) is the one studied in early time by Fermi, Ulam and Pasta [1]. The nature of the stable non-linear excitations of frequencies higher than the cut-off phonon frequency is an interesting problem which was revisited recently by Sievers and Takeno [2]. They found, self-localized anharmonic modes of odd parity with frequency above the Debye cut-off frequency and dependent on the k_2 and k_4 coefficients. Recently, Page [3] has showed that the pure anharmonic hamiltonian i.e. without harmonic interaction ($k_2 = 0$), can be solved and yields two anharmonic modes of odd and even parity with frequencies above the phonon cut-off frequency.

The present study deals with this simplified pure anharmonic hamiltonian ($k_2 = 0$) in ordered and disordered chains. The main results are obtained by numerical simulation of large chains of 16000 atoms. The equation of motion are integrated numerically using either a simple leap-frog algorithm as described in reference [4] or a more sophisticated fourth order symplectic method [5]. Results were found to be independent on the integration method. On a Connection Machine of 16384 processors with single precision hardware Weitek chips and a SUN front-end computer we reached performances of $21 \cdot 10^6$ updates/sec with the leap-frog algorithm. The symplectic solver was about 3 times slower.

Periodic boundary conditions are used and the excitation consist in an initial displacement of unit length of some chosen site (labeled 0). The time steps is 1/50 of the shortest period of vibration and averages were taken over an ensemble of 10 to 200 different samples. Whenever disordered systems were considered, the masses were distributed uniformly around the average mass $m_0 = 250$ with a relative root-mean square deviation :

$$\sigma = \left(\frac{< m_i^2 > - m_0^2}{m_0^2} \right)^{(1/2)} \tag{2}$$

with $\sigma = 11.5\%$.

The characteristic results are represented in the following figures, where the excitations has been applied at $x = 0$. We present the instantaneous energy $E_t(x)$ as a function of the space variable x for different times (in unit of the shortest period of vibration).

1 - Perfect chain

The function $E_t(x)$ is shown at a time ($t = 50$, fig. 1a) where the initial peak of energy has already split into several peaks. At time $t = 1000$ (fig. 1b), a broad packet of the peak fragments is observed. Note that well defined peaks are moving in front of the packet. It will be shown that the r.m.s. of this energy distribution varies linearly in time (cf. figure 4).

2 - Disordered chain

The same plots $E_t(x)$ for now disordered chains reveals a progressive spreading out of the energy in space. Here the energy distribution at $t = 1000$ is decreasing in space, but fluctuates largely for one given sample.

3 - Disordered chain : Ensemble average.

Averages of energy profiles from 200 samples are plotted at different times (fig. 3a). As time increases the energy spreads on the lattice. In fig. 3b we plotted $E_t(x) \cdot \sqrt{t}$ vs. $x/\sqrt{t}$. As time increase the distribution function becomes increasingly well approximated by a gaussian function. For t=10000 data we found that we could express $E_t(x)$ (solid line in fig. 3b) as :

$$E_t(x) = \frac{4.72 \cdot 10^{-2}}{\sqrt{(t)}} exp(-2.18 \cdot 10^{-2} (\frac{x}{\sqrt{(t)}})^2) \tag{3}$$

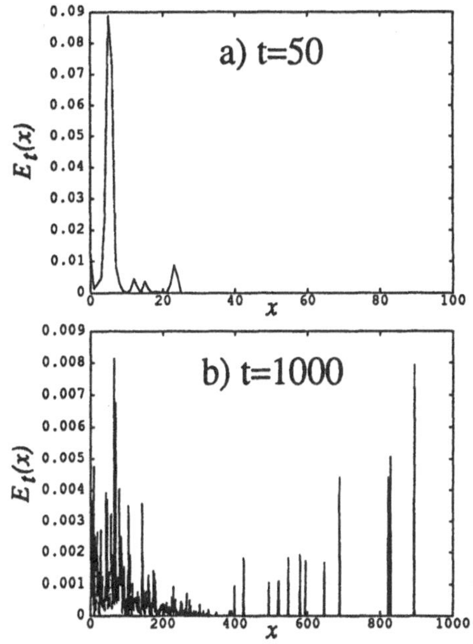

Figure 1. Energy distribution function $E_t(x)$ in the ordered case for a) $t = 50$ and b) $t = 1000$. The horizontal axis is in lattice units, the vertical scale is arbitrary units. The excited sites is site 0. The energy also spread in the negative direction (not shown).

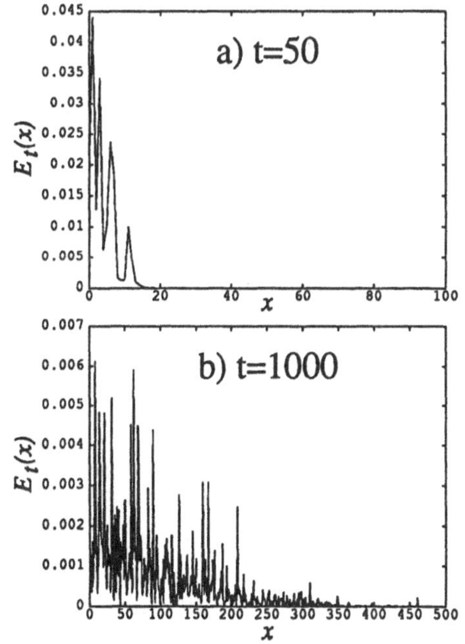

Figure 2. Energy distribution function $E_t(x)$ in the disordered case for a) $t = 50$ and b) $t = 1000$. The energy present large fluctuations with a general tendency to decrease. The characteristic length of decrease grows with time as does the number of peaks.

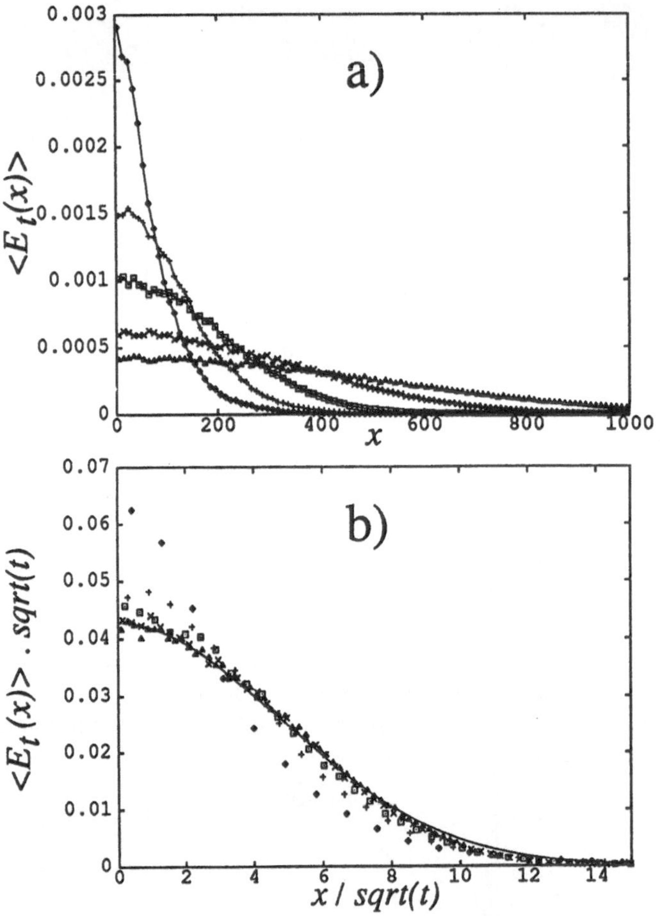

Figure 3. Average energy distribution function $< E_t(x) >$ (a) in the disordered case for $t = 500(\diamond), 1000(+), 2000(\square), 5000(\times)$ and $10000(\triangle)$. Over 200 samples were averaged. In (b) we have plotted $< E_t(x) > \cdot\sqrt{t}$ vs. $x/\sqrt{t}$ and eq. (3) (solid line). As time increase the curves converge toward a gaussian.

4 - Evolution of the energy distribution

The second moment of the energy distribution $< x^2 >$ is plotted as a function of time for three different situations : - the losanges for the perfect chain exhibiting a t^2 variation at large times, - the crosses for an ensemble of 10 disordered chain show a law proportional to time for about two decades, - the square for a pure harmonic disordered chain ($k_4 = 0$) where the localization phenomenon is revealed by the saturation of

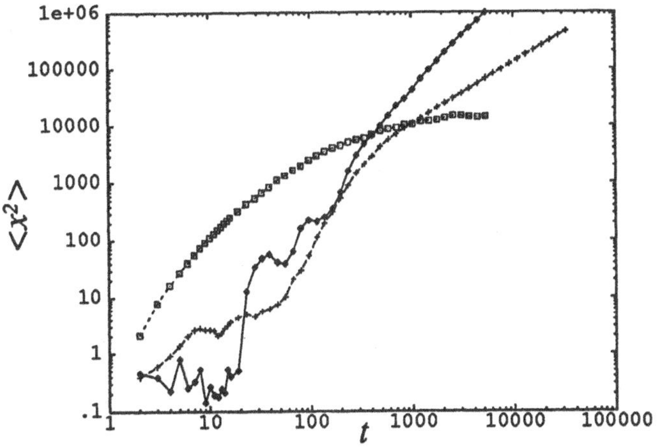

Figure 4. Second moment of the energy distribution function $< x^2 >$ for the harmonic disordered case (□), the anharmonic disordered case (+) and the anharmonic ordered case (◇). The first curve shows how the energy becomes localized due to the disorder in harmonic systems. With anharmonicity the second moment increases linearly with time for ($t > 1000$) as in a diffusion process. In the anharmonic *ordered* case the moment increases as t^2.

$< x^2 >$ at long time. The estimated localization length is obtained from this asymptotic value: $x_0 \approx 100$.

5- Analysis

A preliminary analysis of the observed phenomenon can be formulated in the following terms:

- the energy peaks of the perfect and pure anharmonic chain are basically instable. One observes a spontaneous desintegration a big peaks in fragments of smaller energy. At longer times the packet of fragments moves uniformly in time. This uniform motion reveals an underlying conservation law during the fragmentation of the type "conservation of momentum". This relation comes from the translational invariance property of the perfect chain.

- The apparent "normal diffusion" would come from the fragmentation of the peak excitation on the mass impurity. Let us call R_i, T_i and L_i the fraction of the incident energy which is reflected, transmitted of localized on the mass m_i. The energy conservation law gives : $R_i + T_i + L_i = 1$.

The problem of random fragmentation can be changed into the more conventional problem of random walk of a fictive particle which is reflected, transmitted or immobilized with probability R_i, T_i and L_i. It is not difficult to show [6] that the particle obeys a diffusive law characterized by a diffusion constant $D \propto (R_i/T_i)^{-1}$. Hence, this diffusive motion leads to a well known law $< x^2 >\sim 2Dt$. Since this law is well observed in the simulations, the model of random fragmentation on the impurities is validated. This conceptual frame, in addition to the expected characteristic lengths present in the problem like the interatomic distance and the soliton or peak width, provides us a new length l : the average distance between two random fragmentations. The previous regime of normal diffusion is hence obtained when $l >> \Lambda \sim a$. (Λ is the soliton width). Other interesting regimes could be also considered, particularly when the harmonicity is restored, where an additional characteristic length : the localization length must be taken into account. This more complex regime exhibits anomalous diffusion [4].

References:

[1] E. Fermi, J. Pasta and S. Ulam, Los Alamos Science Laboratory, report no. LA-1940 (1955).
[2] A.J. Sievers and S. Takano, Phys. Rev. Lett., 61, (1988) 970.
[3] J.B. Page, Phys Rev. B, 41, (1990) 7835.
[4] R. Bourbonnais and R. Maynard, Phys. Rev. Lett., 64, (1990) 1397., R. Bourbonnais and R. Maynard, Int. J. of Modern Physics C 1, (1990) 233.
[5] D.B. Duncan, C.H. Walshaw and J.A.D Wattis, $7^t h$ Inter-disciplinary Workshop on Non-Linear Coherent Structure, Dijon June 1991, (this issue).
[6] A. Langenfeld "energy transport in anharmonic and disordered systems" Internal Report, (1990), University Joseph Fourier, Grenoble, France

A NUMERICAL VENTURE INTO THE MENAGERIE OF COHERENT STRUCTURES OF A GENERALIZED BOUSSINESQ EQUATION

C. I. Christov, G. A. Maugin

Laboratoire de Modélisation en Mécanique, CNRS URA 229
Université Pierre et Marie Curie (Paris VI)
Tour 66, 4 Place Jussieu, 75252 PARIS CEDEX 05

1. INTRODUCTION

We consider a generalized Boussinesq equation which is a model for the one-dimensional dynamics of phases in martensitic alloys. The natural difference approximation that coincides with the discrete form of the model is employed and then Newton's quasilinearization of the nonlinear terms is performed. In order to adequately represent the localized solitary-wawe solutions up to 20000 grid points are used in calculations.

Two distinct classes of solutions are found. To the first class belong oscillatory pulses whose envelopes are localized waves. The second class consists of smoother localized solutions that are either kinks or bell-shaped "bumps" depending on the amplitude of the initial condition. The amplitude of a bump decrease with time while its support increases. An appropriate self-similar scaling is found analytically and confirmed by the direct numerical simulations to high accuracy.

2. POSING THE PROBLEM

Following [1],[2],[3],[4] we consider the one-dimensional model of an atomic chain in which the longitudinal displacements couple to the shear strain. Upon introducing relative displacements the Euler-Lagrange equations for variation of the governing functional adopt the form

$$\frac{d^2}{dt^2}S_n = c_T^2(S_{n+1}-2S_n+S_{n-1}) - (S_{n-1}^3-2S_n^3+S_{n+1}^3) + (S_{n+1}^5-2S_n^5+S_{n-1}^5) - \beta(S_{n+2}-4S_{n+1}+6S_n-4S_{n-1}+S_{n-2}), \tag{2.1}$$

whose continuum limit is

$$S_{tt} = c_T^2 S_{\xi\xi} - (S^3)_{\xi\xi} + (S^5)_{\xi\xi} - \alpha S_{\xi\xi\xi\xi}\,. \tag{2.2}$$

In equation (2.1) the spatial variable is defined as $\xi=X/a$ (where a is the distance between the atoms). $X_n=na$ stands for the distance and $\alpha=\beta-0.5\ c_T^2$. The differential form hints at the name - "a generalized Boussinesq equation" in the sense that (2.2) has more complicated nonlinearity than the original Boussinesq equation [5]. The continuum limit provides also the clue how the boundary conditions are to be posed. The natural boundary condition for the Lagrangian functional in differential form is $S_{\xi\xi}=0$. Since the index "1" refers to the first atom of the chain and "N" - to the last one, then for the discrete version of the system the natural boundary conditions can be represented as follows

$$(2.3) \qquad S_0-2S_1+S_2=0\ , \quad S_{N-1}-2S_N+S_{N+1}=0\ .$$

where in order to properly discretize the natural conditions we introduce into consideration two "artificial" members of the chain with indices "0" and "N+1", respectively. We also consider the physically most typical situation when the boundary points are held fixed, namely

$$(2.4) \qquad S_1=0\ , \qquad S_N=0\ .$$

3. ALGORITHM

In order to successfully treat the problem of localized solutions (henceforth called "coherent structures") the first requirement for the algorithm is to be fast enough allowing computations with large number of "grid points" N in order to provide room for a structure to move without interacting with other structures or with the boundaries. The second major requirement is for strong temporal stability in the sense that the different kinds of computational errors do not amplify timewise even for large values of time increments. The last requirement is crucial because some of the properties of the individualized localized solution can be recognized only after very long temporal evolution.

It is convenient to introduce the auxiliary function

$$(3.1) \qquad Q_i = c_T^2 S_i - S_i^3 + S_i^5 - \beta(S_{i-1}-2S_i+S_{i+1})\ , \quad i=1,\ldots N\ .$$

and to recast (2.4) in the form

$$(3.2) \qquad \frac{d^2}{dt^2} S_i = (Q_{i-1}-2Q_i+Q_{i+1})\ , \qquad i=2,\ldots,N-1\ .$$

In terms of function Q_i, the boundary conditions adopt the simple form

$$(3.3) \qquad S_1 = S_N = 0\ , \ Q_1 = Q_N = 0\ .$$

Initial conditions are imposed both for function S and its time derivative, say functions s_i and σ_i. We use the superscript n to denote the current time step on a staggered time mesh $t_n=(n-0.5)\tau$ where τ is the time increment. Then the initial conditions are approximated to second order as follows

$$(3.4) \qquad S_i^0 = s_i - \frac{\tau}{2}\sigma_i\ , \qquad S_i^1 = s_i + \frac{\tau}{2}\sigma_i\ .$$

The main objective in devising the algorithm is to have a stable scheme that would allow us to march with large time increments τ. For this reason we chose a fully implicit scheme. At the time step *(n+1)* we use a consistent Newton's quasilinearization of the terms on the right-hand side of (3.1) according to the formulae:

$$(3.5a)\qquad S_i^3\Big|^{n+1} = 3S_i^{n2}S_i^{n+1} - 2S_i^{n3} + O(\tau^2)\ ,$$

$$(3.5b)\qquad S_i^5\Big|^{n+1} = 5S_i^{n4}S_i^{n+1} - 4S_i^{n5} + O(\tau^2)\ ,$$

$$(3.5c)\qquad S_i^2\Big|^{n+1} = 2S_i^nS_i^{n+1} - S_i^{n2} + O(\tau^2)\ .$$

The strongly implicit scheme requires also that the time derivative of (3.2) be approximated at time step *(n+1)*. At the first step (number *2*) following the two initial steps, one can have only a first-order approximation of the second derivatives over three steps:

$$(3.6a)\qquad \frac{d^2}{dt^2}S_i\Big|^2 = \frac{1}{\tau^2}(S_i^2-2S_i^1+S_i^0) + O(\tau)\ .$$

At all of the next steps *(n>2)* we employ the following four-step scheme with second-order approximation

$$(3.6b)\qquad \frac{d^2}{dt^2}S_i\Big|^{n+1} = \frac{1}{\tau^2}(2S_i^{n+1}-5S_i^n+4S_i^{n-1}-S_i^{n-2}) + O(\tau^2)\ .$$

Introducing the above formulas into (3.1)-(3.2) we arrive at a **coupled** system of difference equations for the two set functions S_i, Q_i, namely

$$(3.7a)\qquad \beta S_{i-1}^{n+1} - (2\beta+c_T^2-3S_i^{n\,2}+5S_i^{n\,4})S_i^{n+1} + \beta S_{i+1}^{n+1} + Q_i^{n+1} = 2S_i^{n3}-4S_i^{n5}\ ,$$

$$(3.7b)\qquad Q_{i-1}^{n+1} - 2Q_i^{n+1} + Q_{i+1}^{n+1} - \frac{2}{\tau^2}S_i^{n+1} = -\frac{1}{\tau^2}(5S_i^n-4S_i^{n-1}+S_i^{n-2})\ .$$

Here the set functions of steps n, $n-1$ and $n-2$ are thought of as being known. Eqs (3.7) are valid for all interior points $i = 2, \dots, N-1$ and are coupled through the boundary conditions (3.3).

The most important feature of the system (3.7) is that upon introducing the composite set function

$$(3.8)\qquad W_{2i-1} \equiv Q_i^{n+1}\ ,\quad W_{2i} \equiv S_i^{n+1}\ ,\qquad i = 1,\dots,N\ ,$$

and after fairly obvious manipulations the said system can be recast as a *five*-diagonal system for the new set function W_k where $1<k<2N$. The band structure allows us to use highly efficient specialized solvers, e.g., the one developed in [6].

IV. RESULTS

To start with we set $c_L=1$, $\beta=1$. It is easily seen that the particular values of these parameters are not so important and by changing the value of the time increment τ we can select most of the principal cases by means of simply rescaling the dependent and independent variables. Due to the strong implicity, the scheme turns out to be stable for a wide range of time increments, $10^{-6}<\tau<10^{6}$, and for a wide range of amplitudes of initial conditions. As it should have been expected the computations with larger τ's led us to the smoother solutions spreading wider in the region under consideration.

The smooth solutions persisted in our calculations when the time increment τ was larger than 5. It goes without saying that we did verify whether the same shape of coherent structure is obtained if the calculation are conducted with different $\tau>5$ (say $\tau=10$ and $\tau=100$). We discovered that what mattered was the interplay between the time increment and the amplitude of the initial condition. When the latter is "moderate" in comparison with τ then the homoclinics of the shape of Airy functions appeares as shown in Fig.1. When the initial amplitude is relatively small the pentic nonlinearity is swithched off even on the earliest stages and then the symmetric homoclinics shown in Fig.2 appeared. Conversely when the amplitude was large enough, then the balance between the two nonlinear terms yielded the kink-shaped structure (heteroclinics) shown in Fig.3.

An interesting feature of the homoclinics is that their shape is not preserved timewise while the heteroclinics (kinks) are stationary patterns. The former decrease in amplitude with time while their support increases and after a sufficiently long time the solution eventually gets on a self-similar track discussed in the next section.

A completely different Universe appeared when the time increments were small enough (say $\tau\leqslant 1$ for $\beta=1$) and allowed development of more complicated "wiggled" shapes. It is more convenient to consider the case $\tau=0.1$, $\beta=0.01$, since the spatial span of the structures is smaller. In the sequence of Figures 4 one sees the development of a "pulse" that has smooth shape in the right-hand side of the interval but in the course of its time evolution spans larger portions of the left-hand side of the interval with its wavy "tail".

The complete classification of the different creatures inhabiting the generilized Boussinesq equation goes far beyond the scope of the present short note. A more systematic account is due elsewhere.

V. THE SELF SIMILAR STAGE

The results of the previous section suggest that for large times some of the solutions tend to adopt a self-similar shape in the sense that their amplitude decreases with time while the length-scale of the support increases. Being reminded that at large times for the decaying solutions one has $S^5 \ll S^3$ we found a self-similar scaling of the following type

$$(5.1)\qquad S = t^{-\alpha}s(\eta), \quad \eta \equiv t^{-\delta}(x-c_T t) \quad \text{where } \alpha = \delta = \frac{1}{3}.$$

Respectively, the equation for the scaled function s reads

$$(5.2) \qquad \frac{2}{3}c_T^2(4s'+\eta s'') = -(s^3)'' - \beta s'''' .$$

Here the primes stand for differentiation with respect to the similarity variable η. It is interesting to note that such self-similar solutions were found both for Burgers' equation [7] and Korteweg-de Vries' equation [8].

It goes beyond the frame of the present work to attempt a direct solution to (5.4). Rather we shall check whether the time dependent solution of the previous section conforms with the asymptotic law (5.1). Let us define the length of support L_s as the distance from the point where the maximum of the structure is situated to the point where the amplitude is 1/100 of the maximum. The definition is appropriate for all of the structures which decay monotonically in the right portion of the region under consideration. In Table 1 we present for the solution from Fig.1 the amplitude and support as a function of dimensionless time after a certain moment of time that we call $t=0$. Next to the column of numerical results we also present in Table 1 the approximation for A and L_s of the type

$$(5.3) \qquad A = a(t+a_0)^{-1/3} , \qquad L_s = b(t+a_0)^{1/3} .$$

where the constants a,b and a_0 are defined so as to provide a best in the least-square sense fit to observations from numerical simulations. It is clearly seen from Table 1 that the asymptotic self-similar powers are in excellent agreement with the data for both the amplitude A and support L_s the difference being less than 1% save the moment $t=0$ which in fact is "too early a moment" to be treated as an asymptotic stage.

TABLE 1. $a_0 = 2.76.10^4$; $a = 1.011$; $b = 63.3$

time	amplitude	approxim.	%	support	approxim.	%
0	0.033063	0.033445	1.14	1866	19139	2.503
20000	0.027840	0.027894	0.19	2281	22948	0.601
40000	0.024881	0.024818	0.255	2587	25792	0.301
60000	0.022767	0.022765	0.0106	2804	28119	0.279
80000	0.021265	0.021257	0.0378	3002	30113	0.308
100000	0.020095	0.020083	0.0597	3183	31873	0.315
120000	0.019139	0.019132	0.0377	3343	33458	0.083
140000	0.018337	0.018338	0.0082	3492	34905	0.042
160000	0.017658	0.017662	0.0244	3632	36241	0.217
180000	0.017066	0.017076	0.0583	3763	37486	0.384
200000	0.016541	0.016560	0.117	3884	38653	0.484

In Fig.5 are shown the rescaled results for the shape of coherent structure. One sees that the similarity is beyond any doubt. The same holds for the second kind of homoclinic solutions (Fig.6) which means that the expanding self- similar solutions are inherent in Boussinesq dynamics.

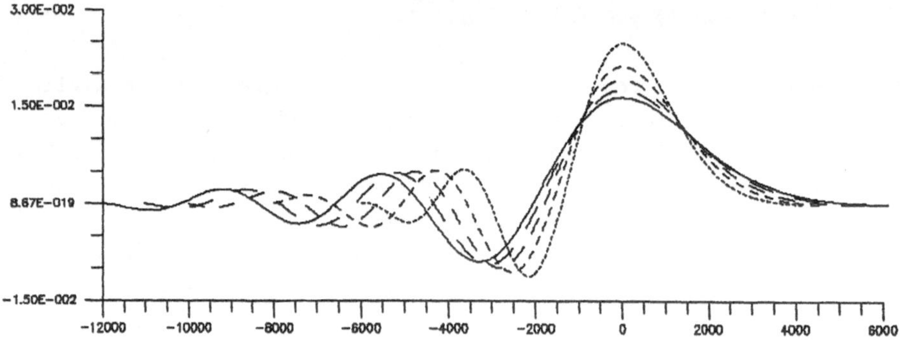

time=40000; time=80000; time=100000;
time=12000; time=140000;

Fig. 1

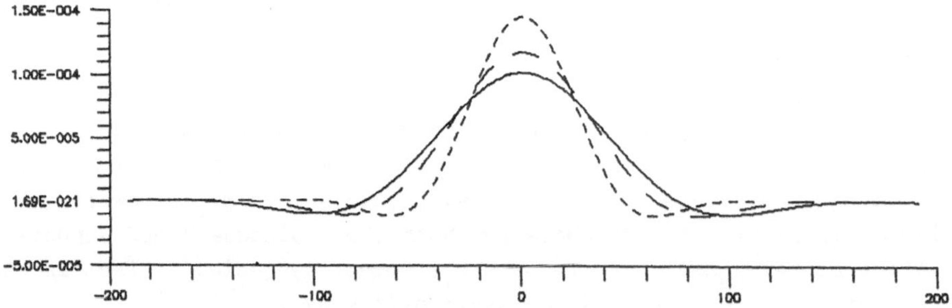

time=51000; —time=200000; time=400000.

Fig. 2

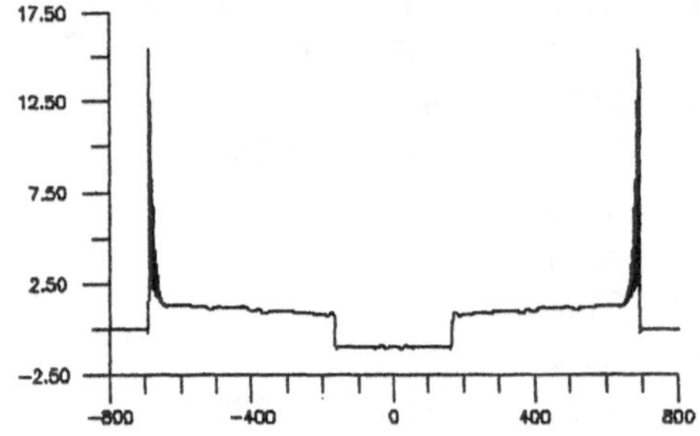

Fig. 3a

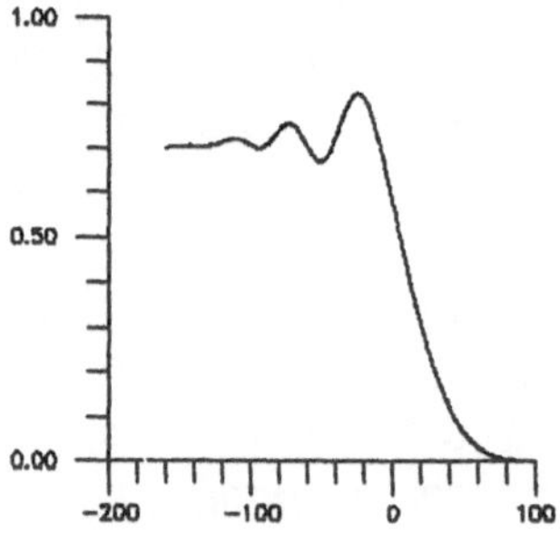

Fig. 3b

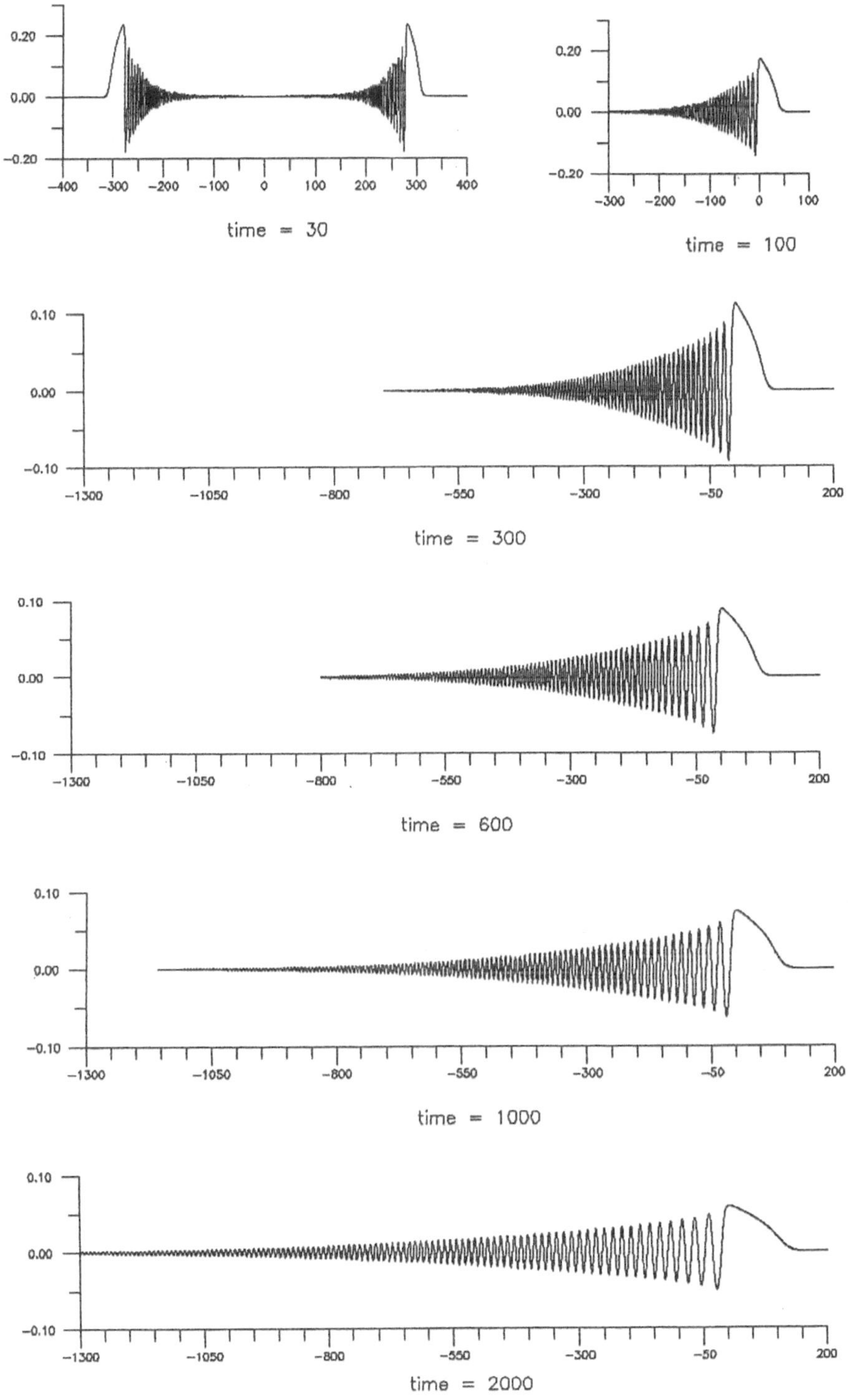

Fig. 4

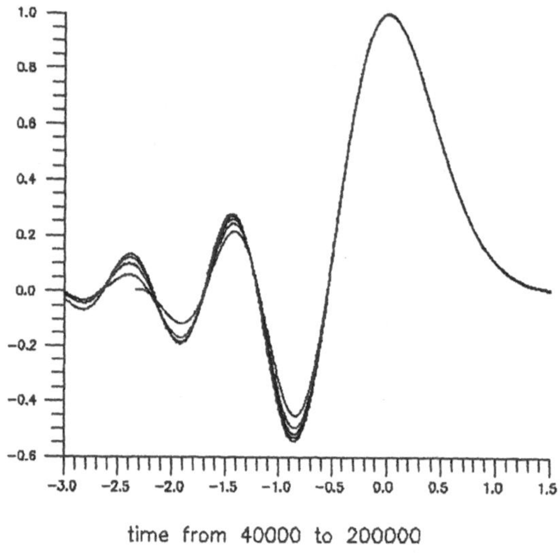

Fig. 5

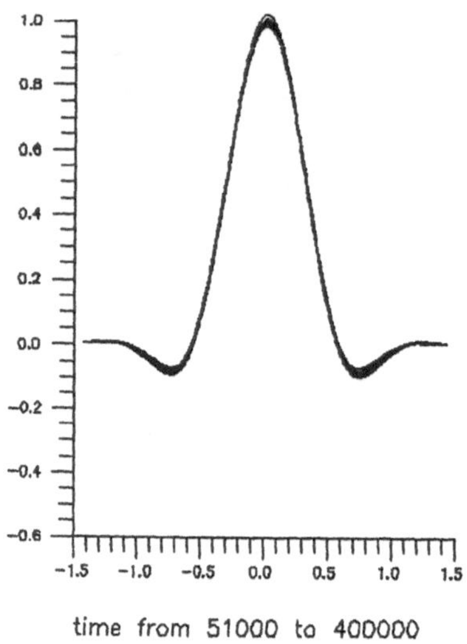

Fig. 6

Acknowledgment. The first author acknowledges a Fellowship from French Ministry of Research and Technology and a partial support from Grant 1052 of Bulgarian Ministry of Science and Higher Education.

LITERATURE

[1] G.A. Maugin, *Non-Classical Continuum Mechanics* (Ed. by R.J.Knops and A.A.Lacey), pp. 272-283, Cambridge University Press, Cambridge (1987)

[2] J. Pouget, *Physical Properties and Thermodynamical Behaviour of Minerals* (Ed. by E. K. Salje), pp. 359 - 401, Riedel, Dordrecht (1988)

[3] J. Pouget, *Continuum Models and Discrete Systems* (Ed. by G.A.Maugin), Longman, vol.1 (1990), pp. 296-312.

[4] G.A. Maugin, S. Cadet, Existence of Solitary Waves in Martensitic Alloys, *Int. J. Engng. Sci.*, 29, No2 (1991), pp.243-258

[5] J. Sander, K. Hutter, On the Development of the Theory of the Solitary wave. A Historical Essay, *Acta Mechanica*, 86 (1991), pp.111-152.

[6] C.I. Christov, An Algorithm for Gaussian Elimination with Pivoting for Multi-diagonal systems, Submitted for publication.

[7] C.I. Christov, On a canonical representation for some stochastic processes with application to turbulence, *Bulg.Acad.Sci., Theor.Appl. Mech.*, 11, No.1 (1980), pp.59-66 (in Russian),reported also in *Continuum Models and Discrete Systems*, (Ed. G. A. Maugin), Longman, vol. 1 (1990), pp. 232-253.

[8] A.C. Newell, Solitons in Mathematics and Physics, SIAM, Philadelphia, 1985

PART IV

TWO-DIMENSIONAL STRUCTURES

SELF-ORGANIZATION AND NONLINEAR DYNAMICS WITH SPATIALLY COHERENT STRUCTURES

K.H. Spatschek, P. Heiermann, E.W. Laedke, V. Naulin, and H. Pietsch
Institut für Theoretische Physik I, Heinrich-Heine-Universität Düsseldorf
D-4000 Düsseldorf, F.R. Germany

Abstract: In near-integrable soliton-bearing systems spatially coherent states can play an important role. In this contribution we briefly review some of the main phenomena for physically relevant situations. We start with the well-known soliton formation in integrable systems which can be interpreted as the first appearence of self-organization in physics. It is shown here that also in non-integrable Hamiltonian systems solitary waves can self-organize. For dissipative systems, the self organization hypothesis is presented and tested for 2d drift-waves. A socalled self-organization instability is found which shows the growth of a spatially coherent (solitary) structure even in the presence of turbulence. The other finding in this respect, the absence of (Anderson) localization in nonlinear disordered systems, is also briefly mentioned. The soliton, as a collective excitation, can overcome individual chaotic motion. A recent result for the proton motion in two Morse-potentials under the influence of oscillations of the heavy ions, is discussed showing the importance of solitons to create ordered structures and collective transport. Nevertheless, solitary waves can also be the constituents of deterministic (temporal) chaos as shown in the final part of this contribution.

1. Self-Organization of spatially coherent structures

The constructive proof [1] of integrability of the 1d KdV-equation can be considered as a milestone in the development of nonlinear physics. As a by-product of the proof, self-organization in the form of stable solitons appears. This is very fascinating and can be considered as an important contribution to the new discipline "synergetics". From the physical point of view the question arises whether this self-organization phenomenon is an artefact of the integrable systems. Integrability can be broken by several means, e.g. higher space dimensions, dissipation, driving, etc. In the following we shall present four examples for self-organization of solitary waves in non-integrable systems. The results follow from numerical simulations, but can be understood by analytical theory.

Let us start with self-organization in KdV-systems. We take as an example the non-integrable 2d KdV equation

$$\partial_t u + u\partial_z u + \partial_z \nabla^2 u = 0 \tag{1}$$

in the Zakharov-Kuznetsov form. Here, $\nabla^2 = \partial_z^2 + \partial_x^2$. It reduces for only one relevant space coordinate z to the celebrated KdV-equation [1]. As has been shown [2], 1d soliton solutions $u_s = 12\eta^2 \operatorname{sech}^2[\eta(z - z_0 - 4\eta^2 t)]$ of (1) are transversely unstable. The growth rate γ_k can be calculated by variational principles to yield

$$\gamma_k^2 = \sup_\varphi \frac{\langle\varphi|\partial_z H_k \partial_z|\varphi\rangle}{\langle\varphi|H_k^{-1}|\varphi\rangle} = \inf_\varphi \frac{\langle\varphi|\partial_z H_k \partial_z H_k \partial_z H_k \partial_z|\varphi\rangle}{\langle\varphi|\partial_z H_k \partial_z|\varphi\rangle} \,. \tag{2}$$

Here $H_k = -\partial_x^2 + 4\eta^2 - u_s + k^2$. The growth rate depends on the transverse wavenumber k; a cut-off appears at $k = k_c \equiv \sqrt{5}\eta$, and the growth rate has its maximum for $k^2 \approx 1.7\eta^2$. In Fig. 1a this dependence is shown by constructing numerically upper and lower bounds from (1). The exact growth rate curve lies within the shaded area. Also the small-k and small-(k_c-k) expansions are shown, respectively. In a 2d numerical simulation [3] we identified this instability and followed its time evolution. A typical result is shown in Fig. 1b. We can interpret this finding in the following

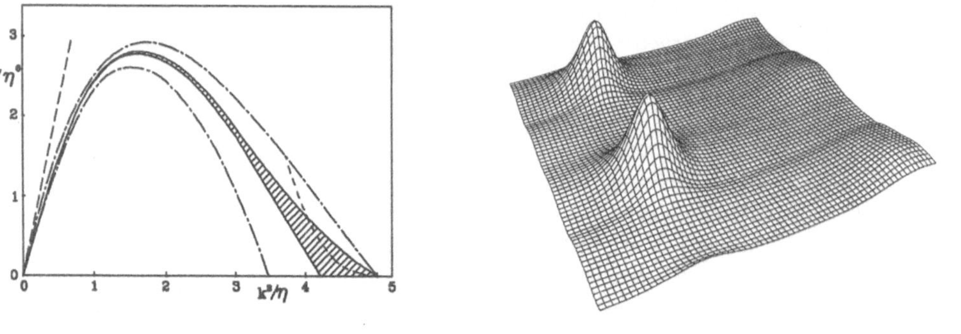

Fig. 1 (a) Transverse instability growth rate γ_k vs. wavenumber k for a 1d KdV soliton.
(b) Appearence of stable 2d KdV-solitons in a numerical simulation of (1). We started with a one-dimensional soliton in the z-direction.

way. In a narrow channel of width d, small k-values cannot occur and the (along the channel) 1d KdV soliton is (transversely) stable. At $d = 2\pi/k_c$ a bifurcation occurs and the 2d KdV solitary wave is the new self-organized state. It is shown in Fig. 1b as one of the humps.

For the 2d soliton solution of (1) a Liapunov functional can be presented [4] in the form $L = L_p\{u\} - L_p\{\bar{u}_M\}$, where

$$L_p\{u\} := \int d^2r[(\nabla u)^2 - \frac{1}{3}u^3 + 4\eta^2 u^2] , \tag{3}$$

and $\bar{u}_M \epsilon S$ belongs to the invariant set S defined with respect to space translations $\vec{\xi}, \bar{u}_M = u_s(\vec{r} - \vec{\xi})$. The functional (3) proves the stability of the 2d stationary localized solitary wave solution of (1). The procedure is standard: for the first variation $\delta L = 0$ and for the second variation $\delta^2 L > 0$ can be shown. [When instead of two space dimensions the three-dimensional case is considered, the stability of a 3d localized solitary wave solution can be proven in a similar manner!]

However, one should be cautious in generalizing these results. If, for example, the case of the cubic nonlinear Schrödinger (NLS) equation

$$i\partial_t q + 2|q|^2 q + \nabla^2 q = 0 \tag{4}$$

is investigated, again the one-dimensional case shows self-organization into 1d solitons $q_s = \eta\,\mathrm{sech}(\eta x)\exp(i\eta^2 t)$. This fact follows from the inverse scattering solutions by Zakharov and Shabat [5]. The soliton solutions are two-dimensionally unstable, with a transverse instability growth rate [6]

$$\gamma_k^2 = \sup_{\varphi} \frac{-\langle\varphi|H_-|\varphi\rangle}{\langle\varphi|H_+^{-1}|\varphi\rangle} = \inf_{\varphi} \frac{-\langle\varphi|H_- H_+ H_-|\varphi\rangle}{\langle\varphi|H_-|\varphi\rangle} . \tag{5}$$

In the second expression, the variation of φ is restricted to the subspace $\langle\varphi|H_-|\varphi\rangle < 0$. Here, the operators H_+ and H_- are defined as $H_+ = -\partial_x^2 - k^2 - 2|q_s|^2 + \eta^2$ and $H_- = H_+ - 4|q_s|^2$, respectively. The cut-off wavenumber is $k_c = \sqrt{3}\eta$. When again considering $d = 2\pi/k$ as the bifurcation parameter,

at $d = 2\pi/k_c$ a bifurcation occurs but instead of a stationary (unstable) 2d Schrödinger soliton a new time-depending (collapsing) solution appears. When the initial state is close to a (unstable) stationary 2d Schrödinger solitary wave, the following theorem can be proven [7]:
Let us assume in 2d that $H\{q\} \leq 0$ holds, where H is the energy functional $H = \int d^2r(|\nabla q|^2 - \frac{1}{2}|q|^4)$. Then, up to translation in space and phase shifts, we can find for every ϵ a δ_ϵ such that

$$\| q(x, t=0) - G(x) \|_{W^{1,2}} \leq \delta_\epsilon \Rightarrow \| q(x,t) - \mu(t) G[\mu(t)x] \|_2 \leq \epsilon \text{ holds,}$$

$$\text{with } \mu(t) \equiv \frac{\|\nabla q\|_2}{\|\nabla G\|_2},\ \mu(t=0) = 1 \text{ and } 0 \leq t < t_c.$$

Here G is the 2d solitary wave solution. It is very interesting to see that, although the stationary solitary wave solution is not stable, the new bifurcating state is connected to the 2d solitary wave solution: it is a solitary wave solution with time-varying parameters, i.e. the width is decreasing with time, leading to a singularity within a finite time. Because of space limitations , we cannot present more details and numerical results here. They will be published elsewhere [7]. We should note that the area of collapsing solutions is a very active one and for arbitrary initial conditions the question of the collapse as an effective dissipation mechanism in plasmas is still open.

Next, we turn to an essentially non-integrable problem (with dissipative and driving terms) to discuss self-organization in nonlinear drift-waves. When dissipative and driving terms are ignored, the basic equation is the Hasegawa-Mima-equation [8] for the normalized electrostatic potential ϕ:

$$\partial_t(1 - \nabla^2)\phi - \kappa_n \partial_y \phi = \hat{z} \times \nabla\phi \cdot \nabla\nabla^2\phi \,. \tag{6}$$

Here κ_n is a normalized density-gradient-coefficient. Equation (5) is non-integrable, but has 2d dipolar vortex solutions. The latter (in general) do not interact elastically, but show a surprising stability against small perturbations. We now generalize (6) by including self-consistently driving and damping terms due to collisions in the same way as done by Kono and Miyashita [9]. In plasma physics the corresponding linear instability is known as the collisional drift instability. Instead of (6) then

$$\partial_t(1 - \nabla^2 - \frac{\kappa_n}{Dk_\|^2}\partial_y)\phi + [-\kappa_n\partial_y + \frac{\kappa_n^2}{Dk_\|^2}\partial_y^2 + \mu\nabla^4]\phi = \hat{z} \times \nabla\phi \cdot \nabla\nabla^2\phi \tag{7}$$

appears. In (7), $D = \Omega_e/\nu_e$ characterizes the collisional contributions and $k_\|$ is an effective parallel wavelength. A numerical simulation [9,10] of (7) shows the self-organization of an arbitrary initial state into a dipolar vortex. The maximum vortex with respect to the (numerically prescribed) box size appears. This end-result is shown in Fig. 2a. For this simulation we started at time $t = 0$ with random noise of low level. The unstable (linear) modes grow, transfer energy via mode-coupling to other modes, and a parametric instability amplifies small-k contributions. This numerical behavior can be understood analytically. A key-role in the interpretation of the final result plays the socalled self-organization hypothesis [11]. It is formulated for nonlinear partial differential equations with dissipation which contain two (or more than two) quadratic (or higher order) conserved quantities in the absence of dissipation. In the case of (6) we shall apply the self-organization hypothesis for the conserved quantities energy $E[\phi] = \int d^2r[(1-\nabla^2)\phi]^2$ and enstrophy $K[\phi] = \int d^2r[(1-\nabla^2)\nabla^2\phi]^2$. The hypothesis is formulated under the following two conditions: (i) There exists a selective dissipation process among the conserved quantities E and K when the dissipation is introduced. That is, one conserved quantity K decays faster than the other E. (ii) The nature of the mode-coupling through the nonlinear terms in the equation is such that the modal cascade in the quantity E is towards small wavenumbers. Then it is assumed (and justified by numerics as, e.g., shown in Fig. 2a) that the following hypothesis holds: The randomly excited field ϕ is expected to reach a quasi-stationary state in which ϕ is described by a deterministic field equation. The latter is obtained by minimizing K within the constraint that E is kept constant: $\delta K - \lambda \delta E = 0$. It is straightforward to show that this variational principle leads to the (quasi-stationary) equation for dipole solutions of (6) [in the absence of dissipation].

We have developed a method and derived concrete equations to explain the self-organization

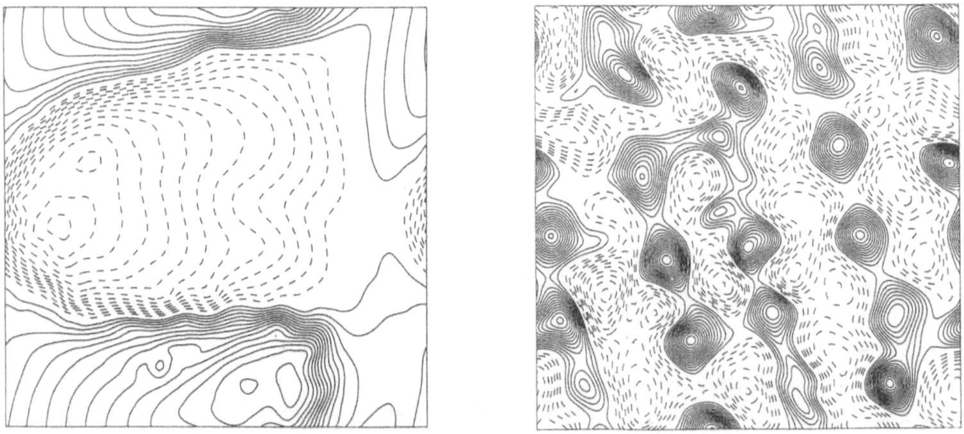

Fig. 2 (a) Self-organization of a big dipolar vortex as a result of numerical simulation of (7). (b) A chain of 2d KdV-type solitary waves appear when (12) is solved.

hypothesis from first principles [12]. Abbreviating the linear operator appearing on the left-hand-side of (7) by $\hat{L}$ we rewrite (7) in the form

$$\hat{L}\phi + \{\phi, \nabla^2\phi\} = 0 \,. \tag{8}$$

Here, $\{...,...\}$ denotes the Poisson-bracket. Next we separate the normalized potential ϕ into a regular (ϕ^R) and a turbulent (ϕ^T) component in the usual way by making use of a turbulent ensemble and the corresponding averaging denoted by $\langle ... \rangle$. Thus when introducing $\phi = \phi^R + \phi^T$ we assume $\langle \phi^T \rangle = 0$. Within this concept we obtain from (8) the coupled equations

$$\hat{L}\phi^R + \{\phi^R, \nabla^2\phi^R\} = -\langle\{\phi^T, \nabla^2\phi^T\}\rangle \,, \tag{9}$$

$$\hat{L}\phi^T + \{\phi^T, \nabla^2\phi^T\} - \langle\{\phi^T, \nabla^2\phi^T\}\rangle = -\{\phi^R, \nabla^2\phi^T\} - \{\phi^T, \nabla^2\phi^R\} \,. \tag{10}$$

In the absence of turbulence ($\phi^T \equiv 0$), the left-hand-side of (9) determines in the usual way the regular structures. On the other hand, in the absence of regular structures, i.e. when the right-hand-side of (10) is zero, the last equation will be similar to that known from (weak) turbulence theory [13]. Linearizing (9) and (10) leads after some tedious algebra [12] to the growth rate

$$\gamma_k = \pi \frac{k k_y^2}{1+k^2} \int_0^\infty k_1^3 W_{k_1} \, dk_1 > 0 \,, \tag{11}$$

where $W_k = \frac{1}{2}(1+k^2)\langle|\phi^T|^2\rangle_k^0$ is the zeroth order turbulent spectral energy density, which has been assumed, in lowest order and for demostration, to be isotropic.

Besides this analytical attempt to justify the self-organization hypothesis we performed a numerical simulation for a slightly different model equation compared to (6). In the presence of a temperature gradient also a KdV-type nonlinearity appears in the basic equation [10]

$$\partial_t(1-\frac{\kappa_n}{Dk_{\|}^2}\partial_y)\phi+[-(u+\kappa_n)\partial_y+\frac{\kappa_n^2}{Dk_{\|}^2}\partial_y^2+\mu\nabla^4]\phi+u\partial_y\nabla^2\phi+\kappa_T\phi\partial_y\phi=\hat{z}\times\nabla\phi\cdot\nabla\nabla^2\phi\,. \quad (12)$$

Here κ_T is the temperature-gradient-coefficient, and we have transformed (with velocity u) into a co-moving frame. The interesting difference with respect to (6) is that (in the dissipationfree case) the relevant conserved quantities change from $E[\phi]$ and $K[\phi]$ to $\tilde{E}[\phi]=\int d^2r[(\nabla\phi)^2-\frac{\kappa_T}{3u}\phi^3]$ and $\tilde{K}[\phi]=\int d^2r\phi^2$, respectively. Now, the self-organization hypothesis yields the variational principle $\delta\tilde{E}-\lambda\delta\tilde{K}=0$ whose solutions are 2d monopole structures of the KdV-type [see Fig. 1b]. And indeed a numerical simulation of (12) confirms this conjecture. In contrast to the single big dipolar vortex for (7) a chain of 2d solitary waves (zonal flow) appears for the model (12) as shown in Fig. 2b. The stability of the 2d solitary waves, even in the presence of the twisting nonlinearity, can be proven, whereas the dipolar vortex is structurally unstable with respect to perturbations in form of a scalar nonlinearity.

2. Self-organized solitary waves as constituents of nonlinear dynamics

We now turn to the question whether solitary waves are "robust". There are three aspects connected with the definition of robustness. The first one is related to linear and nonlinear stability of the exact solutions within the corresponding models. Also the elasticity or inelasticity of collisions falls into this first category. The second one consists of the question whether solitary collective excitations can overcome individual chaotic motion, disturbances due to external fluctuations, etc. The third one is mainly considered here and goes one step further. Can solitary waves (as a whole) behave chaotically in time so that we can consider them as constituents of deterministic chaos?
Let us start with a few remarks with respect to the second aspect. (The first one was already touched in the previous section 1.) A simple example might be helpful. In hydrogen-bonded chains solitary waves are found as the solutions of, e.g., the two-component model [14]

$$\frac{d^2u_n}{dt^2}=u_{n+1}-2u_n+u_{n-1}-\omega_o^2\frac{\partial U(u_n;\rho_n)}{\partial u_n}\,, \quad (13)$$

$$\frac{d^2\rho_n}{dt^2}=\rho_{n+1}-2\rho_n+\rho_{n-1}-\Omega_n^2(\rho_n-\Delta_n)\,. \quad (14)$$

Here, U is in general a double-well potential for the hydrogen-bonded proton; it is created by heavy ions. The potential U is assumed to be a function of two variables: the displacement u_n of the n-th proton from the middle of the hydrogen bridge and the relative displacement ρ_n of the neighbouring heavy ions creating this potential. Solitary wave solutions consist of kinks (or anti-kinks) for the proton displacement and are accompanied by disturbances in the heavy ion sublattice. From the *individual* (proton) point of view, the motion of the particle in an unharmonic potential is driven by the external motion of the heavy ions. Thus, when *collective* solitary wave excitations are not present, the position of the proton can be random. We [15] have verified this statement by a model calculation for the motion of a proton in the superposition of two Morse-potentials created by the two neighbouring heavy ions (see Fig. 3). The equation

$$\frac{d^2u}{dt^2}+\frac{\partial}{\partial u}\tilde{U}(u,\rho)=\gamma\frac{du}{dt} \quad (15)$$

was solved, where γ is a damping decrement and $\tilde{U}$ has the form shown in Fig. 3. The superposition of two Morse-potentials depends on the (normalized) coordinate ρ. For the latter we have assumed an harmonic time-dependence $\rho = \rho_o \sin(\Omega t)$ to simulate the dynamical behavior of the heavy sublattice in the absence of collective solitary wave excitations. As as result, similar to the finding for the Duffing oscillator, chaos can appear. This shows that solitary wave solutions are extremely important (e.g. for transport) since they can override the otherwise chaotic behavior. In this sense they can be called robust.

Another important feature in this respect is the fact that solitary waves can even render an effective transport through random media. As is well-known [16], in linear systems disorder causes Anderson localization, i.e. an exponential decay of the transmission coefficient with the system length. But it has been shown that nonlinearity can lead via soliton formation to an effective transmission mechanism. We have investigated this phenomenon for a similar model as that one originally treated by Caputo et al. [17]. Especially in biological systems, where the environment always causes irregularities in the chain, the formation of solitary waves, their propagation characteristics, and stability are now under active investigation [18].

Here we would like to discuss in more detail the other type of robustness which qualifies solitary

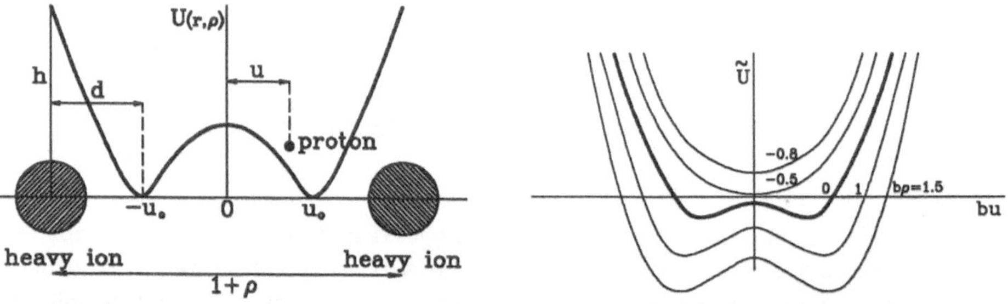

Fig. 3 (a) Motion of a proton in the superposition of two Morse-potentials created by two heavy ions. (b) Changes of the potential with ρ as a parameter.

waves as constituents for nonlinear dynamics, with possible temporal chaos. Let us demonstrate this on the paradigm of a perturbed NLS equation. As has been first demostrated by Nozaki and Bekki [19], for a model of damped nonlinear Langmuir waves driven in a rf capacitor field,

$$i\partial_t q + \partial_x^2 q + 2|q|^2 q = -i\gamma q - iae^{i\omega t}\,, \tag{16}$$

the period-doubling route to temporal chaos occurs for phase-locked solitary waves. Analyzing (16), we can derive the existence condition for a phase-locked solitary wave as $2\gamma\omega^{1/2}/\pi a \ll 1$. The stability of this phase-locked solitary wave was investigated analytically [20]; at finite driving amplitudes (and for fixed damping rate γ and prescribed frequency ω) an instability in form of a Hopf bifurcation takes place and a regulary pulsating solitary wave appears. In a reduced phase-space, the phase-locked solitary wave corresponds to a limit-cycle. With increasing values of driving amplitudes, the system undergoes a series of torus-doubling bifurcations for which the universal Feigenbaum constants $\delta_\infty = 4.6692...$ and $\alpha_\infty = 2.50291...$ could be recovered quite accurately. The situation changes when two space dimensions are taken into account. Then the collapsing solutions can be new attractors as has been discussed in Sec. 1. On the other hand, the whole scenario depends on the form of the "perturbations". If, e.g., we change from (16) to

$$i\partial_t q + \partial_x^2 q + |q|^2 q = -i\alpha q - \beta q - \gamma q^\star \tag{17}$$

or

$$i\partial_t q + \partial_x^2 q + p|q|^2 q = 1 - xq \tag{18}$$

for nonlinear modulated cross-waves in Faraday resonance [21] or radiation in laser irradiated inhomogeneous plasmas [22], respectively, we can find different nonlinear dynamical behaviors with spatial coherence. The first one shows bifurcations into cnoidal-wave-like functions whereas for the second one the quasi-periodic route to temporal chaos occurs. Both models have, in certain parameter regimes, stable solutions [21-23] with spatial coherence; see. Fig. 4. Here, we would like to

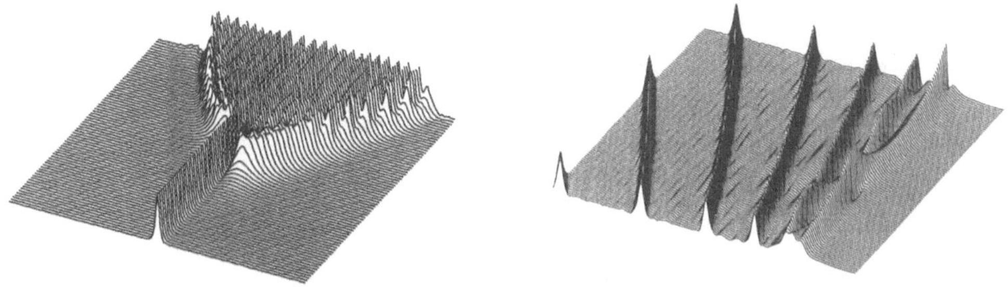

Fig. 4 (a) Appearence of a cnoidal-wave-type stable attractor for $\beta = -1$, $\alpha = 1$, *and* $\gamma = 1.6$ *in (17). (b) Space-time-plot of a solution of (18) for p = 1. A similar regular emission and (accelerated) propagation occurs for* $0.7 \leq p \leq 1.2$.

emphasize a new point in the region of a stable solitary solution to (17). When the driving amplitude is time-modulated [24], i.e. $\gamma = \gamma_o \cos \Omega t$, similar to (16) a phase-locked solitary wave appears which can take part *in toto* in the nonlinear dynamics as a spatially coherent structure. At the first glance, this looks similar to the phenomena detected in (16) and indeed all the tools used there can also be applied here. However, because of the possible bifurcation in space, an interesting interplay between nonlinear dynamics with spatial coherence and simultaneous bifurcation in space can take place. Details will be published elsewhere [25].

3. Summary and conclusions

In this contribution we have given an overview over the possibilities of self-organization and subsequent nonlinear dynamics with spatially coherent structures. The presentation is based on several new and original results which will be published in more details in subsequent publications. The main conclusions are the following: (i) In non-integrable systems stable solitary wave structures are formed by self-organization. (ii) The solitary (and spatially coherent) structures are robust in the sense that they can override individual chaotic behavior and contribute to transport even in disordered systems. (iii) Interesting and generic nonlinear dynamics takes place, with the spatially coherent structures as constituents.

Acknowledgment: This research is supported by the Deutsche Forschungsgemeinschaft through SFB 237.

References

[1] C.S. Gardner, J.M. Greene, M.D. Kruskal, and R.M. Miura, Phys. Rev. Lett. **19**, 1095 (1967).
[2] E.W. Laedke and K.H. Spatschek, J. Plasma Phys. **28**, 469 (1982).
[3] K.H. Spatschek, P. Heiermann, E.W. Laedke, V. Naulin, and H. Pietsch, Proc. 2nd Intern. Toki Conf., Toki, Japan (1990).
[4] E.W. Laedke and K.H. Spatschek, Phys. Fluids **29**, 133 (1986).
[5] V.E. Zakharov and A.B. Shabat, Sov. Phys.-JETP **34**, 62 (1972).
[6] E.W. Laedke and K.H. Spatschek, Phys. Rev. Lett. **41**, 1798 (1978).
[7] E.W. Laedke, R. Blaha, and K.H. Spatschek, J. Math. Phys., in press.
[8] A. Hasegawa and K. Mima, Phys. Fluids **21**, 87 (1978).
[9] M. Kono and E. Miyashita, Phys. Fluids **31**, 326 (1988).
[10] K.H. Spatschek, E.W. Laedke, Chr. Marquardt, S. Musher, and H. Wenk, Phys. Rev. Lett. **64**, 3027 (1990).
[11] A. Hasegawa, Advances in Physics **34**, 1 (1985).
[12] A. Muhm, A.M. Pukhov, K.H. Spatschek, and V.N. Tsytovich, submitted to Comments on Plasma Physics and Controlled Fusion.
[13] V.N. Tsytovich, *Theory of Turbulent Plasma* (Consultants, New York 1977).
[14] A.V. Zolotariuk, St. Pnevmatikos, and A.V. Savin, to appear in Physica D.
[15] R. Grauer, K.H. Spatschek, and A.V. Zolotariuk, to be published.
[16] I.M. Lifshitz, S.A. Gredeskul, and L.A. Pastur, *Introduction to the Theory of Disordered Systems* (Wiley, New York 1988).
[17] J.G. Caputo, A.C. Newell, and M. Shelley, in *Lecture Notes in Physics* **342**, 49 (Springer, Berlin 1989).
[18] O. Kluth, E.W. Laedke, H. Pietsch, K.H. Spatschek, and A.V. Zolotariuk, to be published.
[19] K. Nozaki and N. Bekki, Physica D **21**, 381 (1986).
[20] Th. Eickermann and K.H. Spatschek, in *Inverse Methods in Action*, P.C. Sabatier, ed. (Springer, Berlin 1990), p. 511.
[21] E.W. Laedke and K.H. Spatschek, J. Fluid Mech. **223**, 589 (1991).
[22] K.H. Spatschek, H. Pietsch, and E.W. Laedke, Europhys. Lett. **11**, 625 (1990).
[23] I.V. Barashenkov, M.M. Bogdan, and V.I. Korobov, Europhys. Lett. **15**, 113 (1991).
[24] X.-N. Chen and R.-J. Wei, J. Fluid Mech., to appear.
[25] H. Friedel, E.W. Laedke, and K.H. Spatschek, to be published.

MODULATIONAL INSTABILITY AND TWO-DIMENSIONAL DYNAMICAL STRUCTURES

J. POUGET[(a)] and M. REMOISSENET[(b)]

[(a)]Laboratoire de Modélisation en Mécanique (associé au C.N.R.S.)
Université Pierre et Marie Curie
4 Place Jussieu, 75252 Paris Cédex 05, FRANCE

[(b)]Laboratoire Ondes et Structures Cohérentes
Université de Bourgogne
6 Boulevard Gabriel, 21100 Dijon, FRANCE.

A process of nonlinear structure formation on a two-dimensional lattice is proposed. The basic model consists of a two-dimensional lattice equipped at each node with a molecule or dipole rotating in the lattice plane. The interactions involved in the model are reduced to a periodic potential and nonlinear couplings between first-nearest molecules in the two directions of the lattice. Such a discrete system can be applied to the problem of molecule adsorption on a substrate crystal surface, for instance. The continuum approximation of the model leads to a 2-D sine-Gordon system including nonlinear couplings, which itself can be reduced to a 2-D nonlinear Schrödinger equation in the low amplitude limit. Spatio-temporal structure formation is investigated by means of numerical simulations. These nonlinear structures are caused by modulational instabilities of initial steady states of the two-dimensional system. Moreover, the analogy between the numerically generated patterns and vortex-like excitations in a lattice is also discussed.

1. - INTRODUCTION

Particular interest has been devoted, recently, to the **dynamics of structures on two dimensional nonlinear systems** [1,2]. These structures (dislocations, domain walls, vortices, etc. ....) play an important role in the material properties and they become crutial in nonlinear physics involved in the problem of adsorbates deposited on crystal surfaces [3], in superlattices of ultra thin layers or in large area Josephson jonctions [4], for instance. Here, a particular emphasis is placed on the **dynamical pattern formation mediated by modulational instability** on a two-dimensional Hamiltonian model.

The paper is divided as follows : in Section 2 we introduce our model and show how the basic equation for the lattice model can be reduced to a 2-D nonlinear Schrödinger equation which can exhibit modulational instabilities under certain conditions. In Section 3 we study a particular dynamical regime by means of numerical simulations which then place the role of the modulational instability in evidence for the pattern formation. Then, beyond the instability a self-organisation of 2-D coherent structures takes place.

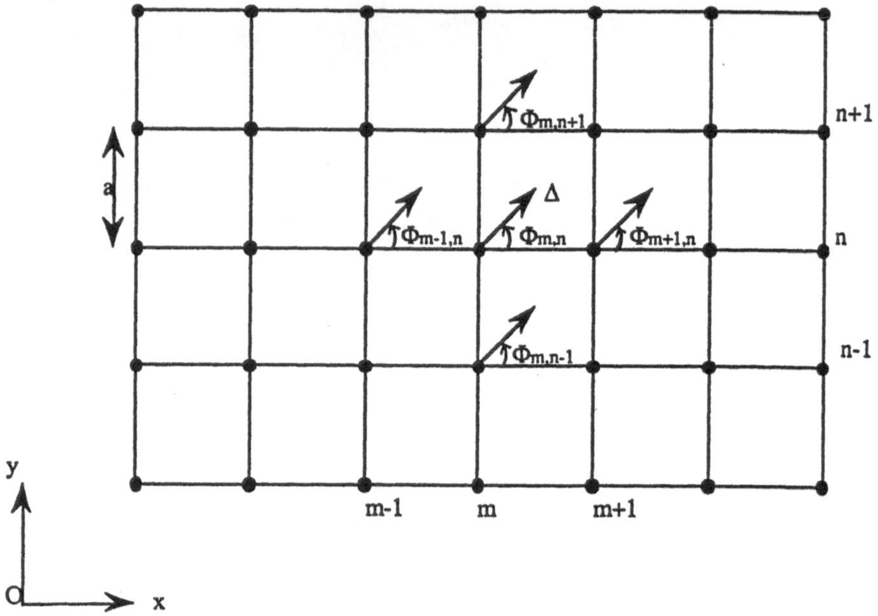

Fig. 1 : *the two – dimensional lattice model equipped, at each node, with rotating molecule (the arrow indicates the molecule orientation).*

2. - THE MODEL

2.1. - Basic discrete equations

The basic model is made of a two-dimensional lattice equipped, at each node, with a **rotator** or **rigid rotating molecule**. Namely, each molecule can rotate in the lattice plane. At site (m,n) the angle of rotation is $\Phi(m,n)$ (see Fig.1). Each molecule interacts **nonlinearly** with its first-nearest neighbors and with a **periodic substrate potential**. Under these conditions the equations of the rotational motion of the molecules can be written as

$$\begin{aligned}\ddot{\Phi}(m,n) = \; & A_L(\Phi(m+1,n) - 2\Phi(m,n) + \Phi(m-1,n)) + A_T(\Phi(m,n+1) - 2\Phi(m,n) \\ & + \Phi(m,n-1)) + B_L\left[(\Phi(m+1,n) - \Phi(m,n))^3 - (\Phi(m,n) - \Phi(m-1,n))^3\right] \\ & + B_T\left[(\Phi(m,n+1) - \Phi(m,n))^3 - (\Phi(m,n) - \Phi(m,n-1))^3\right] \\ & - \omega_0^2 sin(\Phi(m,n)) \quad . \qquad (1)\end{aligned}$$

The inertia of the molecules has been set to unit for ease of presentation. The coefficients A_L and A_T are the linear couplings in the longitudinal and transverse directions while the parameters B_L and B_T are the nonlinear couplings in the longitudinal and transverse directions, respectively. At length, the last term in Eq.(1) is due to the substrate potential where ω_0^2 is the strength of the potential barrier and ω_0 can be interpreted as the frequency of small oscillations in the bottom of the potential wells. Note, if the nonlinear coupling is removed ($B_L = 0$ and $B_T = 0$) Eq.(1) casts in the 2-D Frenkel-Kontorova model (or the 2-D discrete sine-Gordon model [5]). In the following section we restrict our study to the isotropic case, *i.e.* $A_L = A_T = A$ and $B_L = B_T = B$.

2.2. - Continuum approximation

On using a classical procedure we consider the long wave-length limit and reach the continuum approximation of the discrete equations (1) (*i.e.* by expanding in Taylor series $\Phi(m,n)$ in terms of its derivatives about the point $(x = ma, y = na)$). Then Eq.(1) becomes

$$\Phi_{tt} = A\left(\Phi_{xx} + \Phi_{yy}\right) + (A/12)\left(\Phi_{4x} + \Phi_{4y}\right) + B\left(\left(\Phi_x^3\right)_x + \left(\Phi_y^3\right)_y\right) - \omega_0^2 sin(\Phi) \quad , \quad (2)$$

where the variable changes x/a and y/a have been considered (a being the lattice spacing). We notice that if, first, the nonlinear coupling ($B = 0$) and, second, the fourth order derivatives are dropped we recover the usual 2-D sine-Gordon equation [6]. On the other hand, if the substrate potential is removed, Eq.(2) is then somewhat similar to a 2-D Boussinesq equation. Nevertheless, the case for which the substrat potential and nonlinear coupling are both considered is of particular interest for pattern formation.

From Eq.(2), we now look for **plane wave solutions** with a slowly varying envelope of the form

$$\Phi(x,y,t) = \psi(X,Y,T)e^{i(\mathbf{k}.\mathbf{r}-\omega t)} + c.c. \quad , \quad (3)$$

where the wave vector is $\mathbf{k} = (k_L, k_T)$ and ω is the circular frequency of the carrier part of the plane wave and we have also set $\mathbf{r} = (x, y)$. Moreover the envelope ψ is a function of the **slow space and time variables** defined by

$$X = \epsilon x \ , \qquad Y = \epsilon y \qquad \text{and } T = \epsilon t \quad . \quad (4)$$

Where ϵ is a small parameter. In addition, the small amplitude limit has been considered (this allows us to expand the sine function up to the third order with respect to Φ). On inserting (3) into (2) we obtain the **nonlinear dispersion relation**

$$\omega^2 = \omega_0^2 + A\left(k_L^2 + k_T^2\right) - (A/12)\left(k_L^4 + k_T^4\right) + \left(3Bk_L^4 + 3Bk_T^4 - \omega_0^2/2\right)|\psi|^2 \quad . \quad (5)$$

This relation represents the key equation which allows us to reduce Eq.(2) to a 2-D nonlinear Schrödinger equation. It must be noticed that the first three terms in the right hand side of Eq.(5) are the linear part whereas the last term is the nonlinear contribution to the dispersion relation.

2.3. - 2-D nonlinear Schrödinger equation and modulational instability

Now, if we consider slow modulations in space and time of a carrier wave with wave numbers k_{Lc} and k_{Tc}, we can formally expand the dispersion relation (5) around the carrier parameters ($k_L = k_{Lc}$, $k_T = k_{Tc}$ and $|\psi| = 0$) and we arrive at

$$\begin{aligned}
\omega - \omega_c \;=\;& (k_L - k_{Lc})\left(\frac{\partial\omega}{\partial k_L}\right)_c + (k_T - k_{Tc})\left(\frac{\partial\omega}{\partial k_T}\right)_c + \frac{1}{2}(k_L - k_{Lc})^2\left(\frac{\partial^2\omega}{\partial k_L^2}\right)_c \\
& + \frac{1}{2}(k_T - k_{Tc})^2\left(\frac{\partial^2\omega}{\partial k_T^2}\right)_c + (k_L - k_{Lc})(k_T - k_{Tc})\left(\frac{\partial^2\omega}{\partial k_L \partial k_T}\right)_c \\
& + \left(\frac{\partial\omega}{\partial|\psi|^2}\right)_c |\psi|^2 \ .
\end{aligned} \quad (6)$$

The subscript c means that all partial derivatives are taken for the carrier wave features. Then, let the operators $k_L - k_{Lc} : -i\partial/\partial X$, $k_T - k_{Tc} : -i\partial/\partial Y$ and $\omega - \omega_c : i\partial/\partial T$. The frequency

ω_c is the carrier frequency and it is provided by the linear part of the dispersion relation (5). On applying these operators to the amplitude function $\psi(X,Y,T)$, we obtain then the following equation

$$2i\left(\psi_T + v_{gL}\psi_X + v_{gT}\psi_Y\right) + P_1\psi_{XX} + P_2\psi_{YY} + P_3\psi_{XY} + Q|\psi|^2\psi = 0 \quad , \qquad (7.a)$$

where we have set

$$P_1 = \left(A/\omega_c^3\right)\left[\omega_0^2 + Ak_{Tc}^2 - \left(3Ak_{Lc}^4 + Ak_{Tc}^4 + 6Ak_{Lc}^2k_{Tc}^2 + 6\omega_0^2k_{Lc}^2\right)/12\right] \quad , \qquad (7.b)$$

$$P_2 = \left(A/\omega_c^3\right)\left[\omega_0^2 + Ak_{Lc}^2 - \left(3Ak_{Tc}^4 + Ak_{Lc}^4 + 6Ak_{Lc}^2k_{Tc}^2 + 6\omega_0^2k_{Tc}^2\right)/12\right] \quad , \qquad (7.c)$$

$$P_3 = -\,A^2\left(1 - k_{Lc}/6\right)\left(1 - k_{Tc}/6\right)k_{Lc}^2k_{Tc}^2/2\omega_c \quad , \qquad (7.d)$$

$$Q = \left(\omega_0^2 - 6Bk_{Lc}^4 - 6Bk_{Tc}^4\right)/2\omega_c \quad . \qquad (7.e)$$

On considering a frame moving with the wave and using the transformations $\xi = X - v_{gL}T$, $\eta \doteq Y - v_{gT}T$, next (v_{gL} and v_{gT} being the group velocities in the longitudinal and transverse directions), we can rewrite Eq.(7.a) in the standard **2-D nonlinear Schrödinger equation.** The latter equation has been extensively studied especially in plasma physics [7]. With the help of this 2-D nonlinear Schrödinger equation we can investigate the stability of a plane wave traveling on the lattice. A linear analysis of small pertubations of the elementary plane wave solution leads to a criterion of stability or instability maned **modulational instability** [8]. Skipping all the analytic details, the region of instability are given by

$$0 < q_L^2 < 2\psi_0^2 Q/P_1 \;, \qquad 0 < q_T^2 < 2\psi_0^2 Q/P_2 \quad , \qquad (8)$$

where ψ_0 is the constant amplitude of the carrier plane wave, q_L and q_T are the wave numbers of the **perturbation** in the longitudinal and transverse directions, respectively. Accordingly, a perturbation with a wave vector (q_L, q_T) satisfying (8) can trigger instabilities in both directions of the lattice. However, such conditions depend, of course, on the respective signs of the products QP_1 and QP_2.

3. - NUMERICAL SIMULATIONS

We now present the preliminary numerical investigations of the dynamics of structure formation initialized by modulational instabilities. Specifically, an initial carrier wave propagating in the x (or y) direction (5 periods in the propagating direction, the wave number of the carrier wave in x is $k_{Lc} \simeq 0.35$ corresponding to the long wave-length limit) is modulated by adding a random noise of small amplitude to the initial velocity field (the noise is removed afterwards). It is important to emphasize that the numerical simulations are directly performed on the **original lattice model** (see Eq.(1)). A lattice made of 91×70 points is considered and periodic boundary conditions in x and y directions are used for the numerical simulations. Under these conditions the 2-D nonlinear Schrödinger equation describes the dynamical behavior of the system in the very beginning of the instability (at the birth of the instability).

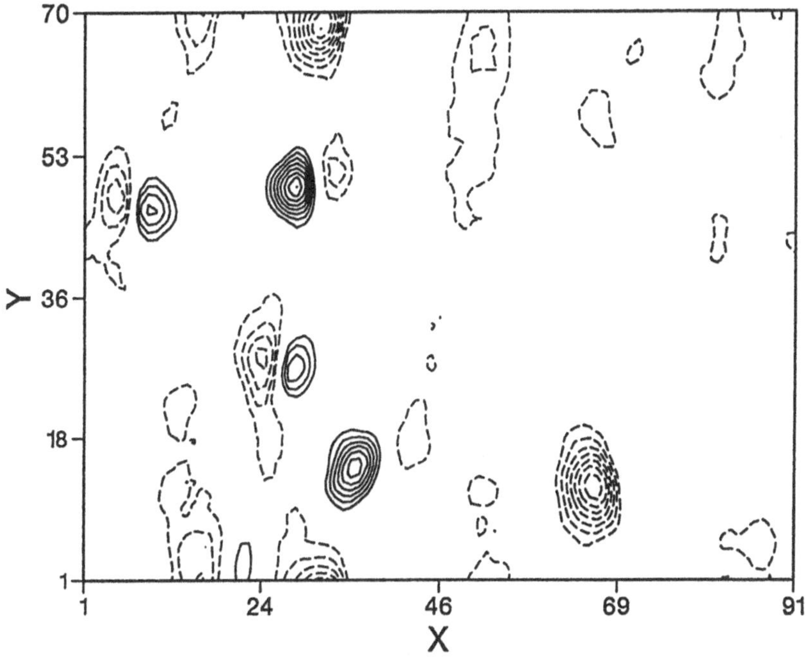

(a)

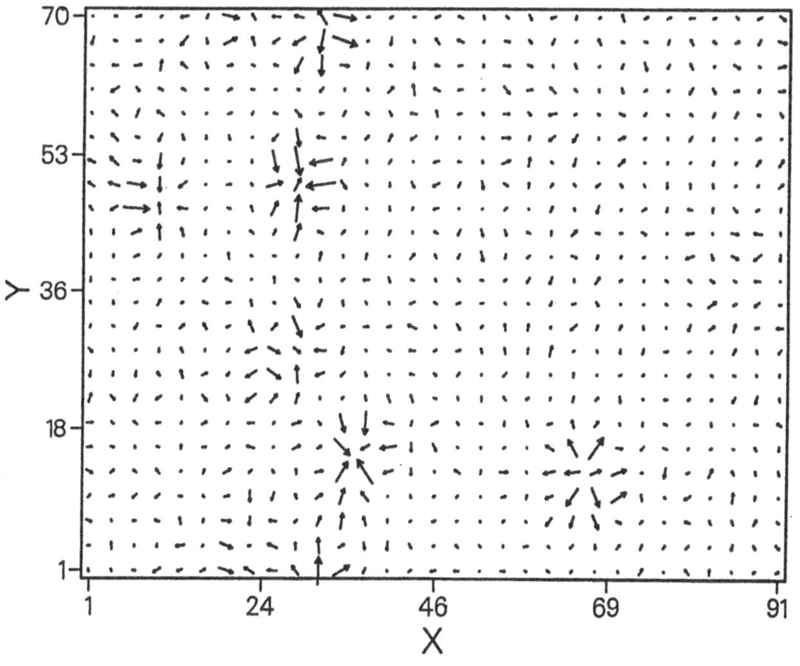

(b)

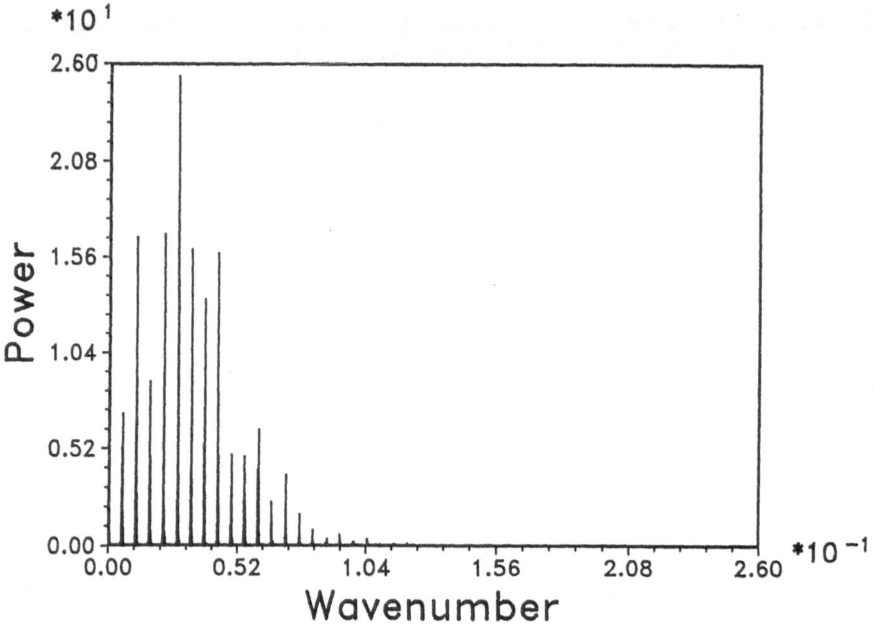

(c)

Fig. 2 : *Numerical simulations of the discrete system Eq.*(1) ($A = 0.1$, $B = 0.2$, $\omega_0^2 = 0.4$) *at* $t = 1200$, (*a*) *contour line graphe for the rotation* (*full line* $\Phi \geq 0$ *and dashed line* $\Phi < 0$), (*b*) *the corresponding pseudo – velocities and* (*c*) *the power spectrum exhibiting additional wave number components due to instabilities*

The most significant results are collected together in Fig.2. Figure 2.a represents the picture of the contour lines for the rotation Φ. This picture exhibits, after a long time, very clear localized structures, in fact these structures are moving on the lattice. We can observe small pertubations due to the phonon radiations but they remain rather weak. On introducing a **"pseudo-velocity"** associated with the rotation Φ (*i.e.* the gradient of Φ) as in perfect fluid hydrodynamics, we can plot a lattice of pseudo-velocity as shown in Fig.2b (each arrow corresponds to the velocity vector). The picture thus numerically generated exhibits sorts of **vortex-like structures**. Finally, the power spectrum of the lattice dynamics is given in Fig.2c showing then additional harmonic components around the high peak corresponding to the carrier frequency. These components are produced by the instabilities. From Fig.2.c, we can compare the range of the instability wave vectors to that given by the conditions (8) and a good agreement is obtained. It is worthwhile noting that, here, in contrast to the pure 2-D nonlinear Schrödinger equation the drastic collapse does not occur [7,9], this emerges from saturation effects because of the substrat potential and discreteness effects.

4. - CONCLUSIONS

We have studied the formation of nonlinear localized structures on a two- dimensional lattice model. We have also shown that these structures are the result of the modulational instabilities of a steady plane wave solution. The most significant idea, which can be underlined in the present work, is that we have been able to reduce the rather complicated dynamics of the lattice to the 2-D nonlinear Schrödinger equation in the long wave-length and small amplitude limits. Although the 2-D nonlinear Schrödinger equation is limited to the birth of the modulational instability, this informs us about the selection mechanism of wave vectors of the instabilities taking place both in longitudinal and transverse directions. In short, the modulational instability is a **natural vehicle** for the nonlinear structure formation. It seems that the characteristic radius of the coherent structures thus produced can be connected with the growth rate of the instabilities as well as the model parameters. This point should be clarified in a further work. In addition, extensions of the study to the specific problems of **vortex-like** and **spiral excitations** will be examined [1,10,11]. Finally, the relative influence of the substrat potential and nonlinear coupling (see the definition of the coefficients of the nonlinear Schrödinger equation, Eqs(7.a)-(7.e)) will be studied in more detail by means of an analytical approach and numerical simulations.

References

[1] **F. FALO, A.R. BISHOP, P.S. LOMDAHL** and **B. HOROVITZ**, *Phys. Rev. B43*, 8081 (1991).

[2] **P. COULLET** and **D. WALGRAEF**, *Europhys. Lett. 10*, 525 (1989).

[3] **I.F. LYUKSYUTOV, A.G. NAUMOVETS** and **Yu.S. VEDULA**, in *Soliton* in *Modern Problems in Condensed Matter Sciences vol. 17* eds S.E. Trullinger, V.E. Zakharov and V.L. Pokrovsky (Elsevier Science Publishers, 1986) p. 605.

[4] **E.M. MASLOV**, *Physica 15D*, 433 (1985).

[5] **J. POUGET, S. AUBRY, A.R. BISHOP** and **P.S. LOMDAHL**, *Phys. Rev. B39*, 9500 (1989).

[6] **O.H. OLSEN, P.S. LOMDAHL, A.R. BISHOP** and **J.C. EILBECK**, *J. Phys. C: Solid State Phys. 18*, L511 (1985).

[7] **V.E. ZAKHAROV**, *Sov. Phys. JETP 35*, 908 (1972).

[8] **T.B. BENJAMIN** and **J.E. FEIR**, *J. Fluid Mech. 27*, 417 (1967).

[9] **E.W. LAEDKE** and **K.H. SPATSCHEK**, in *Differential Geometry Calculus of Variations, and Applications*, ed. G.M. Rassias and T.M. Rassias (Dekker, New York, 1985), p. 335.

[10] **O. HUDAK**, *Phys. Lett. 89A*, 245 (1982).

[11] **Y. ISHIMORI** and **N. MIYAMOTO**, *J. Phys. Soc. Japan 55*, 82 and 3756 (1986).

COMPETITIVE INTERACTIONS AND 2–D STRUCTURES AT FINITE TEMPERATURES

N. Flytzanis† and G. Vlastou–Tsinganos†‡
†Physics Department, University of Crete, Heraklion, Greece
‡Foundation for Research and Technology–Hellas, Heraklion, Greece

Abstract

The phase diagram of a triangular lattice with competitive interactions is obtained at finite temperatures. At high temperatures the existence of an incomplete Devil's staircase points to the existence of incommensurate states near the disorder line. 1–d domain walls and large periodicity 2–d domains are discussed.

1. Introduction

Polytypes [1] is a class of solids, which have the common characteristic to appear in various phases, under small changes in temperature, pressure or other physical parameters. The ground state of such a system may be a periodic structure with a short wavelength (most commonly observed), a long commensurate periodicity, or even incommensurate formations. As examples one can mention the classical structural polytype silicon carbide that presents a large sequence of short period phases, the rare–gas monolayers adsorbed in graphite (2–d systems) that show commensurate and incommensurate phases [2] and the magnetic substance $CeSb$ with an experimentaly observed succesion of long period phases [3].

Although growth kinetics around screw dislocations could contribute to modulated structures, it is commonly accepted that they can exist as stable states in thermodynamic equilibrium with some form of "competition"[4]. This may arise either by antagonistic effective interactions between constituent units , or by competing periodicities. In many cases short range competing interactions give a rather consistent explanation of the multiplicity of phases encountered in polytypes or other structures [1]. Two main categories of models have been proposed to explain the behavior of systems developing modulated structures. These are the discrete variable models, such as the ANNNI type [5] or the clock type models [6] and the continuous variable models of the type of the Frenkel–Kontorova [7,8]. In both the above microscopic class of models the lattice is considered discrete.

In the present work we shall be dealing with a continuous variable model, which can be of one or two degrees of freedom. In the first case, it can represent on one hand a displacement perpendicular to the plane as in the surface reconstruction of the Si (111) surface [9]. On the other hand the variable can be an angle representing rotations, as in an N_2 molecule adsorbed on graphite [10,11] which by lowering the temperature changes to $\sqrt{3} \times \sqrt{3}$ and 2×1 phases. For two degrees of freedom a 2–d model has been proposed [12,13] to explain the three phases of $LiIO_3$.

Very often the competition does not come from the substrate and the interplanar forces, but can be a result of interactions among the atoms as was done in a model used to describe 1–d ferroelectrics [14,15]. In this case for the ground state to be incommensurate one needs at least up to 3rd neighbour interactions. By considering the nearest neighbour strains in the 1–d chain problem, one can transfer to one with a substrate and up to 2nd neighbour interactions.

In the following we shall be presenting a nonlinear model on a triangular lattice, at finite temperature, whose phase diagram exhibits short and long period commensurate phases. In addition, incommensurate phases leave their signature in the form of a "devil's staircase". Also, domain walls are studied, either in 1– or 2–dimensions.

2. Model and Methods

Consider atoms or molecules (of mass M) forming a triangular lattice, having one degree of freedom (displacement or rotation) denoted by $u_{n,m}$. Each particle interacts harmonically with its first and second neighbours with a force constant f_1 and f_2 respectively. An on–site nonlinear double well potential, of the form of ϕ^4 acts on each site ($g_0, g_4 > 0$). The Hamiltonian of the system is the following:

$$\mathcal{H} = \sum_{n,m} \left\{ \frac{1}{2} M \dot{u}_{n,m}^2 + \frac{1}{4} g_4 \left(u_{n,m}^2 - \frac{g_0}{g_4} \right)^2 \right. \\ \left. + \sum_{NN} \frac{1}{2} f_1 (u_{NN} - u_{n,m})^2 + \sum_{NNN} \frac{1}{2} f_2 (u_{NNN} - u_{n,m})^2 \right\}. \tag{1}$$

The last two terms denote summation over nearest neighbours (NN) and next–nearest neighbours (NNN), respectively. The fact that the on–site potential is double well and that $f_1, f_2 \gtrless 0$ arises a conflict between the respective terms in $\mathcal{H}$, and as a result commensurate, incommensurate and even chaotic phases may appear.

Our aim is to construct a phase diagram in the space of the reduced force constants $c_1 = f_1/2g_0$, $c_2 = f_2/2g_0$ (normalized to the frequency of the well minimum) and the temperature. In order to do that one should in principle calculate the full quantum mechanical free energy of the system for every periodicity, as a function of temperature. Since this is an almost impossible task, we confine ourselves to a semi–quantum approximation, based on the Gibbs–Bogoliubov inequality [16]. It gives an upper bound to the free energy in question $\mathcal{F}$, by the use of an auxiliary Hamiltonian $\mathcal{H}_0$.

$$\mathcal{F} \leq \tilde{\mathcal{F}} = \mathcal{F}_0 + < \mathcal{H} - \mathcal{H}_0 >_0 . \tag{2}$$

The upper bound to $\mathcal{F}$ is approximated by $\mathcal{F}_0$ (the free energy corresponding to $\mathcal{H}_0$) plus a correction term. The above expectation value is calculated with the known quantum density matrix ρ_0 of $\mathcal{H}_0$.

The trial Hamiltonian $\mathcal{H}_0$ is chosen in the frame of the independent–site approximation, to be a set of displaced harmonic oscillators, with different frequency at each site.

$$\mathcal{H}_0 = \frac{1}{2} \sum_{n,m} \left[-\frac{\hbar^2}{M} \frac{\partial^2}{\partial u_{n,m}^2} + M\omega_{n,m}^2 (u_{n,m} - \alpha_{n,m})^2 \right] . \tag{3}$$

The density matrix ρ_0 and the free energy $\mathcal{F}_0$ of such a system are known exactly, and therefore the r.h.s of Eq.(2) can be explicitely calculated. The $\alpha_{n,m}$ and $\omega_{n,m}$ are considered as variational parameters and can be obtained by minimizing the free energy.

An optimisation method for the free energy that has proven to be very succesfull is the Monte Carlo Simulated Annealing (MCSA). It has been introduced by Kirkpartick *et al* [17] using the Metropolis Monte Carlo algorithm [18]. It provides an efficient method to determine a global minimum, while with the possibility of climbing over barriers, it avoids being trapped into local minima. The algorithm consists of a basic step, which is used repeatedly in order to simulate a collection of atoms at a given "fictitious" temperature. In each step, an atom is given a small random displacement and the resulting change in energy of the system is computed. The change is accepted unless the energy is higher, in which case the new configuration is accepted, if the probability factor $\exp(-\Delta E/kT)$ is larger than a random number. Here ΔE is the increase in energy from the previous configuration. Thus the system can get out of a local minimum well, and by lowering the "fictitious" temperature, to find the global minimum well.

There are certain parameters involved in this algorithm, as the sequence of lowering the "fictitious" temperature, the number of steps at each temperature, the displacement step, and others, which in practice for finite computer time, must be determined empirically [19], since they strongly depend on the particular form of the function to be minimized. Following the Monte Carlo Simulated Annealing method one proceeds with the Steepest Descent method to locate the exact position of the minimum.

3. Phase Diagram of Periodic Structures and Domain Walls

For the construction of the phase diagram of our model system in the (c_1, c_2, τ) phase space, (τ being the temperature in some convienient units), we have to search at each phase space point for the ground state. For this purpose we compare for each temperature and force constants, the free energies of quasi–one–dimensional periodic structures from 1×1 up to 15×1 and of 2–dimensional ones up to 3×3. The higher 2–dimensional periodic structures were examined by calculating the number of opposite sign first and second neighbours. Depending on the sign of the neighbouring interactions, the vast majority of the 2-d structures have been excluded, and only the ambiguous cases were treated numerically, due to the computing time required. For clarity, we present in

Figure 1 a cross–section of the full phase diagram at constant second neighbour interaction parameter $c_2 = -0.5$. (For other values of c_2 refer to [20]). The notation used to describe one particular configuration is to note the number of consecutive atoms that have the same sign displacement, e.g. a 7×1 structure with signs of displacements as $(++--++-)$ is indicated as $2^3 1$.

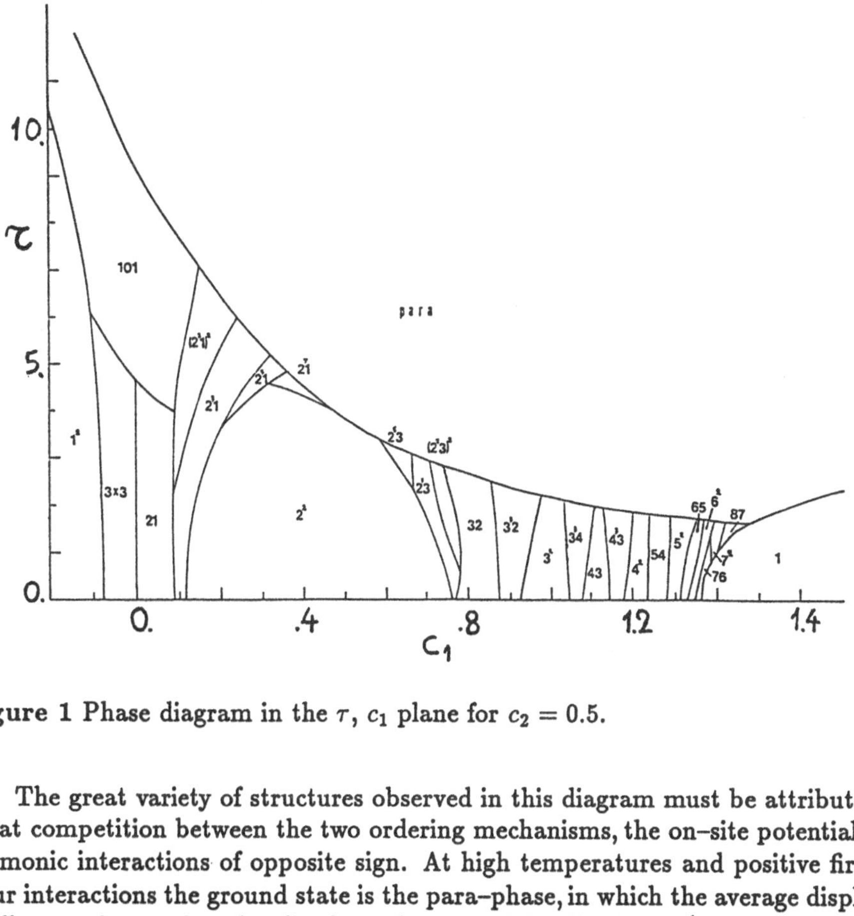

Figure 1 Phase diagram in the τ, c_1 plane for $c_2 = 0.5$.

The great variety of structures observed in this diagram must be attributed to the great competition between the two ordering mechanisms, the on–site potential and the harmonic interactions of opposite sign. At high temperatures and positive first neighbour interactions the ground state is the para–phase, in which the average displacement of all atoms is zero, i.e. they lie above the potential wells of the ϕ^4 potential. It is important to note that an analogous situation in a classical calculation at zero temperature would be unstable.

For both the first and second neighbour interactions being negative, the structures that persist at high temperatures are those that have the highest number of large relative displacements between first and second neighbours. These are the 2×1 and the symmetric 3×1 $(+0-)$.

At lower temperatures let us concentrate on the part of the diagram between the 4×1 and the 1×1 phase. A close look reveals that between two configurations, a third appears with a periodicity the sum of the two. A more quantitative picture of this fact can be obtained by plotting the wavevector $q = 2\pi s/N$ of each state against one of the systems parameters. The number $2s$ accounts for the number of sign changes within one period consisting of N atoms. In the following Figure 2, we are presenting a plot of

s/N versus c_1 for a high temperature. The maximum number of atoms within a period is $N = 30$.

At low temperatures [20], low periodicities occupy large regions of the diagram, but the higher periodicities form a characteristic curve described as a complete devil's staircase. This implies that if there was an infinitely fine grid in c_1 and an infinitely large period searched, then all the rational fractions s/N could be found in the plot. This means that the character of this devil's staircase is complete. At higher temperatures though, and just below the border to the para-phase, the devil's staircase of Fig. 2 appears to have complete and incomplete parts. This is a strong indication of the existence of an incommensurate phase, near the transition to the para-phase. For a more quantitative description of such an incommensurate phase one has to perform discrete mapping techniques [21], which however, are plagued by numerical difficulties since the physically stable states of minimum free energy correspond to unstable orbits of the map.

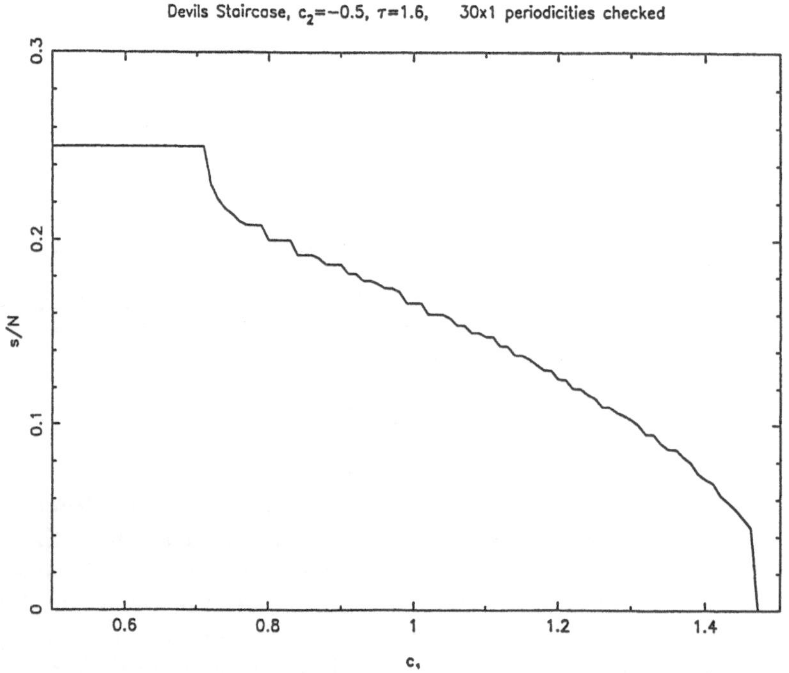

Figure 2 Devil's staircases for fixed $c_2 = -0.5$ and $\tau = 1.6$.

It has been proposed that there is a close relation between the existence of domain wall structures and the incommensurate to commensurate phase transition. Also, it is important to know when and if they can become ground state, in particular on a discrete lattice. On the triangular lattice one can have either 2–d or quasi–one–dimensional walls. The simplest one is the so–called ferro domain wall, of the later kind, that separates two regions with periodicity 1×1, but opposite sign. The method we have studied such formations is the Steepest Descent method, and thus calculated the arrangement of atoms with the lowest free energy.

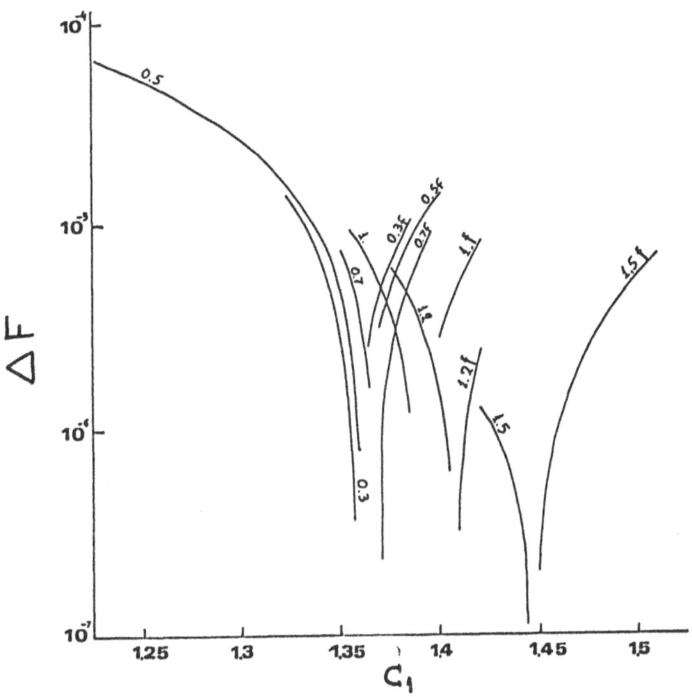

Figure 3 Plot of $\Delta F = (\mathcal{F}_{1\times1}/\mathcal{F}_{wall})-1$ for the ferro wall vs c_1 at $c_2 = -0.5$ and varying $\tau = 0.3, 0.5, 0.7, 1., 1.2, 1.5$. Negative curves are reflected and denoted as "folded" (f).

Let us consider the ferro domain wall free energy in relation to the ferroelectric 1×1 state. It is clear that if the wall free energy becomes lower than the respective 1×1 then the later is no longer the ground state. We shall concentrate in the region between the period 1 and the modulated phases in the phase diagram of Fig. 1. In Figure 3 we make a log–plot of the quantity $(\mathcal{F}_{1\times1}/\mathcal{F}_{wall}) - 1$ as a function of c_1 for various temperatures and fixed c_2. For each temperature there are two curves corresponding to positive and negative (folded) values of the quantity plotted. Moving from higher to lower c_1 values, for the same temperature, the wall free energy becomes lower than 1×1, when the above quantity is positive. This c_1, τ value is near the border line between 1×1 and modulated phases in the diagram. Conversely, when a domain wall becomes energetically favorable it usually means that a modulated phase is the ground state. It was never found, although searched thoroughly a wall being a ground state.

The 2–dimensional walls seem to be better candidates for the ground state due to entropy gain. The pattern shown in Figure 4 is a triangular wall formation, with a 9×9 unit cell, whose short diagonal is a domain wall between two 3×3 structures. Similar to this structure was found to be ground state in an Ising spin system [22]. We have found that such a formation is stable at high temperatures but its free energy is always slightly lower that the 3×3. This leads us to the conclusion that it is a metastable state which could become gound state if additional interactions were important.

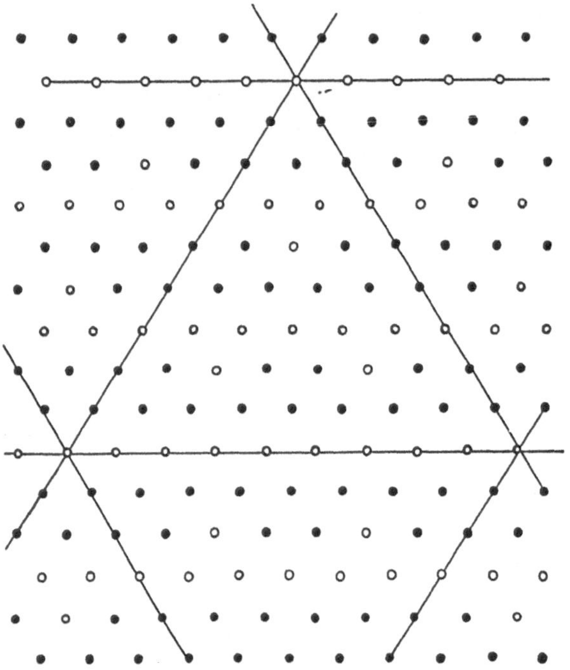

Figure 4 Triangular domain structure (9×9) consisted by the 3×3 asymmetric phase and 2×1.

We have extended the calculations to a two degrees of freedom system and applied it to $LiIO_3$. The phase diagram is in good agreement with the experimentally observed $\alpha \rightarrow \gamma$ transition and previous numerical classical equilibrium calculations. Of course the zero point energy introduces even at $T = 0$ a region in the phase digram, where the para–phase is the ground state [13].

4. Concluding Remarks

In summary, using a semi–quantum variational principle based on the Gibbs–Bogoliubov inequality and the MCSA optimization method, we can obtain the phase diagram for a 2–d model with competitive interactions at finite temperatures. The phase diagram shows some similarities with 1–d ANNNI models except that a number of 2–d structures appear. Sequences of ground states given in the phase diagram have also been observed in $CeSb$ [1] by varying a physical parameter like temperature.

The inclusion of weak longer range interactions does not change the phase diagram, while for strong we can expect significant changes. The degree of changes depends on n_i, i.e. the number of ith neighbours with large relative displacements and of course the surface symmetry. This is evident also from lattice gas models of surface reconstruction, where as a function of overlayer coverage, different range interactions come into play and the corresponding ground state can be very different. Assymetries in the interactions can also influence the ground state. These effects are under investigation along with non convex interplanar nonlinear interactions and interactions that include strain gradient terms.

References

1. J.M. Yeomans 1988 in "Solid State Physics" Vol 41, eds H. Ehrenreich and D. Turnbull (Academic Press, London) P. 151
2. P. Bak, D. Mukamel, J. Villain and K. Wentowska 1979 Phys. Rev B *19* 1710
3. P. Fisher, B. Lebech, G. Meier, B.D. Rainford and O. Vogt 1978 J. Phys. C *11* 345
4. F. Axel and S. Aubry 1981 J. Phys. C *14* 5433
5. M.E. Fisher and W. Selke 1980 Phys. Rev. Lett. *44* 1502
6. D.A. Huse 1981 Phys. Rev *B 24* 5180
7. Y. Frenkel and T. Kontorova 1938 Zh. Eksp. Theor. Phys. 8 *89* 1340
8. F.Y. Frank and J.H. van der Merwe 1949 Proc. R. Soc. A *198* 205
9. J. Kanamori and M. Okamoto 1985 J. Phys. Soc. Japan *54* 4636
10. C.R. Fuselier, N.S. Gillis and J.C. Raich 1978 Solid. St. Com *25* 747
11. O.G. Mouritsen and A.J. Berlinsky 1982 Phys. Rev. Lett. *48* 181
12. E. Coquet, M. Peyrard and H. Büttner 1988 J. Phys. C *21* 4895
13. G. Vlastou–Tsinganos 1991 PhD Thesis, University of Crete
14. T. Janssen and J.A. Tjon 1982 Phys. Rev. B *25* 3767
15. Y. Ishibashi 1991 J. Phys. Soc. Japan *60* 212
16. R.P. Feynman and A.R. Hibbs "Quantum Mechanics and Path Integrals (McGraw-Hill, New York, 1965)
17. S. Kirkpatrick, C.D.Jr. Gelattand M.P. Vecchi 1983 Science *220* 671
18. N. Metropolis, A. Rosenbluth, M. Rosenbluth, A. Teller and E. Teller 1953 J. Chem. Phys. *21* 1087
19. G. Vlastou-Tsinganos, N. Flytzanis and H. Büttner 1990 J. Phys. A *23* 225
20. G. Vlastou-Tsinganos, N. Flytzanis and H. Büttner 1990 J. Phys. A *23* 4553
21. T. Janssen 1986 "Incommensurate Phases in Dielectrics" Vol 1, eds R.Bline and A.P.Levanyuk (Amsterdam: North Holland)
22. K. Nakanishi and H. Shiba 1982 J. Phys. Soc. Japan *51* 2089

INTERACTIONS OF SOLITONS in (2+1) DIMENSIONS

Bernard Piette
Wojciech J. Zakrzewski

*Department of Mathematical Sciences,
University of Durham, Durham DH1 3LE, England*

Abstract: We consider instanton solutions of the CP^N models in two Euclidean dimensions as solitons of the same models in (2+1) dimensions. We find that, in general, the solitons tend to shrink so to stabilise them we add special potential and skyrme-like terms. We show that in head-on collisions the solitons scatter at 90° to the direction of their original motion and that they also undergo a shift along their trajectories.

1. INTRODUCTION

Consider the scattering of two solitons in any relativistic (1+1) dimensional model, such as the well known Sine-Gordon model. Imagine that one soliton moves with velocity v_1, the other with velocity v_2 and that, at some time $t = t_0$, they are well separated and their separation is δ. If the velocities of the solitons are such that they approach each other then after a certain length of time they will interact (and at that time the description of the system in terms of two isolated solitons ceases to be applicable) but then, some time later, they will emerge from the scattering region and later on will become well separated again. Looking at the positions of the solitons at this time, we find that each is at a place different from where it would have reached, had there been no interaction between them. We see that one effect of their interaction is to shift the solitons along their trajectories, the direction and the magnitude of this shift being determined by the strength of the interaction. All this is well known and has been observed in many models.

Let us now increase the spatial dimension by one and look at the scattering of solitons in (2+1) dimensions. What would be the corresponding properties? To answer this question we observe, that on purely kinematical grounds, there are more possibilities. As solitons correspond to extended structures, we see, that as they approach each other they can either experience:

1) a head on,

2) a small impact parameter or

3) a large impact parameter collision.

In the latter case, if the impact parameter is larger than the size of each soliton (in particular if there are no net forces acting on the solitons) we would expect the solitons to pass each other experiencing only small perturbations due to their interactions.

The most interesting are clearly head-on collisions and the scattering at very small impact parameters. So what happens with solitons in such scatterings? To try to answer this question we have to decide what we mean by a soliton in (2+1) dimensions. Clearly, we would expect it to represent at each value of time a localised but spatially extended structure of finite energy. So what model should we use and will the results depend on its choice?

Most of the properties of solitons in (1+1) dimensions are associated with the integrability of the underlying theory. Unfortunately, although several integrable models in (2+1) dimensions are known, none of them is relativistically covariant. But why should we consider a relativistically covariant model? There are many reasons; let us mention only that static extended structures arise often in many such theories and, in particular, many properties of the physical proton follow quite naturally from its description in terms of such an extended structure in a phenomenologically successful "Skyrme model of the proton" (of course, in a relativistic model in (3+1) dimensions). So if we want to consider solitons in a relativistic model we cannot rely on the integrability of the model for the properties of their scattering.

The simplest relativistic model in which we can study various properties of solitons in (2+1) dimensions is the S^2 σ model, also called the CP^1 model, which involves one real vector field of 3 components, $\vec{\phi} \equiv (\phi^1, \phi^2, \phi^3)$. In (2+1) dimensions $\vec{\phi}$ is a function of the space-time coordinates (t, x, y) which we will also write as (x^0, x^1, x^2). The model is defined by the Lagrangian density

$$\mathcal{L} = \tfrac{1}{4}(\partial^\mu \vec{\phi}) \cdot (\partial_\mu \vec{\phi}) \,, \tag{1}$$

together with the constraint $\vec{\phi} \cdot \vec{\phi} = 1$, *i.e.* $\vec{\phi}$ lies on a unit sphere S^2_ϕ. In (1) the Greek indices take values $0, 1, 2$ and label space-time coordinates, and ∂_μ denotes partial differentiation with respect to x^μ. Note that we have set the velocity of light, c, equal to unity, so that in all our calculations we can use dimensionless quantities. The Euler-Lagrange equations derived from (1) are

$$\partial^\mu \partial_\mu \vec{\phi} + (\partial^\mu \vec{\phi} \cdot \partial_\mu \vec{\phi}) \vec{\phi} = \vec{0} \,. \tag{2}$$

For boundary conditions we take

$$\vec{\phi}(r, \theta, t) \to \vec{\phi}_0(t) \quad \text{as} \quad r \to \infty \,, \tag{3}$$

where (r, θ) are polar coordinates and where $\vec{\phi}_0$ is independent of the polar angle θ. In two Euclidean dimensions (*i.e.* taking $\vec{\phi}$ to be independent of time) this condition ensures finiteness of the action and in (2+1) dimensions it leads to a finite potential energy.

It is convenient to express the $\vec{\phi}$ fields in terms of their stereographic projection onto the complex plane W

$$\phi^1 = \frac{W + W^\star}{1 + |W|^2}, \quad \phi^2 = i\frac{W - W^\star}{1 + |W|^2}, \quad \phi^3 = \frac{1 - |W|^2}{1 + |W|^2}. \tag{4}$$

The W formulation is very useful, because it is in this formulation that the static solutions take their simplest form; namely, as originally shown by Belavin and Polyakov[1] and Woo,[2] they are given by W being any rational function of either $x + iy$ or of $x - iy$. In this formulation the Lagrangian density is given by

$$\mathcal{L} = \frac{\partial_\mu W \partial^\mu W^\star}{(1 + |W|^2)^2}, \tag{5}$$

where * denotes complex conjugation and the equations of motion are given by

$$\partial_t^2 W = \frac{2W^\star((\partial_t W)^2 - (\partial_x W)^2 - (\partial_y W)^2)}{1 + |W|^2} + \partial_x^2 W + \partial_y^2 W. \tag{6}$$

The simplest nontrivial static solution of (6) is given by

$$W = \lambda \frac{z - a}{z - b}, \tag{7}$$

where $z = x + iy$ and a, b and λ are arbitrary complex numbers. It is easy to calculate the energy density, E, corresponding to the static solution (7). We find

$$E = \frac{8|\lambda|^2 \, |a - b|^2}{(|z - b|^2 + |\lambda|^2 \, |z - a|^2)^2} \tag{8}$$

and so we see that the extended structure of this solution has a bell-like shape, with its position and size determined by

$$\frac{a|\lambda|^2 + b}{|\lambda|^2 + 1} \quad \text{and} \quad \frac{|\lambda|^2 \, |a - b|^2}{(|\lambda|^2 + 1)^2}$$

respectively. Taking the limit $\lambda \to \infty$, $b \to \infty$ while keeping their ratio fixed (and still called λ) allows us to consider $W = \lambda(x + iy - a)$ as our candidate for a soliton of "size" λ, which is positioned at a. In the same sense $W = \lambda\frac{(x+iy-a)(x+iy+a)}{2a}$, provides us with a field configuration describing two such solitons (positioned at $\pm a$).

These solutions, strictly speaking, can be considered as soliton configurations for the (2+1) dimensional model only if they are stable and do not desintegrate when we consider their time evolution. So what is their evolution? We performed several numerical studies using a 4th order Runge-Kutta method of simulating time evolution. The calculations were performed on fixed lattices which varied from 201×201 to 512×512, with lattice spacing $\delta x = \delta y = 0.02$. The time step was 0.01. We used fixed boundary conditions for most of our simulations involving the $\vec{\phi}$ fields (with some absorption on the boundaries) and the extrapolated boundary conditions for the simulations involving the W_i fields. We have tested our results by changing the lattice size and varying the boundary conditions and we are reasonably confident as to their validity.

2. CP^1 RESULTS

The obtained results reveal that when the solitons are sent initially at zero impact parameter they scatter at 90°. At the same time, however, they are unstable in the sense that they tend to change their size. We have analysed this problem in some detail[3] [4] . The observed instability is due to the fact that the pure S^2 model (has no intrinsic scale and so admits the existence of solitons of arbitrary size. Hence under small perturbations the solitons can either expand indefinitely or shrink to become infinitely tall spikes of zero width. Our simulations have shown that this is exactly what happens in this model. In fact, as soon as the solitons are purturbed, *e.g.* start moving, they start shrinking. This is true not only in the full simulation of the model but also[6] in the approximation to the full simulation provided by the so-called "collective coordinate" approach in which the evolution is approximated by the geodesic motion on the manifold of static solutions.

However, even though the extended structures shrink, we can still look at their scattering properties as they become very spiky only some time after emerging from the scattering region. We can lengthen this time by letting them expand as they move towards each other (in this case they shrink less as they come out). In all cases the scattering proceeds through the same intermediate stages; first the extended structures come towards each other, then when they are close together they form a ring and finally they emerge out of the ring at 90° to the original direction of motion. We can follow their trajectory in the x, y plane; an example of a typical simulation is shown in fig 1a; in fig 1b we show a plot of the time dependence of the relative distance between the extended structures. Looking at fig 1a we see that when the solitons are close together it is really difficult to talk about their positions (they are really in the shape of a ring); when we think of their speeds and the positions they started at and they find themselves at when they have emerged from the interaction region we see that, in analogy to what happens in (1+1) dimensions, their positions are translated forward along their trajectories. How much are they shifted along their trajectories? This question is difficult to answer as, due to the ring structure of their intermediate state, we do not understand their trajectories. What happens when the two solitons are on top of each other? Such a configuration would correspond to $W = \mu(x + iy - a)^2$. As its easy to see the energy density of such a configuration is in the shape of a ring centered at $x + iy = a$. Thus it would seem natural to assume that the two solitons come on top of each other before they scatter at 90°.

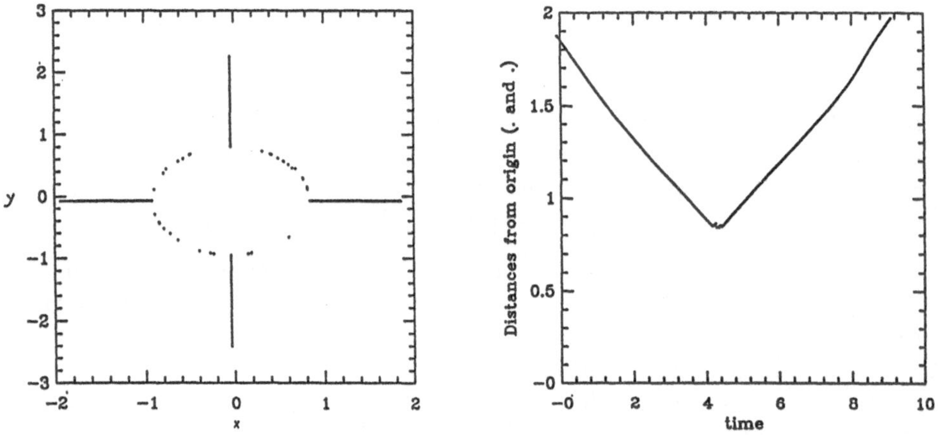

To study the importance of the ring formation and/or the shrinking of solitons we decided to go beyond the simplest S^2 model. To overcome the shrinking one has to break the conformal invariance of the model. This can be done by introducing a "Skyrme-like" and a potential term to the Lagrangian. Such a model was introduced in ref. [5] and used for many simulations. The "Skyrme-like" term of that model was later shown to be unique[7] so that the only arbitrariness resides in the form of the potential term. Of course, the form of the potential term does effect the details of the scattering, but all our simulations have shown[8] that the gross features are always the same. The introduction of the new terms generates some forces between the solitons; in particular, if we stick to the model of ref [5] these forces are repulsive and the only static solution corresponds to one soliton of a fixed "size" (whose value is determined by the parameters of the additional terms). The existence of the repulsive forces introduces a "critical" velocity into the model; in head-on collisions below a certain value of the velocity the solitons scatter back to back, above it they scatter at 90°. If one assumes that the scattering at 90° proceeds through the intermediate stage of the solitons "coming on top of each other" one can estimate the value of this critical velocity by performing an energy balance; such a calculation can be found in ref [9] where it was shown that this estimate agrees well with the results obtained in numerical simulations.

Similar results have also been found in the models based on other potentials[8] . In all cases, at sufficiently high velocities (to overcome all the repulsive forces), the elementary head-on collisions between two solitons exibit the 90° scattering. At lower velocities, and in interactions involving more solitons the scattering properties are more complicated. For lack of space we will present the results of our investigations of these cases elsewhere.

3. CP^2 MODEL

To go beyond the formation of a ring, when the solitons are on top of each other, we have to consider a model with a larger target manifold space; the simplest such model is the CP^2 model. This model involves two complex W fields (like the W field of the CP^1 model) and the Lagrangian density is given by

$$\mathcal{L} = \frac{\partial_\mu W_1 \partial^\mu W_1^\star + \partial_\mu W_2 \partial^\mu W_2^\star + (W_1\partial_\mu W_2 - W_2\partial_\mu W_1)(W_1\partial^\mu W_2 - W_2\partial^\mu W_1)^\star}{(1+|W_1|^2+|W_2|^2)^2}, \tag{9}$$

and the equations of motion are given by

$$\begin{aligned} \partial_t^2 W_1 &= \frac{2W_1^\star\Big((\partial_t W_1)^2 - (\partial_x W_1)^2 - (\partial_y W_1)^2\Big)}{1+|W_1|^2+|W_2|^2} \\ &+2W_2^\star\Big(\frac{(\partial_t W_1)(\partial_t W_2) - (\partial_x W_1)(\partial_x W_2) - (\partial_y W_1)(\partial_y W_2)}{1+|W_1|^2+|W_2|^2}\Big) + \partial_x^2 W_1 + \partial_y^2 W_1 , \end{aligned} \tag{10}$$

and a similar equation for W_2, obtained from (10) by the interchange $(1 \leftrightarrow 2)$.

It is easy to see that $W_1 = \lambda z^2$, $W_2 = \mu z$ is a static solution of the equations of motion and describes two solitons on top of each other (and located at $z = 0$). For a general choice of the parameters λ and μ the energy density of the configuration has a ring-like structure (like in the CP^1 case); however, when the parameters μ and λ satisfy $\mu^2 = \sqrt{2}\lambda$ the energy density takes the shape of a single peak (*i.e.* the ring becomes a peak). We can displace the solitons initially by choosing $W_1 = \lambda(z^2 - a)$ for some reasonable value of a, and then taking W_2 as above with $\mu^2 =$

$\sqrt{2}\lambda$, set the two solitons moving towards each other by starting the simulation off with $\frac{dW_1}{dt} = aV$, $\frac{dW_2}{dt} = 0$. With such an initial value problem the solitons are set to expand as well so that when they emerge out of the interaction region they do not shrink too fast.

We have performed many simulations corresponding to different values of the the initial velocity V. All our simulations showed a 90° scattering. Moreover, they also showed a shift along the trajectory as seen initially in the CP^1 case. In fig2 we display typical trajectories of our solitons and in fig3 a,b and c we show the time dependence of the distance between the solitons for simulations started with three values of V. We clearly see a shift along the trajectory which is similar to the one observed in the CP^1 case except that this time the interpretation is easier (our picture suggests that as the solitons are close together they speed up and then come on top of each other where they spend some time after which they separate and gradually, as they leave the interaction region, they regain their initial speed). Clearly this is only a qualitative picture of their interaction; when they are close together they loose their identity, and like in the (1+1) dimensional case, it makes little sense of talking about their trajectories. Moreover, as is easy to check, the shift along the trajectories does not depend on V (and in the case of the simulations shown in fig3. its value is $\delta = 1$, if we assume that the solitons go through the origin).

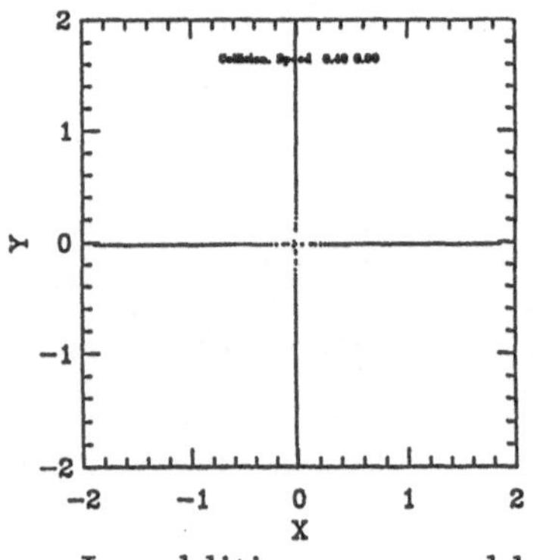

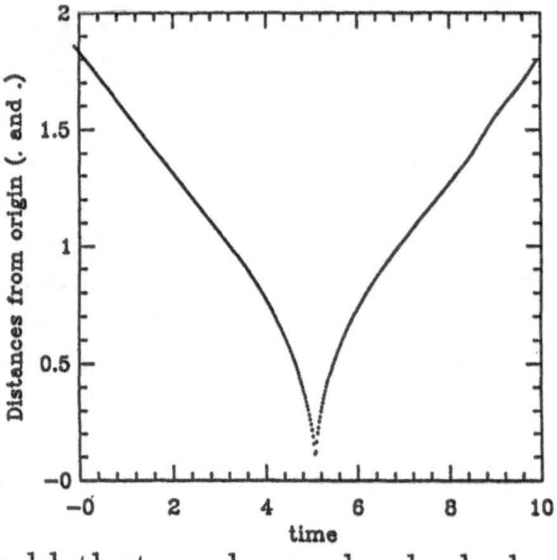

In addition we would like to add that we have also looked at some field configurations corresponding to one soliton and one antisoliton. As the forces between them are attractive, placed some distance apart, solitons and antisolitons move towards each other and then annihilate into pure radiation. The angular dependence of the outgoing radiation is not uniform; most of it is, again, sent at out at 90° to the direction of their final approach (just before the annihilation). This has also been observed in some earlier simulations[5] .

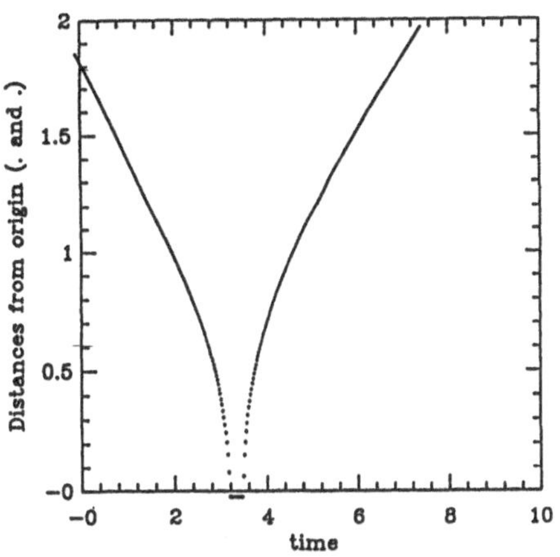

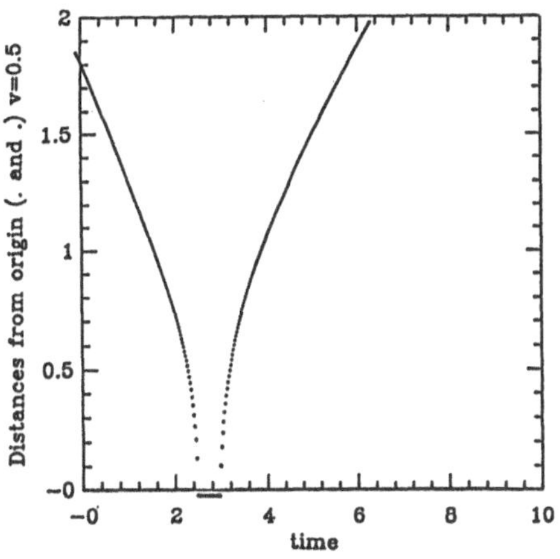

Thus we see that our solitons in (2+1) dimensions behave very much like real solitons. In the scattering processes they preserve their shape and although some radiation effects are present these effects are always very small. In conclusion, we see that the modified S^2 model, although non-integrable, is almost integrable in that it has many features in common with many integrable models. Most differences or deviations are rather small. As most physically relevant models are not integrable our results suggest that the results found in some integrable models should not be dissmissed as not relevant; it is quite likely that some of these results may also hold in models which, strictly speaking, are not integrable but whose deviations from integrability are rather small.

REFERENCES

1. A.A. Belavin and A.M. Polyakov, *JETP Lett*, **22**, 245 (1975)
2. G. Woo *J. Math. Phys.*, **18**, 1264 (1977)
3. R.A. Leese, M. Peyrard and W.J. Zakrzewski - *Nonlinearity*, **3**, 387 (1990)
4. W.J. Zakrzewski - Soliton-like Scattering in the $O(3)$ σ model in (2+1) Dimensions - *Nonlinearity* to appear (1990)
5. R.A. Leese, M. Peyrard and W.J. Zakrzewski Soliton Scatterings in Some Relativistic Models in (2+1) Dimensions - *Nonlinearity*, **3**, 773 (1990), M. Peyrard, B. Piette and W.J. Zakrzewski - Soliton Scattering in the Skyrme Model in (2+1) Dimensions 1. and 2., Durham University preprints *DTP-90/37* and 39 (1990)
6. R.A. Leese - *Nucl. Phys.* **B**, **B334**, 33, (1990)
7. J.M. Izquierdo, B. Piette, M.S.S. Rashid and W.J. Zakrzewski - Models with Solitons in (2+1) Dimensions, Durham University preprint DTP-91/21 (1991)
8. B. Piette and W.J. Zakrzewski - in preparation
9. B. Piette and W.J. Zakrzewski - Skyrmions and Their Scattering in (2+1) Dimensions, Durham preprint *DTP-91/11*, 1991

SPIRAL WAVES IN EXCITABLE MEDIA

Pierre Pelcé and Jiong Sun

Laboratoire de Recherche en Combustion
Université de Provence, St Jerome
13397 Marseille Cedex 13, France

INTRODUCTION

The problem of the formation of spiral waves in two-dimensional excitable media is one of the classical problems of nonequilibrium pattern-forming system which remains unsolved. It is clear now , from numerical simulations, that reaction diffusion sytems are a good physical basis for the explanation of the formation and of the characteristics of the spiral waves. However, the mathematical understanding of these solutions is far from being complete. A first problem is to determine the characteristics of the steadily rotating spiral wave , rotational frequency and radius of the unexcited circle lying at the center of the spiral. A second problem is the stability of the spiral wave. In a range of control parameters, an instability mode called meandering is found. Meandering of spiral waves is a significant deviation of the spiral tip from circular trajectories and is now well confimed by experiments [1],[2]. Meandering is also observed in numerical simulations of standard reaction-diffusion models showing that the explanation of this fascinating behaviour must be found in the frame of these reduced systems. Contrary to experiments transition to meandering is well defined. It is a supercritical Hopf bifurcation [3].

In order to understand these phenomena, the simplest model from which one can start consists of a set of two coupled reaction diffusion equations with two very different time scales, one for the trigger variable c_1 which varies on the fast scale, the other for the recovery variable c_2 , on the slow scale. Numerical simulations of this model exhibit steadily rotating fields , with a well determined frequency ω, the region where the gradient of c_1 is sharp being located on a spiral shape. In order to understand these solutions and determine the frequency of rotation of the spiral, one simplifies the model and assimilates the regions where c_1 varies rapidly to a closed moving contour which delimits an excited region. Outside this contour , in the refractory region , the slow recovery variable decays until the occurence of another excitation. Moreover, aroud the center of rotation of the spiral lies an unexcited region with a well determined radius r_0 [4],[5],[6].

In a first part we describe this simplified model and discuss how it can be determined from the set of two coupled reaction-diffusion equations. In a second part, we study the solutions of the model corresponding to steadily rotating spirals. We determine uniquely the angular velocity ω and the hole of radius r_0 as a function of the control parameters of the model, i.e. ε the ratio between fast reaction time and refractory time and δ the excitability [6]. In a third part we discuss the stability of the spiral waves in a limiting case where a first analysis can be performed [7].

THE MODEL

The basis of W.F.M. is the classical piece-wise linear model for the following reaction-diffusion system:

$$\varepsilon\frac{\partial c_1}{\partial t}=\varepsilon^2\Delta c_1+f(c_1,c_2) \qquad (1)$$

$$\frac{\partial c_2}{\partial t}=c_1-\delta \qquad (2)$$

where $f(c_1, c_2)$ is the piece-wise linear function drawn on Fig.1:

$$f(c_1,c_2)=\begin{cases}-c_1-c_2+1\ , & 0<c_1\\ -c_1-c_2-1\ , & c_1<0\end{cases} \qquad (3)$$

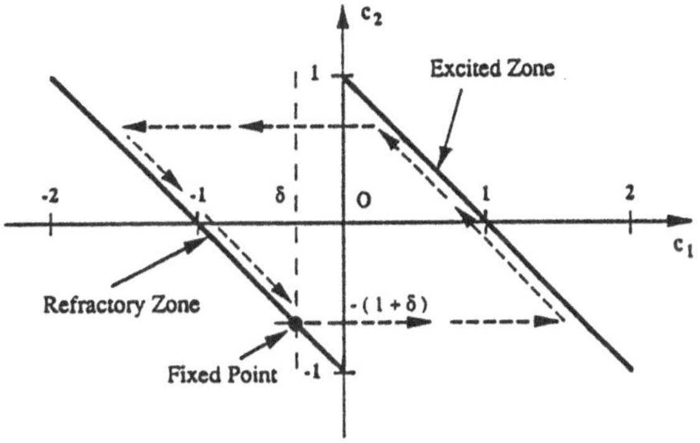

Fig.1 Piecewise linear model.

and δ a real and negative number larger than -1 which characterizes the excitability of the system. Here, $\varepsilon = \tau / T$ is the ratio between fast reaction time τ and the refractory time T. Time is scaled with T and lengths by the diffusive length $(D\tau)^{1/2} / \varepsilon$, where D is the diffusive coefficient of the trigger variable c_1. It is assumed here that only the trigger variable can diffuse in the system.

It is well known that when ε is small , the reaction-diffusion system admits one-dimensional solutions corresponding to the propagation of a sequence of stable pulses. Each pulse of the sequence is characterized by two waves (the front and the back) propagating with the same velocity separated by an excited region. In each wave , the concentration of the recovery variable c_2 can be assumed constant . The propagation velocity $c(c_2)$ of the wave can be found as a function of c_2 after integration of eqn.1 in a frame moving with constant velocity $c(c_2)$, in the new space variable $\xi = (x - c(c_2)t)/\varepsilon$. In the case of the piece-wise linear model (3), this velocity is found as

$$c(c_2) = -\frac{2c_2}{\sqrt{1-c_2^2}} + O(\varepsilon) \qquad (4)$$

As introduced in Zykov [8] book p.5, an important quantity caracterizing the front is the maximum relative rate of increase of c_1, $E_{max} / A = |(\partial c_1 / \partial t)|_{max} / A$, where A is the amplitude of the wave. In the case of the simple piece-wise linear model defined above, this quantity is found as $|c_2| / \varepsilon$. Its inverse corresponds to the transit time of the wave. In the case of the single pulse $c_2 = -(1 + \delta)$, sothat $E_{max} / A = |1 + \delta| / \varepsilon$. When the pulse is isolated, the duration of the pulse (transit time of the excited region) is found as $D = -\mathrm{Log}\,|\delta|$. Experimentalists are more familiar with the quantities E_{max} / A and D . The model is defined with the two independant parameters ε and δ. In the following we will use alternatively these two kinds of parameters.

When the pulse propagates in an inhomogeneous field c_2, the equiconcentration waves of the trigger variable c_1 become distorted and the normal velocity (4) of the waves are modified by transverse concentration fluxes. It is well known that when the curvature radius of the wave is large compared to its thickness, the normal velocity of the wave is modified proportionally to the curvature as

$$\vec{v}.\vec{n} = c(c_2) - \varepsilon\kappa + O(\varepsilon) \qquad (5)$$

where $c(c_2)$ is determined by relation (4). We will now explain the model studied in the paper. We call this model " Wave Front Interaction Model " (W.F.M.) since it describes the motion of two fronts , the front and the back , which move with a normal velocity given by relation (5).These two waves interact with a relaxational field c_2 which satisfies the equation

$$\frac{\partial c_{2\pm}}{\partial t} = -c_{2\pm} \pm 1 - \delta \qquad (6)$$

where the subscript + (resp. -) means that the relaxational field c_2 is calculated in the excited region (resp. refractory region). This last equation is simply deduced from relation (2) with the additional relation $c_1 = \pm 1 - c_2$ which holds respectively in excited and refractory regions. The common end point of the two curves must have a zero normal velocity. This model appears as a free-boundary problem in the spirit of the one proposed by Fife [3] and Tyson and Keener [4] , but simpler and thus more tractable since the diffusion coefficient of the recovery variable is assumed here to be zero. It contains two control parameters , the ratio between fast reaction time and refractory time ε , a real positive number and the excitability δ, a real negative number larger than -1. It is clear from the beginning that a first validity condition of the model is that the two waves will be well separated, which imposes that $0 < \varepsilon \ll 1$. Secondly, that the curvature radius of the wave is much larger than the front thickness, which implies $c \ll 1$. From relation (4), this condition can be satisfied if δ goes to -1.

STEADILY ROTATING SPIRALS

We look for solutions of W.F.M. corresponding to clockwise spirals rotating at constant angular velocity ω , around a hole of radius r_0. Consider a polar coordinate system (r, θ) rotating with constant angular velocity ω. Then , the shapes of the front and the back , respectively $\theta_F(r)$ and $\theta_B(r)$ satisfy eqn.5 , i.e.

$$\frac{\omega r}{\sqrt{1+\Psi_{F,B}^{2}}} = \pm c\,(c_{2F,B}) - \varepsilon\left(\frac{\frac{d\Psi_{F,B}}{dr}}{\sqrt{1+\Psi_{F,B}^{2}}^{\,3}} + \frac{\Psi_{F,B}}{r\sqrt{1+\Psi_{F,B}^{2}}}\right) \qquad (7)$$

Here , $\Psi_{F,B} = r\, d\,\theta_{F,B}(r)/dr$, the sign + (resp. -) corresponds to the subscript F (resp. B). and c (c_2) is determined by relation (4). The two curves meet tangentially to the circle of radius r_0 sothat $\Psi_F(r_0) = -\Psi_B(r_0) = -\infty$, or from eqn. (7) their curvature reaches their critical value $\kappa_{cr} = c\,(c_{2F,B})/\varepsilon$. At large distance from the tip, front and back behave as phase-shifted Archimedean spirals, i.e. $\theta_B(r) = kr = \theta_F(r) +$ constant , where k is a positive number . In each part of the domain delimited by the curves, the concentration of the recovery variable c_2 satisfies

$$\omega\frac{dc_{2\pm}}{d\theta} = -c_{2\pm} \pm 1 - \delta \qquad (8)$$

where the subscripts + and - are respectively associated to the excited and refractory regions. On the front , $c_{2+}(\theta_F(r)) = c_{2-}(\theta_F(r)) = c_{2F}$ and on the back, $c_{2+}(\theta_B(r)) = c_{2-}(\theta_B(r)) = c_{2B}$.

From Eqn.(8), one can deduce c_{2F} and c_{2B} as :

$$c_{2F} = 1 - \delta + \frac{2\left(\exp\left(\frac{\theta_B - \theta_F - 2\pi}{\omega}\right) - 1\right)}{\left(1 - \exp -\frac{2\pi}{\omega}\right)} \qquad (9)$$

$$c_{2B} = 1 - \delta + \frac{2\left(\exp -\frac{2\pi}{\omega} - \exp\left(\frac{\theta_F - \theta_B}{\omega}\right)\right)}{\left(1 - \exp -\frac{2\pi}{\omega}\right)} \qquad (10)$$

Numerical Results

For the details of the resolution of the system composed by eqns. (7), (9),(10) see the full paper by Pelcé and Sun [6]. For convenience we introduce the radial coordinate scaled with the tip radius $R = r\,\kappa_{tip} = r\,c/\varepsilon$.

Angular velocity ω and hole radius r_0

Hole radius r_0

The smaller is the ratio between fast reaction time and refractory time, the smaller is the radius of the circle around which the spiral tip rotates at constant angular velocity Fig.2. This curve diverges at the limiting value $\varepsilon_{max} = 6.8$. This means that solutions for steadily rotating spirals exist only for values of ε less than 6.8. For this particular value of δ, large hole radii are found for values of ε which are not small so that results obtained from W.F.M. may differ from the one obtained from the complete reaction-diffusion system. On the other hand solutions corresponding to small radii which are obtained for small values of ε may be in good agreement with the one obtained from complete simulations of the reaction-diffusion system.

Angular velocity ω

It is a decreasing function of ε Fig.3.

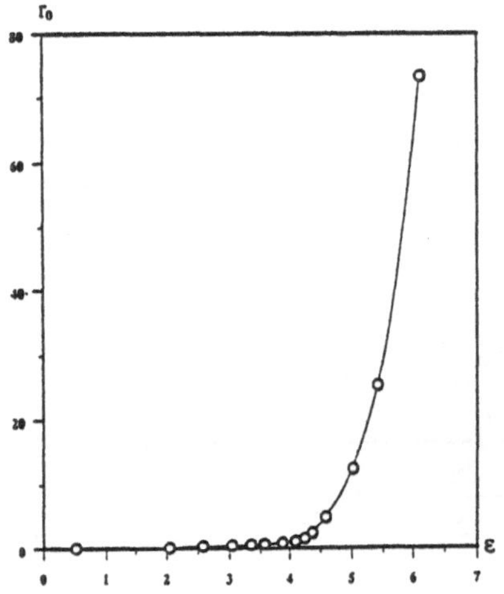

Fig.2 Hole radius r_0 as a function of ε (δ = - 0.1).

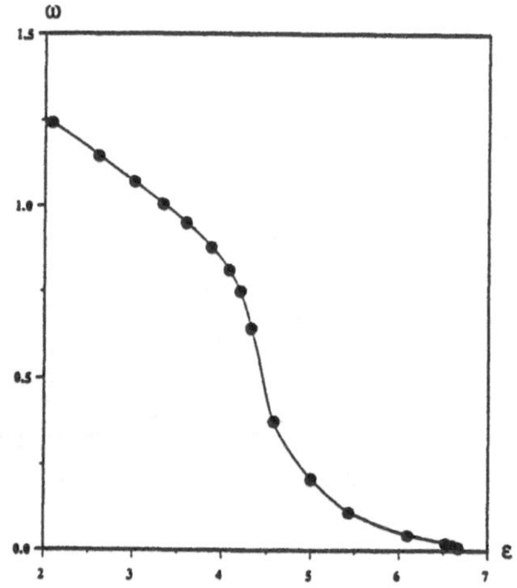

Fig.3 Angular velocity ω as a function of ε (δ = - 0.1).

The spiral shape

Spiral shapes are drawn on Fig.4 for R_0 = 1. and R_0 = 10. for the same value of the excitability δ = -.1.

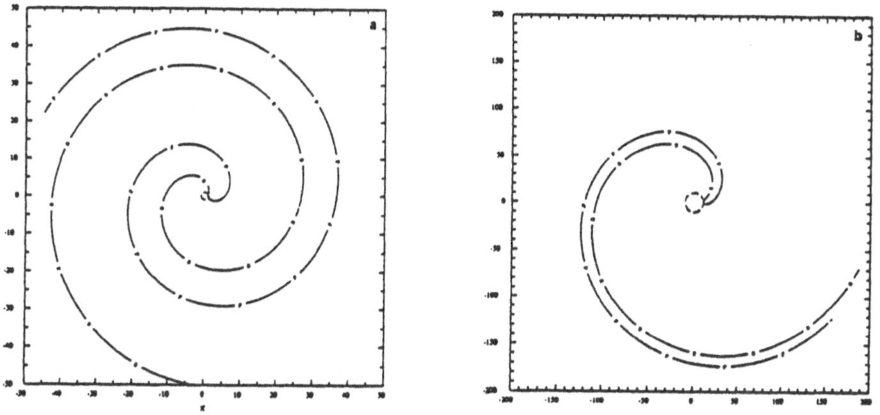

Fig.4 Shapes of spirals (δ = -.1): left R_0 = 1. ; right R_0 = 10

Steadily rotating spirals in the diagram (E_{max} /A , D)

It is more convenient to draw the steady state curves for rotating spirals in the diagram (maximum relative rate of increase of the trigger variable c_1, pulse duration D) (Fig.7). This kind of diagram is well discussed in the Zykov book [8] and very useful as far as tip meandering problem is posed in the cardiologist context. For this, we take different values of ε and δ , determine the corresponding values of E_{max} /A and D and compute the hole radius R_0 for these values in a similar way as what was done in section II.a) . Then, curves of equiradius are drawn in the diagram (E_{max} /A, D) . As was found for the approximate solutions of Zykov [8], for a fixed hole radius, the maximum relative rate of increase of the trigger variable c_1 decreases when the pulse duration D increases. On this diagram, we draw the "validity line" of W.F.M. which limits the region where $0 < \varepsilon < .5$.

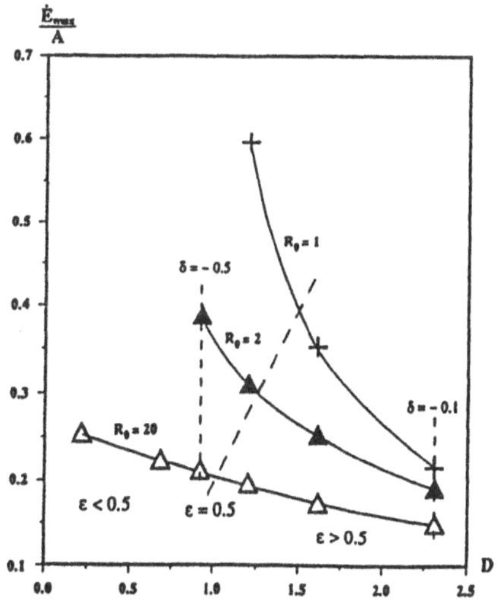

Fig.7 Steady state rotating spirals in the (E_{max} / A, D) diagram. The dashed lines limits the region where $0 < \varepsilon < .5$.

FIRST APPROACH FOR A LINEAR STABILITY ANALYSIS.

Before to perform the whole linear stability analysis involving perturbations of the two waves, it is more convenient to analyse a simpler situation [7]. It was observed from the analysis of the steady states that for an excitability δ slightly larger than a minimum value δ_{min} (ε) the radius of the central unexcited region (hole radius) is very large [8],[9]. Below this value spiral wave retracts and no steady rotation is possible. When the spiral tip rotates on a trajectory of large curvature radius, the time between two consecutive excitations is very large and the refractory fiels has time to come back to its steady value. Thus the front can be considered as moving in an unexcited medium at equilibrium. It is shown [8] that in this case, the structure of the wave is that of a pulse moving with a well determined normal velocity c terminating on a free end with zero normal velocity and constant tangential velocity c. It is known from experiments and numerical simulations that in this case, the uniform spiral rotation is stable. Even in this simple situation it is not evident to explain the reason for this stability. If at large distance from the tip it appears clear that perturbations of the spiral shape must be smoothed out because of the stabilizing effect of the curvature on the normal velocity, it is not so clear why , if the tip penetrates in the unexcited region (Fig.8) , it is repelled towards the steady tip trajectory.

Thus, we perform a linear stability analysis of the uniform steady rotation of an opened curve moving with a normal velocity linear function of the curvature and with a free end moving with zero normal velocity and constant tangential velocity c. As expected, we show that this uniform rotation is stable , i.e., all the eigenvalues of the stability spectrum are negative (Fig.9). Furthermore, as it is often the case for linear stability problems in semi-infinite space, this spectrum is found to be discrete.

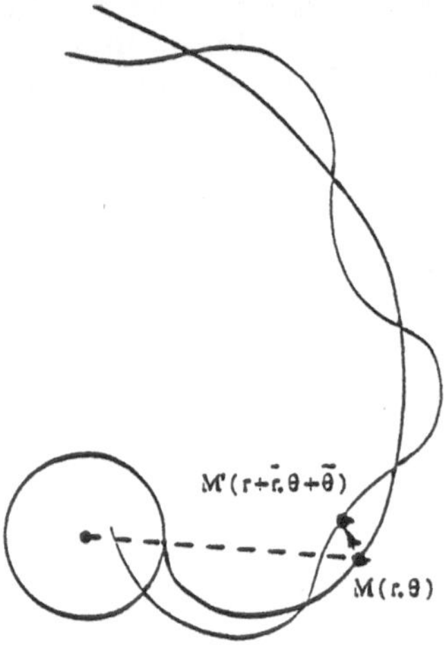

Fig.8: Sketch of a perturbation of the steadily rotating spiral.

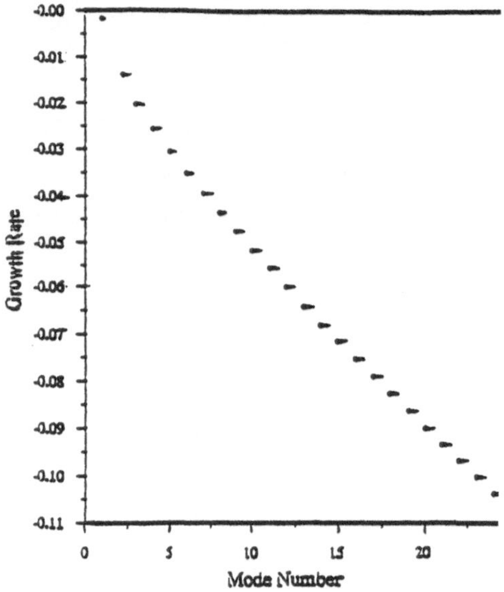

Fig.9: Spectum of growth rates

Acknowledgments: This research is supported by a collaborative research grant NATO.

REFERENCES

[1] T. PLESSER, S. MULLER and B. HESS, (1990) J.Phys.Chem.94, 7501.
[2] G.S. SKINNER and H.L. SWINNEY, (1990) To appear in Physica D.
[3] A. KARMA, (1990) Phys.Rev.Lett. 65, 2824.
[4] P.C. FIFE, (1984) in Non-Equilibrium Dynamics in Chemical systems, C.Vidal and A.Pacault eds, Springer Verlag.
[5] J.J. TYSON and J.P. KEENER, Physica D 32, 327.
[6] P.PELCE and J.SUN, (1991) Physica D 48, 353.
[7] P.PELCE and J.SUN, (1991) Submitted to Phys.Rev.A. Rapid.Com.
[8] V.S. ZYKOV, (1988) Modelling of wave processes in excitable media (Manchester University Press, Manchester).
[9] A.KARMA, (1991) Phys.Rev.Lett. To appear

KADOMTSEV-PETVIASHVILI AND (2+1)-DIMENSIONAL BURGERS EQUATIONS IN THE BÉNARD PROBLEM

R.A. KRAENKEL, S.M. KURCBART, J.G. PEREIRA
Instituto de Física Teórica
Universidade Estadual Paulista
0145-SAO PAULO, SP, BRESIL

and

M.A. MANNA
Laboratoire de Physique Mathématique
Université MONTPELLIER II, Sciences & Techniques du Languedoc
34095-MONTPELLIER CEDEX, FRANCE

ABSTRACT. We study the surface perturbation of a viscous fluid adequately heated from below. We shown that under appropriate perturbations to the static solution the system exhibit oscillatory instabilities governed by the Kadomtsev-Petviashvili equation or by the (2+1) dimensional Burgers equation.

I. INTRODUCTION

The system formed by a fluid heated from below, the so called Bénard problem, has been along the years a standard model for many studies in fluid dynamics [1]. In most cases, however, the main interest has been concentrated in convection phenomena. While considering the same system, our concern here will be quite different since we shall be interested in the study of surface waves and in situations for which the Rayleigh number R is well below that determined by the onset of convection. Furthermore, we shall only consider systems for which the upper boundary is a two-dimensional surface. Yet, we shall restrict ourselves to the study of long surface waves, on whose description the slow variables play a very important role. By using appropriated slow space and time variables we shown that nearly one-dimensional undamped waves, described by the Kadomtsev-Petviashvili[2] equation, may propagate in a shallow viscous fluid, provided the Rayleigh number of the system satisfy the condition $R = 30$. This extend the result obtained by Alfaro and Depassier[3], in (1+1) dimensions. Furthermore, it will be shown that changing appropriately the perturbation scaling and the slow variables, the evolution equation governing the surface displacement is the (2+1)-dimensional Burgers[4] equations provided the Rayleigh number satisfy the condition $R \neq 30$.

II. THE SYSTEM

Let us consider a Bénard system consisting of a fluid bounded below by a plane stress-free perfect thermally conducting medium at $z=0$ and temperature $T = T_b$ and above by a free surface which, at rest, lies at $z=d$. The depth d is such that the buoyancy effect is predominant when compared to the influence of the surface tension. This is the reason we assume a vanishing surface tension. The equations governing the hydrodynamical flow of a viscous fluid can be simplified considerably by using the Boussinesq approximation. The origin of this simplification is the smallness of the coefficient of thermal expansion γ. As for mostly situations of practical occurence γ is indeed small, ranging usually from $10^{-3}$to 10^{-4} , this approximation does not impose severe restrictions from the physical point of view. In this approximation, the equations describing the motion of a fluid are given by

$$\vec{\nabla}.\vec{V} = 0 \tag{1}$$

$$\rho_0 \frac{d\vec{V}}{dt} = - \vec{\nabla}p + \mu \nabla^2 \vec{V} + \vec{g}\rho \tag{2}$$

$$\frac{dT}{dt} = \kappa \nabla^2 T \tag{3}$$

$$\rho = \rho_0[1 - \gamma(T-T_0)] \tag{4}$$

where $d/dt = \partial/\partial t + \vec{V}.\vec{\nabla}$ is the convective derivative, $\vec{V} = (u,v,w)$ is the fluid velocity, and p is the pressure. The viscosity μ, thermal diffusivity κ, and coefficient of thermal expansion γ, are constant.T_0 and ρ_0 are a reference temperature and density, respectively.

On the upper free surface $z= d+\eta(x,y,t)$ the boundary conditions are[5]

$$\eta_t + u\eta_x + v\eta_y = w \tag{5}$$

$$(p-p_a)\eta_x-\mu \{2 u_x\eta_x - (u_z+w_x) + (u_y+v_x) \eta_y\} = 0 \tag{6}$$

$$p-p_a + \mu \{(w_x + u_z)\eta_x - 2w_z + (w_y+v_z)\eta_y\} = 0 \tag{7}$$

$$(p-p_a)\eta_y - \mu \{(v_x+u_y)\eta_x - (v_z+w_y) + 2v_y\eta_y\} = 0 \tag{8}$$

and

$$\hat{n} \cdot \vec{\nabla} T = - \frac{F}{k} \tag{9}$$

where $\hat{n}$ is the unit vector normal to the free surface, given by

$$\hat{n} = (-\eta_x, -\eta_y, 1)/N \quad ; \quad N = \left(1 + \eta_x^2 + \eta_y^2\right)^{1/2},$$

F is the normal heat flux, k is the thermal conductivity, and p_a is a constant pressure exerted on the upper free surface.

An important point is the dynamical boundary condition to be satisfied at the lower plane. We suppose that the sliding resistance between two portions of the fluid is much greater than between the fluid and the plane[6]. Under this condition, it is reasonable to assume a stress-free lower surface, which implies[1]

$$w = u_z = v_z = 0 \tag{10}$$

for $z = 0$.

The static solution to these equations depends only on the coordinate z and is given by

$$T_s = T_0 - \frac{F}{k}(z-d) \quad , \quad \rho_s = \rho_0\left[1 + \frac{\gamma F}{k}(z-d)\right]$$

$$p_s = p_a - g\,\rho_0 \left[(z-d) + \frac{\gamma F}{2k}(z-d)^2\right] \quad .$$

In order to get the dimensionless form of the equations, boundary conditions and static solutions, we adopt d as unit of length, d^2/κ as unit of time, $\rho_0 d^3$ as unit of mass, ant Fd/k as unit of temperature. Furthermore, we introduce three dimensionless parameters : the Prandtl number $\sigma = \mu/\rho_0\kappa$, the Rayleigh number $R = \rho_0 g\gamma F d^4/k\kappa\mu$ and the Galileo number $G = gd^3\,\rho_0^2/\mu^2$.

III. THE KADOMTSEV-PETVIASHVILI EQUATION

In order to obtain the Kadomtsev-Petviashvili (KP) equation we need introduce the following slow variables

$$\xi = \epsilon\,(x - ct) \quad ,$$
$$\zeta = \epsilon^2 y \quad ,$$
$$\tau = \epsilon^3 t \quad ,$$

with ϵ a small parameter, chosen so that the amplitude η of the surface perturbations is $O(\epsilon^2)$:

$$\eta = \epsilon^2\left(\eta_0 + \epsilon\eta_1 + \epsilon^2\eta_2 + \ldots\right) \quad .$$

Introducing now the expansions

$$u = \epsilon^2 \left(u_0 + \epsilon u_1 + \epsilon^2 u_2 + \ldots\right) \quad ,$$

$$v = \epsilon^3 \left(v_0 + \epsilon v_1 + \epsilon^2 v_2 + \ldots\right) \quad ,$$

$$w = \epsilon^3 \left(w_0 + \epsilon w_1 + \epsilon^2 w_2 + \ldots\right) \quad ,$$

$$P - P_s = P_0 + \epsilon\, P_1 + \epsilon^2\, P_2 + \ldots \quad ,$$

$$T - T_s = \theta_0 + \epsilon\, \theta_1 + \epsilon^2 \theta_2 + \ldots \quad ,$$

where all quantities are dimensionless, we can obtain an order by order solution to the equations (1)-(10). In order ϵ^0, the solution is given by

$$\theta_0 = P_0 = 0 \quad .$$

In order ϵ^1 it is

$$\theta_1 = P_1 = 0 \quad , \quad \eta_0 = f(\xi,\zeta,\tau)/c \quad ,$$

$$u_0 = f(\xi,\zeta,\tau), \; w_0 = -z\; f(\xi,\zeta,\tau) \quad ,$$

with $f(\xi,\zeta,\tau)$ an arbitrary function. In order ϵ^2 we find

$$\theta_2 = 0 \quad , \quad P_2 = \sigma^2 G\, \eta_0 \quad , \quad \eta_1 = g(\xi,\zeta,\tau)/c \quad ,$$

$$u_1 = g(\xi,\zeta,\tau) \quad , \quad v_0 = r(\xi,\zeta,\tau) \quad ,$$

$$w_1 = -\, z\; g_\xi(\xi,\zeta,\tau) \quad ,$$

with $g(\xi,\zeta,\tau)$ and $r(\xi,\zeta,\tau)$ arbitrary functions. At this order, the solubility condition implies

$$c^2 = \sigma^2\, G \quad .$$

In the next order the solution is given by

$$\theta_3 = \frac{1}{6}\, f_\xi\, (z^3 - 3z) \quad ,$$

$$P_3 = \frac{1}{24}\, \sigma\, R\; f_\xi(z^4 - 6z^2 + 5) - 2\sigma\, f_\xi + \sigma^2\, G\, \eta_1 \quad ,$$

$$u_2 = \frac{1}{24} f_{\xi\xi}(z^6 - 15 z^4 + 39z^2) + h(\xi,\zeta,\tau) \quad ,$$

$$v_1 = s(\xi,\zeta,\tau) \quad ,$$

$$w_2 = - \frac{1}{168} f_{\xi\xi\xi}(z^7 - 21 z^5 + 91 z^3) - z(h_\xi + r_\zeta) \quad ,$$

with $h(\xi,\zeta,\tau)$ and $s(\xi,\zeta,\tau)$ arbitrary functions. The solubility condition in this order determines the critical Rayleigh number $R_c = 30$. On the other hand, the boundary conditions yield now two relations among the arbitrary functions $f(\xi,\zeta,\tau)$, $h(\xi,\zeta,\tau)$ and $r(\xi,\zeta,\tau)$:

$$ch_\xi - c^2(\eta_2)_\xi = - f_\tau - 2 ff_\xi - \frac{71}{168} cf_{\xi\xi\xi} - cr_\zeta \quad , \tag{11}$$

$$r_\xi = f_\zeta \quad . \tag{12}$$

Finally, in order ϵ^4 the expression for θ_4, P_4 and the boundary conditions, yields the relation

$$ch_\xi - c^2(\eta_2)_\xi = f_\tau + \frac{30 + \sigma G}{\sigma G} ff_\xi + \frac{272\sigma - 15}{168} cf_{\xi\xi\xi} \quad . \tag{13}$$

The requirement of compatibility of eqs. (11)-(13) provides an evolution equations for the function $f(\xi,\zeta,\tau)$:

$$\left(f_\tau + \frac{3(10 + \sigma G)}{2\sigma G} ff_\xi + c\left(\frac{17}{21}\sigma + \frac{1}{6}\right) f_{\xi\xi\xi}\right)_\xi = - \frac{1}{2} cf_{\zeta\zeta} \quad .$$

By transforming the variables as

$$\xi \longrightarrow \Lambda^{-1} \xi \ , \ \zeta \longrightarrow \left(\frac{c}{6\Lambda}\right)^{1/2} \zeta \quad , \quad f \longrightarrow - \frac{4G\sigma}{(10+G\sigma)\Lambda} f \quad ,$$

with

$$\Lambda = \left[c \left(\frac{17}{21}\sigma + \frac{1}{6}\right)\right]^{-1/3}$$

we get

$$(f_\tau - 6 ff_\xi + f_{\xi\xi\xi})_\xi + 3 f_{\zeta\zeta} = 0 \quad . \tag{14}$$

which is the KP equation[7].

The solitary-wave solution of (14) is given by[8]

$$f(\xi,\zeta,\tau) = A\ \mathrm{sech}^2\left\{\left[\frac{1}{2}(m-\ell)\ \xi - \frac{1}{2}(m^2-\ell^2)(6/\Lambda c)^{1/2}\zeta - 2(m^3-\ell^3)\ \tau/\Lambda\right]\Lambda-\beta\right\} \qquad (15)$$

with m and ℓ real numbers, $\beta = \frac{1}{2}\ \ell n\left(-\frac{m}{\ell}\right)$, and

$$A = \frac{2G\sigma}{\Lambda(10 + G\sigma)}\ (m-\ell)^2 \quad .$$

If $\ell \neq -m$, eq. (15) represents an oblique solitary wave which moves at a certain angle to the ξ-axis and does not decrease along the direction defined by the equation

$$\xi = (\ell+m)(6/\Lambda c)^{1/2}\zeta \quad ,$$

as $\xi, \zeta \longrightarrow \infty$. If, however, $\ell = -m$, the solution (15) is converted into the Korteweg-de Vries soliton. In this way we see that, the solitary-wave solution of the KP equation describes a wave whose pattern consists of a horizontal streamline. These solitary waves, associated with oscillatory instabilities, are sustained by the adverse temperature gradient applied and just come into play because the amount of energy released by buoyancy exactly compensates the amount dissipated by viscosity.

IV. THE (2+1)-DIMENSIONAL BURGERS EQUATION

Here, we shall obtain a non-linear evolution of the surface displacement governed by the (2+1)-dimensional Burgers equation. To this end let us consider surface perturbation and slow variables given by

$$\eta = \epsilon\ \left(\eta_0 + \epsilon\ \eta_1 + \epsilon^2\ \eta_2\right) \quad ,$$

$$\xi = \epsilon(x - ct) \quad ,$$

$$\zeta = \epsilon^{3/2}y \quad ,$$

$$\tau = \epsilon^2\ t \quad .$$

Furthermore, we introduce the expansions

$$u = \epsilon\ \left(u_0 + \epsilon\ u_1 + \epsilon^2\ u_2 + \ldots\right) \quad ,$$

$$v = \epsilon^{3/2}\left(v_0 + \epsilon\ v_1 + \epsilon^2\ v_2 + \ldots\right) \quad ,$$

$$w = \epsilon^2\left(w_0 + \epsilon\ w_1 + \epsilon^2 w_2 + \ldots\right) \quad ,$$

$$P-P_s = P_0 + \epsilon\ P_1 + \epsilon^2\ P_2 + \ldots \quad ,$$

$$T-T_s = \theta_0 + \epsilon\, \theta_1 + \epsilon^2\, \theta_2 + \ldots \quad .$$

In the lowest order, the solution of equation (1)-(10) is given by

$$\theta_0 = 0 \quad , \quad P_0 = 0 \quad , \quad u_0 = f(\xi,\zeta,\tau) \quad ,$$

$$w_0 = -z\, f_\xi(\xi,\zeta,\tau) \quad ,$$

with $f(\xi,\zeta,\tau)$ an arbitrary function. In the next order it is

$$\eta_0 = \frac{1}{c} f \quad , \quad v_0 = h(\xi,\zeta,\tau) \quad ,$$

$$\theta_1 = 0 \quad , \quad P_1 = G\sigma^{\,2}\eta_0 \quad ,$$

$$u_1 = g(\xi,\zeta,\tau) \quad , \quad w_1 = -z(g_\xi + h_\zeta) \quad .$$

with $g(\xi,\zeta,\tau)$ and $h(\xi,\zeta,\tau)$ arbitrary functions.
In the order ϵ^2 we get

$$\theta_2 = \frac{1}{6} f_\xi (z^3 - 3z) \ , \ P_2 = \left[\frac{R\sigma}{24}\ (z^4 - 6z^2 + 5) - 2\sigma\right] f_\xi + \frac{R\sigma}{2}\eta_0^2 + G\sigma^2\, \eta_1 .$$

The boundary condition on the upper free surface yield the equation

$$c^2\eta_{1\xi} - cg_\xi = f_\tau + 2\, ff_\xi + c\, h_\zeta \quad . \tag{16}$$

At this order, there appear a solubility condition giving

$$c^2 = G\sigma^2 \quad .$$

In the next order the expressions for θ_3, P_3 and the remaining boundary condition yield the equation

$$- c^2\, \eta_{1\xi} + cg_\xi = f_\tau + \left[1 + \frac{R}{G\sigma}\right] ff_\xi - \sigma \left[4 - \frac{2R}{15}\right] f_{\xi\xi} , \tag{17}$$

as well as a relation between the arbitrary functions $f(\xi,\zeta,\tau)$ and $h(\xi,\zeta,\tau)$

$$f_\zeta = h_\xi \quad . \tag{18}$$

The requirement of compatibility of Eqs (16) (17) and (18) provides an evolution equation for f :

$$\left[f_\tau + \left(\frac{3G\sigma + R}{2G\sigma}\right) ff_\xi + \nu\, f_{\xi\xi}\right] = -\frac{c}{2} f_{\zeta\zeta} \quad ,$$

where

$$\nu = 2\sigma - \frac{R\sigma}{15} \quad .$$

By transforming f according to

$$f \longrightarrow \frac{2G\sigma}{3G\sigma+R} f \quad ,$$

we obtain the equation

$$(f_\tau + ff_\xi - \nu f_{\xi\xi})_\xi = -\frac{c}{2} f_{\zeta\zeta} \quad . \tag{19}$$

This is the (2+1)-dimensional Burgers equation. The complete integrability of (19) is currently under investigation[9]. Eq (19) has a progressive wave solution of the form

$$f(\Lambda) = f(\mathbb{A}\xi + \mathbb{B}\zeta - \mathbb{C}\tau) \quad ,$$

whose explicit form is

$$f(\Lambda) = \frac{(3G\sigma+R)(2\mathbb{C}-c\mathbb{B}^2)}{4G\sigma} \left\{1 - \text{tgh}\left[\frac{(2\mathbb{C}-c\mathbb{B}^2)}{4\nu}(\Lambda-\Lambda_0)\right]\right\} \quad .$$

where $\mathbb{A},\mathbb{B},\mathbb{C}$ are constants depending on the initial condition. It represents a nearly one-dimensional Burgers shock wave. This shock wave is propagated without any overtaking and it is substained by the adverse temperature gradient provided the Rayleigh number of the system satisfy the condition $R \neq 30$. As the Rayleigh number approaches the point $R = 30$, the coefficient ν approaches to zero, nonlinearity is not compensated by dissipation and the wave begins to break.
It should be noticed, however, that this solution is not valid for an arbitrary large R, since in this region other phenomena which are not considered by our approach, may take place.

V. CONCLUSIONS

We have extended a (2+1) dimensions the result that a solitary wave may propagate in a viscous fluid subject to an adverse temperature gradient. We found that the surface displacement obey the Kadomtsev-Petviashvili equation.

On the other hand we shown that a much larger surface perturbation will be governed by the (2+1) dimensional Burgers equation. We predict the existence of a nearly one dimensional kink, provides the Rayleigh number satisfy the condition $R \neq 30$.

REFERENCES

1. S. CHANDRASEKHAR, Hydrodynamics and Hydromagnetic Stability (Clarendon, Oxford, 1955).

2. B.B. KADOMTSEV and V.I. PETVIASHVILI, Soviet Physics Dokl. 15, (1970), 539.

3. C.M. ALFARO and M.C. DEPASSIER, Physics Review Letters 62, (1989), 2597.

4. M. BERTUCCELLI, P. PANTANO and T. BRUGARINO, Lettere Nuovo Cimento, 37, (1987), 433.

5. J.V. WEHAUSEN and E.V. LAITONE in Encyclopaedia of Physics, Vol. 9, Ed. S. Flugge (Springer, Berlin, 1960).

6. G.G. STOKES, Trans. Cambridge Philos. Soc. 8, (1845), 278 ; reprinted in : Mathematical and Physical Papers. Vol. 1 (Johnson Reprint Corporation, New-York, (1966), 75).

7. S.M. KURCBART, M.A. MANNA, J.G. PEREIRA and A.N. GARAZO, Physical Letters A. 148, (1990), 53.

8. M. JAULENT, M.A. MANNA and L.M. ALONSO, Inverse Problems 5, (1989) 573.

9. G.M. WEBB and C.P. ZAUK, Physics Letters A 150, (1990), 14.

PART V

THEORETICAL PHYSICS

NON-LINEARITY AND COHERENCE IN MODELS OF SUPERCONDUCTIVITY

J.M. Dixon
Department of Physics, University of Warwick
Coventry, CV4 7AL, U.K.
and
J. A. Tuszyński
Department of Physics, University of Alberta
Edmonton, Alberta, T6G 2J1, Canada

Theoretical investigations of standard low-temperature superconductivity frequently proceed in one of two main ways. The first begins with the well-known BCS Hamiltonian which may be written as

$$H^{BCS} = \sum_{\underline{k}\sigma} \omega_{\underline{k}}\, a^{\dagger}_{\underline{k}\sigma} a_{\underline{k}\sigma} + \sum_{\substack{\underline{k},\underline{\ell},\underline{m} \\ \sigma,\sigma'}} \Delta_{\underline{k}\,\underline{\ell}\,\underline{m}}\, a^{\dagger}_{\underline{k}\sigma} a^{\dagger}_{\underline{k}\sigma'} a_{\underline{m}\sigma'} a_{\underline{k}+\underline{\ell}-\underline{m}\sigma} \,. \qquad (1)$$

In equation (1) the one-body part of the Hamiltonian is assumed diagonal in the plane-wave basis, each vector being labelled by a wave-vector $\underline{k}$, $\underline{\ell}$ or $\underline{m}$. The two-body term is assumed to be attractive in nature, the vectors on annihilators and creators being arranged to conserve linear momentum. The labels σ and σ' denote the components of spin, the total spin on each particle being $S = 1/2$. One standard approach is to use Bogoliubov-Valatin transformations [1] to diagonalise H^{BCS}, so that the effective one-body term exhibits explicitly a gap representing the minimum energy required to create an excitation in the system. Both this transformation and subsequent approximations are guided by the knowledge, obtained originally variationally [2], that in the ground state electrons are formed into Cooper pairs with their wavevectors and spin components equal and opposite.

The second approach is to take a Landau-Ginzburg (LG) view and write down a free energy density expansion [3] in the form

$$G_S(\underline{r}) = G_N(\underline{r}) + A(T)\,|\psi(\underline{r})|^2 + \frac{1}{2} C|\psi(\underline{r})|^4 + \frac{\hbar^2}{2m^*}|\nabla\psi(\underline{r})|^2 \;, \qquad (2)$$

where $A(T) = \overline{a}\ (T-T_c)$, $\overline{a}$ being a constant and $T-T_c$ denoting the temperature difference from the critical temperature $T=T_c$. Here, $\psi(\underline{r})$ is the order parameter and $|\psi(\underline{r})|^2$ measures the density of superconducting electron pairs and hence the local degree of superconductivity. $G_N(\underline{r})$ is the free energy density of the normal state, m* is an effective mass and C a parameter which is assumed to be such that $C>0$. At this stage, despite the work of Gor'kov [4], this route is usually considered to be phenomenological but at the time this was put forward it constituted a break through in the understanding of critical phenomena [5]. Thus, the first approach uses a number of approximations to incorporate Cooper pairs and an energy gap and the second appears to be phenomenological. However, the present authors [6] have recently developed a novel approach, initially without spin components, using the Hamiltonian

$$H_{TD} = \sum_{\underline{k},\underline{\ell}} \omega_{\underline{k},\underline{\ell}} q_{\underline{k}}^{\dagger} q_{\underline{\ell}} + \sum_{\underline{k},\underline{\ell},\underline{m}} \Delta_{\underline{k},\underline{\ell},\underline{m}} q_{\underline{k}}^{\dagger} q_{\underline{\ell}}^{\dagger} q_{\underline{m}} q_{\underline{k}+\underline{\ell}-\underline{m}} \ . \tag{3}$$

The idea was to first write down Heisenberg's equations of motion for the annihilators and creators in equation (3). One notices immediately that these rate equations have the same form whether the q_η or $(q_\eta^\dagger)$'s describe Bosons or Fermions. The next step is to define a quantum field by

$$\psi(\underline{r}) = \Omega^{-1/2} \sum_{\underline{k}} e^{-i\underline{k}\cdot\underline{r}} q_{\underline{k}} \quad , \tag{4}$$

and to rewrite the equations of motion in terms of this field alone. This latter part of the procedure is not easy because the parameters or matrix elements in (3) are functions of the subscripted wave-vectors. However, if these elements are expanded as a series about some suitable point in $\underline{k}$-space then clearly the equations of motion may be written in terms of ψ, ψ^* and their gradients. In general, there will be an infinite number of terms as a consequence of this procedure. If the point in $\underline{k}$-space is chosen as a critical point of the system, then it is well known [7] that close to this point the order parameter ψ will be predominantly classical, corrections being very small and of order $\hbar$. It is then only necessary to perform the expansion in $\underline{k}$-space up to second order because Renormalisation Group Theory tells us that we only need to include terms up to ψ^n in the Hamiltonian, in N-dimensional space time [8], where $n = 2N/(N-2) = 4$ – or terms in ψ^3 or its equivalent in the equations of motion. Thus, the form of the second order equations is virtually exact and higher order terms merely redress those in lower orders. This general equation of motion takes the highly non-linear form [6]

$$i\hbar\partial_t\psi = \lambda_o\psi + i\underline{\lambda}_1\cdot(\nabla\psi) - \frac{1}{2}\sum_{i,j}(\underline{\lambda}_2)_{ij}\,\partial^2_{x_ix_j}\psi + v_2\psi^\dagger\psi\psi$$

$$+\,\Omega\,[i\,(\nabla_{\underline{n}}f)_o\cdot\psi^\dagger\psi\nabla\psi + i\,(\nabla_{\underline{m}}f)_o\cdot\psi^\dagger(\nabla\psi)\psi - i\,(\nabla_{\underline{k}}f)_o\cdot(\nabla\psi^\dagger)\psi\psi]$$

$$+\frac{\Omega}{2!}\sum_{i,j}(\partial^2_{m_i m_j} f)_o \{\psi^\dagger [-\partial^2_{x_i x_j}\psi - 2i(\partial_{x_i}\psi)m_j^o + \psi m_i^o m_j^o]\psi$$

$$+\psi^\dagger\psi[-\partial^2_{x_i x_j}\psi - 2i(\partial_{x_i}\psi)\eta_j^o + \eta_i^o\eta_j^o\psi] + 2\psi^\dagger[(i\partial_{x_i}\psi - m_i^o\psi)(i\partial_{x_j}\psi - \eta_j^o\psi)]\}. \quad (5)$$

What is perhaps so surprising is that, if H_{TD} is itself written in terms of the field ψ a standard Landau-Ginzburg form is retrieved. Furthermore, terms in $\psi(\nabla\psi^\dagger)$, $(\nabla^2\psi^\dagger)\psi$, $(\nabla^2\psi^\dagger)\psi^\dagger\psi\psi$ and $(\nabla\psi^\dagger)(\nabla\psi^\dagger)\psi\psi$, which were actually considered and investigated by Landau, also appear! The equations of motion for ψ turn out to be the Euler-Lagrange equations obtained by finding the extrema, relative to ψ^*, of the Hamiltonian functional, equation (3) being considered as a Hamiltonian density when written in terms of the classical field.

What, therefore, is so significant is that in the first approach above it is not necessary to make approximations; we can retain the full generality of the second quantised form and the second approach is not phenomenological at all but a work of true genius.

So far this new approach has not incorporated spin at all. At this stage this will be omitted and to focus attention on standard low temperature superconductors we put

$$\omega_{\underline{k},\underline{\ell}} = \left(\frac{\hbar^2 k^2}{2m^*} - E_F\right)\delta_{\underline{k},\underline{\ell}} \quad , \quad (5a)$$

and assume

$$\Delta_{\underline{k},\underline{\ell},\underline{m}} = -V_o\,\delta_E \quad , \quad (5b)$$

as is customarily done in BCS-type theories. In equation (5) E_F is the Fermi energy, V_o is a constant energy and the symbol δ_E is zero unless the kinetic energies are within $\hbar\omega_D$ (where ω_D is the Debye angular frequency). Using standard methods in quantum field theory [7], near the critical point we describe the field ψ by

$$\psi = \phi + \Lambda \quad ,$$

where ϕ is a large classical envelope and Λ is the smaller quantum component and, as a first approximation in the equations of motion, drop the quantum component Λ. In zeroth order the equations of motion reduce to

$$i\hbar\partial_t\phi = -E_F\,\phi + \frac{\hbar^2}{2m^*}\nabla^2\phi - 2\Omega V_o\phi^*\phi\phi \quad , \quad (6)$$

which is a non-linear Schrödinger equation in three-dimensional space. This equation is integrable in 1+1 dimensions, as is well known, and among its solutions are stable solitons. It has also been studied in depth using the Symmetry Reduction method [8]. Analytical results appear to be only found when the level surfaces of the symmetry variable ξ correspond to planar, cylindrical and spherical manifolds. Asymptotic behaviour for some of these geometries corresponds to solutions of the quasi-linear equation and for this reason we confine our attention to these. Writing the classical field ϕ as

$$\phi = \exp\left(+i\frac{Et}{\hbar}\right)\exp(i\chi)\,\eta \quad , \tag{7}$$

substituting into (6), and separating real and imaginary parts results in

$$A\eta + C\eta^3 - \frac{\hbar^2}{2m^*}\nabla^2\eta + \frac{\hbar^2}{2m^*\eta^3}C_1^3 = 0 \tag{8}$$

where $A = E_F - E$, $C = 2\Omega V_0$ and C_1 defines the magnitude of the superconducting current, $\underline{j}_s$.

When there are <u>no currents</u>, corresponding to the case when $C_1=0$, there are two main categories of solution to consider. Firstly, when $m^*>0$ the lowest exact solution is a mean field and if $A>0$ this describes the normal state whereas if $A<0$ we obtain the ordered superconducting state. In the case when $A<0$, just below the energy of the normal phase, one finds, with this approach, a discrete ladder of one-, two- etc. up to N-soliton states. If one assumes that the physical situation may be described by a nearly free soliton gas, then one can show that the N-soliton condensate is separated from the normal phase by a gap Δ. If, as is the case in the LG picture, the constant A is temperature dependent through the usual relation

$$A = \bar{a}\,(T - T_c) \quad , \tag{9}$$

then we find that the gap scales with temperature as

$$\Delta \sim (T - T_c)^{1/2} \quad , \tag{10}$$

exactly as it does for standard low temperature superconductors! In fact, between the mean field energy and disordered phase one finds a continuum of snoidal and dnoidal (of the Jacobi elliptic type) waves, the latter representing thermodynamically stable fluctuations. As the temperature approaches the critical point from above the normal phase becomes destabilised so that for $T<T_c$ soliton

energy levels become available to the system and dnoidal waves execute small oscillations about the mean superconducting energy.

Still describing the situation when $m^*>0$, but when superconducting currents $\underline{j}_s \neq 0$ are present, we may still solve the non-linear equation of motion exactly and above T_c all the solutions become unstable. However, below T_c, $\underline{j}_s$ breaks the topological solitons (or kinks) into pairs of bumps and the elliptic solutions become distorted. At a critical value of the current $\underline{j}_c$ both elliptic waves and bump solitons below them disappear and the superconducting state is destroyed. Remarkably, this critical condition enables us to deduce the correct scaling of the current with temperature as

$$\underline{j}_s \sim |T - T_c|^{3/2} \quad . \tag{11}$$

This is again as in standard superconductors [3].

The other main case is when $m^*<0$. This may arise due to band structure effects, re-dressing from higher order terms but in any case can happen if transport is by holes in an approximately two-dimensional system [9] as is the situation for high-temperature superconductivity. This situation is very different from that for $m^*>0$ and results in a ground state which is strongly modulated in space (cn-waves) and has a critical current whose square has a cubic dependences on $T\text{-}T_c$ [10].

All the results above are consequences of the model in the absence of spin so it is legitimate to ask what difference its inclusion would produce. Following a similar procedure as outlined above, the starting point would be the analogue of equation (3) with spin components introduced, namely

$$H_1 = \sum_{\substack{\underline{k},\underline{\ell} \\ \sigma}} \omega_{\underline{k},\underline{\ell}}\, q^{\dagger}_{\underline{k}\sigma} q_{\underline{\ell}\sigma} + \sum_{\underline{k},\underline{\ell},\sigma',\sigma} \Delta_{\underline{k},\underline{\ell},\underline{m}}\, q^{\dagger}_{\underline{k}\sigma} q^{\dagger}_{\underline{\ell}\sigma'} q_{\underline{m}\sigma'} q_{(\underline{k}+\underline{\ell}-\underline{m})\sigma} \quad . \tag{12}$$

In equation (12) we have assumed for simplicity that the one- and two-body operators from which ω and Δ arise do not depend on spin. The labels σ, σ' refer to the components of spin, not necessarily for a spin $S=1/2$, and $q_{k\sigma}$ or $(q^{\dagger}_{\underline{k}\sigma})$'s can refer to either Bosons or Fermions at this stage. If we take the case where the spins do refer to electrons, as an example, and denote spin components $\sigma = +1/2$ by '+' and $\sigma = -1/2$ by '−', then the Heisenberg equation of motion for $q_{\eta +}$ is given by

$$i\hbar\partial_t q_{\underline{n}+} = \sum_{\underline{k}} \omega_{\underline{n},\underline{k}}\, q_{\underline{k}+} + \sum_{\underline{k},\underline{m}} \{\Delta_{\underline{n}\,\underline{k}\,\underline{m}} q^{\dagger}_{\underline{k}+} q_{\underline{m}+} q_{(\underline{n}+\underline{k}-\underline{m})+} - \Delta_{\underline{k}\,\underline{n}\,\underline{m}} q^{\dagger}_{\underline{k}+} q_{\underline{m}+} q_{(\underline{k}+\underline{n}-\underline{m})+}$$

$$+ \Delta_{\underline{n}\,\underline{k}\,\underline{m}} q^{\dagger}_{\underline{k}-} q_{\underline{m}-} q_{(\underline{n}+\underline{k}-\underline{m})+} - \Delta_{\underline{k}\,\underline{n}\,\underline{m}} q^{\dagger}_{\underline{k}+} q_{\underline{m}-} q_{(\underline{k}+\underline{n}-\underline{m})+}\} \quad , \qquad (13)$$

with a corresponding equation for $i\hbar\,\partial_t q_{\underline{n}-}$, obtained from (13) by interchanging '+' and '−'. These latter two equations may now be replaced by field equations involving, in this simple example, two spin-dependent fields defined by

$$\psi_+(\underline{r}) = \Omega^{-1/2} \sum_{\underline{k}} e^{-i\underline{k}\cdot\underline{r}}\, q_{\underline{k}+} \;, \qquad \psi_-(\underline{r}) = \Omega^{-1/2} \sum_{\underline{k}} e^{-i\underline{k}\cdot\underline{r}}\, q_{\underline{k}-} \qquad (14)$$

using a similar procedure to that in the spinless field case. The field equation for ψ_+ may be obtained by making the following replacement in equation (5)

a) $q_{k\sigma} \rightarrow \psi_\sigma$

b) $k_x q_{k\sigma} \rightarrow \partial_{x_1}\psi_\sigma$ (15)

c) $\psi^{\dagger}\psi\,\nabla\psi \rightarrow \psi^{\dagger}_{+}\psi_+\psi_+ + \frac{1}{2}\psi^{\dagger}_{-}\psi_-\nabla\psi_+ + \frac{1}{2}\psi^{\dagger}_{-}\psi_+\nabla\psi_-$

d) $\psi^{\dagger}[\partial^2_{x_ix_j}\psi]\psi \rightarrow \psi^{\dagger}_{+}[\partial^2_{x_ix_j}\psi_+]\psi_+ + \frac{1}{2}\psi^{\dagger}_{-}[\partial^2_{x_ix_j}\psi_-]\psi_+ \;\frac{1}{2}\psi^{\dagger}_{-}[\partial^2_{x_ix_j}\psi_+]\psi_-$

and similar replacements by permutation of co-ordinates. An equation for the ψ_- field may also be obtained in this way but with '+' replaced by '−' in c) and d) of (15) and vice versa. A full calculation will confirm the relationships in (15). What one notices on writing down these equations of motion is that those terms which couple the ψ_+ and ψ_- fields together may all be written in terms of brackets of the form

$$[\psi_-\psi_+ + \psi_+\psi_-] \quad . \qquad (16)$$

However, these all vanish due to the commutation relations for the quantum fields! Hence the equation of motion for ψ_+ becomes identical in form to the spinless case, with ψ_+ replaced by ψ. In a similar way the equation for ψ_- is also, on putting $\psi_- \rightarrow \psi$, exactly the same as in the spinless case. Hence, for Bosons with spin S=0 and Fermions with S=1/2 the equations of motion are of the <u>same</u> form, despite the fact that they apparently contain very complicated couplings between the spin component fields!

Even when the spin is greater than one-half for Fermions, a very similar argument, when the one-body and two-body terms involve sums over spin components, may be used to show that, for Fermions, for any total spin, the equation of motion for a field associated with one component of the spin is identical to that for any other component and has the same form as in the spinless case. This does take rather a lot of tedious algebra but is a remarkable result. The Boson case for $S \neq 0$, is a little more complicated but in this, although the equation of motion for any spin component has the same form, it is not identical to the spinless case and the two-body interaction terms become scaled by the spin degeneracy, $2S+1$. Thus, this result might lead to higher critical temperatures if the mechanism for superconductivity involved quasi-particles with a larger effective spin. In this connection we remark that the transition temperature for ^4He is much higher than in ^3He, provided we compare the two under the same conditions of pressure so this would agree with our deliberation above [11].

What we have presented in this paper is a nonlinear field-theoretic approach to superconductivity which is applicable to situations where a single phase exists. Depending on the case we may obtain a normal phase, a homogeneous superconducting phase, or indeed a host of modulated superconducting phases with periodically distributed regions of high and low concentration of superconducting charge. The regions with low concentrations of charge density will be easily penetrable by magnetic flux lines and may result in the formation of a flux lattice. Thus, this type of description appears fairly suitable for both type I and type II superconductors. However, in the case of the new ceramic type of superconductor the spatial ordering may be somewhat more complicated and the role of defects, twin boundaries and structural disorder should not be underestimated. Admittedly then, the model we have presented in this work would only be suitable for a single grain and interactions between the neighbouring domains of superconductivity would have to be modelled independently through the inclusion of Lawrence-Doniach terms, for example. We have made a first step in the direction of extending the present approach to embrace ceramic superconductors [12]. Interesting phenomena appear to be already incorporated, for instance the associated vortex arrays form from the order parameter phases of the domains and may or may not be commensurate with the structure of the island or domain envelopes. This may explain the experimentally observed glassy features of these materials [13].

REFERENCES

[1] P.L. Taylor, A Quantum Approach to the Solid State (Prentice-Hall, Englewood Cliffs, NJ, 1970).

[2] J. Bardeen, L.N. Cooper and J.R. Schrieffer, Phys. Rev. 106 (1957) 179.
[3] R.H. White and T. Geballe, Solid State Phys., Long-Range Order in Solids (Academic Press, New York, 1979).
[4] L.P. Gor'kov, Sov. Phys. J.E.T.P. 9 (1959) 1364.
[5] L.D. Landau and E.M. Lifshitz, Statistical Mechanics (Pergamon, Oxford, 1969).
[6] a) J.A. Tuszyński and J.M. Dixon, J. Phys. A: Math. Gen. 22 (1989) 4877.
b) J.M. Dixon and J.A. Tuszyński, J. Phys. A: Math. Gen. 22 (1989) 4895.
[7] R. Jackiw, Rev. Mod. Phys. 49 (1977) 681.
[8] L. Gagnon and P. Winternitz, J. Phys. A 21 (1988) 1493.
[9] L. Eaves, University of Nottingham, U.K., private communication (1991).
[10] J.A. Tuszyński and J.M. Dixon, Physica C 161 (1989) 687.
[11] S.J. Putterman, Superfluid Hydrodynamics (Elsevier, New York, 1974).
[12] M. Otwinowski, J.A. Tuszyński and J.M. Dixon, J. Phys.: Condensed Matter 2 (1990) 6381.
[13] I. Morgenstern, K.A. Müller and J.G. Bednorz, Physica C 153-155 (1988) 67.

CHAOTIC POLAROINC and BIPOLARONIC STATES IN COUPLED ELECTRON-PHONON SYSTEMS

Serge AUBRY, Gilles ABRAMOVICI and Jean-Luc RAIMBAULT
Laboratoire Léon Brillouin[1] , CEN Saclay -91191-Gif-Sur Yvette Cédex (France)

For large electron-phonon coupling, many models describing the electron-phonon coupling have a limit called anti-integrable with trivially chaotic states. For example, let us consider the Holstein Hamiltonian which is the sum of three terms which are

(1-a) $$H = H_k + H_{ep} + H_p$$

1- H_k is an electronic Hamiltonian

(1-b) $$H_k = -T \sum_{<i,j>,\sigma} c^+_{i,\sigma} c_{j,\sigma}$$

which corresponds to a band of electrons (described with standard Fermion operators $c^+_{i,\sigma}$ and $c_{i,\sigma}$) propagating on an arbitrary lattice $\mathbb{L}$ periodic or not (<i,j> denotes neighboring sites and σ the spin of the electron which can be $\uparrow$ or $\downarrow$)

2- H_p is the phonon Hamiltonian

(1-c) $$H_p = \sum_i \hbar \omega_0 \left(a^+_i a_i + \frac{1}{2} \right)$$

which corresponds to Einstein oscillators with frequency ω_0 at the sites i of the lattice $\mathbb{L}$ (a^+_i and a_i are standard Boson operators) and

3- H_{ep} is the electron-phonon coupling Hamiltonian

(1-d) $$H_{ep} = g \sum_i n_i \left(a^+_i + a_i \right)$$

---

[1] Laboratoire commun CEA-CNRS

where

(1-e) $$\mathbf{n}_i = c_{i\uparrow}^{+} c_{i\uparrow} + c_{i\downarrow}^{+} c_{i\downarrow}$$

It couples the electronic density operator $\mathbf{n}_i$ at site i with the oscillator position. Choosing for this operator (in appropriate units) as

(2-a) $$u_n = \frac{\hbar \omega_0}{4g} (a_n^{+} + a_n)$$

and its conjugate momentum

(2-b) $$p_n = \frac{2g}{\hbar \omega_0} i (a_n^{+} - a_n)$$

and choosing

(2-c) $$E_0 = \frac{8 g^2}{\hbar \omega_0}$$

as the unit of energy, this Hamiltonian **H** becomes the sum of three terms

(3-a) $$\hat{H} = \frac{H}{E_0} = H_{AI} + t H_K + \beta H_Q$$

where

(3-b) $$H_{AI} = \sum_i \frac{1}{2} \left(u_i^2 + n_i u_i \right)$$

(3-c) $$H_K = -\frac{1}{2} \sum_{<i,j>, \sigma} c_{i,\sigma}^{+} c_{j,\sigma}$$

(3-d) $$H_Q = \frac{1}{2} \sum_i p_i^2$$

This Hamiltonian contains two dimensionless parameters which are

(4-a) $$t = \frac{T \hbar\omega_0}{4 g^2} > 0$$

and

(4-b) $$\beta = \frac{1}{4} \left(\frac{\hbar \omega_0}{2g} \right)^4$$ respectively.

This new formulation of the initial Hamiltonian is specially interesting for making appearent two important physical limits which are the (standard) adiabatic limit and the anti-integrable limit. Up to now, the importance of this second limit for understanding the physical behavior of the model, was not recognized in the litterature although it has been already considered implicitly but unperfectly in few articles.

Adiabatic Limit: The standard adiabatic limit is obtained when β is zero. This approximation is valid when

(5-a) $$g >> \hbar\omega_0$$

that is for large enough electron coupling g. In that case, the effect of the quantum lattice fluctuations is negligible. At the adiabatic limit, the problem becomes variational since operator $\mathbf{u}_i$ commutes with the Hamiltonian and can be considered as a scalar $u_i = <\mathbf{u}_i>$. Note that the adiabatic Hamiltonian obtained for β=0, is equivalent to the standard meanfield Hamiltonian which is the selfconsistent Hamiltonian obtained by replacing operator $\mathbf{u}_i\mathbf{n}_i$ by $\mathbf{u}_i<\mathbf{n}_i> + <\mathbf{u}_i> \mathbf{n}_i - <\mathbf{u}_i><\mathbf{n}_i> = \mathbf{u}_i\, n_i + u_i\, \mathbf{n}_i - u_i\, n_i$ in (3-b). This approximation neglects the fluctuation $(\mathbf{u}_i - u_i)(\mathbf{n}_i - n_i)$ which becomes zero only when β=0.

For the adiabatic Hamiltonian, the eigenstates are characterized by the atomic configurations $\{u_i\}$ which are obtained as local minima of the variational form

(5-b) $$\Phi(\{u_n\}) = \sum_\nu \sigma_\nu E_\nu(\{u_n\}) + \sum_i \frac{1}{2} u_i^2$$

where $E_\nu(\{u_n\}))$ are the eigenenergies of the tight-binding Schroedinger equation

(6-a) $$- t \; (\bar{\bar{\Delta}}\Psi^\nu)_n + u_n \Psi_n^\nu = E_\nu(\{u_i\}) \, \Psi_n^\nu$$

($(\bar{\bar{\Delta}}\Psi)_n = \sum_{m \mathcal{V} n} \Psi_m$ is the sum over the neighboring sites m to n on the lattice denoted $m\mathcal{V}n$).This eigenequation also determines the electronic eigenstates $\{\Psi_n^\nu\}$ for the atomic configuration $\{u_n\}$. σ_ν is the population number of state ν. We have

(6-b) $$\sigma_\nu = < c^+_{\nu,\uparrow} c_{\nu,\uparrow} > + < c^+_{\nu,\downarrow} c_{\nu,\downarrow} > = 0 , 1 \text{ or } \frac{1}{2}$$

It can be chosen according to the standard Fermi rule

(7-a) $$\sigma_\nu = 1 \qquad \text{for} \qquad E_\nu(\{u_n\}) \leq E_F$$

and

(7-b) $$\sigma_\nu = 0 \qquad \text{for} \qquad E_\nu(\{u_n\}) > E_F$$

when the electrons are in the ground-state determined by the lattice potential. We also consider situations where the electrons are in excited states as in polaronic or mixed polaronic-bipolaronic states.

Anti-Integrable Limit: When considering large electron-phonon coupling g, the relevant limit is obtained for t=0 (and β=0). This condition is fulfilled when

(8-a) $$g >> \sqrt{T\ \hbar\omega_0}$$

the model becomes trivially soluble since the Hamiltonian **H** (or $\hat{\mathbf{H}}$) commutes with the electronic density operators $c^+_{i\uparrow} c_{i\uparrow}$ and $c^+_{i\downarrow} c_{i\downarrow}$ for spins ↑ and ↓. For any arbitrary pseudo-spin configurations $\{\sigma_{i\uparrow}\}$ and $\{\sigma_{i\downarrow}\}$ with

(8-b) $$\sigma_{i\uparrow} = 0 \text{ or } 1 \quad \text{and} \qquad \sigma_{i\downarrow} = 0 \text{ or } 1$$

there exists an eigenstate of the "anti-integrable" Hamiltonian such that

(8-c) $$<c^+_{i\uparrow} c_{i\uparrow}> = \sigma_{i\uparrow} \qquad \text{and} \qquad <c^+_{i\downarrow} c_{i\downarrow}> = \sigma_{i\downarrow}$$

These eigenstates are periodic, quasi-periodic, chaotic or else as the pseudo-spin configurations $\{\sigma_{i\uparrow}\}$ and $\{\sigma_{i\downarrow}\}$ which label it. By analogy with dynamical systems where the anti-integrable limit is a limit where the chaotic behavior is perfect, we called this limit anti-integrable[1,2].

For t=0, when a single electron is present at site i, there exists a lattice distortion at this site. The electron associated with the lattice distortion is called a *polaron*. This object is magnetic because of the free spin of the electron. When two electrons with opposite spins are present at site i, this pair of electrons associated with the corresponding lattice distortion is called a *bipolaron*. This object is non magnetic.

Thus, the eigenstates determined by pseudospin configurations fulfilling $\{\sigma_{i\uparrow}\}=\{\sigma_{i\downarrow}\}$ are bipolaronic configurations. Those where $\sigma_{i\uparrow} \times \sigma_{i\downarrow} = 0$ for all i, are polaronic configurations. In the general case, the eigenstates determined by $\{\sigma_{i\uparrow}\}$ and $\{\sigma_{i\downarrow}\}$ are mixed polaronic-bipolaronic configurations.

We have proven through tedious mathematics[3], that these chaotic polaronic and bipolaronic states obtained trivially for t=0, survives to perturbations in t, that is for large enough electron-phonon coupling. This proof holds very generally for most arbitrary lattices which could be random or not (except for exponential lattices such as Bethe lattices), also in the presence of an arbirary uniform magnetic field (not represented in the model described here).

More precisely, for each eigenstate $\Psi(\{\sigma_{i\uparrow}\},\{\sigma_{i\downarrow}\})$ at the anti-integrable limit, there exists a positive non zero number t_c (independant of the system size) such that for $|t| < t_c$, there exists an eigenstates $\Psi(\{\sigma_{i\uparrow}\},\{\sigma_{i\downarrow}\}, t)$ of the adiabatic Holstein model (with $t\neq0$ and $\beta=0$) which depends uniformly continuously on the parameter t and such that for t=0, we have $\Psi(\{\sigma_{i\uparrow}\},\{\sigma_{i\downarrow}\},0) = \Psi(\{\sigma_{i\uparrow}\},\{\sigma_{i\downarrow}\})$.This exact result means that for large enough electron-phonon coupling, there exists infinitely many local minima of the adiabatic energy.

Since they originate by continuity from the anti-integrable limit, these states are also called bipolaronic, polaronic or mixed polaronic-bipolaronic structures. However, note that these polarons and bipolarons become interacting "particles" for $t\neq0$

For larger t, (that is for smaller electron-phonon coupling), these structures disappear through complex cascade of bifurcations qualitatively similar to those observed for dynamical systems.

The real bounds which can be observed numerically for t_c are much larger than the bounds obtained rigorously. Although these last bounds were found independent of the pseudo-spin configuration, the real bound in t of these configurations depends on the choice of the pseudo-spin configurations $\{\sigma_{i\uparrow}\}$ and $\{\sigma_{i\downarrow}\}$. In fact the dimensionless parameter t_c ranges not very far from unity which proves for many systems the physical relevance of the domain of existence for these polaronic and bipolaronic states.

Many other exact results concerning the properties of these states are obtained. All these structures are proven to be insulating with both a non vanishing electronic gap and a non-vanishing phonon gap. These structures are "charge defectible" which means that non uniform distribution of the electronic charges are metastable as in true insulators unlike metals and semi-conductors. The local perturbations due for example to impurities or to extra-electrons decay exponentially at long distance.

For large enough t, the ground-state of the adiabatic Holstein model, is proven to correspond to an ordered bipolaronic structure which may be commensurate, incommensurate (or else?) and thus is a *bipolaronic charge density wave*. It is characterized by a periodic, quasi-periodic (or else) ordering of the pseudo-spin configuration $\{\sigma_{i\uparrow}\} = \{\sigma_{i\downarrow}\}$. Therefore, the other bipolaronic and mixed polaronic-bipolaronic structures correspond to low energy configurational excitations of this ordered ground-state. Again, this exact result holds at any dimension and for most lattices.

For small magnetic field, the ground-state remains a bipolaronic structure but for a large enough magnetic field, the ground-state is proven to become a mixed polaronic-bipolaronic state which is magnetic, through complex and unexplored cascades of transitions.

The role of the dimensionless parameter β describing the quantum lattice fluctuations which is not strictly zero but small, has been analysed[5]. It has been shown that their effect are negligible for the polaronic and bipolaronic structures providing that the phonon gap remains large enough. By contrast, it has been proven[5-d] that the existence of a gapless phonon mode (e.g. phason), breaks down the validity of the Born-Oppenheimer approximation on which the standard theory of Charge Density Waves (CDW) is based. In that case, superconductivity is conjectured. As a result, all CDW's should be viewed as Bipolaronic CDW's, that is an incommensurate ordering of bipolarons. The physical consequences of this description sharply differ on many points from those of the widely admitted standard Peierls-Frohlich theory of CDW[6] and could provide a new fruitful basis for understanding many experiments in CDW systems which up to now received inconsistent theoretical interpretations[7].

In summary, this work confirms and extends early numerical studies and conjectures on the Peierls instability in this model in one dimension model where a transition by "breaking of analyticity" toward a bipolaronic CDW was already observed[4]. This recently obtained exact result should break down some wrong ideas concerning the effects of large electron-phonon coupling which are widely spread in the litterature. It opens a new direction for understanding globally the properties of Charge Density Waves Systems and perhaps in further stages, other structural problems concerning for example magnetic structures.

[1] S. AUBRY and G. ABRAMOVICI (1990) *Chaotic Trajectories in the Standard Map: The Concept of Anti-Integrability"* Physica **D43** 199-219 and Erratum (in press) and S.AUBRY (1991) *The Concept of Anti-Integrability: Definition, Theorems and Applications* Proceeding of the IMA Workshop on *Twist Mappings and their Applications* March 12-16, 1990 Minneapolis, USA in press

[2] S. AUBRY, J.P. GOSSO, G.ABRAMOVICI, J.L. RAIMBAULT and P. QUEMERAIS (1991) *Effective Discommensurations in the Incommensurate Ground-States of the Extended Frenkel-Kontorowa Models* Physica **D47** 461-497

[3] S.AUBRY, G.ABRAMOVICI and J.L. RAIMBAULT (1991) *Chaotic Polaronic and Bipolaronic States in the Adiabatic Holstein Model* prepint Submitted to Journal of Statistical Physics

[4] S. AUBRY and P. QUEMERAIS (1989) in *Low Dimensional Electronic Properties of Molybdenum Bronzes and Oxides* 295-405 Ed.Claire SCHLENKER Kluwer Academic Publishers Group

[5] (a) S.AUBRY, P.QUEMERAIS (1989) in *Singular Behavior & Nonlinear Dynamics* Ed. St. and Sp. Pnevmatikos and T. Bountis, Riedel pp.342-363 (b) S.AUBRY, G.ABRAMOVICI, D.FEINBERG, P.QUEMERAIS and J.L. RAIMBAULT (1989) in *Non-linear Coherent Structures in Physics, Mechanics and Biological Systems* Lectures Notes in Physics (Springer) **353** pp.103-116 (c) S.AUBRY, P.QUEMERAIS and J.L. RAIMBAULT (1990) in Proceeding of Third European Conference on *Low Dimensional Conductors and Superconductors* Fisica **21** Supp.3, 98-101 and Fisica **21** Supp.3, 106-108 Ed. S. Barisić (d) S. AUBRY (1991) *Bipolaronic Charge Density Waves* Proceeding of *Microscopic Aspects of Non-Linearity in Condensed Physics* NATO A.R.W., Florence (Italy) June 1990

[6] P.A.LEE, T.M. RICE and P.W. ANDERSON (1973) *Phys.Rev.Lett.* **31**, 462

[7] see for example P.MONCEAU *Recent Developments in Charge Density Wave Systems* in *Application of Statistical and Field Theory Methods to Condensed Matter* p.357 Ed D.BAERISWYL et al Plenum NY (1990) and refs. therein.

Chaotic motion of solitons in the PDE model of long Josephson junctions

G.Rotoli and G.Filatrella

Department of Physics, University of Salerno
I-84081 Baronissi (SA), Italy.

The Josephson junction [1] is an extremely interesting solid-state device both for applied physics and non-linear physics. It consists simply of a sandwich of two superconductive films separated by a thin layer, generally of few Å, of an insulator. Depending on the thickness of the insulator the 'macroscopic' wave functions of the two superconductors overlap in the insulator region leading to a coupling that constitutes the essence of Josephson effect [2]. The phase difference $\phi(x,t)$ between the two wave functions can show, temporally and/or spatially, a highly correlated behavior. In particular it can be shown, by means of a simple electrical model [3], that the phase satisfies a partial differential equation (PDE) known as the Perturbed Sine-Gordon equation. This equation, if only one spatial dimension is longer than the so-called Josephson penetration depth λ_J and in the so-called 'inline' configuration, is:

$$\phi_{tt} - \phi_{xx} + \sin\phi = \alpha\phi_t - \beta\phi_{xxt}. \tag{1}$$

We have in addition two time-dependent boundary conditions for a current biased junction irradiated by a microwave field [4,5]:

$$\phi_x(0,t) + \beta\phi_{xt}(0,t) = -\chi + \eta(t), \tag{2a}$$

$$\phi_x(l,t) + \beta\phi_{xt}(l,t) = \chi + \eta(t). \tag{2b}$$

In these formulas all the distances are normalized to λ_J, and times to $\omega_J = \bar{c}/\lambda_J$, the plasma frequency, where $\bar{c}$ is the speed of light in the junction. $\eta(t)$ is the normalized external magnetic field at the edges of the junction, α and β are loss parameters, and χ is the normalized bias-current supplied to the junction.

It is well known that Eq.(1) admits solitonic solutions [1,4]; in the context of Josephson junctions such solitons are named 'fluxons' because they carry a flux quantum $h/2e$. In the absence of any time dependent signal, $\eta(t) = 0$, the fluxons propagate back and forth in the junction, the energy dissipated in the propagation being replenished by the bias current at the boundary χ. Thus the frequency of the fluxon ω_{free} is dependent on the bias current χ via a non-linear relation. From the second Josephson equation it is possible to link the frequency of the fluxon to the d.c. voltage at the end of the junction; the branches obtained on the current-voltage (I-V) d.c. characteristics are called Zero Field Steps (to recall the absence of any external field, also static), and correspond to the propagation of one or more fluxons; we often refer to them simply as the unperturbed I-V curves. If we write for the external signal the form $\eta(t) = \eta_0 \sin(\omega_s t)$ [4,5] a single fluxon can be forced, for some values of the bias current, to oscillate in phase with this signal at the same frequency ω_s, i.e. the soliton is phase-locked to the external signal (the values of the bias current for which we have this effect are called phase-locking range). More generally a phase-locking state is obtained if after m periods of fluxon and n periods of the external signal the same phase-relation is reached, i.e. the frequencies of the signal and of the fluxon are in a fractional ratio [6]. Thus we have the following link between the soliton frequency and d.c. voltage at the end of the junction (in normalized units) [6]:

$$V_{n,m} = \frac{2n}{m}\omega_s. \tag{3}$$

In the phase-locking range of bias-current this phenomenon gives rise to constant-voltage steps on the I-V characteristics of the junction. Steps on I-V characteristic were observed in several experiments [7]; in many cases this observation is referred to the $m = 1, n = 1$ case (fundamental frequency of the junction), that has, as can be demonstrated (see Ref. [6]) the biggest phase-locking range. Experimentally subharmonic steps, $m > 1$, appear to be [8] much smaller, and, in general, are difficult to observe, though the theory predicts phase-locking ranges smaller than the case of the fundamental frequency, but surely in principle not unobservable. As we will see this experimental difficulty can be ascribed to chaotic behavior in long junctions pumped with a subharmonic frequency.

Kautz [9] first observed chaotic behavior in small (with respect to λ_J) Josephson junctions both numerically and experimentally. For such a junction equation (1) reduces to the equation of a forced pendulum; the short junction can be locked to the external signal (again producing constant-voltage steps on the I-V curves), but for suitable values of the amplitude and frequency of the external signal the phase-lock becomes unstable and a period two solution appears. Since now a frequency of the system is not well defined Eq. (3) can be substituted by

$$V = \frac{4\pi}{\langle T_p \rangle_{ave}}, \tag{4}$$

where T_p is the period of pendulum oscillations. Continuing to increase the amplitude of the signal the system undergoes a standard bifurcation cascade until a chaotic behavior is reached. It is interesting to note that constant-voltage steps on the I-V curves again exist also if the pendulum motion is chaotic for a short range of values of amplitude of the external signal after the beginning of the chaotic region in the parameter space [9]; sometimes this phenomenon is referred to as 'phase-locking chaos' or 'frequency locking' [5,10], but we prefer the more correct name of 'voltage locking'.

In the long junction case chaotic behavior has been observed numerically in many cases [11,12,13]. In all these cases the chaos is both spatial and temporal (turbulence), i.e. it appears essentially as a higher-dimensional phenomenon clearly related to the infinite degrees of freedom of the full equation (1).

On the other hand we can proceed in another direction: applying the McLaughlin-Scott [14] perturbative approach we can treat the motion of a single fluxon in the junction as the motion of a relativistic particle in the interior of the junction. Giving to this particle-fluxon a sharp energy supply at the boundary [15] we can, if the frequency of energy supply is in a fractional relation to the time of flight of fluxon T_k in the junction, phase-lock the soliton to the external field. This method is called 'map approach' after [6] because we can write, solving the equation of motion of the particle-fluxon in the junction, an (analytic) bidimensional map directly in terms of time of flight T_k and energy y_k (here the label indicates the k-reflection of the fluxon at one edge of the junction). We do not write explicity the map here; the interested reader can find it in ref. [16]. Obviously if we know the times of flights of the fluxon, we can obtain the d.c. voltage by means of the formula (4) merely identifying T_p with T_k cfr. [6]. If the external signal makes n periods while the fluxon makes m oscillations in the junction, the Eq. (4) reduces to Eq. (3). Since the map is analytic an analysis of the stability of phase-locked solutions can be easily conducted: if the amplitude of the external signal is sufficiently high again phase-locked solutions are unstable and 2-times of flight solution sets in. Then continuing to increase the amplitude, via a bifurcation cascade, the system becomes chaotic, i.e. fluxons travel in the junction with a chaotic distribution of times of flight.

The features of this chaotic motion are studied in detail in the ref. [10,17], here we limit ourselves to report the main aspects only:

i) the amplitude of the external signal necessary to arrive at the chaos decreases with m, i.e. chaotic states appear mainly on subharmonic steps $m \geq 3$;

ii) in terms of the bias current χ the chaos begins to develop on the unperturbed I-V characteristic at the center of the phase-locking step;

iii) increasing the losses, α and β, the stability of the system is increased;

iv) the system appears to be 'voltage locked', i.e. though the motion of fluxons is chaotic, the average of time of flight in Eq.(4) is such that V remains unchanged at a phase-locked value, so on the I-V characteristic the step remains vertical;

v) continuing to increase the amplitude of the external signal the system produces very long TOFs, that consequently annihilate the fluxon after a certain value $\eta_{0,a}$.

It is natural that the low-dimensionality of this chaos is an intrinsic feature of the perturbative approach used to reduce a PDE to an ODE equation. In fact the hypothesis of a single particle-fluxon propagating in the junction is essential in the perturbative approach to obtain all the above mentioned results. Nothing can be asserted about the full equation (1), i.e. we cannot say that (1) shows this type of low-dimensional chaos except by directly attacking it.

Our procedure was first to choose a region of parameter space where the map shows chaotic phenomena; then we have integrated numerically Eq.(1). The method used is based on the reduction of Eq.(1) to a system of ordinary differential equations (ODE's), then integrating the ODE's using some standard method (or methods). In our case, to obtain a very careful integration, we have used two different spatial discretization formulas (3-point and 5-point formulas within the junction) and two different methods to integrate the ODE system (a simple and fast Predictor-Corrector method [18] and a Bulirsh-Stoer method [19]). The results appear to be the same, within the discretization and time-step induced errors.

Using the map prediction we search for chaotic dynamics on a junction of normalized length $l = 10$, pumped with a frequency of $\Omega = 0.4$ on the $m = 3$ subharmonic. Biasing the junction at center of phase-locking step, $\chi = 0.493$, with an external amplitude of $\eta_0 = 0.100$ the fluxon appears to be phase-locked to the external signal. In Fig.(1) we report the voltage peaks at two edges of the junction signaling the reflection of the fluxon at the edges. The time interval between two successive peaks yields the TOF of the soliton in the junctions, from Fig.(1) we see that this time is constant and equal to $T = 23.56$, which implies a phase-locking voltage of $V = 0.2666 = (2/3)\,0.4$, as follows by Eq.s (3) and (4).

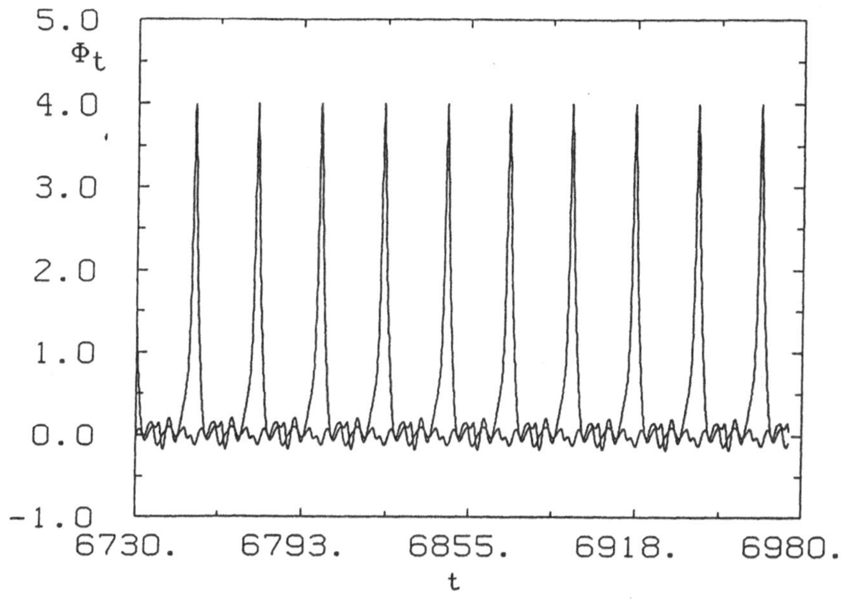

Fig.(1)

The small undulations following the reflection peaks can be ascribed to external signal effects when no soliton is present at the edges. The error, deduced from the numerical data, in the estimation of TOF is of the order of magnitude of the time-step, which is $\Delta t = 0.01$; thus we can resolve TOFs separated by more than this time-step (this fixes the above reported number of significant digits in the voltage). In Fig.(2) we have increased the value of η_0 to 0.150: as is clearly seen we have now a 2-TOFs solution in the junction. The two TOFs adjust themselves to make again $V = 0.2666$. Continuing to increase the amplitude of the external signal the 2-TOFs solution becomes unstable and the dynamics evolves via a bifurcation cascade to a chaotic state, that is shown in Fig.(3) for a field $\eta_0 = 0.190$. Though the motion is chaotic the system is again 'voltage locked' to the external signal: in fact the numerical evaluation of Eq.(4) over $n_r \sim 10^3$ reflections gives for the voltage $V = 0.2667$.

We stress that n_r is a relatively small number with respect to what might be used in a 'map approach', but in the PDE system it has to be considered a good goal, in view of the long integration time of the Eq.(1) (the number of the time steps in a typical run is $\sim 10^6$). However in the PDE system this situation is not permanent, further increase in the amplitude of the external signal leads to the loss of 'voltage locking': in fact for a field $\eta_0 = 0.260$, we obtain $V = 0.2080$, i.e. also the voltage assumes chaotic values and on the I-V characteristic the step ceases to be vertical. The bifurcation-cascade for the PDE approach is shown in Fig.(4). In Fig.(4) we have marked with an arrow the positions of η_0 relative to Fig.s (1) and (2). The dotted line is the approximate separation between the 'voltage locked' region and chaotic voltage region, occurring at an amplitude $\eta_{0,s} \sim 0.195$. After $\eta_{0,a} \sim 0.270$ the fluxon motion is unstable and the fluxon is annihilated.

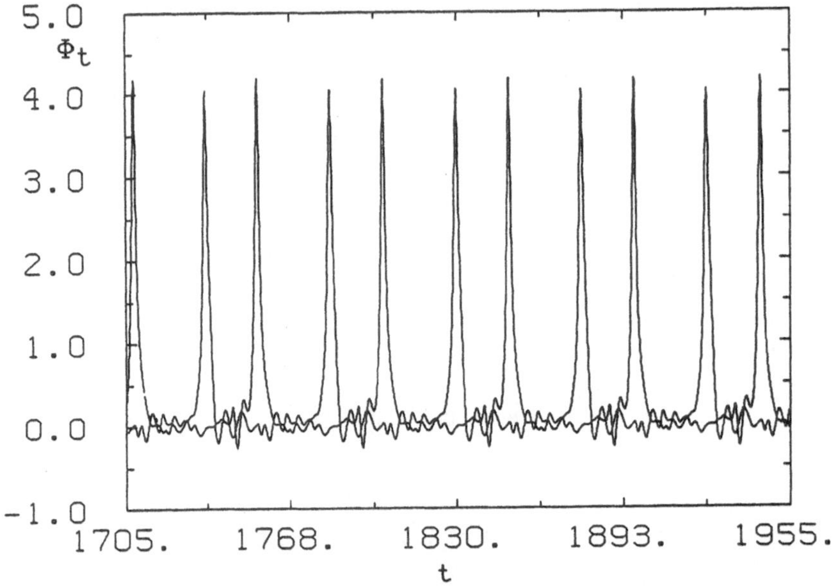

Fig.(2)

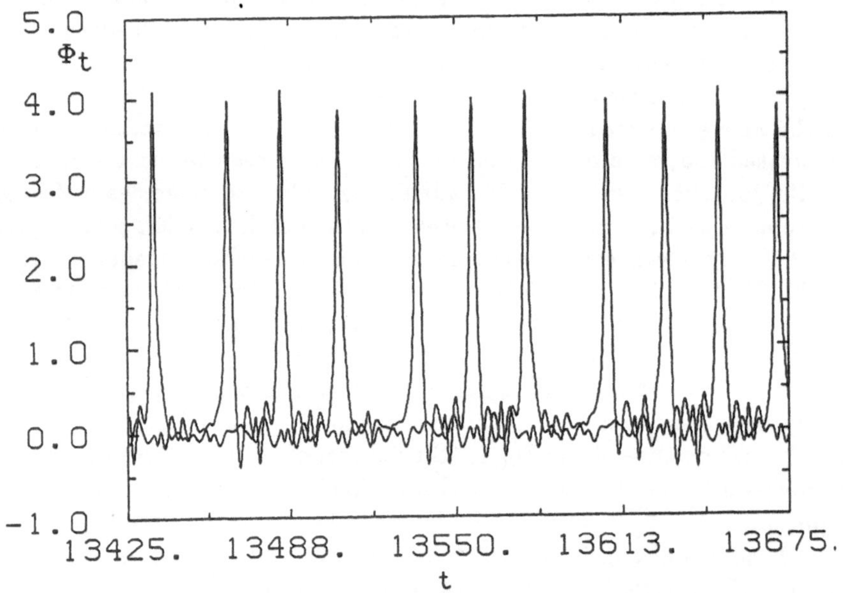

Fig.(3)

An estimate of the Feigenbaum ratio:

$$f = \lim_{n\to\infty} \frac{\eta_{0,n+1} - \eta_{0,n}}{\eta_{0,n} - \eta_{0,n-1}}$$

where $\eta_{0,n}$ is the value of nth-bifurcation parameter, for the PDE data gives $f \sim 6.1$.

Comparision of Fig.(4) with the 'map approach' bifurcation [18], obtained with the same parameters, gives two quantitative differences. The first is that in the PDE model the phase-locking appears to be more stable, i.e. the value of the first bifurcation in the PDE approach is $\sim$ 20% higher than in the map case; this can be partially explained with the differences in boundary conditions in the two methods: in the map context the external signal is an 'abrupt' delta function whereas in the PDE it is a 'smeared' sinusoidal signal, which implies that a strict comparison between the two amplitudes is not possible.

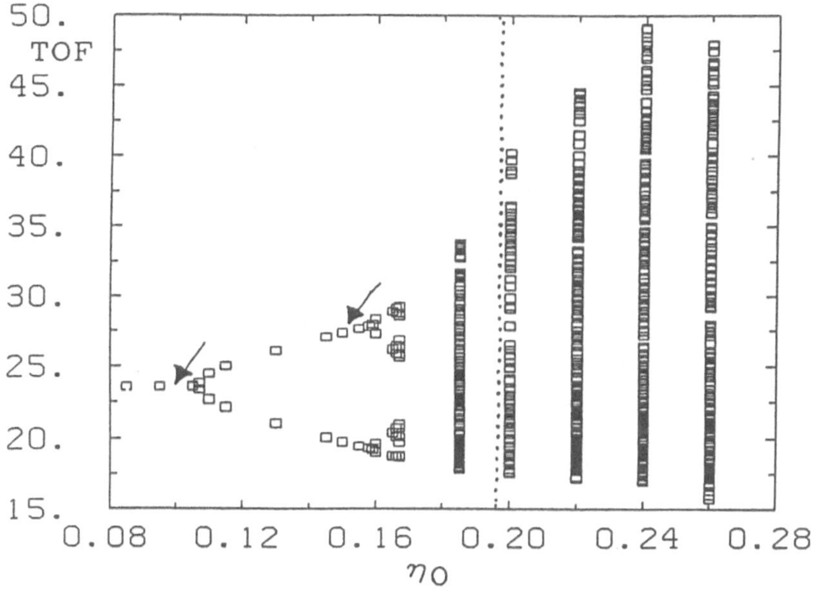

Fig.(4)

Moreover the bifurcation tree in the PDE is 'longer', i.e. the interval of η_0 in which the fluxon is stable (including the chaotic regime) is about 2 times the same interval with the map. On the other hand a comparison between map and PDE strange attractors shows that both objects are structurally very similar [20] confirming that the map approach grasps the substantial dynamics of Eq.(1). The main features of this chaotic motion of fluxon (PDE low-dimensional chaos) can be summarized as follows:

i) also here the amplitude of the external signal necessary to arrive at chaos decreases with m, i.e. in all runs chaotic states appear only on subharmonic steps $m \geq 3$; on the steps at the fundamental frequency an amplitude of $\eta_0 \sim 0.5$ in general destroys the 1-fluxon solution, but up to such an amplitude neither chaotic motion nor bifurcations are exibited;

ii) in terms of the bias current χ the chaos begins to appear on the unperturbed I-V characteristic at the center of the phase-locking step;

iii) increasing the losses, α and β, the stability of the system is increased;

iv) the system appears to be 'voltage locked' until an amplitude $\eta_{0,s}$, i.e. though the motion of fluxons is chaotic, the average of time of flight in Eq.(4) is such that V remains unchanged at the phase-locking value; after $\eta_{0,s}$ the 'voltage locking' state is lost;

v) a further increase of the amplitude of the external signal produces very long TOFs, that consequently annihilate the fluxon beyond a certain value $\eta_{0,a}$.

It is interesting to note that the value of $\eta_{0,s}$, depends upon the bias current χ, but the minimun is not at the center of the step, which implies that deviations from the voltage begin to develop in the low bias part of the step. A possible explanation could be the overlapping of attraction basins of more subharmonics [21].

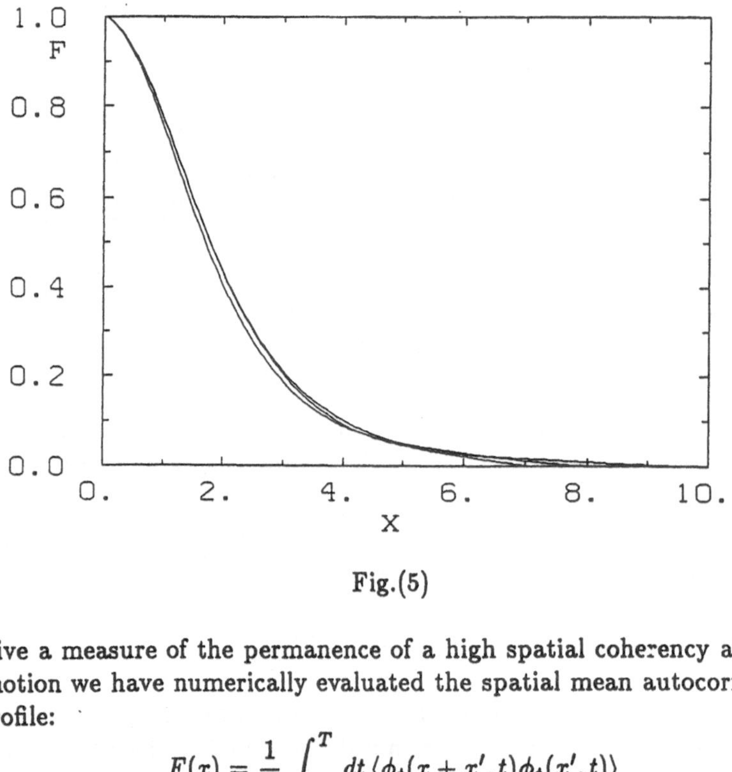

Fig.(5)

To give a measure of the permanence of a high spatial coherency also during the chaotic motion we have numerically evaluated the spatial mean autocorrelation of the soliton profile:

$$F(x) = \frac{1}{T} \int_0^T dt \, \langle \phi_t(x + x', t) \phi_t(x', t) \rangle \tag{5}$$

For a well-defined fluxon lineshape we expect correlation to extend over a distance of the order of λ_J, i.e. the typical distance of variation of the phase in the junction. A first result of this calculation is shown in Fig.(5), where we report Eq.(5) for three typical cases: phase-locking, and two chaotic solutions. As is evident from the figure correlation extends over a distance of the order of λ_J in all the cases. In Eq.(5) T was chosen of order of 10^2 periods. The small decrease in the spatial correlation of $\eta_0 = 0.190$ curves is within the accurancy of the estimate of integral (5). Thus we conclude that in the entire parameter interval the fluxon remains highly correlated though its motion is chaotic, confirming the existence of a low-dimensional chaotic attractor for the full PDE system governed by Eq.(1).

Acknowledgement

We wish to thank R.D. Parmentier for illuminating comments and for a critical reading of the manuscript. We had several interesting discussions with G. Costabile, S. Pagano, N.F. Pedersen, and M. Salerno, to whom goes our gratitude. This work is in partly sponsored by the Progetto Finalizzato "Tecnologie Superconduttive e Criogeniche" of the Italian CNR.

Bibliography

[1] R.D.Parmentier, in *Solitons in Action* , edited by K.Lonngren and A. Scott, Wiley, New York 1978, p.173.
[2] B.D.Josephson, Phys. Lett. 1, 251, 1962.
[3] S.Pagano, PhD Thesis, Tech. Univ. of Denmark report No.S42, 1987.
[4] A.Davidson and N.F.Pedersen, Phys.Rev.B41, 178, 1990.
[5] G.Rotoli, G.Costabile and R.D.Parmentier, Phys.Rev.B41,
[6] M.Salerno, M.R.Samuelsen, G.Filatrella, S.Pagano, and R.D.Parmentier, Phys.Rev.B41, 6641, 1990.
[7] G.Costabile, R.Monaco and S.Pagano, J. Appl. Phys. 63, 5406, 1988.
[8] J.J.Chang, Phys.Rev.B34, 6137, 1986.
[9] R.L.Kautz, IEEE Trans. Magn. MAG-19, 465 (1983); R.L.Kautz and R.Monaco, J.Appl.Phys. 57, 875, 1985.
[10] M.Salerno, Phys.Lett.A144, 453, 1990.
[11] M.Octavio Phys.Rev.B29, 1231, 1984.
[12] A.R.Bishop, K.Fesser, P.S.Lomdahl, W.C.Kerr, and S.E.Trullinger, Phys.Rev.Lett. 50, 1095, 1983.
[13] M.Cirillo, to be published in Journ.Appl.Phys.
[14] D.W.McLaughlin and A.C.Scott, Phys.Rev.A18, 1652, 1978.
[15] O.A. Levring, N.F. Pedersen, ams M.R. Samuelsen, J. Appl. Phys. 54, 987 (1983).
[16] G.Filatrella, G. Rotoli, and R.D. Parmentier, Phys. Lett. A148, 122, 1990.
[17] G.Rotoli and G.Filatrella to be published Phys.Lett.A 1990.
[18] W.E.Milne, *Numerical solutions of differential equations* , Wiley, New York 1953, chap.2.
[19] W.H.Press, B.P.Flannery, S.A.Teukolsky, and W.T.Vetterling, *Numerical Recipes* (Cambridge University press, Cambridge, 1986) , Chap. 15.
[20] G.Filatrella, N.Grønbech-Jensen, R.Monaco, S.Pagano, R.D.Parmentier, N.F.Pedersen, G.Rotoli, M.Salerno and M.R.Samuelsen, Proc. of Workshop "Nonlinear Superconductivity Electronics and Josephson Devices", eds. G.Costabile, A.Davidson, S.Pagano, N.F.Pedersen, and M.Russo, Capri, sept. 1990 in print Plenum Press.
[21] G.Filatrella and G.Rotoli, Nato-Asi conference "Chaos: theory and practice", Patras, Greece, July 1991.

NONLINEAR STRUCTURE OF PHASE MOTION FROM THE STUDY OF DIFFERENTIAL EQUATIONS NEAR RESONANT TORI

Michel Planat
Laboratoire de Physique et Métrologie des Oscillateurs
associé à l'Université de Franche-Comté-Besançon
32, avenue de l'Observatoire - 25000 Besançon

1. Introduction

Nonlinear dynamics undergoes a period of explosive growth today. However some federative concepts have emerged : solitons for nonlinear waves, strange attractors for dissipative phenomena, resonant tori in hamiltonien theory, scaling laws and fractals in many physical systems from low dimensional chaos to fully developed turbulence. In particular, nonlinear oscillator differential equations provided a source of inspiration for perturbation methods in the early work of Düffing, Van der Pol and Rayleigh, then for bifurcation and modern qualitative analysis in the work of Poincaré, Birkhoff and others russian mathematiciens including Liapourov, Andronov and more recently Kolmogorov and Arnold.[1,2] Most of the modern approach turns around the idea of periodic orbits and resonances in hamiltonien systems. From Kolmogorov-Arnold-Moser (KAM) theory we expect to preserve the structural stability of nonresonant tori against small hamiltonien conservative perturbations.[3] Perturbed resonant trajectories were first studied by Poincaré who predicted the phenomenon of splitting of separatrices and local stochasticity for two degrees of freedom (2-tori). One popular example of Poincaré-Birkhoff theory of resonance is given by the so-called Henon mapping as a model of the motion of stars in the gravitational field of the galaxy.[4,5] For n-tori ($n > 2$) a random wandering in the resonant regions between invariant tori (Arnold diffusion and stochastic webs) was predicted.[3] More recently the minimal dimensionality for stochastic webs to form in hamiltonien systems reached its lower bound ($n = 1.5$), the half degree of freedom refering to a time periodic perturbation.[6]

In this paper we study nonlinear oscillator problems in the spirit of modern theory of resonance. We emphasize that the concept of winding number and Arnold tongues is the most relevant to study structural stability of resonances. We derive a criterium for the numerical plot of Arnold tongues and apply it to Arnold's circle map and others two dimensional mappings.[7] Then we find an explicit form of Poincaré surface of section closed to resonances of oscillator differential systems. All these mappings are found nonhamiltonien so that current knowledge about the nature of resonance cannot be used.

Phase mapping for Mathieu equation is found one dimensional : correspondingly the amplitude-phase mapping is critical between elliptic and hyperbolic behavior. We find regions in parameter space where the frequency locking is reached through intermittent phase locking. It is associated to a structure of the parameter space in localized chaotic regions and escaping regions.[7]

Mappings for the Morse oscillator, Van der Pol and Düffing equation leads to a new concept of phase turbulence. At the high amplitude edge of the Arnold tongue, a transition layer is found where the dynamics changes from stable to chaotic through intermittent phase and frequency jumps. Phase becomes chaotic through doubling of jumps leading to a probability distribution resembling the Cauchy density. On the other hand, in the transition region, the winding number (or frequency) shows jumps of amplitude ℓ scaled approximately as ℓ^{-1}. They obey a 1/f power spectral density.

2. Structural stability for oscillator mappings

According to hamiltonien topology the trajectory corresponding to an n-dimensional integrable system lies on an n-dimensional torus. This allows to write the solution explicitely in terms of action-angle variables. The coordinates are the most useful when we consider the n-1 dimensional surface of section of the torus. Recent work has focused at the simplest case of a 2-torus because the surface of section has 1 dimensional form $\theta_{n+1} = \theta_n + \Omega$ (mod 1) where $\Omega = \omega_1/\omega_2$; ω_1 and ω_2 are the two frequencies for the trajectory winding around the torus and θ_n is the angle (or phase) at the nth cross section of the trajectory.

Let now consider the nonlinear phase motion closed to a resonance in an ordinary pendulum. It can be approximated as[8,9]

$$\theta_{n+1} = \theta_n + \Omega - \frac{c}{2\pi} \sin 2\pi\theta_n \ (mod\, 1) \qquad (1)$$

To study eq. (1) we introduce the rotation (or winding number, or mean frequency) with the relation

$$w = \lim_{n\to\infty} (\theta_n - \theta_o)/n \qquad (2)$$

In the unperturbed case, $w = \Omega$ so that the curve $w = f(\Omega)$ is just the straight line ; no structurally stable regions is seen since even a slight variation of parameter Ω leads to a new resonance. When $c \neq 0$, the $w = f(\Omega)$ curve has small steps near each resonance $\Omega = p/q$ (p and $q \in$ Q) and forms a devil's staircase. There are also steps in the $w = g(c)$ curve. Each region with a constant winding number is a so-called Arnold tongue. At $c = 1$, the devil's staircase is complete and has a fractal dimension $D = 0.87$.[9] When $c \geq 1$, locked regions begin to overlap and this is the threshlod for chaotic trajectories to exist in phase space. The concept of rotation intervall was introduced to account for structural stability above the chaotic limit.[10]

In order to obtain a numerical plot of Arnold tongues, we derived the following criterium : given a small real number ε and a big integer n, we admit that two points Ω, c and $\Omega + \Delta\Omega$, $c + \Delta c$ belongs to the same Arnold tongue if

$$|w^{n}_{\Omega+\Delta\Omega,\, c+\Delta c} - w^{n}_{\Omega,c}| < \varepsilon \tag{3}$$

where w^n means n iterations of the mapping. This definition implies the nonexponential divergence of initially closed trajectories. It also implies the connexity of sets in real space and can be used to derive the Mandelbrot's set.[7] Arnold tongues for mapping (1) are given in Fig. 1.

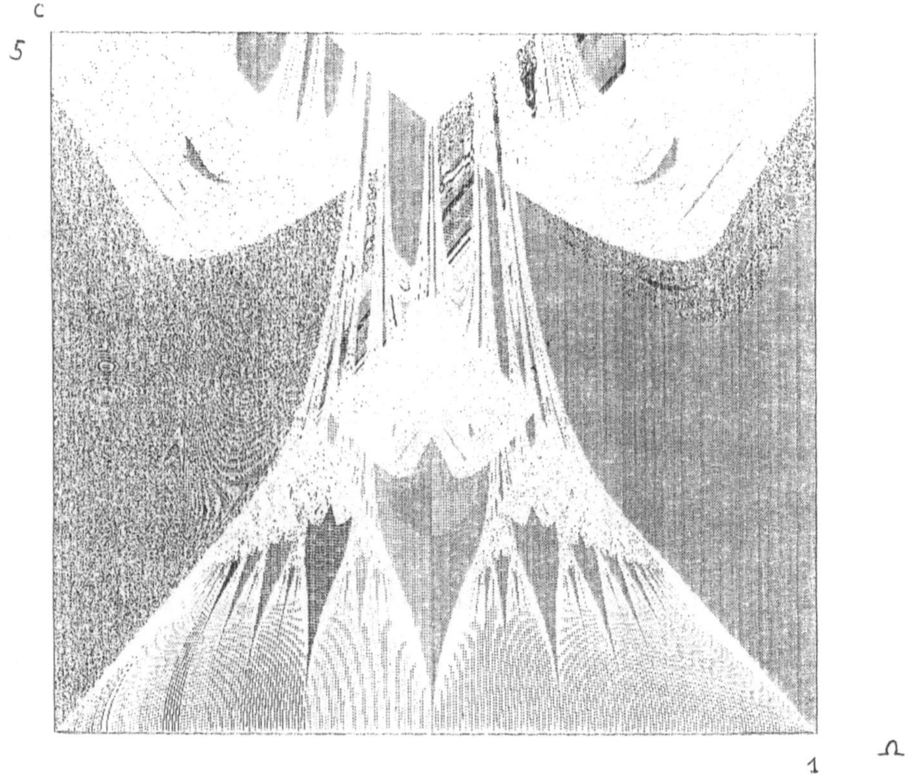

Fig. 1 : *Arnold tongues for the circle map (3)*

In the figure we choosed $n = 250$ and the color code was choosen equal to $[1/w]$ where [] means the integer part. Arnold tongues are a useful tool to study the qualitative phase dynamics. As an example, we found intermittent phase locked states closed to the boundary between the stable 0/1 region and the chaotic region ($\Omega = 0.33$; $c = 3.7$). At this point the winding number (or frequency) has an approximate 1/f power spectral density. Intermittency will be also found at the transition to chaos in oscillator differential systems.

3. Closed to resonances Poincaré mappings of oscillator differential systems

KAM theory predicts structural stability closed to resonances of an hamiltonien nondegenerate mapping. At resonances Poincaré-Birkhoff theory predicts overlap of secondary resonances and local stochasticity in a layer enclosing the separatrix between resonances. KAM theory fails when the unperturbed system is linear because the degenary condition is fullfilled ; this is the case of Zaslavsky mapping which leads to stochastic web and Arnold diffusion.[6] We will encounter another failure of the conditions for KAM theory to apply : Poincaré mappings for oscillator differential systems may not to be hamiltonien. We will show that the transition to chaos for these mappings is characterized by intermittent phase and frequency jumps obeying scaling laws.

To derive these Poincaré mappings we assume the closed proximity of a resonance. Let consider the unperturbed periodic signal $x(t) = r(t)\, exp\,(i\theta(t))$ with non dimensional frequency $\Omega = \omega_1/\omega_2$, where ω_1 and ω_2 are the unperturbed frequencies on the resonant torus. At each crossing of the Poincaré section for the perturbed torus, the trajectory will undergo a phase jump $\Delta\theta$ and an amplitude jump Δr. They are calculated by assuming that the amplitude is constant $\rho = \rho_0$ and the phase vary linearly in time $\theta = \omega t$ during each period between two crossings. The jumps process is then generalized between the nth and $n+1^{th}$ the crossing leading to the explicit relation for the Poincaré section.

The method will be applied here to Morse, Van der Pol and Düffing oscillators with main emphasis on the phase motion. The Morse potential provides a useful basis for interpreting the vibrational spectra of diatomic molecules. In conjonction with a sinusoidal force it allows to explain stochasticity and dissociation of these molecules. Van der Pol equation is a useful model of relaxation oscillators ; Düffing equation provides a model to account for amplitude-frequency effect and bistability observed in high quality piezoelectric resonators. The closed to resonances phase behavior of these three models is similar and leads to phase and frequency jumps obeying scaling laws.

The hamiltonien for the **periodically driven Morse oscillator is**

$$H = \frac{v^2}{2m} + D\left(1 - e^{-a(u-u_e)}\right)^2 - \varepsilon(u - u_e)\cos(\omega t + \Phi_0) \tag{4}$$

where u is the internuclear distance, u_e is the equilibrium distance, v is the momentum, m the reduced mass, D the dissociation energy, a is the potential parameter and ε the excitation amplitude.[11]

Hamilton's equations are written in dimension less variables as

$$\begin{aligned} x' &= (1 - x)y \\ y' &= -(1 - x)x + A\cos(\Omega\tau + \Phi_0) \end{aligned} \tag{5}$$

with the new parameters $x = 1 - e^{-\alpha(u-u_e)}$, $y = v / \sqrt{2Dm}$; the new time scale is $\tau = \omega_0 t$ with $\omega_0 = \alpha\sqrt{2D/m}$, $\Omega = \omega/\omega_0$ and the excitation amplitude is $A = \varepsilon / (2D\alpha)$.

Rewritting eq. (5) in polar coordinates $x = r\cos\Phi$, $y = r\sin\Phi$ we get

$$\begin{aligned} r' &= A \sin\Phi \cos(\Omega\tau + \Phi_0) \\ \phi' &= -1 + r\sin^2\Phi\cos\Phi + r\cos^3\Phi + (A/r)\cos\Phi\cos(\Omega\tau + \Phi_0) \end{aligned} \tag{6}$$

From now we assume the closed proximity to a resonance. Using the relations

$$r = r_0 \,, \quad \Phi = \Omega\tau \quad when \quad 0 < \tau < 2\pi/\Omega \tag{7}$$

and integrating during the first period of oscillation we get amplitude jump Δr and phase jump $\Delta\phi$. The jump process from the nth to the $n+1^{th}$ period gives the Poincaré section in the following form

$$\begin{aligned} r_{n+1} &= r_n + \Omega^{-1} c \sin(2\pi\theta_n) \\ \theta_{n+1} &= \theta_n + \Omega^{-1}\left[1 - \frac{c}{2\pi r_{n+1}}\cos(2\pi\theta_n)\right] (mod\, 1) \end{aligned} \tag{8}$$

where we have used the mod 1 phase $\theta = -\Phi/2\pi$ and $c = \pi/A$. The jacobian matrix $\mathcal{A}$ of the linearized system at the fixed points $2\pi\theta_0 = 0$ or π is such that $\det\mathcal{A} = 1$ and $\mathrm{tr}\mathcal{A} = 2 + \Omega^{-2}c^2/r_0^2$ with $\Omega^{-1}(1 \mp c/(2\pi r_0)) = m \in \mathbb{N}$. The map is conservative and both fixed points are unstable.

Van der Pol equation is written in dimensionless form as

$$x'' = -x + \varepsilon x'(1 - \alpha x^2) + A\cos(\Omega\tau + \Phi_0) \tag{9}$$

where $\Omega = \omega/\omega_0$ and $\tau = \omega_0 t$. With the same notations as above, its Poincaré section takes the form

$$\begin{aligned} r_{n+1} &= r_n\,[1 + \pi\varepsilon\,\Omega^{-1}(1 - \alpha r_n^2/4)] + c\,\Omega^{-1}\sin(2\pi\theta_n) \\ \theta_{n+1} &= \theta_n + \Omega^{-1}\left[1 - \frac{1}{2\pi}\frac{c}{r_{n+1}}\cos(2\pi\theta_n)\right] (mod\, 1) \end{aligned} \tag{10}$$

where $c = \pi A$. It is interesting to observe that for an autooscillating motion at amplitude $r_n = 2/\sqrt{\alpha}$ the Morse eq. mapping (8) is found. In subsequent analysis we will restrict for this case.

Düffing equation is written in dimensionless form as

$$x'' + \mu x' + x + \delta x^3 = x_0 \cos(\Omega\tau + \Phi_0) \tag{11}$$

With the same notations, its Poincaré section takes the form

$$s_{n+1} = s_n\,[1 - \varepsilon(\Omega)] + c\,\Omega^{-1}\,(\sin 2\pi\theta_n)$$
$$\theta_{n+1} = \theta_n + \Omega^{-1}\,(1 + s_{n+1}^2) - \frac{c\,\Omega^{-1}}{2\pi\, s_{n+1}}\cos(2\pi\theta_n)\ (mod\,1) \tag{12}$$

with $s = K_1 r$; $K_1 = 1/2\,(3\,\delta/2)^{1/2}$; $\varepsilon(\Omega) = 2\pi^2/(Q\Omega)$ (Q is the quality factor of the resonator) and $c = \pi a_1 K_1$.

In the case of high Q resonators, $\varepsilon(\Omega) \ll 1$ so that the mapping is conservative. Period 1 fixed points (θ_0, s_0) are given by

$$2\pi\theta_o = \begin{matrix} 0 \\ \pi \end{matrix}$$
$$\Omega^{-1}\,[1 + s_o^2 \mp a_1 K_1 / (2 s_o)] = m \in IN \tag{13}$$

By linearizing the map at the fixed points we find tr $\mathcal{A} > 2$ for the fixed point $\theta_0 = 0$. The fixed point $2\pi\theta_0 = \pi$ may be stable if $|\mathrm{tr}\mathcal{A}| < 2$. By contrast to the Morse oscillator, inclusion of the square term s^2_{n+1} leads to linearly bistable states. The point $\theta_0 = 0$ corresponds to the normal use of a piezoelectric resonator and will be the only one studied further.

4. Phase mappings of oscillator differential systems

In the small amplitude limit $r_n = 0$ all mappings (8), (10), (12) have the one dimensional form

$$\theta_{n+1} = \theta_n + \Omega^{-1} - \frac{1}{2\pi}\, cotg\,(2\pi\theta_n)\ (mod\,1) \tag{14}$$

Plotting the winding number (2) versus the frequency ratio Ω^{-1} we find none structurally stable regions. We plotted the phase jump $\Delta\theta_n = \theta_{n+1} - \theta_n$ (mod Ω^{-1}) (Fig. 2a) and the winding number (Fig. 2b) versus the number of iterations ($n < 25\,550$; the time signal curve is at the bottom of plots). Jumps of arbitrary amplitude are found. This behavior is reminiscent of Cauchy type processes.[12] Going back to the modulo 2π phase $\Phi = -2\pi\,\theta$, the phase jump $X = \Delta\Phi$ obeys the equation

$$X = -\,cotg\,\Phi = tg\,(\Phi - \frac{\pi}{2}) \tag{15}$$

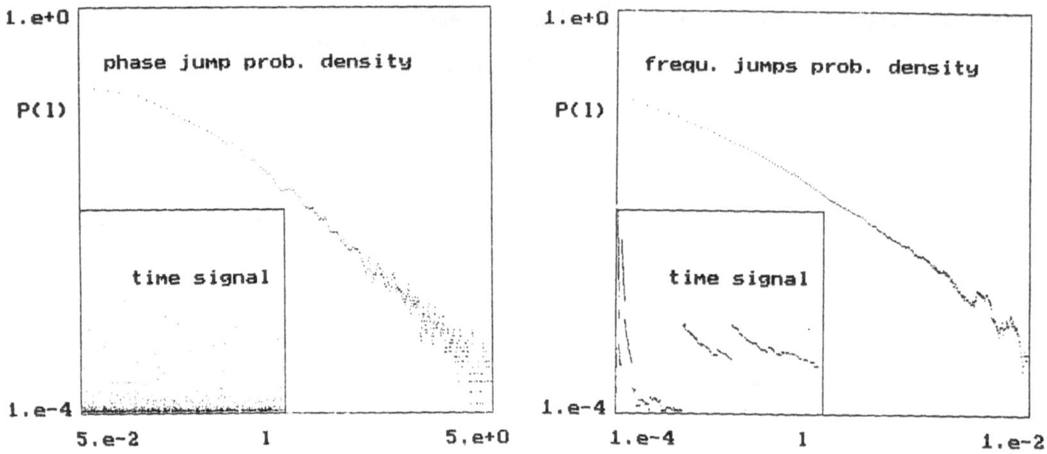

(a) (b)

Fig. 2 : *Phase and frequency jumps for mapping (14)*

From Fig. 2a, phase jumps seems to evolve randomly and probability theory is useful to describe the motion. Eq. (15) corresponds to an experiment in which a vertical mirror evolves freely and randomly with an angle Φ around its axis and projects a horizontal light ray at a distance X from its starting position. If Φ is assumed uniformly distributed between 0 and π, the distribution function is $F(X) = \Phi/\pi = 1/2 + (1/\pi) \arctan X$ and the probability density is $f(X) = F'(X) = 1/(\pi(1+X^2))$. The normal density $(1/\pi)\exp(-X^2)$ is recovered only at small distances X. The X axis is approached so slowly that no expectation value or second moment should be expected.

We defined the probability density $P(\ell)$ as the probability that X makes a jump of magnitude ℓ as usual

$$P(\ell)\,d\ell = prob. \quad (\ell < X \leq \ell + d\ell)\, d\ell \tag{16}$$

and plotted this density in log.log scales when X is the phase jump (Fig. 2a) or the frequency jump (Fig. 2b). As shown in Fig. 2a, the probability density approximates the Cauchy density. On the other hand the probability density of frequency jumps approximately scales as ℓ^{-1}.

This scaling picture gives rise to a power law frequency dependence of the Fourier spectrum. The phase jump amplitude ℓ and the life time τ are easily derived by differentiation of the resonance condition (7) to get $\ell = \Omega\tau$ so that the scaling law becomes $P(\ell) = P(\Omega\tau)$. As is well known this distribution of lifetimes leads to the power frequency spectrum with Fourier frequency $\bar{\omega}$ as follows

$$S(\bar{\omega}) = \int \frac{\tau}{1 + \bar{\omega}^2 \tau^2} P(\tau)\, d\tau \tag{17}$$

Since $P(\tau) = \tau^{-\alpha}$, we get $S(\bar{\omega}) = \bar{\omega}^{-2+\alpha}$ with $\alpha \simeq 2$ for phase jumps (white spectrum) and $\alpha \simeq 1$ for frequency jumps ($1/\omega$ spectrum). Numerical experiments confirms this conjecture.

Mapping (14) approximates the phase motion when the amplitude of selfsustained oscillations $r = r_0$ is small. For nonsmall amplitudes r_0 the motion can be studied in parameter scales Ω^{-1} and $q = c\,\Omega^{-1} / r_0$ with the following mapping

$$\theta_{n+1} = \theta_n + \Omega^{-1} - \left(\frac{q}{2\pi}\right) \cos 2\pi\theta_n \,/\, (1 + q \sin 2\pi\theta_n) \;\; (mod\, 1) \tag{18}$$

Arnold tongues for this mapping are given in Fig. 3. We studied numerically the transition from frequency locked to chaotic regions by using a one dimensional bifurcation diagram for the amplitude ℓ of a phase jump versus the nonlinear parameter q (Fig. 4a with $\Omega^{-1} = 1$ and Fig. 4b with $\Omega^{-1} = (\sqrt{5} - 1)/2$).

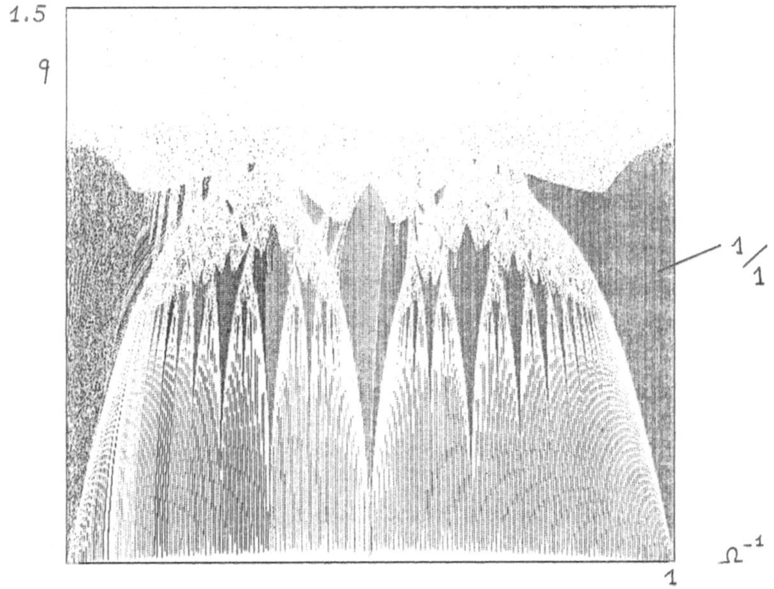

Fig. 3 : *Arnold tongues for the Van der Pol eq. mapping closed to a limit cycle (Eq. 18) in nondimensional parameters* $q = c\,\Omega^{-1} / r_0$, $\Omega^{-1} = \omega_0/\omega$

This transition is shown on Fig. 3 for the main tongue 1/1. We observe an increase of the phase jump amplitude until at $0.79 < q < 0.82$ the phase jump remains constant. When $0.82 < q < 0.83$ four phase jumps are allowed. When $0.83 < q < 1$ a doubling cascade is seen. Phase jump probability density shows no large scale correlation in this region. At q = 1, an abrupt change of behavior happens, large scale jumps are seen and probability densities described in cotg mapping (14) are found.

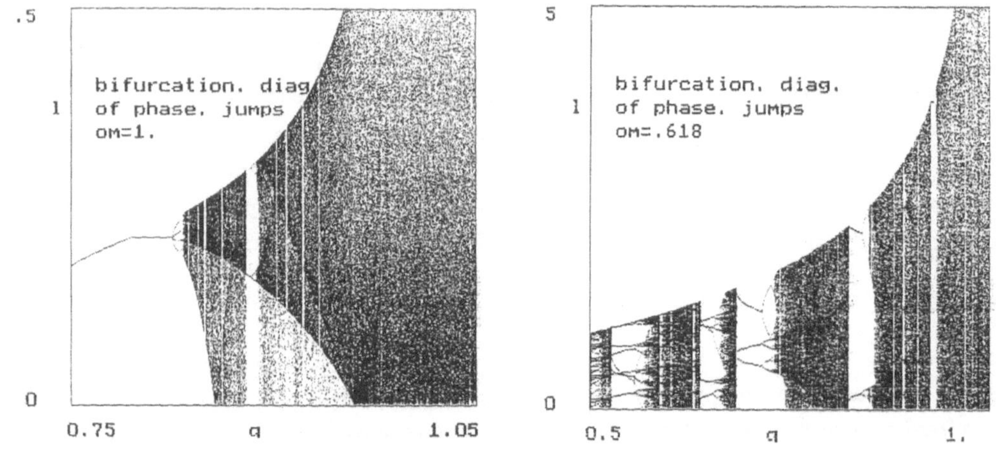

(a) (b)

Fig. 4 : *Bifurcation diagram for the amplitude of phase jump in mapping (18)*

For Düffing equation, nonlinear stability at the fixed point $\theta_0 = 0$ has been studied by assuming that the excitation frequency is such that $\Omega^{-1} = m \in /N$. From (13), $s_0 = (1/2\, a_1 K_1)^{1/3}$. Introducing the new parameter $\bar{q} = q/s_0$, the phase mapping in (12) becomes

$$\theta_{n+1} = \theta_n + \Omega^{-1} + \frac{\bar{q}}{2\pi}\,[1 - \cos 2\pi\theta_n \,/\, (1 + \bar{q}\, \sin 2\pi\theta_n)] \;(mod\, 1) \tag{19}$$

Eq. (19) differs from (18) by the phase shift $\bar{q}/2\pi$. The net effect of this term is to tilt Arnold tongues towards the low winding number region. The transition region between stable and unstable states extends slightly with minor changes.

5. Conclusion and perspectives

The analysis can be persued further by looking at the effect of amplitude-phase coupling on Arnold tongues. The transition to chaos has some spectacular consequences in the intermittent behavior of trajectories in phase space. One example is given for the Morse oscillator at the resonance 1/1 (Fig. 5a) : at starting times the trajectory switches intermittently between the two wells ; then it may escape from the central region and collide erratically at edges of the phase space. For incommensurate frequencies the phase space shows a wavelike trajectory (Fig. 5b).

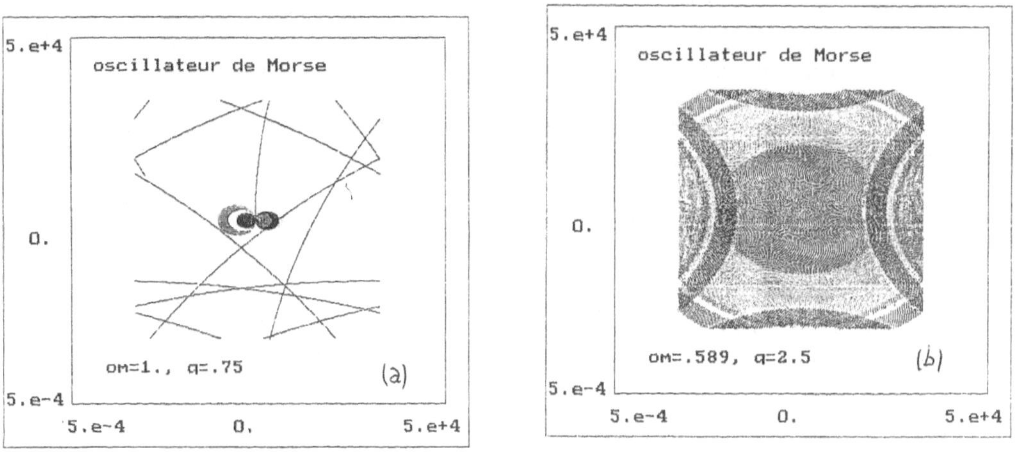

Fig. 5 : *Trajectories in phase space for the Morse eq. mapping (8)*

As a conclusion, a new picture of resonance emerges from the study of Poincaré section in oscillator differential equations. Conditions of KAM theory are not filled since perturbations are nonhamiltonien. For sustained oscillations a transition to chaos through intermittent phase jump doubling leading to Cauchy type density and 1/f frequency noise is found. Numerous applications can be expected : stability of piezoelectric clocks, models of dissociation in complex molecules or celestial mechanics.

Acknowledgements : The author wish to thank J. Miehe for helpful discussions.

References

[1] J. Guckenheimer and P. Holmes, "Nonlinear oscillations, dynamical systems and bifurcation of vector fields", Springer Verlag, N.Y. (1983).

[2] A.H. Nayfeh, D.T. Mook, "Nonlinear oscillations", John Wiley & Sons, N.Y. (1979).

[3] V.I. Arnold, "Mathematical methods of classical mechanics", Springer Verlag, N.Y. (1978).

[4] A.J. Lichtenberg, M.A. Liebermann, "Regular and stochastic motion", Springer Verlag, N.Y. (1983).

[5] M.C. Gutzwiller, "Chaos in classical and quantum mechanics", Springer Verlag, N.Y. (1990).

[6] A. Chernikov, R. Sagdeev and G. Zaslavsky, "Physics Today, nov. 1988, p. 27.

[7] M. Planat, in preparation.

[8] B.V. Chirikov, Physics Reports 52, 264 (1979).

[9] P. Cvitanovic, M.H. Jensen, L.P. Kadanoff and I. Procaccia, Phys. Rev. Lett. 55, 343 (1985).

[10] R.S. Mackay and C. Tresser, Physica 19D, 206 (1986).

[11] Y. Gu and J.M. Yuan, Phys. Rev. A 36, 3788 (1987).

[12] W. Feller, "An introduction to probability theory and its applications", John Wiley & Sons, vol. II, p. 51.

NOISE INDUCED BIFURCATIONS IN SIMPLE NONLINEAR MODELS

Karen Lippert, Konrad Schiele, Ulrich Behn and Adolf Kühnel
Sektion Physik der Universität Leipzig
Augustusplatz 10, O-7010 Leipzig Germany

Abstract

For two generalizations of the Stratonovich model with Gaussian white noise (GWN) and dichotomous Markovian process (DMP) the stationary probability density and moments are calculated. The bifurcation pattern of the stationary solution can be changed qualitativly by varying the deterministic or the noise parameters. We show cases where noise induced states and a subcritical bifurcation can be detected only from the knowledge of the variance.

1. Introduction

We consider systems far from equilibrium described by a simple nonlinear differential equation with few parameters and a noise term,

$$\dot{x} = f(x) + g(x)\cdot R_t \quad . \tag{1}$$

R_t denotes the noise. If the stationary solution x_s of the deterministic equation f(x)=0 bifurcates one observes similarities with phenomena in equilibrium phase transitions associating the position of the maxima of the stationary probability density with the order parameter. For R_t being a GWN ξ_t (with $\langle \xi_t \rangle = 0$, $\langle \xi_t \xi_{t'} \rangle = 2D\cdot\delta(t-t')$) or a DMP I_t (with $\langle I_t \rangle = 0$, $\langle I_t I_{t'} \rangle = \Delta^2 \exp\{-2\alpha|t-t'|\}$) the stationary probability distribution is calculated explicitly following standard methods [1-10]. For multiplicative coupling of the noise with the linear term the bifurcation point is shifted. For suitable chosen f(x) and g(x) the bifurcation type can be controlled by deterministic as well as by noise parameters.

For two generalizations of the Stratonovich model we show in which way the bifurcation pattern can be changed completely shifting the deterministic and the noise parameters and ask wether these changes can be detected already from the knowledge of the first and second moments. This is of interest for more complicated models wich do not allow to determine the stationary probability density.

2. Crossover from Supercritical to Subcritical Bifurcation Driven by Noise

The model is given by the equation

$$\dot{x} = ax - x^3 + (x - bx^3)\cdot R_t \quad , b > 0 \ . \tag{2.1}$$

If the control parameter *a* changes its sign the deterministic part of (2.1) describes a supercritical (forward) bifurcation, cf. Fig.1.

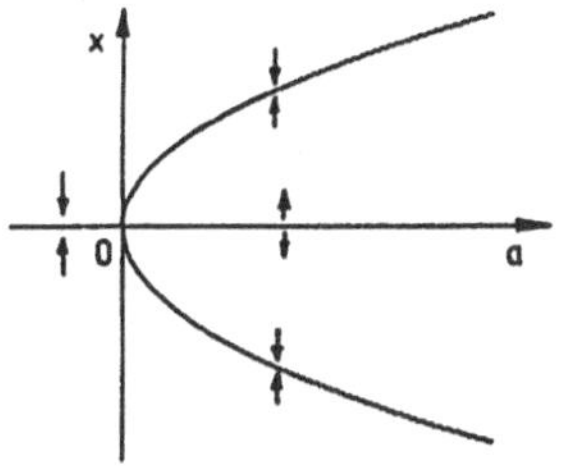

Fig.1. Bifurcation of the stationary solution of Eq.(2.1) in the deterministic case. The arrows show the direction of the flow.

2.1. The Case of GWN

The analysis of the flow yields a certain region in the (x,a)-plane (dashed in Fig. 2) which cannot be left once reached, i.e. this region is the support of the stationary probability density.

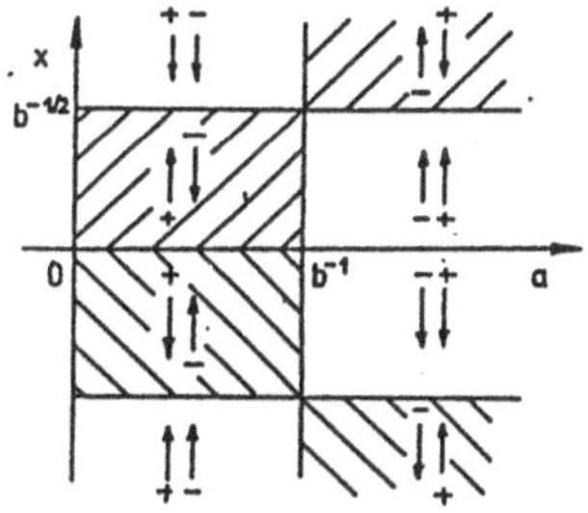

Fig.2. Support of the stationary probability density (2.2)

For a < 0 the stationary probability density P_s is degenerated to $\delta(x)$ like for the Stratonovich model. For a > 0 the stationary probability density and the moments are given by

$$P_s(x|x_o > 0) = N\cdot x^{\frac{a}{D}-1}\cdot|1-bx^2|^{-\frac{a}{2D}-1}\cdot\exp\left\{\frac{1}{2D}(a-b^{-1})/(1-bx^2)\right\} \ , \tag{2.2}$$

$$
\langle x^q \rangle = \begin{cases} \dfrac{\Gamma\left(\frac{a}{2D}+\frac{q}{2}\right)}{\Gamma\left(\frac{a}{2D}\right)}\, b^{-q/2} \left(\dfrac{1-ab}{2Db}\right)^{a/2} \Psi\left(\frac{a}{2D}+\frac{q}{2}\, ,\, \frac{a}{2D}+1\, ;\, \frac{1}{2D}(b^{-1}-a)\right) , & a<b^{-1} \\ \\ b^{-q/2}\, \dfrac{\Psi\left(1-\frac{q}{2}\, ,\, \frac{a}{2D}+1\, ;\, \frac{1}{2D}(a-b^{-1})\right)}{\Psi\left(1,\, \frac{a}{2D}+1\, ;\, \frac{1}{2D}(a-b^{-1})\right)} , & a>b^{-1}. \end{cases} \tag{2.3}
$$

$\Psi(a,b;x)$ denotes the degenerated hypergeometric function [12]. Obviously in the case $a > b^{-1}$ moments of order $q \geq 2$ diverge. The extreme values of P_s obey the equation $3b^2Dx^4 - (4bD - 1)\cdot x^2 + D - a = 0$. Noise induced states are possible iff $4bD>1$. The regions of different qualitative behaviour of P_s are shown in Fig. 3.

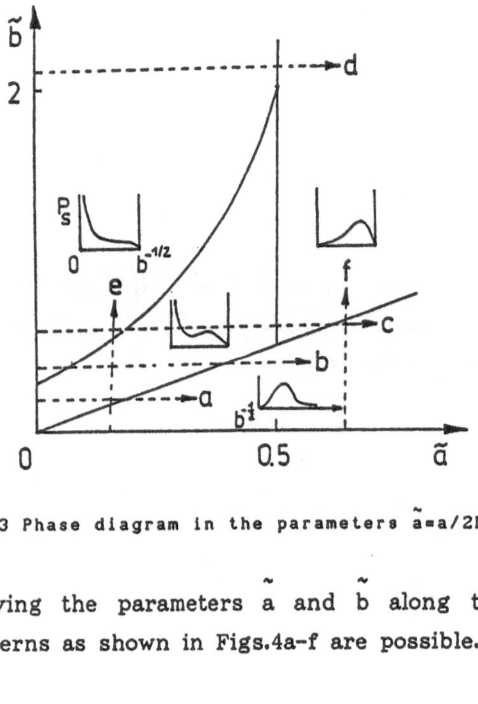

Fig.3 Phase diagram in the parameters $\tilde{a}=a/2D$, $\tilde{b}=1/2bD$.

Varying the parameters $\tilde{a}$ and $\tilde{b}$ along the arrows in Fig.3 bifurcation patterns as shown in Figs.4a-f are possible.

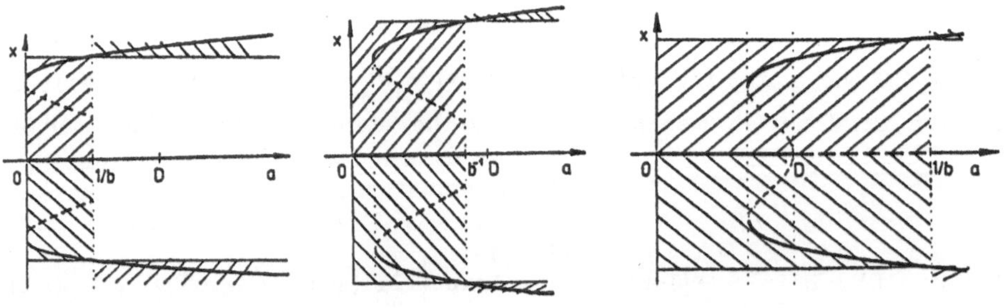

Fig.4a-c. Bifurcation patterns for varying $\tilde{a}$ along arrows a-c in Fig.3.

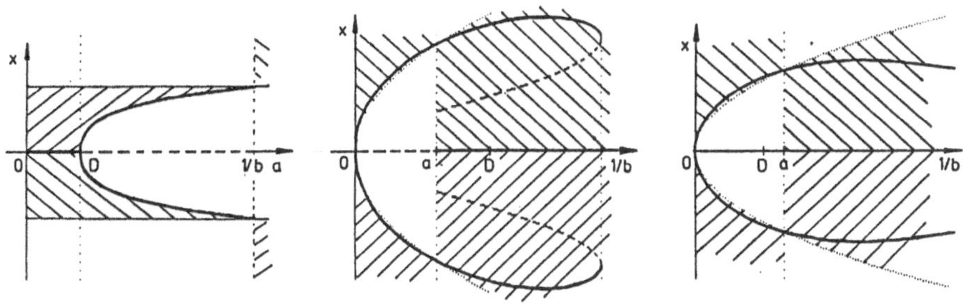

Fig.4d-f. Bifurcation patterns along arrows d-f in Fig.3.

We discuss now the shape of P_s for $a < b^{-1}$. For subcritical bifurcations P_s is bimodal in a certain parameter region. For supercritical parameters P_s is monomodal. The peak becomes sharper with increasing *a* so that the variance should decrease. It is evident that in this case the maximum of the variance must be located at values of *a* smaller than D, i.e. a subcritical maximum of the variance indicates a subcritical bifurcation. These qualitative arguments are supported by the numerical results shown in Fig.5. To distinguish between monomodal and bimodal behaviour one may also consider other integral quantities of P_s, e.g. the entropy.

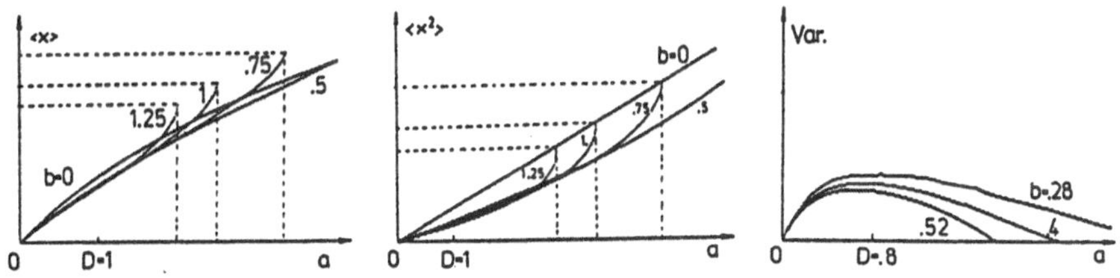

Fig.5. Moments for model (2.1).

2.2. The Case of DMP

The DMP $I_t = \Delta(-1)^{N_t}$, (N_t is a Poisson process) jumps only between two states. To determine the support of the stationary probability density one considers the flow for the realizations $+\Delta$ and $-\Delta$. The solutions of the equation $f(x) \pm \Delta g(x) = 0$ form the borders in between the system is pushed to and fro by the DMP [10]. There are two different possibilities for the shape of the support of P_s shown in Fig.6.

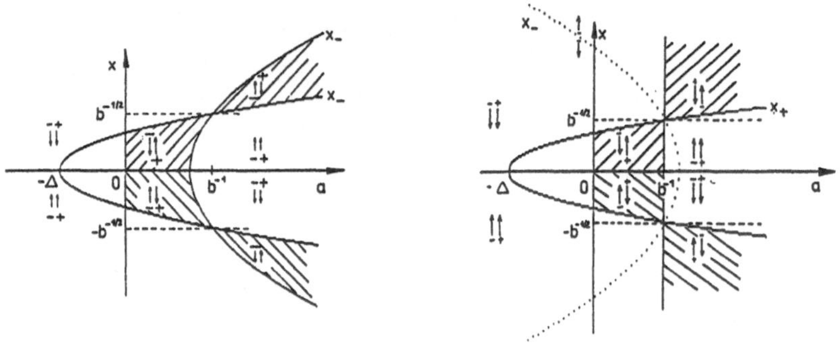

Fig.6. Support of the stationary density (2.4) for $b<1/\Delta$ (a), and $b>1/\Delta$ (b)

In the GWN limit Fig. 6b transformes into Fig. 2. P_s and the moments are calculated as

$$P_s(x|x_o > 0) = N\cdot|1-bx^2|\cdot x^{-2(\lambda+\mu)-1}\cdot|x^2-x_+^2|^{\lambda-1}\cdot|x^2-x_-^2|^{\mu-1} \quad , \tag{2.4}$$

$$\langle x^q \rangle = \begin{cases} x_+^q \dfrac{B(\lambda,-\lambda-\mu+q/2)}{B(\lambda,-\lambda-\mu)}\cdot\dfrac{F\left(q/2-1,\lambda,q/2-\mu;z\right) - x_+^2 b\,\frac{\lambda+\mu-1}{\mu-q/2}\cdot F\left(q/2,\lambda,q/2-\mu;z\right)}{1 + x_+^2/x_-^2\cdot\frac{\lambda}{\mu} - bx_+^2\cdot\frac{\lambda+\mu}{\mu}} \\ \text{with } z=x_+^2/x_-^2 \text{ for } a < \min\{\Delta,1/b\} \\ x_-^q\cdot\dfrac{F\left(\lambda+\mu+1-q/2,\mu,\lambda+\mu;z\right) - bx_-^2 F\left(\lambda+\mu-q/2,\mu,\lambda+\mu;z\right)}{F\left(\lambda+\mu+1,\mu,\lambda+\mu;z\right) - bx_-^2 F\left(\lambda+\mu,\mu,\lambda+\mu;z\right)} \\ \text{with } z=(x_-^2-x_+^2)/x_-^2 \text{ for } \Delta < a < 1/b \\ x_+^q\cdot\dfrac{bx_+^2 F\left(\lambda+\mu-q/2,\lambda,\lambda+\mu;z\right) - F\left(\lambda+\mu+1-q/2,\lambda,\lambda+\mu;z\right)}{bx_+^2 F\left(\lambda+\mu,\lambda,\lambda+\mu;z\right) - F\left(\lambda+\mu+1,\lambda,\lambda+\mu;z\right)} \\ \text{with } z=(x_+^2-x_-^2)/x_+^2 \text{ for } \Delta < 1/b < a \quad , \end{cases} \tag{2.5}$$

where we use the notations $2\lambda=\alpha/(a+\Delta)$, $2\mu=\alpha/(a-\Delta)$ and $x_\pm^2=(a\pm\Delta)/(1\pm b\Delta)$. $F(a,b,c;x)$ is the hypergeometric function and $B(x,y)$ the beta function. For $a < 0$ P_s in (2.4) is not normalizible. For this parameter region $P_s = \delta(x)$ as shown in [9]. The possibilities for the qualitative shape of P_s are the same as for GWN but the phase diagramm is much more complicated due to the additional parameter.The extreme values of P_s obey a cubic equation in $y=x^2$
$3b(b^2\Delta^2-1)y^3 + (2\alpha b+2ab+5-7b^2\Delta^2)y^2 + (5b\Delta^2-2\alpha b-2\alpha-6a+a^2b)y + 2\alpha a+a^2-\Delta^2 = 0.$

3. Crossover from Supercritical to Subcritical Bifurcation Driven by the Control Parameter

We consider another generalization of the Stratonovich model [10,11],

$$\dot{x} = ax + 2bx^3 - x^5 + x \cdot R_t \tag{3.1}$$

Depending on the sign of the control parameter b the stationary deterministic part of (3.1) describes a supercritical bifurcation $(b < 0)$ or a subcritical bifurcation $(b > 0)$, cf. Figs. 7a,b.

Fig.7. Supercritical and subcritical bifurcation of the stationary solution of Eq.(3.1) in the deterministic case.

In the following we focus our interest on the case of a subcritical bifurcation.

3.1 The Case of GWN

Depending on the initial value $x_0 > 0$ $(x_0 < 0)$ the whole upper (lower) half plane is the support of P_s. For P_s and the moments we obtain

$$P_s(x|x_o > 0) = N \cdot x^{\frac{a}{D} - 1} \cdot \exp\left\{ -\frac{1}{D}\left(\frac{x^4}{4} - bx^2 \right) \right\} \quad , \tag{3.2}$$

$$\langle x^q \rangle = \frac{\Gamma\left(\frac{a}{2D} + \frac{q}{2} \right)}{\Gamma\left(\frac{a}{2D} \right)} (2D)^{-q/4} \frac{\mathfrak{D}_{-a/2D-q/2}\left(-b\sqrt{D/2} \right)}{\mathfrak{D}_{-a/2D}\left(-b\sqrt{D/2} \right)} \quad , \tag{3.3}$$

where $\mathfrak{D}_\nu(x)$ denotes the parabolic cylinder function [12].
The extreme values of P_s are given by $x^2 = b \pm \sqrt{b^2 + a - 1/D}$. The phase diagram is shown in Fig. 8.

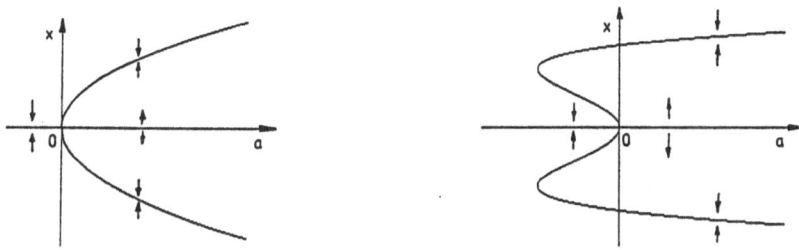

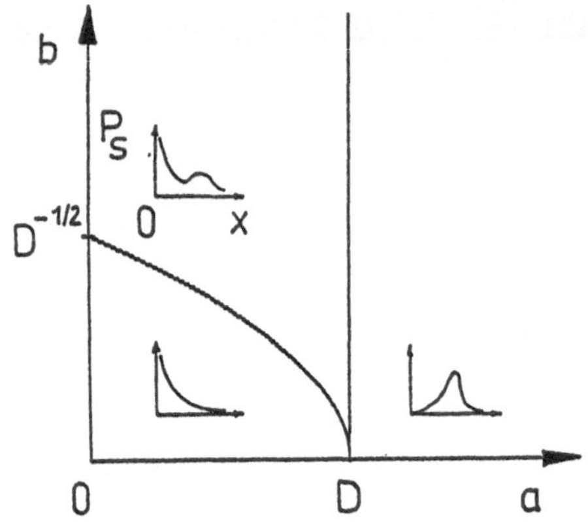

Fig.8.Phase diagram

The bifurcation patterns are similar to those of Figs. 4a, c and d. In Fig. 9 we show the moments and variance as functions of a.

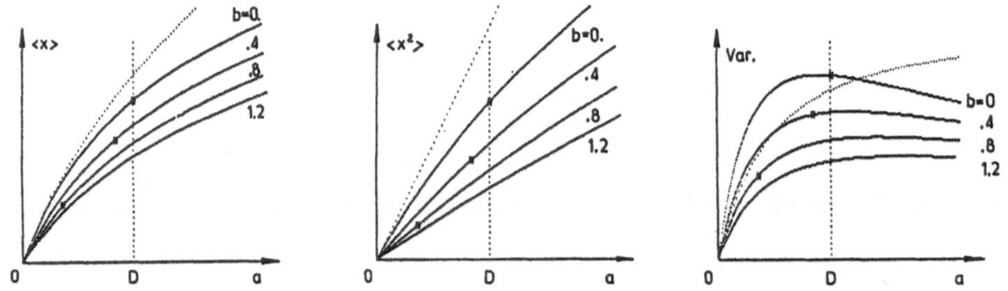

Fig.9. Moments for model (3.1). The dotted lines correspond to the Stratonovich model.

3.2. The Case of DMP

After a certain time the system is trapped between the boundaries x_{++} and x_{+-}, where $x_{\sigma\sigma'} = b+\sigma\sqrt{b^2+a+\sigma'\Delta}$. The shape of the support is shown in Fig.10.

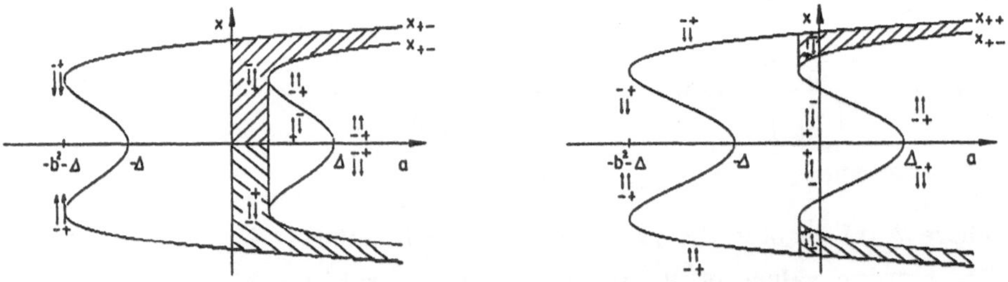

Fig.10. Support of the stationary probability density (3.4) for $b^2>\Delta$ (a) and $b^2<\Delta$ (b).

The stationary probability density is given by

$$P_s(x|x_0 > 0) \sim x^{-1} \prod_{\substack{\sigma \pm 1 \\ \sigma' \pm 1}} x^{-2\lambda_\sigma} \cdot \left(x^4 - 2x_{\sigma\sigma'}^2 x^2 + 2bx_{\sigma\sigma'}^2 + a + \sigma'\Delta \right)^{\chi} \tag{3.4}$$

with the exponents $2\lambda_\sigma = \alpha/(a+\sigma\Delta)$ and $\chi = \alpha/4(a+\sigma'\Delta+bx_{\sigma\sigma'}^2)-1$. The extreme values of P_s obey a quartic equation in $y = x^2$

$$9y^4 - 28by^3 + (20b^2 - 18a - 2\alpha)y^2 + (4\alpha b + 12ab)y + 2\alpha a + a^2 - \Delta^2 = 0$$

The bifurcation of the extreme values occurs at $a_c = \min\left(\Delta - b^2, -\alpha + \sqrt{\alpha^2 + \Delta^2}\right)$. Due to the four parameters of the model the phase diagramm is complicated again. In analogy to the situation in Section 2 different bifurcation patterns may be obtained changing the parameters in an appropriate way. The moments can be expressed by series expansions. A detailed discussion is however feasible only for fixing special values of the parameters.

References

[1] Kitahara,K.; Horsthemke,W.; Lefever,R.: Phys.Lett. 70A (1979) 377

[2] Kitahara,K.; Horsthemke,W.; Lefever,R.; Inaba,Y.: Prog.Theor.Phys. 64 (1980) 1233

[3] Klyatskin,V.I.: Izv.VUZ Radiofiz. (UdSSR) 20 (1977) 562

[4] Pawula,R.F.: IEEE Trans.Inf.Theory 13 (1967) 33

[5] Pawula,R.F.: Internat.J.Control 25 (1977) 238

[6] Pomeau,Y.: J.Stat.Phys. 24 (1981) 189

[7] Sancho,J.M.; San Miguel,M.: Prog.Theor.Phys.69 (1983) 1085

[8] Behn,U.;Schiele,K.;Teubel,A.:Wiss.Z.KMU Math.-Naturwiss.R. 34 (1985) 6

[9] Teubel,A.; Behn,U.; Kühnel,A.: Z.Phys.B Condensed Matter 71 (1988) 393

[10] Behn,U.; Schiele,K.: Dubna-Preprint EE 17-84-300 (1984)

[11] Stratonovich,R.L.: *Topics in the theory of random noise.* Gordon and Breach New York 1967 Vol.2

[12] Abramowitz,M.;Stegun,I.A. (eds) :*Pocketbook of mathematical functions.* Thun, Frankfurt/Main Deutsch 1984

Coherent Behaviour of Single Degrees of Freedom in an Order-to-Chaos Transition

A. Campa, A. Giansanti and A. Tenenbaum

Dipartimento di Fisica
Università di Roma "La Sapienza"
Piazzale A. Moro 2, 00185 Roma (Italy)

Introduction

A quite relevant theme in biological physics is the coherent energy transduction at the macromolecular level. One of the main theoretical problems in this field is the construction of a realistic model for nondissipative intramolecular energy transfer, through a nonlinear coupling among different degrees of freedom (DOFs).

In the recent past a soliton model, originally proposed by Davydov, has been extensively studied [1] in the general context of energy transport in proteins. We contributed [2,3,4] to the subject with a molecular dynamics study of acetanilide (ACN), a model for α-helical regions in proteins. In our work on ACN chains we have investigated the dynamics of the DOFs involved in the soliton generation and propagation. If a soliton travels along such a chain, one may have an ordered dynamics in a limited region of space, and for a limited time; the ordered region would move along the chain with the soliton. Because of the need to recognize those vibrations that can sustain ordered motions over adequate time scales, in a background of chaotic uncorrelated motions, we have elaborated new diagnostic tools to analyse the dynamical coherence of each DOF in a complex molecular system.

In a highly chaotic regime one does not expect to observe qualitatively different behaviours among the different DOFs: equipartition of energy holds and memory of the initial conditions is rapidly lost. However, in the transition from a fully chaotic regime to a regime dominated by ordered motions, one expects to find a mixed situation where part of the system may have a degree of chaoticity which is different from the rest of the system, and which could also vary in time. The usual indicators of order and chaos (e.g. Lyapunov spectra [5], fractal dimensions [6], spectral entropies [7]), which give a global information on the system and are based on asymptotic time scale estimates, would not be useful, e.g., in identifying in the complex structure of a molecular chain the DOFs involved in a soliton. In this work we propose new indicators: partial Lyapunov exponents, computed from the dynamics of the tangent space vector associated with a given dynamical system. The finite time analysis of the growth rate of the components of this vector will give the required knowledge on the chaoticity of the single DOFs. In order to test our diagnostics we study here a simple dynamical system exhibiting a large range of characteristic frequencies.

The model

Our model consists of a system of five nonlinearly coupled linear oscillators. The hamiltonian of the model is given by:

$$H = \frac{1}{2}\sum_{i=1}^{5}\left(p_i^2 + \omega_i^2 q_i^2\right) + \frac{1}{2}\sum_{i\neq j}^{1,5} q_i^2 q_j^2. \tag{1}$$

In one case, which we call the single gap system (SG), the values of the frequencies are the following: $\omega_1 = 1$, $\omega_2 = \frac{\pi}{2}$, $\omega_3 = 2$, $\omega_4 = 10e$, and $\omega_5 = 30$; in another case, which we call the double gap system (DG), we put $\omega_5 = 90$. The choice of the last two frequencies, an order of magnitude higher than the first three, is aimed at showing the influence of a broad range of frequencies with a gap on the dynamics of the various DOFs.

We have numerically integrated both the equations of motion and the variation equations of motion with the central difference algorithm. In our case the second order equations of motion are:

$$\ddot{q}_i = -\omega_i^2 q_i - 2\sum_{j(\neq i)}^{1,5} q_j^2 q_i \qquad i = 1,\ldots,5. \tag{2}$$

The equations of motion in the tangent space (variation equations) are related to the stability of the orbits; they are usually written as first order equations, obtainable from the hamiltonian equations of the system. If $\dot{x}_i = F_i(\{x_j\})$ are the hamiltonian equations of motion, where x_i is one of the q_i or one of the p_i, then the equations of motion of the tangent space vector $\vec{y}$ corresponding to the variation of $\vec{x}$ are given by:

$$\dot{y}_i = \sum_k \frac{\partial F_i(\{x_j\})}{\partial x_k} y_k \tag{3}$$

where the factor of y_k is computed along the trajectory $\vec{x}(t)$. Here we write (3) as second order equations:

$$\ddot{\xi}_i = -\left(\omega_i^2 + 2\sum_{j(\neq i)}^{1,5} q_j^2(t)\right)\xi_i - 4\sum_{j(\neq i)}^{1,5} q_i(t)q_j(t)\xi_j, \qquad \eta_i \equiv \dot{\xi}_i,$$

$$i = 1,\ldots,5, \tag{4}$$

where $(\xi_1,\ldots,\xi_5,\eta_1,\ldots,\eta_5) \equiv \vec{y}$, with ξ_i and η_i corresponding to the variations of q_i and p_i respectively.

From eqs. (4) one can compute the Lyapunov spectrum in the usual way [5]; in particular, the maximum Lyapunov exponent is given by:

$$\lambda_M = \lim_{t\to+\infty} \lambda(t), \qquad \lambda(t) = \frac{1}{t} Log\frac{|\vec{y}(t)|}{|\vec{y}(0)|} \tag{5}$$

where $\vec{y}(0)$ is the initial vector in the tangent space, taken randomly. We have defined new partial Lyapunov exponents (PLEs), referring to single DOFs:

$$\lambda_i = \lim_{t \to +\infty} \lambda_i(t), \qquad \lambda_i(t) = \frac{1}{2t} Log \frac{\omega_i^2 \xi_i^2(t) + \eta_i^2(t)}{\omega_i^2 \xi_i^2(0) + \eta_i^2(0)}, \qquad i = 1, \ldots, 5. \tag{6}$$

We want to remark the following point. It is a necessary condition that at least one of the λ_i is equal to λ_M, the maximum Lyapunov exponent defined in (5); but, a priori, some of them could be smaller. However, a generic initial vector in the tangent space will expand in modulus like $e^{\lambda_M t}$ when $t \to \infty$ with probability one. Therefore, also λ_i, for each i, should be equal to λ_M with probability one. However, this is true only in the asymptotic limit; it is not to be expected that the $\lambda_i(t)$s, at finite times, should be equal. As a matter of fact, their differences are a central point of our investigation. It turns out that, in the transition region between ordered and chaotic motions, there are still significant differences in the values of the $\lambda_i(t)$s for times three or four orders of magnitude higher than the characteristic periods of the oscillators. This is the manifestation of a qualitative diversity in the behaviour of the single DOFs. We have studied the differences among the $\lambda_i(t)$s by computing the quantities $\delta_i(t) = (\lambda(t) - \lambda_i(t))/\lambda_M$, and we have defined a characteristic "coherence time" for each DOF through:

$$\tau_i = \frac{1}{T} \int_0^T t \delta_i(t) dt. \tag{7}$$

Maximum Lyapunov exponents

We have studied the two versions of our model at different values of ϵ, the energy per DOF, or energy density. We have computed the maximum Lyapunov exponent (λ_M) to find out whether a transition in the dynamics takes place and the energy range in which this happens. In Fig. 1 we show λ_M at different energy densities for the SG version of our model. It is evident from the sharp change in the slope of $Log(\lambda_M)$ that such a transition takes place indeed [8], and is located between $\epsilon = 0.9$ and $\epsilon = 1$. Below this threshold the slope is strongly positive: λ_M changes by four orders of magnitude, passing from $\epsilon = 0.7$ to $\epsilon = 1$. Above the threshold the slope becomes small. The value of λ_M for the DG version at $\epsilon = 0.7$ is 1.04×10^{-3}, much larger than the corresponding value in the SG version (0.79×10^{-4}). This point will be discussed in the last section.

In the frame of the KAM theorem [9] it is interesting to contrast the values of λ_M with the anharmonicity of the system. We have computed the ratio R of the anharmonic to the total harmonic energy, and we report in Table I its values at the different energies that we have simulated. It is interesting to note that at $\epsilon = 0.6$, where the system SG gives $\lambda_M = 0$, the anharmonicity R is still as high as 4.7% . This shows that for this system - with a significant gap in the frequency spectrum - the ratio of the perturbed part of the hamiltonian to the harmonic part may reach quite relevant values before the KAM tori begin being destroyed.

TABLE I

ϵ	0.6	0.7	0.8	0.9	1.0	1.5	2.0
$R(\%)$	4.7	5.1	5.5	4.5	5.8	6.4	7.4

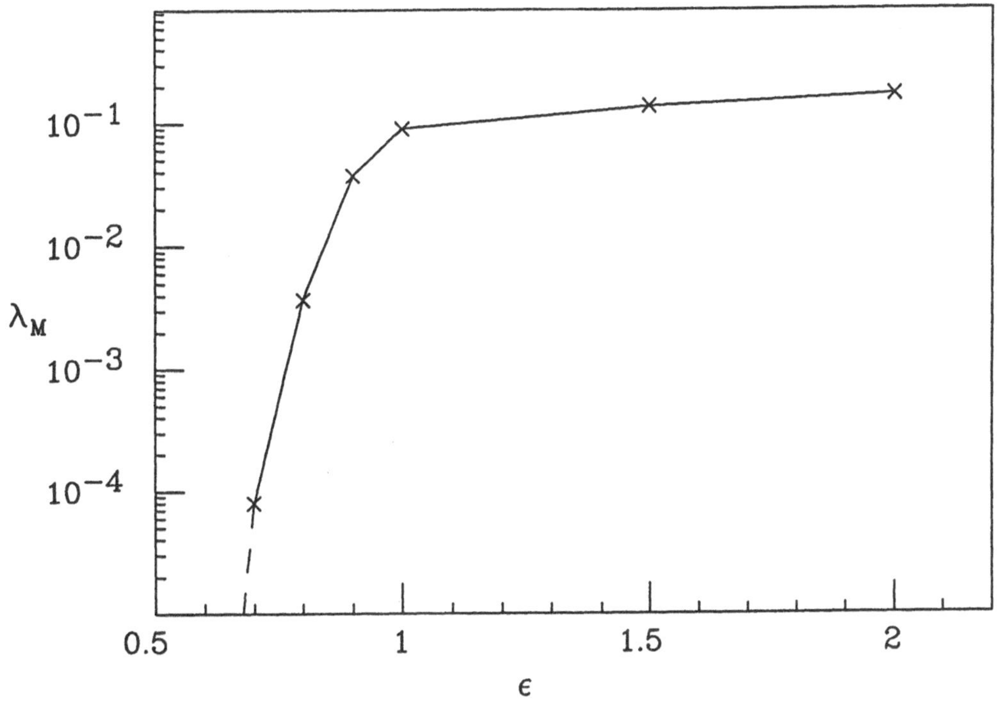

Fig. 1. Maximum Lyapunov exponent vs. energy density (SG case).

Correlation functions

The statistical properties of the dynamics of the single DOFs may be characterized by the autocorrelation functions (ACFs). We have collected in Fig. 2 these functions for the harmonic energies of the different DOFs for the SG case, at total energy density $\epsilon = 0.9$.

Some of the graphs are symmetric with respect to the zero line. This happens because in Fig. 2 only maxima and minima over groups of 100 computed points are plotted, in order to have clear graphs over large times. In the case of an ACF rapidly oscillating around zero this appears as a symmetrical graph. This way of plotting is sufficient if one is interested in studying the decay of the ACFs. From this point of view, they exhibit quite different patterns: there is a clear distinction between the high-

and low-frequency DOFs at that energy; while the high-frequency ACFs show a short decay time, the low-frequency ones exhibit a large decay time.

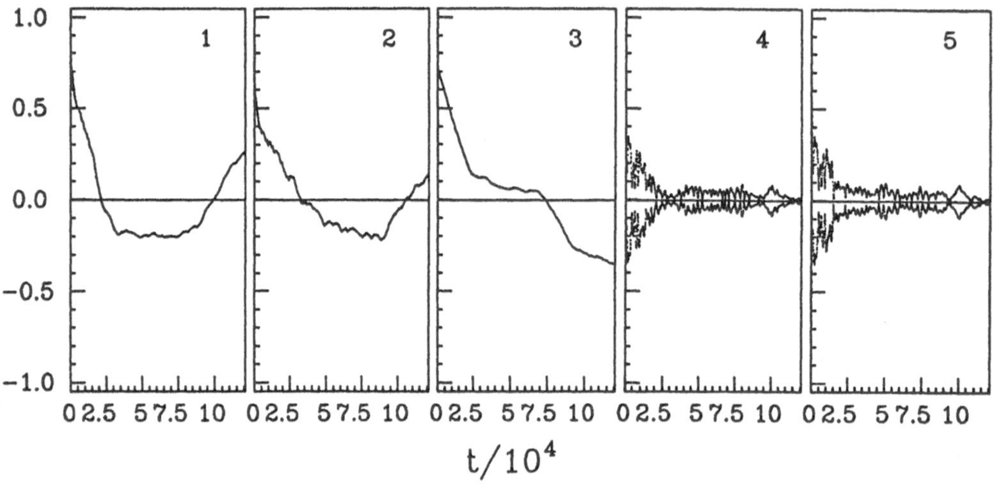

Fig. 2. Autocorrelation functions of the harmonic energy for the five degrees of freedom (SG case), at $\epsilon = 0.9$.

Partial Lyapunov Exponents

We have computed the PLEs for the five degrees of freedom of our system at energies near to the transition point individuated by λ_M. In Fig. 3 we show a graph with five curves corresponding to the DOFs for the SG case, at $\epsilon = 0.9$. It is clear that, because of the gap, the curves behave in very different ways; a group, corresponding to the DOFs with small eigenfrequencies, oscillates around zero; a second group, corresponding to the large frequencies, decays to zero only over large times. It should be noted here that the dynamical behaviour of the individual DOFs, as derived from the PLEs, is not related in a systematic way to the complementary information that can be deduced from the ACFs. Thus, while the PLEs of the high-frequency DOFs show a slower divergence of those DOFs for nearby trajectories (i.e., a localized ordered behaviour) than for the low-frequency DOFs, the corresponding ACFs at the same energy give a different indication, i.e., that the autocorrelation time is shorter for the high frequencies.

Using (7), one finds $\tau_4 = 182$ and $\tau_5 = 192$. The values computed for the low-frequency DOFs are not significant because the variance of the integrand in (7) is so large (in particular, larger than τ_i) that τ_i itself loses its meaning, as it could be expected looking at Fig. 3. The variance of the integrand in (7) for $i = 4,5$ is small, which shows that the corresponding functions $\delta_i(t)$ are close to hyperbola with the x-axis as asymptote.

A point has to be remarked. One could try to assign a characteristic decay time to the ACFs shown in the previous section. It is evident, inspecting Fig. 2, that the shape of the ACFs in most cases does not allow to define a characteristic time of a decay process. Indeed, either the ACFs are correlated over times which are even greater than the whole simulation time (DOFs 1, 2 and 3), or the structure is quite irregular (DOFs 4 and 5). At best, in the last case, one could roughly identify in the ACFs a superposition of two decaying processes with completely different characteristic times [10].

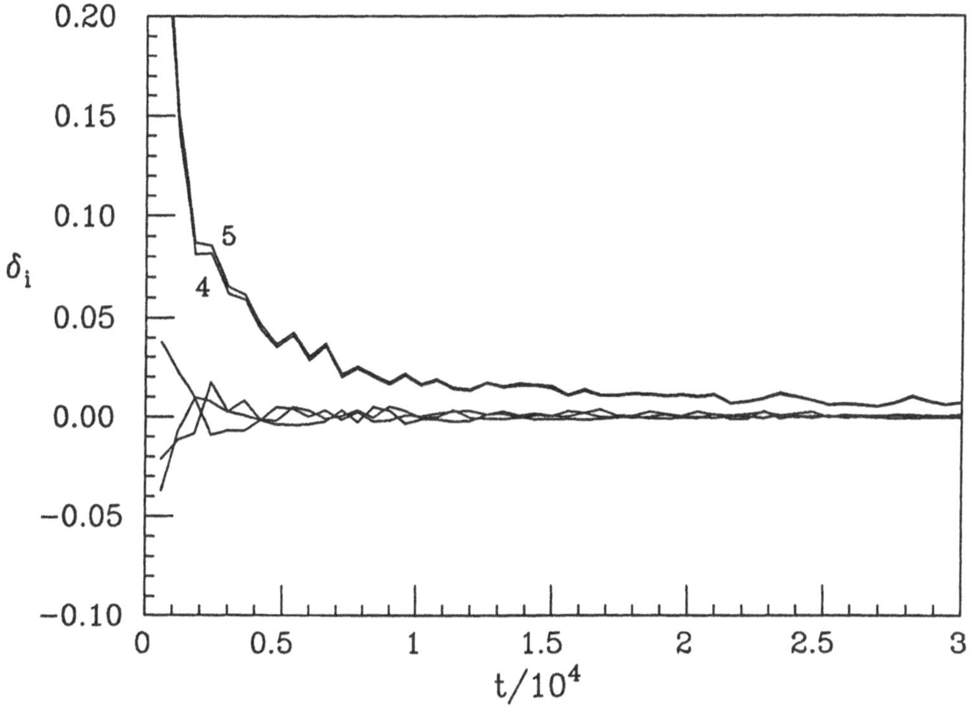

Fig. 3. Functions δ_i vs. time at $\epsilon = 0.9$ (SG case).

We have computed the LPEs also for the DG version at $\epsilon = 0.7$. The LPEs are shown in Fig. 4, using the same method as in Fig. 3. One can see that in this version, where the high frequencies are separated, the corresponding $\delta_i(t)$s are also separated, giving rise to quite different τ_is: $\tau_4 = 5400$, $\tau_5 = 7800$. Each τ_i of the high-frequency DOFs turns out to be approximately inversely proportional to λ_M, when ϵ is changed; the proportionality constant is approximately the same in both versions of the model for $i = 4$. More generally, looking at all the results, this constant seems to depend only on the frequency of each DOF.

Discussion and Comments

As noted before, the DG version has a λ_M which is much larger, at $\epsilon = 0.7$, than the λ_M of the SG version. That is, the system with the larger frequency gap is more chaotic. This shows how a prediction of the qualitative features of the dynamics of a system based simply on an inspection of the frequencies (or frequency differences) of the DOFs may be misleading. On the other hand, the characteristic coherence times τ_i here introduced are able to single out the behaviour of each DOF also in a complex situation, in which a clear definition of a correlation time through the ACFs may not be possible (as shown in Fig. 2).

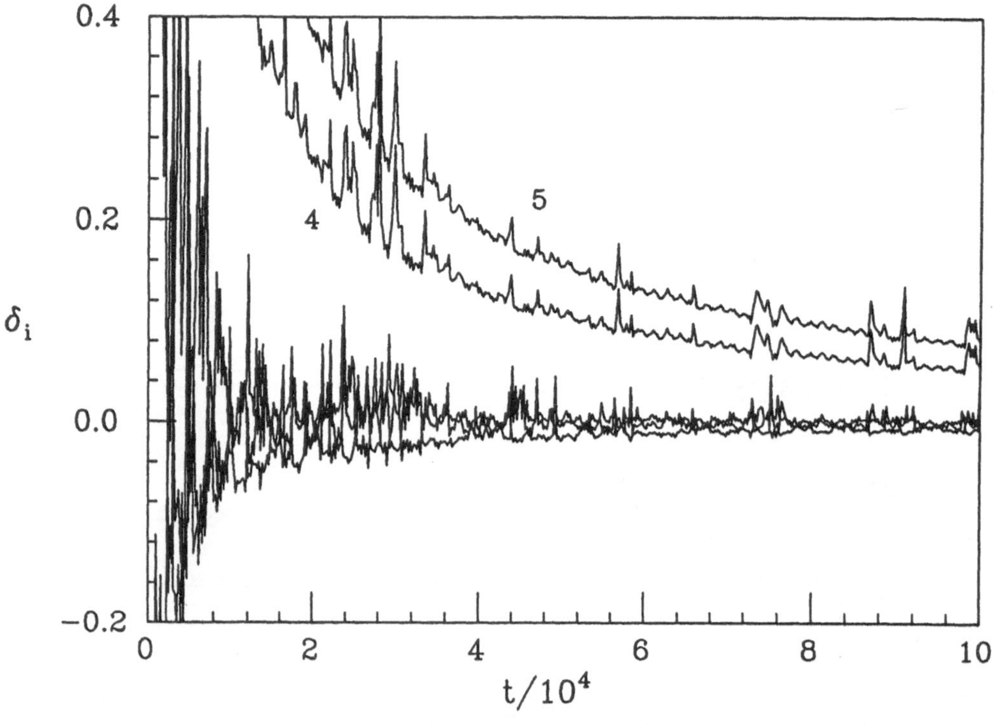

Fig. 4. Functions δ_i vs. time at $\epsilon = 0.7$ (DG case).

The shape of the curves given in Fig. 4 allows also to gain insight into the intermittent character of the dynamics of a single DOF. The peaks which accompany the time decrease of the curves indicate that the corresponding DOF have just had transient phases of more coherent dynamics, i.e., of lower values of the corresponding $\lambda_i(t)$. If one computed the $\lambda_i(t)$s over time intervals comparable with the average width of the peaks, one whould find a step-wise pattern, with short intervals of very coherent dynamics for the single DOFs. The coincidence of these peaks in different curves gives a clear indication of a cross-correlation of the coherent character in the dynamics of the DOFs.

In conclusion, we would like to stress an important feature of the new diagnostics we have introduced. When simulating a real system, one has always to use simplified models; it is therefore necessary to ask oneself if the results derived from the computer experiments are reliable. In this context the question of structural stability of the equations of the model is of utmost importance. As this question is usually very difficult to answer for complex systems, one can think of using the PLEs to check whether the DOFs which are more relevant in the description of the phenomena of interest in the simulation have long coherence times. If this were the case, one could argue that these DOFs would also be structurally stable for small changes introduced in the equations of motion, in the same way in which they keep coherence for small perturbations of their trajectory.

More fundamentally, the computation of the coherence times of the DOFs in a realistic system, say in a biological macromolecule, could be a clue to understand how certain specific DOFs are able to perform ordered dynamical sequences over time intervals which are orders of magnitude larger than their characteristic period, while the others simply vibrate in a disordered, thermal way.

References

[1] See P. L. Christiansen and A. C. Scott, eds., *Davydov's soliton revisited: Self-trapping of vibrational energy in proteins*, (Plenum Publishing Co., New York, 1990).
[2] A. Campa, A. Giansanti and A. Tenenbaum, *Phys. Rev.* **B36**, 4394 (1987).
[3] A. Tenenbaum, A. Campa and A. Giansanti, *Phys. Lett.* **A121**, 126 (1987).
[4] A. Giansanti, A. Campa, D. Levi, O. Ragnisco and A. Tenenbaum, in Ref. [1], p. 439.
[5] G. Benettin, L. Galgani and J. M. Strelcyn, *Phys. Rev.* **A14**, 2338 (1976).
[6] E. Ott, *Rev. Mod. Phys.* **53**, 655 (1981).
[7] R.Livi, M. Ruffo, M. Sparpaglione and A. Vulpiani *Phys. Rev.* **A31**, 1039 (1985).
[8] M. Pettini and M. Landolfi, *Phys. Rev.* **A41**, 768 (1990).
[9] See, for example, G. Gallavotti: *The Elements of Mechanics* (Springer, New York, 1983); p. 466-470.
[10] G. Benettin and A. Tenenbaum, *Phys. Rev.* **A28**, 3020 (1983).

Dissipation in Quantum Field Theory

E. Celeghini[(1)], M. Rasetti[(2)] and G. Vitiello[(3)]

(1) Dipartimento di Fisica dell'Universitá and I.N.F.N., Firenze , Italy

(2) Dipartimento di Fisica and I.N.F.M., Politecnico di Torino, Italy

(3) Dipartimento di Fisica, Universitá di Salerno and I.N.F.N., Sezione di Napoli, Italy

A major difficulty that appears in the study of dissipative systems in Quantum Mechanics is that the canonical commutation relations (*CCR*) are not preserved by time evolution due just to damping terms. Then one introduces fluctuating forces in order to preserve the quantum mechanical consistency, namely the canonical structure. Another way to handle the problem is to start from the beginning with an Hamiltonian which describes the system, the bath and the system-bath interaction. Subsequently, one eliminates the bath variables which originate both damping and fluctuations, thus obtaining the reduced density matrix.

Purpose of the present paper is to discuss some aspects of dissipation in Quantum Field Theory (*QFT*) resorting just to the relatively simple example of the damped harmonic oscillator, whose classical equation of motion is

$$m\ddot{x} + \gamma\dot{x} + \kappa x = 0 \quad . \tag{1}$$

Contrary to the more traditional attitude by which one tries to accomodate the non-unitary character of time-evolution implied by dissipation in the familiar framework of the unitary operator algebra, our task here is to provide a picture of non-unitary evolution at a quantum level without forcing or reducing it to the framework of unitary operator algebra. In the following we show that in *QFT* there is enough room to produce such a picture if one takes advantage of the existence of infinitely many unitarily inequivalent representations of the *CCR*. As a prize for such an unconvenctional attitude, the statistical nature of dissipative phenomena naturally emerges from our formalism and some light is shed on the question of if and where the arrow of time can be find at microscopic level.

In order to deal with an isolated system, as the canonical quantization scheme requires, a procedure of doubling of the phase-space dimension is necessary[1]. Thus the lagrangian for system (1) is written as

$$L = m\dot{x}\dot{y} + \frac{1}{2}\gamma(x\dot{y} - \dot{x}y) - \kappa xy \quad , \tag{2}$$

where y denotes the position variable for the *doubled* system. Intuitively one expects the y variable to grow as rapidly as the x solution decays: in this sense y may be thought of as describing an effective degree of freedom for the heat bath to which the system (1) is coupled. (1) is obtained by varying (2) with respect to y, whereas variation with respect to x gives indeed $m\ddot{y} - \gamma\dot{y} + \kappa y = 0$, which appears in fact to be the *time reversed* ($\gamma \to -\gamma$) of (1). The canonical momenta p_x and p_y (the collection of dynamical variables $\{x, p_x, y, p_y\}$ spans the new phase-space) are then given by $p_x \equiv \frac{\partial L}{\partial \dot{x}} = m\dot{y} - \frac{1}{2}\gamma y$; $p_y \equiv \frac{\partial L}{\partial \dot{y}} = m\dot{x} - \frac{1}{2}\gamma x$. Canonical quantization may then be performed by introducing the commutators $[x, p_x] = i\hbar = [y, p_y]$, $[x, y] = 0 = [p_x, p_y]$, and the corresponding sets of annihilation and creation operators $a, a^\dagger, b$ and $b^\dagger$. Performing the linear canonical transformation $A \equiv \frac{1}{\sqrt{2}}(a+b)$, $B \equiv \frac{1}{\sqrt{2}}(a-b)$, the quantum hamiltonian is obtained

$$\begin{aligned} \mathcal{H} &= \mathcal{H}_0 + \mathcal{H}_I \quad , \\ \mathcal{H}_0 = \hbar\Omega(A^\dagger A - B^\dagger B) \quad &, \quad \mathcal{H}_I = i\hbar\Gamma(A^\dagger B^\dagger - AB) \quad , \end{aligned} \tag{3}$$

where $\Gamma \equiv \frac{\gamma}{2m}$ is the decay constant for the classical variable $x(t)$. It is easy to realize that the dynamical group structure associated with our system of coupled quantum oscillators is that of $SU(1,1)$. The two mode realization of the algebra $su(1,1)$ is indeed generated by

$$J_+ = A^\dagger B^\dagger \quad , \quad J_- = J_+^\dagger = AB \quad , \quad J_3 = \frac{1}{2}(A^\dagger A + B^\dagger B + 1) \quad , \tag{4}$$

corresponding to the Casimir operator $\mathcal{C}$ defined as: $\mathcal{C}^2 \equiv \frac{1}{4} + J_3^2 - \frac{1}{2}(J_+J- + J_-J_+)$ $= \frac{1}{4}(A^\dagger A - B^\dagger B)^2$. Note that $[\mathcal{H}_0, \mathcal{H}_I] = 0$, as $\mathcal{H}_0$ is in the center of the dynamical algebra.

Let us denote by $\{|n_A, n_B>\}$ the set of simultaneous eigenvectors of $A^\dagger A$ and $B^\dagger B$, with n_A, n_B non-negative integers. Let the initial state be the *vacuum* $|n_A = 0, n_B = 0> \equiv |0>$, such that $A|0> = 0 = B|0>$, its time-evolution is given by

$$\begin{aligned} |0(t)> = \exp\left(-it\frac{\mathcal{H}}{\hbar}\right)|0> &= \exp\left(-it\frac{\mathcal{H}_I}{\hbar}\right)|0> \\ = \frac{1}{\cosh(\Gamma t)} &\exp\left(\tanh(\Gamma t) J_+\right)|0> \quad , \end{aligned} \tag{5}$$

namely a two-mode Glauber coherent state [1, 2, 3] (*i.e.* a generalized coherent state for $su(1,1)$). Eq. (5) shows that at every time t the state $|0(t)>$ has unit norm, however as $t \to \infty$ it gives rise to an asymptotic state which is orthogonal to the initial state $|0> : <0(t)|0(t)> = 1$ and

$$\lim_{t\to\infty} <0(t)|0> = \lim_{t\to\infty} \exp\left(-\ln\cosh(\Gamma t)\right) \to 0 \quad . \tag{6}$$

Eq. (6) expresses the instability (*decay*) of the vacuum under time evolution operator $\mathcal{U} \equiv \exp\left(-it\frac{\mathcal{H}_I}{\hbar}\right)$. We reach thus the conclusion that as an effect of damping (recall that $\mathcal{H}_I \to 0$ as $\gamma \to 0$) the time-evolution generator $\mathcal{H}_I$ leads outside the original Hilbert space of states. As the ordinary quantization procedure requires that a definite representation of the *CCR* should be used in order to describe the system under study, we say that the time-evolution generator of the damped oscillator, by leading outside the original Hilbert space, produces *unquantization*. This is an obstruction which is to be bypassed in producing a canonical scheme. One should notice that so far the problem was tackled within the framework of Quantum Mechanics, namely within a scheme were the von Neumann theorem only allows unitarily equivalent representations of *CCR*. In what follows, we intend to show how the pathology exhibited by (6) can be controlled if one operates in a different scheme, giving up in particular the condition of finiteness of the number of degrees of freedom. This, in turn, is obviously equivalent to moving to a second quantization scheme, *i.e.* to a *QFT*, where the infinite number of degrees of freedom allows the coexistence of infinitely many unitarily inequivalent representations of the *CCR*'s. The most straightforward extension to *QFT* of the hamiltonian given by eq. (3), describing an (infinite) collection of damped harmonic oscillators, is

$$\mathcal{H} = \mathcal{H}_0 + \mathcal{H}_I \quad ,$$
$$\mathcal{H}_0 = \sum_\kappa \hbar\Omega_\kappa (A_\kappa^\dagger A_\kappa - B_\kappa^\dagger B_\kappa) \quad , \quad \mathcal{H}_I = i\sum_\kappa \hbar\Gamma_\kappa (A_\kappa^\dagger B_\kappa^\dagger - A_\kappa B_\kappa) \quad , \tag{7}$$

where κ labels the field degrees of freedom, *e.g.* spatial momentum. As usual, the computational strategy is now to work at finite volume of the system V, and to perform at the end the limit $V \to \infty$. The commutation relations are:

$$[\, A_\kappa, A_\lambda^\dagger \,] = \delta_{\kappa,\lambda} = [\, B_\kappa, B_\lambda^\dagger \,] \quad ; \quad [\, A_\kappa, B_\lambda^\dagger \,] = 0 = [\, A_\kappa, B_\lambda \,] \quad . \tag{8}$$

We still have $[\, \mathcal{H}_0, \mathcal{H}_I \,] = 0$, and corresponding to eq. (5) we have (formally, at finite volume V),

$$|0(t)> = \prod_\kappa \frac{1}{\cosh(\Gamma_\kappa t)} \exp\left(\tanh(\Gamma_\kappa t) J_+^{(\kappa)}\right)|0> \quad , \tag{9}$$

with $J_+^{(\kappa)} \equiv A_\kappa^\dagger B_\kappa^\dagger$. Moreover $<0(t)|0(t)> = 1 \;\; \forall t$, and

$$<0(t)|0> = \exp\left(-\sum_\kappa \ln\cosh(\Gamma_\kappa t)\right) \quad ; \tag{10}$$

which shows how, provided $\sum_\kappa \Gamma_\kappa > 0$,

$$\lim_{t\to\infty} <0(t)|0> \propto \lim_{t\to\infty} \exp\left(-t\sum_\kappa \Gamma_\kappa\right) = 0 \quad . \tag{11}$$

Now, using the customary continuous limit relation $\sum_\kappa \mapsto \frac{V}{(2\pi)^3}\int d^3\kappa$, in the infinite-volume limit we have (for $\int d^3\kappa\, \Gamma_\kappa$ finite and positive)

$$\begin{aligned} &<0(t)|0> \underset{V\to\infty}{\longrightarrow} 0 \quad \forall t \quad , \\ &<0(t)|0(t')> \underset{V\to\infty}{\longrightarrow} 0 \quad \forall t, t' \quad , \quad t \neq t' \quad . \end{aligned} \tag{12}$$

We notice that the time-evolution transformations

$$\begin{aligned} A_\kappa \mapsto A_\kappa(t) &= e^{-i\frac{t}{\hbar}\mathcal{H}_I} A_\kappa e^{i\frac{t}{\hbar}\mathcal{H}_I} = A_\kappa \cosh(\Gamma_\kappa t) - B_\kappa^\dagger \sinh(\Gamma_\kappa t) \quad ; \\ B_\kappa \mapsto B_\kappa(t) &= e^{-i\frac{t}{\hbar}\mathcal{H}_I} B_\kappa e^{i\frac{t}{\hbar}\mathcal{H}_I} = -A_\kappa^\dagger \sinh(\Gamma_\kappa t) + B_\kappa \cosh(\Gamma_\kappa t) \quad , \end{aligned} \tag{13}$$

and their hermitian conjugates can each be implemented, for every κ, as inner automorphism for the algebra $su(1,1)_\kappa$. Such an automorphism is nothing but the well known Bogolubov transformations. The transformations (13) are canonical, as they preserve the CCR (8). So, for each t we have a copy $\{A_\kappa(t), A_\kappa^\dagger(t), B_\kappa(t), B_\kappa^\dagger(t);\ |0(t)>\ |\forall\kappa\}$ of the original algebra and of its highest weight vector $\{A_\kappa, A_\kappa^\dagger, B_\kappa, B_\kappa^\dagger;\ |0>\ |\forall\kappa\}$, induced by the time evolution operator. The time evolution operator can therefore be thought of as a generator of the group of automorphisms of $\bigoplus_\kappa su(1,1)_\kappa$ parametrized by time t.

It is very important to point out that the various copies need not to be unitarily equivalent representations of the CCR's; as a matter of fact, they do become unitarily inequivalent in the infinite-volume limit, as shown in (12). This implies that the automorphisms (13) are defined up to arbitrary intertwining operators; in fact, one should more accurately say that the dynamical algebra $\mathcal{X}$ is given globally by the doubly-continuous direct sum $\bigoplus_t \bigoplus_\kappa \mathcal{A}^{\kappa(t)} \circ su(1,1)_\kappa$, where $\mathcal{A}^{\kappa(t)}$ denotes the intertwining operator [4, 5, 8] connecting the representations realizable at time t.

As a direct check, one can easily verify how at each time t one has

$$A_\kappa(t)|0(t)>= 0 = B_\kappa(t)|0(t)> \quad , \quad \forall t \quad .$$

The number of modes of type A_κ is given, at each instant t by

$$\mathcal{N}_{A_\kappa} \equiv < 0(t)|A_\kappa^\dagger A_\kappa|0(t)>= \sinh^2(\Gamma_\kappa t) \quad ; \tag{14}$$

and similarly for the modes of type B_κ.

The state $|0(t)>$ on the other hand may be written as:

$$\begin{aligned} |0(t)> &= \exp\left(-\frac{1}{2}\mathcal{S}_A\right) \exp\left(\sum_\kappa A_\kappa^\dagger B_\kappa^\dagger\right)|0> \\ &= \exp\left(-\frac{1}{2}\mathcal{S}_B\right) \exp\left(\sum_\kappa B_\kappa^\dagger A_\kappa^\dagger\right)|0> \quad , \end{aligned} \tag{15}$$

with

$$\mathcal{S}_A \equiv -\sum_\kappa \left\{A_\kappa^\dagger A_\kappa \ln \sinh^2(\Gamma_\kappa t) - A_\kappa A_\kappa^\dagger \ln \cosh^2(\Gamma_\kappa t)\right\} \quad ; \tag{16}$$

and $\mathcal{S}_B$ given by the same expression with B_κ and $B_\kappa^\dagger$ replacing A_κ and $A_\kappa^\dagger$, respectively. As A_κ's and B_κ's commute, due to (15) we shall simply write $\mathcal{S}$ for either $\mathcal{S}_A$ or $\mathcal{S}_B$.

From (15) one derives the expansion $|0(t)>=\sum_{n\geq 0}\sqrt{W_n(t)}\,|n,n>$, where n denotes the multi-index $\{n_\kappa\}$, with $\sum_{n\geq 0} W_n(t)=1$,

$$< 0(t)|\mathcal{S}|0(t)>= -\sum_{n\geq 0} W_n(t)\ln W_n(t) \quad . \tag{17}$$

Eq. (17) leads us therefore to interpreting $\mathcal{S}$ as the *entropy* for the dissipative system. On the other hand we have, for the time variation of $|0(t)>$

$$\frac{\partial}{\partial t}|0(t)>= -\frac{i}{\hbar}\mathcal{H}_I|0(t)> \quad . \tag{18}$$

Use of eqs. (7) and (16) shows then that

$$\frac{\partial}{\partial t}|0(t)>= -\left(\frac{1}{2}\frac{\partial\mathcal{S}}{\partial t}\right)|0(t)> \quad , \tag{19}$$

which may also be derived directly from (15). Equation (19) shows that $i\left(\frac{1}{2}\hbar\frac{\partial\mathcal{S}}{\partial t}\right)$ is the generator of time-translations, namely time-evolution is controlled by the entropy variations. It appears to us suggestive that for a dissipative quantum system the same operator that controls time evolution could be interpreted as defining a dynamical variable whose expectation value is formally an entropy: we conjecture that the connection between these features of $\mathcal{S}$ reflects correctly the irreversibility of time evolution characteristic of dissipative motion[6]. Damping (or, more generally, dissipation) implies indeed the choice of a privileged direction in time evolution (*time arrow*) with a consequent breaking of time-reversal invariance.

We also observe that $< 0(t)|\mathcal{S}|0(t)>$ grows monotonically with t from value 0 at $t=0$ to infinity at $t=\infty$; *i.e.* the entropy for both A and B increases as the system evolves in time towards the stability condition at $t=\infty$. Moreover the difference $\mathcal{S}_A-\mathcal{S}_B$ of the A- and B-entropies is constant in time: $[\mathcal{S}_A-\mathcal{S}_B,\mathcal{H}]=0$. Since the B-particles are the holes for the A-particles, $\mathcal{S}_A-\mathcal{S}_B$ turns out to be, in fact, the (conserved) entropy for the complete system.

In conclusion, the system in its evolution runs over a variety of representations of the CCR's which are unitarily inequivalent to eac h other for t $\neq$ t' in the infinite-volume limit. It is in fact the non-unitary character of time-evolution implied by damping which is recovered, in a consistent scheme, in the unitary inequivalence among representations at different times in the infinite-volume limit.

Also, the statistical nature of dissipative phenomena naturally emerges from our formalism, even though no statistical concepts were introduced a priori: for example, the *entropy operator* enter s the picture as time evolution generator. It is therefore an interesting question asking ourselves whether and how such statistical features may actually be related to thermal concepts. Let us clarify this point in the following discussion.

Let us focus, for the sake of definiteness, on the A-modes, and introduce the functional

$$\mathcal{F}_A \equiv < 0(t)|\Big(\mathcal{H}_A-\frac{1}{\beta}\mathcal{S}_A\Big)|0(t)> \quad . \tag{20}$$

β is a strictly positive function of time to be determined; $\mathcal{H}_A$ is the part of $\mathcal{H}_0$ relative to the A-modes only, namely $\mathcal{H}_A = \sum_\kappa \hbar\Omega_\kappa A_\kappa^\dagger A_\kappa$. We write $\vartheta_\kappa \equiv \Gamma_\kappa t$, and look for the values of ϑ_κ rendering $\mathcal{F}_A$ stationary:

$$\frac{\partial \mathcal{F}_A}{\partial \vartheta_\kappa} = 0 \quad ; \quad \forall \kappa \quad . \tag{21}$$

Condition (21) is clearly a stability condition to be satisfied for each representation. Setting $E_\kappa \equiv \hbar\Omega_\kappa$, it gives

$$\beta(t)E_\kappa = -\ln \tanh^2(\vartheta_\kappa) \quad . \tag{22}$$

From (22) we have then

$$\mathcal{N}_{A_\kappa}(t) = \sinh^2(\Gamma_\kappa t) = \frac{1}{e^{\beta(t)E_\kappa} - 1} \quad ; \tag{23}$$

which is the Bose distribution for A_κ at time t provided we assume $\beta(t)$ to represent the inverse temperature $\beta(t) = \frac{1}{k_B T(t)}$ at time t (k_B denotes the Boltzmann constant). This allows us to recognize $\{|0(t)>\}$ as a representation of the *CCR*'s at finite temperature, equivalent – up to an arbitrary choice of the temperature scale, which is a differentiable function of time – with the Thermo-Field Dynamics representation $\{|0(\beta)>\}$ of Umezawa and Takahashi[7].

We can now therefore interpret $\mathcal{F}_A$ as the free energy and $\mathcal{N}_A$ as the average number of activated A-modes at the temperature defined by time t through the function $\beta(t)$. In function of the time, the change in the energy $E_A \equiv \sum_\kappa E_\kappa \mathcal{N}_{A_\kappa}$ is given by the relation $dE_A = \sum_\kappa E_\kappa \dot{\mathcal{N}}_{A_\kappa} dt$; as one should expect, since the time evolution induces transitions over different representations, which in turn imply changes in the number of activated modes. Time derivative of $\mathcal{N}_{A_\kappa}$ is of course obtained from eq. (14). We can also compute the change in entropy [8]

$$dS_A = \frac{\partial}{\partial t}\Big(< 0(t)|S_A|0(t) >\Big) \;=\; \beta \sum_\kappa \hbar\Omega_\kappa \dot{\mathcal{N}}_{A_\kappa}(t)dt \;=\; \beta dE_A(t) \quad . \tag{24}$$

Eq. (24) shows that

$$dE_A - \frac{1}{\beta} dS_A = 0 \quad . \tag{25}$$

When changes in inverse temperature are slow namely $\frac{\partial \beta}{\partial t} = -\frac{1}{k_B T^2}\frac{\partial T}{\partial t} \approx 0$ (which is the case for adiabatic variations of temperature, at T high enough), eq. (25) can be obtained directly by minimizing the free energy (20) $d\mathcal{F}_A = dE_A - \frac{1}{\beta} dS_A = 0$; which – by reference to conventional thermodynamics – allows us to recognize E_A as the internal energy of the system. It also expresses the first principle of thermodynamics for a system coupled with environment at constant temperature and in absence of mechanical work.

We may also define as usual heat as $dQ = \frac{1}{\beta} dS$. We thus see that the change in time $d\mathcal{N}_A$ of particles condensed in the vacuum turns out into heat dissipation dQ.

Let us close this paper with few remarks[8]. The total Hamiltonian (7) is invariant under the transformations generated by $J_2 = \bigoplus_\kappa J_2^{(\kappa)}$. The vacuum however is not invariant under J_2 (see eq.(9)) in the infinite volume limit. Moreover, at each time t, the representation $|0(t)>$ may be characterized by the expectation value in the state $|0(t)>$ of, *e.g.*, $J_3^{(\kappa)} - \frac{1}{2}$: thus the total number of particles $n_A + n_B = 2n$ can be taken as an order parameter. Therefore, at each time t the symmetry under J_2 transformations is spontaneously broken. On the other hand, $\mathcal{H}_I$ is proportional to J_2. Thus, in addition to breakdown of time-reversal (discrete) symmetry, already mentioned above, we also have spontaneous breakdown of time translation (continuous) symmetry.

In other words we led dissipation (i.e. energy non-conservation), as it has been described in this paper, to an effect of breakdown of time translation and time-reversal symmetry. It is an interesting question asking which is the mode playing the role of the Goldstone mode related with the breakdown of continuous time translation symmetry: we observe that since $n_A - n_B$ is constant in time, the condensation (annihilation and/or creation) of AB-pairs does not contributes to the vacuum energy so that AB-pair may play the role of Goldstone mode.

From the point of view of boson condensation, time evolution in the presence of damping may be then thought of as a sort of continuous transition among different phases, each phase corresponding, at time t, to the representation $|0(t)>$ and characterized by the value of the order parameter at the same time t. The damped oscillator thus provides an archetype of system undergoing continuous phase transition.

We also observe that in perturbation theory a basic role is played by the adiabatic hypotesis by which the interaction may be switched off in the infinite time limit. It is such a possibility which allows the definition and the introduction of non-interacting fields. In the case of damped oscillator the switching off of the interaction in the infinite time limit is not possible since time evolution is intrinsically non-unitary and the adiabatic hypotesis thus fails. As a matter of fact we have seen that the set of annihilation and creation operators changes at each time and the same concept of non-interacting field thus loses meaning. We thus conclude that damping and dissipation require a non-perturbative approach and perturbation methods may be used only for local (in time and temperature variables) fluctuations.

Finally it has been recently shown[10] that the squeezed states of light entering quantum optics[11] can be identified, up to elements of the group $\mathcal{G}$ of automorphisms of $su(1,1)$, with the states of the damped quantum harmonic oscillator.

References

[1] E. Celeghini,M. Rasetti and G.Vitiello, in Proc. 2nd Workshop on *Thermal Field Theories and Their Applications*, H.Ezawa, T.Aritmitsu and Y.Hashimoto Eds. (Elsevier, Amsterdam, 1991)

[2] M. Rasetti, *Intl. J. Mod. Phys.* **13** (1973) 425

[3] J.R. Klauder and E.C. Sudarshan, *Fundamentals of Quantum Optics* (Benjamin, New York, 1968)
[4] G. Lindblad and B. Nagel, *Ann. Inst. H. Poincaré* **XIII A** (1970) 27
[5] E.M. Stein and A.W. Knapp, *Ann. Math.* **93** (1971) 489
[6] S. De Filippo and G. Vitiello, *Lett. Nuovo Cimento* **19** (1977) 92
[7] Y. Takahashi and H. Umezawa, *Collective Phenomena* **2** (1975) 55
[8] I. Hardman, H. Umezawa and Y. Yamanaka, *Thermal Energy in QFT*, Edmonton University preprint (1990)
[9] E. Celeghini, M. Rasetti and G. Vitiello *Florence University preprint* DFF 135/5/'91
[10] E. Celeghini, M. Rasetti, M. Tarlini and G. Vitiello, *Mod. Phys. Lett.* **3 B** (1989) 1213
[11] D. Stoler, *Phys. Rev.* **D 1** (1970) 3217

Coherence and Quantum Groups

E. Celeghini < celeghini@fi.infn.it >
Dip. di Fisica dell'Universitá and Sez. I.N.F.N., Firenze, Italy

Quantum groups have shown in recent years to be an exceptionally promising and rich structure whereby one can expect a growing wealth of new results in statistical mechanics and quantum field theory [1]. Stemmed out of the algebraic structure dictated by integrability conditions for a class of integrable systems, quantum groups can be intuitively thought of as the deformation (determined by the quantum Yang-Baxter equations and the R-matrix they involve) of the universal enveloping algebra of some given Lie algebra $\mathcal{L}$ of dynamical variables yet preserving the associativity properties of $\mathcal{L}$ (we consider here quantum groups as synonymous of Hopf algebras, disregarding more general definitions).

On a different side, an increasing deal of attention has been devoted to the notion of coherence in optics [2]. In particular a strong interest has been devoted to squeezed states those, constructed as generalized coherent states for suitable algebras, may describe a wide class of systems characterized by reduced quantum fluctuations.

Recently, and unexpectedly, the two fast developing fields of research have been connected and quantum groups have been introduced in the context of quantum optics. [3].

An interesting role in the description of the physical features of this new model – which represents situations in which the interaction between atom and radiation field is intensity dependent – is played by a set of states recently introduced by [4].

In the present note the two concepts of quantum groups and squeezing are discussed together taking into account the ideas of ref. [4] and it is shown that they can already be bridged in the simple setting provided by a subtle q-deformation of the usual coherent states.

So, to reach our goal we have to refrase the usual approach to coherent states in such a way that it can be "quantized" i.e. we have to stress the Lie algebra (or superalgebra, as we shell see) aspects of the usual creation and annihilation operators. The Fock space, from this point of view, is simply a representation of this structure.

When the "right" algebraic aspects has been identified, we know, at least in principle, from the general properties of quantum groups how to perform the job: the representation (i.e. the Fock space) will remain unmodified while the generators change in function of a parameter (usually called "q", for "quantum"). This means that the quantum operators are elements of the universal enveloping algebra of the original ones; to be more precise they are the same of the corresponding non quantum ones, but multiplied for integer functions of the Cartan subalgebra and of the Casimir operators.

The first and simpler idea would be to deform the group $H(1)$:

$$[a,\bar{a}] = H, \quad [N,a] = -a, \quad [N,\bar{a}] = \bar{a}, \quad [H,\cdot] = 0 \,. \tag{1}$$

The quantum groups are new mathematical objects and their general theory is, quite far from complete, but semisimple quantum groups are well known and, so, the best way for our job is to generalize to quantum groups the concept of contraction [5]. It can be shown that the contaction is more or less the same with the basic difference that the "q" parameter also is involved in the singular transformation.

Conforming to this idea, we define the following transformation on the four generators of $U(2)_q \equiv SU(2)_q \otimes U(1)$ and on the parameter $z = \log q$.

$$\begin{pmatrix} a_q \\ \bar{a}_q \\ N \\ H \\ w \end{pmatrix} = \begin{pmatrix} \varepsilon & 0 & 0 & 0 & 0 \\ 0 & \varepsilon & 0 & 0 & 0 \\ 0 & 0 & -1 & \varepsilon^{-2} & 0 \\ 0 & 0 & 0 & 2 & 0 \\ 0 & 0 & 0 & 0 & \varepsilon^{-2} \end{pmatrix} \begin{pmatrix} J_+^q \\ J_-^q \\ J_3 \\ K \\ z \end{pmatrix} \tag{2}$$

where K is the $U(1)$ generator; explicit and simple calculations show that this is exactly the singular transformation – beside the part related to z – giving (1) from $u(2)$.

The transformation (2) is applied to $U(2)_q$ (letting $e^z = q$) i.e. to the structure that have the algebraic relations:

$$[J_3^q, J_\pm] = \pm J_\pm, \qquad [J_+^q, J_-^q] = [2J_3^q]_q \equiv \frac{\mathrm{sh}(zJ_3^q)}{\mathrm{sh}(z/2)} \qquad [K, \cdot] = 0,$$

the coalgebra:

$$\begin{aligned} \Delta J_\pm &= e^{-zJ_3^q/2} \otimes J_\pm + J_\pm \otimes e^{zJ_3^q/2} , \\ \Delta J_3^q &= \mathbf{1} \otimes J_3^q + J_3^q \otimes \mathbf{1} , \\ \Delta K &= \mathbf{1} \otimes K + K \otimes \mathbf{1} , \end{aligned}$$

and counit and antipode:

$$\gamma(J_\pm) = -e^{zJ_3^q/2} \, J_\pm \, e^{-zJ_3^q/2} , \qquad \gamma(J_3^q) = -J_3^q , \qquad \gamma(K) = -K$$
$$\epsilon(\mathbf{1}) = 1, \qquad \epsilon(J_\pm) = \epsilon(J_3^q) = \epsilon(K) = 0 ,$$

we get taking the limit $\varepsilon \to 0$:

$$[a_q, \bar{a}_q] = \frac{\mathrm{sh}(wH/2)}{w/2}, \qquad [N, a_q] = -a_q, \qquad [N, \bar{a}_q] = \bar{a}_q, \qquad [H, \cdot] = 0 .$$

end

$$\Delta a_q = e^{-wH/4} \otimes a_q + a_q \otimes e^{wH/4} , \qquad \Delta \bar{a}_q = e^{-wH/4} \otimes \bar{a}_q + \bar{a}_q \otimes e^{wH/4} , \tag{3}$$

$$\Delta N = \mathbf{1} \otimes N + N \otimes \mathbf{1} , \qquad \Delta H = \mathbf{1} \otimes H + H \otimes \mathbf{1} .$$

Counit and antipode are:

$$\gamma(a_q) = -a_q\,, \qquad \gamma(\bar{a}_q) = -\bar{a}_q\,, \qquad \gamma(N) = -N\,, \qquad \gamma(H) = -H$$
$$\epsilon(1) = 1, \qquad \epsilon(a_q) = \epsilon(\bar{a}_q) = \epsilon(N) = \epsilon(H) = 0\,.$$

We note, in passing, that the contraction is a very powerful technique and it can be used also to derive the universal R-matrix of the quantum algebra we obtained (let us call it $H(1)_q$) from the one of $U(2)_q$. Explicit calculations give:

$$R = e^{-(w/2)(H \otimes N + N \otimes H)} \sum_{k=0}^{\infty} \frac{1}{k!} \left(w^{1/2} e^{wH/4}\, a_q\right)^k \otimes \left(w^{1/2} e^{-wH/4}\, \bar{a}_q\right)^k.$$

In this way, the quantified of the algebra (1) has been studied in all aspects. This is very interesting, particularly because it is the first example of non semisimple quantum group studied in all details but cannot help in building deformed coherent states: at the algebra level $H(1)_q$ is completely equivalent to an $H(1)$ in which only the Planck constant has rescaled as: $H' = \dfrac{\mathrm{sh}(wH/2)}{w/2}$. So, disregarding this irrelevant rescaling, the q-oscillators of $H(1)_q$ are the same of the ones of $H(1)$.

The play is different if instead of the algebra (1) we quantize $osp(1|2)$. With a suitable choice of the generators it can be written:

$$\{a, \bar{a}\} = 4M, \qquad [M, a] = -\frac{1}{2}a, \qquad [M, \bar{a}] = \frac{1}{2}\bar{a}.$$

This superalgebra looks unfamiliar but, as a matter of fact, is nothing else that the 3 generators structure of the creation annihilation operators, as can be easy seen defining the number operator as $N \equiv 4M - \frac{1}{2}$:

$$\{a, \bar{a}\} = 2N + 1, \qquad [N, a] = -a, \qquad [N, \bar{a}] = \bar{a}.$$

There is no room to discuss here the details of this quantization [6]. It results:

$$\{a_q, \bar{a}_q\} = \frac{\mathrm{sh}(uM)}{\mathrm{sh}u/4}, \qquad [M, a_q] = -\frac{1}{2}a_q, \qquad [M, \bar{a}_q] = \frac{1}{2}\bar{a}_q \tag{4}$$

The coproduct is easily obtained following [6] and is indeed different from (3):

$$\Delta a_q = e^{-uM/2} \otimes a_q + a_q \otimes e^{uM/2}\,, \qquad \Delta \bar{a}_q = e^{-uM/2} \otimes \bar{a}_q + \bar{a}_q \otimes e^{uM/2}\,,$$

$$\Delta M = 1 \otimes M + M \otimes 1\,, \quad .$$

while counit and antipode are:

$$\gamma(a_q) = -e^{-u/4}\, a_q\,, \quad \gamma(\bar{a}_q) = -e^{u/4}\, \bar{a}_q\,, \quad \gamma(M) = -M\,,$$
$$\epsilon(1) = 1, \quad \epsilon(a_q) = \epsilon(\bar{a}_q) = \epsilon(M) = 0\,.$$

These relations turn $\mathcal{S} \equiv osp_q(1|2)$ (sometimes called also $B_q(0|1)$) into a $\mathbf{Z}_2$-graded Hopf algebra. In showing this one has to take into account the graded multiplication and comultiplication on $\mathcal{S}\otimes\mathcal{S}$ [7]. For instance, if b and c are homogeneous elements in $\mathcal{S}$, then the product on $\mathcal{S}\otimes\mathcal{S}$ is defined as:

$$(a \otimes b)(c \otimes d) = (-1)^{p(b)p(c)}(ac \otimes bd)\,,$$

where $p(b), p(c) \in \mathbf{Z}_2$ are the degrees of b and c respectively.

It is amusing that we have found the bosonic operators into a graded algebra i.e. in a structure more suitable for describing the fermion ones, but this structure admitting the q-expansion has many unexpected features. On this point we limit ourselves to quote that it allows to generalize the well known Jordan-Schwinger realization of $SU(2)$ to $SU(2)_q$: the generators of $SU(2)_q$ are written as bilinear in the creation and annihilation operators of $osp_q(1|2)$ exactly as the generators of $SU(2)$ are written as bilinear in the usual creation annihilation operators [8]:

$$J_+^q = a_q^1 \bar{a}_q^2, \quad J_-^q = a_q^2 \bar{a}_q^1, \quad J_3^q = \frac{1}{2}(a_q^1 \bar{a}_q^1 - a_q^2 \bar{a}_q^2).$$

Therefore, because $osp_q(1|2)$ is a promising q-deformed structure at all levels, including algebra, we can attempt to rewrite the usual theory of coherent states in its scheme.

As a first step, we have to equip first the realization of $\mathcal{S}$ with the notion of hermiticity that has been lost in the quantification.

Indeed, as stressed in general, the Fock space is not modified in the quantum expansion and so $\mathcal{S}$ and consequently N can be identified with the customary Fock space and its number operator but a_q and $\bar{a}_q$ are different from a and $\bar{a}$ end are no more hermitian conjugate: conjugation of (4) leads indeed to

$$a_q^\dagger = \bar{a}_{q^*}\, [\chi_q(N)]^{-1} \quad , \quad \bar{a}_q^\dagger = \chi_q(N) a_{q^*} \quad ,$$

where $\chi_q(N)$ is a function to be determined by suitable self-consistency conditions. A convenient, natural assumption to fix it is that $a_q, a_q^\dagger$ are canonically conjugate with each other, *i.e.* that in the corresponding sector $\mathcal{S}$ be identical with the usual Weyl-Heisenberg algebra, namely $[\,a_q, a_q^\dagger\,] = 1$. This condition induces the choice

$$\chi_q(N) = \frac{([N+1]_{q^*}[N+1]_q)^{\frac{1}{2}}}{N+1} \quad ,$$

where $[\,n\,]_q \equiv \dfrac{q^{\frac{1}{2}n} - q^{-\frac{1}{2}n}}{q^{\frac{1}{2}} - q^{-\frac{1}{2}}} = \dfrac{\sinh(\frac{1}{2}nz)}{\sinh(\frac{1}{2}z)}$, with $z \equiv \ln q$. One has as well

$a_{q^*} = a_q \left(\dfrac{[N]_{q^*}}{[N]_q}\right)^{\frac{1}{2}}$ and $\bar{a}_{q^*} = \left(\dfrac{[N]_{q^*}}{[N]_q}\right)^{\frac{1}{2}} \bar{a}_q$.

Notice that as $q \to 1$, everything collapse on the customary bosonic realization of the Weyl-Heisenberg algebra.

On the states

$$a_q|n> = \left(\frac{[n]_q}{[n]_{q^*}}\right)^{\frac{1}{4}} n^{\frac{1}{2}}|n-1> \quad ,$$
$$\bar{a}_q|n> = \left(\frac{[n+1]_q}{[n+1]_{q^*}}\right)^{\frac{1}{4}} \left(\frac{[n+1]_q[n+1]_{q^*}}{n+1}\right)^{\frac{1}{2}} |n+1> \quad ,$$

whence $\bar{a}_q a_q|n> = [n]_q|n>$, $a_q^\dagger a_q|n> = n|n>$. Note that $a_q^\dagger a_q = a^\dagger a = N$, because a_q end $a_q^\dagger$ differ from a and a_q only for a phase. $|0>$ is the highest weight vector of $\mathcal{S}$: $a|0> = 0$.

We can at this point define the *quantum* analog of position (Q_q) and momentum (P_q) operators :

$$P_q \equiv i\left(\frac{m\hbar\omega}{2}\right)^{\frac{1}{2}} (\bar{a}_q - \bar{a}_q^\dagger) \quad ,$$
$$Q_q \equiv \left(\frac{\hbar}{2m\omega}\right)^{\frac{1}{2}} (\bar{a}_q + \bar{a}_q^\dagger) \quad . \tag{5}$$

Q_q and P_q are hermitian and have commutation relation

$$[Q_q, P_q] = i\hbar_q \equiv i\hbar(1 + \frac{1}{4}\operatorname{Re}(z^2)N(N+1) + \mathcal{O}(z^4)) \quad . \tag{6}$$

and give rise to a *quantum* version of the quantized harmonic oscillator, with hamiltonian

$$H_q \equiv \frac{1}{2m}{P_q}^2 + \frac{1}{2}m\omega^2{Q_q}^2 = \frac{1}{2}\hbar\omega\left(\bar{a}_q^\dagger\bar{a}_q + \bar{a}_q\bar{a}_q^\dagger\right)$$
$$= \frac{1}{2}\hbar\omega\left(1 - \frac{1}{2}\operatorname{Re}(z^2)(n+1)^2 + \mathcal{O}(z^4)\right) \tag{7}$$

Note also that eq. (7) suggests a possible experimental test of the q-effects through the spectrum of the quantum harmonic oscillator for large N.

We are now ready to define now the coherent states $\{|\alpha; q> |\alpha, q \in \mathbb{C}\}$ by

$$a_q|\alpha; q> = \alpha|\alpha; q> \quad ,$$

it is straightforward to check that

$$|\alpha; q> = \mathcal{N}(|\alpha|)\exp_q(\alpha\bar{a}_q)|0> \quad ,$$

where $\exp_q(x) \equiv \sum_{n=0}^{\infty} \frac{x^n}{[n]_q!}$ is the *quantum* version of the exponential function ($[n+1]_q! \equiv [n+1]_q[n]_q!$), and $\mathcal{N}(|\alpha|)$ is a normalization factor which with the above choice turns out to be independent of q : $\mathcal{N}(|\alpha|) = \exp(-\frac{1}{2}|\alpha|^2)$.

Then also,

$$|\alpha;q>=\exp(-\frac{1}{2}|\alpha|^2)\sum_{n=0}^{\infty}\frac{\alpha^n}{\sqrt{n!}}\left(\frac{[n]_{q^*}!}{[n]_q!}\right)^{\frac{1}{4}}|n> \quad .$$

It is worth noticing here that : *a)* in the limit $q\to 1$, $|\alpha;q>$ turns into a customary Glauber coherent state, *b)* the definitions adopted give for $|\alpha;q>$ a form which is somewhat different from that of ref.[4] (who, incidentally, give their definition for $q\in\mathbb{R}$); a difference with interesting consequences.

In order to show that the coherent states above defined $\{|\alpha;q>\}$ are squeezed, i.e. that their uncertainty in Q_q or in P_q or in both is smaller of $\frac{1}{2}\hbar$ (or $\frac{1}{2}\hbar_q$, we don't discuss the details here) we have now to evaluate the variances of Q_q and P_q. This can be done numerically. For $|q|\simeq 1$ and $|\alpha|\sim 1$ there are plenty of interesting situations with astonishing squeezing especially when q is near to the unit circle.

To conclude let us stress the fundamental result of the report: the quantum group coherent states seem to be, in general, the natural candidate to describe squeezed quantum states of matter.

References

[1] M. Jimbo, *Lett. Math. Phys.* **10** (1985) 63 ; **11** (1986) 247 (1986)

[2] D. Stoler, *Phys. Rev.* **D1** (1970) 3217
H.P. Yuen, *Phys. Rev.* **A13** (1976) 2226

[3] M. Chaichian, D. Ellinas, and P. Kulish, *Phys. Rev. Lett.* **65** (1990) 980

[4] L.C. Biedenharn, *J. Phys.* **A22** (1989) L873
A.J. MacFarlane, *J. Phys.* **A22** (1989) 4581

[5] E.Celeghini, R.Giachetti, M.Tarlini and E.Sorace *J. Math. Phys.* **31** (1990) 2548; **32** (1991) 1155; **32** (1991) 1159

[6] P.P.Kulish and N.Yu.Reshetikhin, *Lett. Math. Phys.* **18** (1989) 143

[7] Yu.I.Manin *Quantum groups and non-commutative geometry* Preprint Montreal University, CRM-1561 (1988)

[8] E. Celeghini, T.D. Palev, and M. Tarlini, *Mod. Phys. Lett.* **B5** (1991) 187

[9] E.Celeghini, M.Rasetti and G.Vitiello *Phys. Rev. Lett* **66** (1991) 2056

EXACT PERIODIC SOLUTIONS FOR A CLASS OF MULTISPEED DISCRETE BOLTZMANN MODELS

H. Cornille
Spht CE Saclay, 91191 Gif-sur Yvette, France

ABSTRACT

Only for multispeed discrete Boltzmann models can we obtain well-defined temperature. Recently different hierarchies of multispeed, multidimensional ($d > 1$) discrete models have been characterized by their $(1 + 1)$-dimensional Pde along one axis. Here, for the simplest hierarchy with five independent densities and two speeds which are 1 and either $\sqrt{d}$ or $\sqrt{2}$, we construct $(1+1)-$dimensional periodic solutions. The physical corresponding models are the planar square $8v_i, d = 2$ model and two three-dimensional $14v_i, d = 3$ models(one of them being the Cabannes model).

1. INTRODUCTION

For unispeed discrete Boltzmann models (discrete velocities v_i and densities N_i) the energy E and the mass M are propotional so that the temperature $T = 2E/M - V^2$ is ill-defined and cannot be distinguished from the mean velocity V. On the contrary for multispeed discrete Boltzmann models, E and M are independent, all conservation laws are satisfied and we can define and study the temperature.

For the unidimensional $4v_i$, $d = 1$ model with two speeds $1, 2$ and two different masses for the particles, exact solutions have been found[1].

For multidimensional models with $d > 1$ many multispeed models for single gas (same mass for the particles) are known. We can classify these models following the restriction of their Pde along one x-coordinate axis. It was found that different hierarchies exist which , for each class, are defined by the same set of independent densities and a system of Pde differing only by the dimension d of the space[2]. For instance the $8v_i$ square model[3] with the x-axis along either the medians or the diagonals of a square and the two $14v_i$ cubic[4] Cabannes models belong to two different hierarchies and differ only by the two dimensional $d = 2$ or 3 values. Then we can study the classes of d-dependent exact solutions[2] for each hierarchy and find common properties for models which belong to different spatial dimensions. We notice also the connections between discrete kinetic models[5] and their partners in lattice gas models[5].

Height different hierarchies have been found: Class I with two speeds 1 and either $\sqrt{d}$ or $\sqrt{2}$ which include the square $8v_i$ and two $14v_i$ models. ClassII with three speeds $0, 1, \sqrt{2}$ including the square $9v_i$ model. ClassIII with two speeds 2 and either $\sqrt{d}$ or $\sqrt{2}$, still including the square $8v_i$ and two $14v_i$ models. ClassIV with speeds $0, 2, \sqrt{2}$ includes another $9v_i$ model. Class V with $1, \sqrt{2}$ speeds begins with the popular $18v_i$model[5] and ClassVI with $0, 1, \sqrt{2}$ speeds has the $19v_i$ model associated to the $18v_i$ one. Classes VII and VIII have also the $18v_i, 19v_i$ models but with one speed equal to 2 instead of 1.

For all these hierarchies of models, exact shock waves solutions have been found. For the independent densities N_i they are of the type:

$$N_i = n_{0i} + n_i/(1+e^{\gamma\eta}),\ \eta = x - \varsigma t$$

Up to now only for Class I, other exact solutions have been found.They are $(1+1)$-dimensional solutions which are the superposition of two similarity waves, either real or complex conjugate:

$$N_i = n_{0i} + n_{1i}/D_1 + n_{2i}/D_2,\ D_i = 1 + d_i e^{\gamma_i \eta_i},\ \eta_i = x - \varsigma_i$$

$$N_i = n_{0i} + n_i/\Delta + n_i^*/\Delta^*, \Delta = 1 + \delta e^{\rho t + i\gamma x},\ \delta = \text{real const.} \qquad (1,1)$$

Periodic solutions were first obtained for the unispeed Broadwell model[6] and such solutions exist also for one $14v_i$ Cabannes model[7]. Here we present periodic solutions for the Class I hierarchy which include both the planar $8v_i, d = 2$ model and two three dimensional $d = 3$, $14v_i$ models. For Class I we notice also that $(1+1)$-dimensional shock waves have recently been found[8].

2. CLASS I HIERARCHY OF MODELS

For Class I there exist 5 independent densities $(N_1, M_4), M_2, (N_2, M_1)$ with coordinates $-1, 0, 1$ along the x-axis and two collision terms C, D

$$C = \overline{c}(N_2M_4 - N_1M_1), D = \overline{d}d_c(M_1M_4 - M_2^2), \overline{d} = 2/d, d_c > 0 \text{ arbitrary} \qquad (2.1a)$$

for two d-dependent subclasses (i) and (ii). We write down the system of five nonlinear equations which include three linear ones equivalent to the mass,momentum and energy conservation laws.We define $p_\pm = \partial_t \pm \partial_x$:

$$p_-N_1 = -p_+N_2 = C, M_{2t} = D, p_+M_1 = -(d-1)D + d_*C, p_-M_4 = -(d-1)D - d_*C$$

$$p_-N_1 + p_+N_2 = p_+M_1 + p_-M_4 + 2(D-1)M_{2t} = p_+M_1 - p_-M_4 - 2d_*p_-N_1 = 0 \quad (2.1b)$$

We define the macroscopic quantities: mass$M$, momentum J, mean velocity $V = J/M$ and energy$E$ from which we can construct nontrivial temperature $T = 2E/M - V^2$:

$$M = M_1 + M_4 + 2(d-1)M_2 + d_*(N_1 + N_2),\ J = M_1 - M_4 + d_*(N_2 - N_1)$$

$$E = (M_1 + M_4)/2 + (d-1)M_2 + d_*d_{**}(N_1 + N_2), (i)d_{**} = d/2, (ii)d_{**} = 1 \qquad (2.2)$$

For the subclass (i) we have $(2^d + 2d)v_i$,the speeds are $1, \sqrt{d}$, and $d_* = 2^{d-1}$,$\overline{c} = \sqrt{(d+3)}/2$. For the subclass (ii): $2(3d-2)v_i$,the speeds are $1, \sqrt{2}$, and $2(d-1) = d_*$,$2\overline{c} = \sqrt{5}$. For the d-dimensional coordinates let us write $(x_1 = x, x_2, ..., x_d)$ for the velocities of the different models. The three physical models (see fig.1) $d \leq 3$ are:

$$d = 2, 8v_i:\ N_i\ (\pm1, \pm1),\ M_i\ (\pm1, 0), (0, \pm1) \text{both (i) and (ii)}$$

$$d = 3, 14v_i, (i): N_i\ (\pm1, \pm1, \pm1),\ M_i(\pm1, 0, 0), (0, \pm1, 0), (0, 0, \pm1) \text{ Cabannes model}$$

$$d = 3, 14v_i, (ii): N_i\ (\pm1, \pm1, 0), (\pm1, 0, \pm1),\ M_i(\pm1, 0, 0), (0, \pm1, 0), (0, 0, \pm1)$$

For the $d = 2, d = 3(i), d = 3(ii)$ models we see that N_1 (and N_2) are the densities associated respectively to $2, 4, 4$ velocities in the space. For both models M_1 and M_4 correspond to only one velocity while M_2 is associated with respectively $2, 4, 4$ velocities. For these models three types of collisions can occur: between particles of speed 1, between particles of speed $\sqrt{d}$ (or speed$\sqrt{2}$ for subclass (ii)) and collisions $(1, \sqrt{d}) \longleftrightarrow (1, \sqrt{d})$. However the second subclass (ii) with speeds $1, \sqrt{2}$ is more interesting because we can add another velocity ($d = 2, 9v_i$ and $d = 3, 15v_i$) with speed 0 and then mixed collisions are possible: $(1, 1) \longleftrightarrow (0, \sqrt{2})$, (Class II hierarchy).

3. PERIODIC SOLUTIONS

3.1*Algebraic Determination of the Solutions for the General d case*

For the 5 independent densities we seek positive periodic solutions of the type(1.1):

$$N_1(x,t) = n_{01} + n_1/\Delta + n_1^*/\Delta^*, \; N_2(x,t) = N_1(-x,t)$$

$$M_1(x,t) = m_{01} + 2Re(m_1/\Delta), M_4(x,t) = M_1(-x,t), \; M_2 = m_{01} + m_2 2Re(1/\Delta) \quad (3.1)$$

with δ, γ and ρ in (1.1) and m_2 in (3.1) real while $n_1 = n_{1R} + in_{1I} = |n_1| \, e^{i\theta}$,$m_1 = m_{1I} + m_{1R}$ in (3.1) are complex. We notice that $N_2 = N_1$, $M_4 = M_1$ at $x = 0$ so that these solutions satisfy a specular reflection boundary condition at a wall $x = 0$.
We substitute (3.1) into (2.1) and both from the three linear relations and two nonlinear ones (with linear terms p_N_1,M_{2t}) we find:

$$n_{1R}\rho + n_{1I}\gamma = 0, \; m_{1R}\rho - m_{1I}\gamma + \rho(d-1)m_2 = 0$$

$$m_{1R}^2 - m_{1I}^2 = m_2^2, \; m_{1I}\rho + m_{1R}\gamma = d_*(\rho n_{1I} - \gamma n_{1R})$$

$$\rho n_{1I} - \gamma n_{1R} = -2\bar{c}(n_{1R}m_{1I} + n_{1I}m_{1R}) = 2\bar{c}(n_{01}m_{1I} + m_{01}n_{1I})$$

$$\rho m_2 = \bar{d}d_c(|m_1|^2 - m_2^2) = 2\bar{d}d_c m_{01}(m_2 - m_{1R}) \quad (3.2)$$

For ten real parameters $|n_1|, \theta, m_{1R}, m_{1I}, m_2, \rho, \gamma, n_{01}, m_{01}, d_c$ we have eight real relations leaving two arbitrary parameters chosen to be $n_{01} > 0$ as the scaling parameter and d_c or equivalently $\alpha = \bar{c}/\bar{d}d_c$. We introduce scaled parameters $\overline{m}_{1I} = m_{1I}/m_{1R}$, $\overline{m}_2 = m_2/m_{1R}$ and after some trivial algebra rewrite (3.2):

$$\overline{m}_{1I}^2 = 1 - \overline{m}_2^2 = \alpha(d-1)\overline{m}_2^2 \sin^2\theta, \; \overline{m}_{1I} = -\tan\theta(1 + (d-1)\overline{m}_2)$$

$$|n_1| = -n_{01}(1 - (d-1)\overline{m}_2)/\cos\theta(1 + d\overline{m}_2), \; m_{1R} = d_*|\overline{n}_1|/(\overline{m}_{1I}\sin\theta - \cos\theta)$$

$$1 + \overline{m}_2 + m_{01}/m_{1R} = 0, \; \rho = 2\bar{d}d_c m_{1R}\overline{m}_{1I}^2/\overline{m}_2, \; \gamma = -\rho/\tan\theta \quad (3.3)$$

We define $\eta_1 = \pm 1, \eta_2 = \pm 1, {\eta_i}^2 = 1$ and from the first three (3.3) relations we get:

$$\overline{m}_2 = 1/(1 - d + \eta_1\sqrt{\alpha(d-1)}\cos\theta)$$

$$\cos^2\theta - \eta_1\cos\theta\sqrt{(d-1)/\alpha} - 1/2 + d(d-2)/2\alpha(d-1) = 0 \quad (3.4)$$

At the present stage we only give the algebraic construction of the solutions without discussing the possible d_c or α intervals for which the solutions exist and are positive. From d, α given we first obtain $\cos\theta$:

$$2\cos\theta = \eta_1\sqrt{(d-1)/\alpha} + \eta_2\sqrt{2 + (1 + d(2-d))/\alpha(d-1)} \qquad (3.5)$$

giving $\overline{m}_2$ with (3.4) and successively $\overline{m}_{1I}, |n_1|, m_{1R}, \rho, \gamma, m_{01}$ with (3.3) and $n_{01} > 0$ arbitrary. Then with the scaled parameters we multiply by m_{1R} and reconstruct the original ones m_1, m_2. For the complete positivity discussion we study successively $d = 2$ and 3. We recall[6] that for solutions with $\rho > 0$ it is sufficient for positive N_i that $n_{01} > 0$ as well as $m_{01} > 0$ for positive M_i. If these properties hold then the densities are positive when the time is infinite and for positivity at finite time it is sufficient to choose sufficiently large δ in the complex denominators Δ. We recall also that for the $d = 3$ subclass (i), Cabannes model, then Cabannes and Tiem[7] have previously found periodic solutions.

3.2 *Positive Solutions for the $d = 2, 8v_i$, $d_* = 2$, $d_c = \sqrt{5}/2\alpha$ Planar model*

Theorem 1: For the d=2 case, the sufficient positivity conditions $\rho > 0$, $m_{01} > 0$ are satisfied if: $\eta_1 = -1$, $\eta_2 = 1$ and $0 < \alpha < 4$ or equivalently for the cross-section $d_c > \sqrt{5}/8$.

The solution $(3.3-4-5)$ can be written down analytically:$\gamma\tan = -\rho$ and

$$0 < \cos\theta = (-1 + \sqrt{2\alpha+1}/(2\sqrt{\alpha}) < 1, \ |n_1| = 2n_{01}\sqrt{\alpha}/(3 - \sqrt{2\alpha+1}) > 0 \qquad (3.6)$$

$$m_{01} = |n_1|(-1 + \sqrt{2\alpha+1}/(\cos\theta + \sqrt{\alpha}) > 0, \ m_1 = -2|n_1|(1 + \sqrt{\alpha}e^{-i\theta})/(\sqrt{\alpha} + \cos\theta)$$

$$m_2 = 2|n_1|/(\sqrt{\alpha} + \cos\theta), \ \rho = 4\overline{d}d_c\sqrt{\alpha}(\sqrt{2\alpha+1} - 1 + \alpha)|n_1|/(\sqrt{2\alpha+1} - 1 + 2\alpha)$$

The condition $\alpha < 4$ arises from $|n_1| > 0$.

3.3 *Positive Solutions for the two $d = 3, 14v_i$, $d_* = 4$, Three Dimensional models*

Theorem 2: For the two d=3 case, the sufficient positivity conditions $\rho > 0, m_{01} > 0$ are satisfied if: $\eta_1 = -1$, $\eta_2 = 1$ and $(3 - \sqrt{3})/2 < \alpha < 9/2$ or equivalent conditions for the cross-sections with $d_c = 3\sqrt{6}/4\alpha$ for the (i) model and $d_c = 3\sqrt{5}/4\alpha$ for the (ii) model.

In the (i) case we have $\overline{c}/\overline{d} = 3\sqrt{6}/4$ and $\overline{c}/\overline{d} = 3\sqrt{5}/4$ in the (ii) one.The solutions can still be written down analytically.

$$-1 < \cos\theta = (-1 + \sqrt{\alpha - 1/2})/\sqrt{2\alpha} < 1, \ |n_1| = n_{01}\sqrt{2\alpha}/(2 - \sqrt{\alpha - 1/2}) \qquad (3.7)$$

$$m_2 = 2\sqrt{2}|n_1|/(\sqrt{\alpha} + \sqrt{2}\cos\theta), \ m_1 = -4(\sqrt{2} + \sqrt{\alpha}e^{-i\theta})/(\sqrt{2}\cos\theta + \sqrt{\alpha})$$

$$m_{01} = 2\sqrt{2\alpha(\alpha - 1/2)}|n_1|/(\alpha - 1 + \sqrt{\alpha - 1/2}) > 0$$

$$\rho = \overline{d}d_c|n_1|8\sqrt{2\alpha}\alpha\sin^2\theta/(\alpha - 1 + \sqrt{\alpha - 1/2}) > 0, \ \gamma\tan\theta = -\rho$$

The conditions on α arise from $\cos\theta$ real, $\rho > 0$, $m_{01} > 0$ and $|n_1| > 0$.

3.4 *Macroscopic Quantities Associated to the* $d \leq 3$ *models*

$$d = 2 : M = M_1 + M_4 + 2(M_2 + N_1 + N_2), \;\; J = M_1 - M_4 + 2(N_2 - N_1)$$

$$E = (M_1 + M_4)/2 + M_2 + 2(N_1 + N_2)$$

$$d = 3 \; (i)\text{and} \; (ii) : M = M_1 + M_4 + 4(M_2 + N_1 + N_2), J = M_1 - M_4 + 4(N_2 - N_1)$$

$$(i)E = (M_1 + M_4)/2 + 2M_2 + 6(N_1 + N_2), \;\; (ii)E = (M_1 + M_4)/2 + 2M_2 + 4(N_1 + N_2)$$

3.5 *Numerical Calculations (fig.2)*

We present the curves of the temperature for the $d = 2$ and $d = 3$ subclass (ii) models corresponding to the same arbitrary parameters values: $n_{01} = 1$ and $d_c = 1$. In order that the microscopic densities N_i and M_i be non negative we choose $\delta = 7.8$ for the $d = 2$ model (see fig.3) and $\delta = 4.5$ for the $d = 3$ one. The temperatures are plotted for $\gamma x/2\pi$ varying in the interval $(0, 1)$ and we notice the symmetry with respect to the 0.5 value.

REFERENCES

1. H. Cornille and Y.H. Qian J. Stat. Phys. 61,683,1990; Y.H. Qian, Thesis, Paris ENS, January 1990.

2. H. Cornille to appear in J.Math.Phys 1991; Physics Letters A vol.154,339,1991, Aspects of Nonlinear Dynamics Brussels December 1990, Springer-Verlag 1991; Advances in Kinetic Theory and Continuum Mechanics ed. R. Gatignol Springer-Verlag 1991, p.109.

3. D. d'Humieres, P. Lallemand and U. Frisch Europhys. Let. 2,291,1986

4. H. Cabannes J. Mecan. 14,703,1975,Mech.Research Commun. 10,317,1983

5. For a review paper see D. d'Humieres in Proceed. in Physics 46, ed. P. Manneville Springer-Verlag 1989, p186; Discrete Kinetic Theory, Lattice Gas Dynamics ed. R. Monaco, World Scientific 1988; R. Gatignol Lecture Notes in Physics 37, Springer 1975; T Platkowski and R. Illner SIAM Rev.30,213,1988; S.Chen, M. Lee, M. Zhao and G.D. Doolen Physica D37,42,1989; B.T. Nadiga, J.E. Broadwell and B. Sturtevant Rarefied Gas Dynamics, Progr.Astronaut. vol.116, ed. E.P. Muntz 1988 p.155.

6. H. Cornille Partially Integrable Eqs. in Physics ed R. Conte , Kluwer Academic Publishers, the Netherlands 1990, p39; Lecture Notes in Mathematics 1460, Springer-Verlag 1988 ed. G. Toscani V.Boffi and S.Rionero p.70; Rigorous Methods in Particle Physics ed. S. Ciulli, Springer-Verlag vol.119, 1990,p.168.

7. H. Cabannes and D. Tiem Complex Syst. 1,574, 1987

8. H. Cornille the results will be presented in Nonlinear Dispersive Waves ed. Debnath.

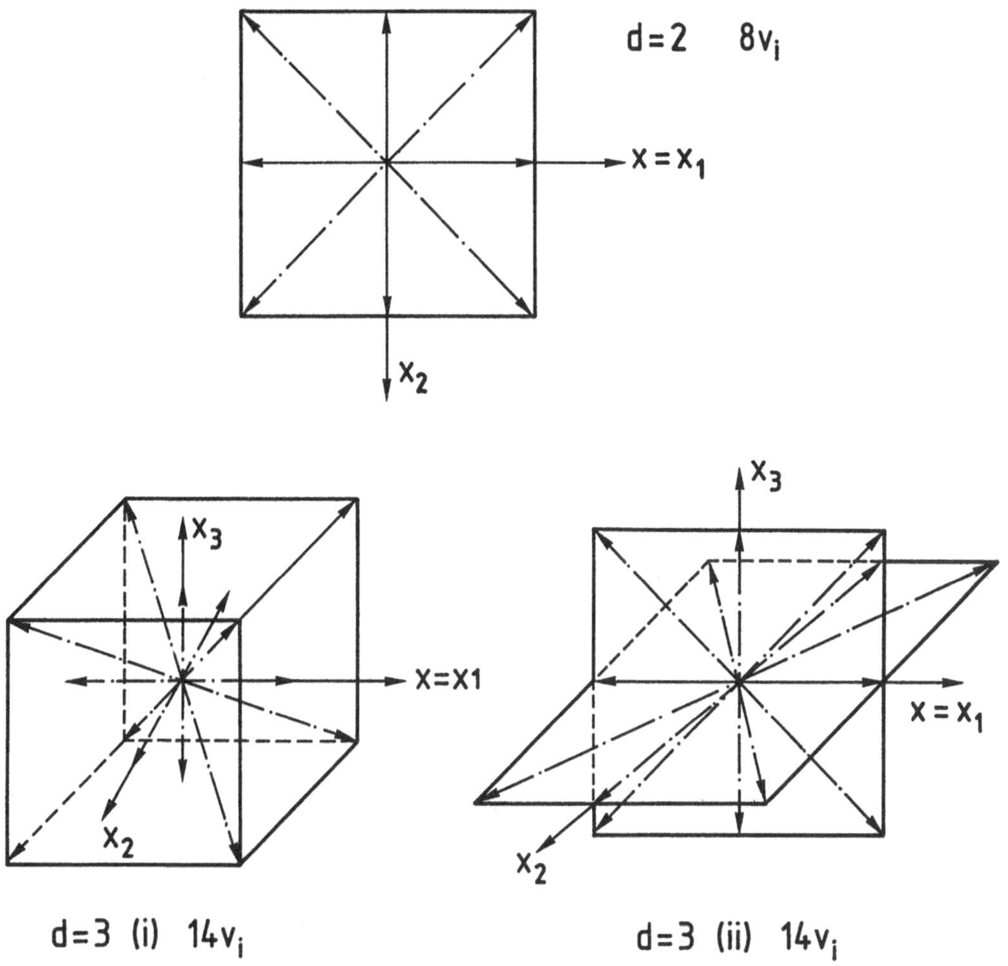

Fig. 1 - d⩽3 class I models

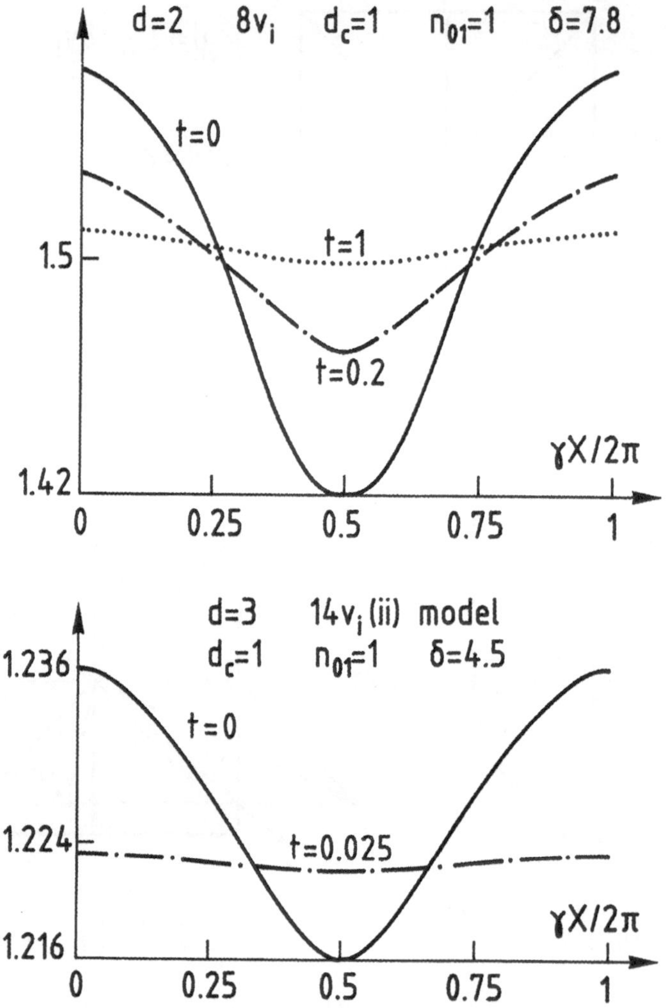

Fig. 2 - Temperature for periodic solutions of class I

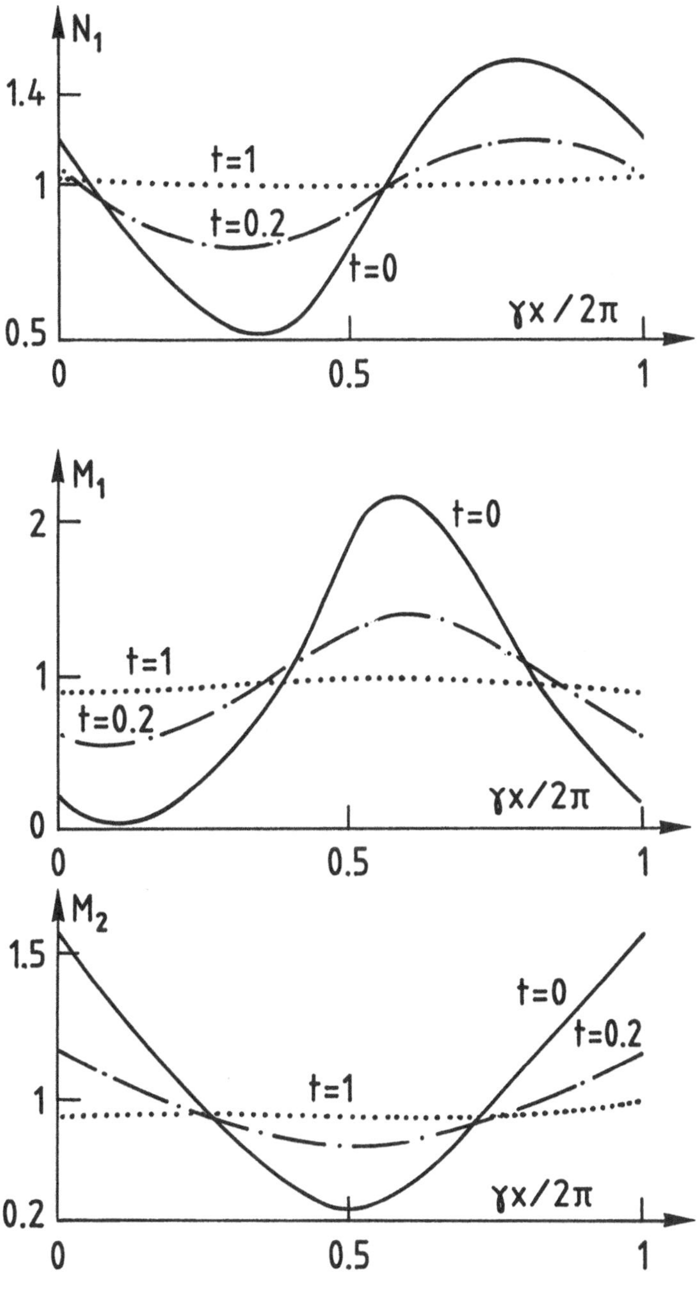

Fig. 3 - Densities for periodic solutions of the d=2 $8v_i$ model
$d_c=1$ $n_{01}=1$ $\delta=7.8$

PART VI

MATHEMATICAL METHODS

COLLECTIVE COORDINATES BY A VARIATIONAL APPROACH: PROBLEMS FOR SINE GORDON AND Φ^4 MODELS

J. G. Caputo
LESP, INSA and URA CNRS 230
BP 8, 76131 Mont-Saint-Aignan cedex, France

N. Flytzanis
Department of Physics, University of Crete
Heraklion, Greece

The method of collective-coordinates obtained by a variational approach is examined for the sine-Gordon and ϕ^4 models. It is shown that the evolution equations for the collective coordinates can be ill-defined because they are obtained by projecting on a null vector. New ansatzes that do not have this problem are presented.

I. Introduction

A certain number of non-linear partial differential equations can be solved exactly by the method of inverse scattering [1]. When they are perturbed the evolution of the solution can be obtained by perturbation methods [2]. Parameters of the unperturbed problem become slowly varying because of the perturbation.

In many other cases however only a few analytical solutions are known and the equation only has a few conserved quantities so that the mathematical structure of the problem is much more limited. In those situations to obtain the evolution of collective coordinates a variational method can be used. This method introduced by Rice and Mele [3] in the study of lightly doped poly-acetylene has had some remarkable sucesses for the sine-Gordon model [4]. For kink-antikink collisions in the sine-Gordon equation, Legrand guided by an algebraic identity gave a collective coordinate ansatz which he used for the study of a perturbed breather [5]. This ode description compared well with the pde solution despite of a sign error in the Lagrangian. The correct sign as will be shown below introduces a mathematical singularity that cannot be removed. The shape mode coordinate blows up when the breather is "flat" because the Lagrange equations are obtained through a projection on a vector which becomes zero at that point.

The paper is derived from [6]. It is organised as follows. Section II presents the variational procedure in the sine-Gordon case. Section III explains the ill-definition of the collective coordinates for kink-antikink collisions in sine-Gordon and ϕ^4. Section IV shows how to fix things and introduces new ansatzes.

II. The variational approach for the sine-Gordon equation

Consider the perturbed sine-Gordon equation:

$$\phi_{tt} - \phi_{xx} + \sin\phi = F(\phi, \phi_t, t) \tag{1}$$

When $F = 0$, equation (1) can be integrated exactly and two well-known solutions are [1] : the breather

$$\phi_B(x,t) = 4arctan(\frac{k_B \sin(\omega_B t + \phi_B)}{\omega_B \cosh(k_B x)}) \tag{2}$$

where $k_B^2 = 1 - \omega_B^2$ and the kink-antikink

$$\phi_{K\bar{K}}(x,t) = 4arctan(\frac{\sinh(\gamma_L u t)}{u \cosh(\gamma_L x)}) \tag{3}$$

where $\gamma_L = \frac{1}{\sqrt{1-u^2}}$ is the Lorenz factor. Formulas (2) and (3) can be seen as special cases of:

$$\phi(x, y(t), k(t)) = 4arctan(\frac{\sinh(y(t))}{\cosh(k(t)x)}) \tag{4}$$

where y(t) and k(t) have different expressions depending on whether the solution is a kink-antikink or a breather. This ansatz relies on an algebraic identity between the sum of a soliton and antisoliton profiles and the expression in the rhs of (4) -2π. It was put forth by Legrand [5]. The collective coordinate approach is to assume that the solution has again the form (4) when the perturbation is present. In order to derive the evolution equations for y and k, one could plug (4) into (1) but there would still be an x dependance. Instead, a variational approach is used. Equation (1) with $F = 0$ can be derived from the following Lagrangian density:

$$l = \frac{1}{2}{\phi_t}^2 - \frac{1}{2}{\phi_x}^2 - (1 - cos\phi) \tag{5}$$

by writing that the variation of $\int\int l dx dt$ is zero. Assuming that the solution follows (4), the evolution of y and k is then obtained from the Lagrangian $L(y, \dot{y}, k, \dot{k}, t) = \int l dx$ where expression (4) for ϕ is used to compute (5), by writing the usual Lagrange equations. The terms in the perturbation that cannot be incorporated in the Lagrangian density such as the damping are treated separately [5]. If

$$F = \epsilon sin\omega t - \delta\phi_t \tag{6}$$

the equations of motion are :

$$P = \partial\Lambda/\partial\dot{y} \tag{7.a}$$

$$Q = \partial\Lambda/\partial\dot{k} \tag{7.b}$$

$$\frac{dP}{dt} + \delta P = \partial\Lambda/\partial y \tag{7.c}$$

$$\frac{dQ}{dt} + \delta Q = \partial\Lambda/\partial k \tag{7.d}$$

where

$$\Lambda = \dot{y}^2\frac{8}{k}(1+\frac{2y}{\sinh 2y}) - \dot{y}\dot{k}16\frac{y}{k^2} + \dot{k}^2\frac{2}{3k^3}[(\pi^2+4y^2)(1-\frac{2y}{\sinh 2y}) + 8y^2]$$
$$-8k(1-\frac{2y}{\sinh 2y}) - \frac{8}{k}(\tanh y)^2(1+\frac{2y}{\sinh 2y}) + 4\pi\frac{y}{k}\epsilon\sin\omega t \tag{7.e}$$

is the Lagrangian and P (Q) is the momentum associated to y (k) respectively.

In the original paper of Legrand, a sign error was made in the coefficient of the k_t^2 term in Λ. It was written to be $(\pi^2+4y^2)(1+\frac{2y}{\sinh 2y}) + 8y^2$. We will see that this mistake completely hid the real behavior of the solution. All the terms in Λ except for the coefficient of $\dot{y}^2$ become 0 when $y = 0$. So only the y kinetic energy remains when $y = 0$. Therefore numerical problems are to be expected for k when y is small.
Equations (7) can be integrated numerically by transforming them into a system of first order differential equations using as variables (y,k,P,Q). The equations for $\dot{y}$ and $\dot{k}$ come from the definition of P and Q and are obtained by inverting the system:

$$\begin{pmatrix} P \\ Q \end{pmatrix} = \begin{pmatrix} 16\frac{\alpha}{k} & -16\frac{y}{k^2} \\ -16\frac{y}{k^2} & \frac{4}{3k^3}[\gamma(2-\alpha)+8y^2] \end{pmatrix} \begin{pmatrix} \dot{y} \\ \dot{k} \end{pmatrix} \tag{8}$$
$$\alpha = 1 + \frac{2y}{\sinh 2y} \quad \gamma = \pi^2 + 4y^2$$

The equations for $\dot{P}$ and $\dot{Q}$ are given by (7.c) and (7.d). Even in the absence of perturbations problems are to be expected when $y = 0$ because the rhs. matrix of (8) becomes non-invertible. To study that case δ and ϵ are set to 0. A breather initial condition was chosen and its evolution monitored through the variables y and k.
All the standard ode integrators tried failed to get across $y = 0$ and maintain reasonable accuracy. A finite difference energy conserving scheme suggested by Luiz Vazquez [7] showed the same behavior. This situation is also observed by Flesch in his study of the kink-antikink collisions in the ϕ^4 model [8] even though he uses an algebraic differential equation solver which is a completely different method. From these numerical results it becomes clear that equations (7) are ill-defined when $y = 0$. It turns out that this feature can be predicted from the ansatz (4).

Before going into the detail of the explaination it is worth mentioning that the agreement Legrand finds between the numerical solution of the pde and his collective coordinates odes is for the y variable only. There is very little interplay between the y and k variables except for when y=0. Computing $(y(t), k(t))$ with the wrong sign in (7.e) and a pure breather initial condition with $k = 0.1$ leads to a value of $\dot{k}$ about 10^{-3} clearly non zero but still small. The $\dot{k}$ terms in the Lagrangian are very small. Furthermore all the comparisons are shown with the perturbation term present. One situation displayed is a collision and annihilation due to damping; because of the damping it would be hard to see the difference with the pde. In the other situation where the kink and antikink pass through each other the comparison would point out the mistake. Therefore it is important in these collective coordinate studies to check the agreement of the pde with the behavior of all the variables.

III. Analysis of the singularity: the projection argument

The Lagrange equations derived from :

$$L(y, \dot{y}, k, \dot{k}, t) = \int l(\phi_t, \phi_x, \phi, t)dx \tag{9}$$

have a solution that is undetermined when $y = 0$. Legrand [5] examined in detail the procedure in which (4) is substituted into (9). The spatio-temporal dependance of ϕ in (4) can be written as: $\phi = \phi(x, a_i(t))$ where the a_i are the collective coordinates. The Lagrange equations from (9):

$$\frac{\partial L}{\partial a_i} - \frac{d}{dt}(\frac{\partial L}{\partial \dot{a}_i}) = 0 \tag{10}$$

imply using the fact that $\frac{\partial \phi_t}{\partial \dot{a}_i} = \frac{\partial \phi}{\partial a_i}$, integrating by parts with respect to x, and assuming that $\frac{\partial \phi}{\partial a_i}$ vanishes for $x \to \pm\infty$:

$$\int_{-\infty}^{+\infty} dx[\frac{\partial l}{\partial \phi} - \frac{\partial}{\partial t}(\frac{\partial l}{\partial \phi_t}) - \frac{\partial}{\partial x}(\frac{\partial l}{\partial \phi_x})]\frac{\partial \phi}{\partial a_i} = 0 \tag{11}$$

Equation (11) is the key to the understanding of the ill-definition of the equations. The expression in brackets is the left hand-side of the evolution equation as obtained from the condition that the spatio-temporal variation of l is stationary. If l is the Lagrangian density associated to the sine-Gordon equation then the lhs of (11) is identically zero for an exact solution of the sine-Gordon equation. Therefore (11) can be seen as a projection of the sine-Gordon operator onto the mode $\partial\phi/\partial a_i$.

For the (y, k) variables the first mode is non-zero for all values of y but the second mode is zero when $y = 0$ so that the Lagrange equation for the evolution of k is automatically satisfied no matter what the dependance of y and k on time is. For $y = 0$, the projection is done on a zero mode, therefore giving no information. Note that this problem only occurs for a breather or kink-antikink ansatz and not for a pure kink. It has been shown in [6] that all ansatz of the type:

$$\phi(x,y(t),k(t)) = 4arctan(\frac{\sinh[f(\,y(t)\,,k(t)\,)]}{\cosh(g(\,y(t)\,,k(t)\,)x\,)}h(\,y(t)\,,k(t)\,)) \tag{12}$$

where f can be zero cause the y and k evolution equations to be ill-defined. This is because the partial derivatives $\frac{\partial\phi}{\partial k}$ and $\frac{\partial\phi}{\partial y}$ become proportional when f=0.

Let us now consider the case of the ϕ^4 model [1]:

$$\phi_{tt} - \phi_{xx} - (\phi - \phi^3) = 0 \tag{13}$$

which can be seen as a generalisation of an equation derived from (1) in which the sine is replaced by the two first terms of its expansion. Even though the equation is not integrable an exact solution is known: the kink (or antikink):

$$\phi(x,t) = \pm\tanh[\frac{1}{\sqrt(2)}\frac{(x-vt)}{\sqrt{1-v^2}} + \xi] \tag{14}$$

To describe collisions between kinks for this model collective coordinates have been used. In particular Flesch [8] used an ansatz:

$$\phi(x,x_0(t),y_0(t)) = 1 - \tanh[\frac{y_0(x-x_0)}{\sqrt{2}}] + \tanh[\frac{y_0(x+x_0)}{\sqrt{2}}] \tag{15}$$

where x_0 is the center of mass variable and y_0 is the inverse of the width of the kinks. The Lagrangian he obtains is very similar to (7.e) in the sense that when x_0 goes through 0 the coefficients of the terms in $\dot{y_0}^2$ and $\dot{x_0}\dot{y_0}$ got to zero. In fact

$$\phi_{y_0} = -(sech[\frac{y_0(x-x_0)}{\sqrt{2}}])^2\frac{(x-x_0)}{\sqrt{2}} + (sech[\frac{y_0(x+x_0)}{\sqrt{2}}])^2\frac{(x+x_0)}{\sqrt{2}} \tag{16}$$

is zero for $x_0 = 0$ so that again the projection is done on a null vector. The problem cannot be fixed by introducing $\dot{x_0}$ in the ansatz or by adding a radiation term because the projection on the y_0 mode is the problem. Exactly as for the (y,k) variables, the numerical simulations show that $\dot{y_0}$ blows up when $x_0 = 0$. From the hamiltonian Flesch shows [8] that $\dot{y_0}$ necessarily goes to ∞ when $x_0 = 0$ unless $\dot{x_0}$ is such that $\frac{1}{2}m_1\dot{x_0}^2 + V - H_0 = 0$ where V is the potential energy and H_0 the total energy.
Prior to the study of Flesch, Jeyadev and Schrieffer [9] had done a collective coordinate study of the ϕ^4model using an ansatz derived from the Lorenz transformed solution of the linearised ϕ^4 equation around a static kink. Because of the complication in the calculations they simplified the expression. Again they obtained a blow-up for the shape mode [6].

To complete the picture of the singularity it is shown in [6] for the sine-Gordon model that the position y and the width k of a kink can be modulated in terms of collective coordinates and that this is not the case for a breather when y is close to zero. To be more specific, consider the normal modes of the collective coordinate equations in the sine-Gordon case and ϕ^4. For both models the translation mode associated to a zero frequency is non zero both for a kink and a kink-antikink pair (or breather in the sine-Gordon case). On the contrary the shape mode which corresponds to a non-zero frequency is non-zero for the kink alone and can become zero for the kink-antikink pair.

IV. New ansatzes for sine Gordon and ϕ^4

Because the ill definition of the equations is due to the fact that the Lagrange equations are obtained by a projection on a null vector, a way to fix things is not to do a simple projection anymore. Introducing an $\dot{a}_i$ dependancy in the ansatz leads to a Lagrangian $L(a_i, \dot{a}_i, \ddot{a}_i)$ The Lagrange equation for a_i is now:

$$\frac{\partial L}{\partial a_i} - \frac{d}{dt}(\frac{\partial L}{\partial \dot{a}_i}) + \frac{d^2}{dt^2}(\frac{\partial L}{\partial \ddot{a}_i}) = 0 \tag{17}$$

Using the same sort of arguments as in the previous section it can be shown [6] that the Lagrange equation (17) is equivalent to:

$$\int_{-\infty}^{+\infty} dx[\Xi \frac{\partial \phi}{\partial a_i} - \frac{\partial}{\partial t}(\Xi \frac{\partial \phi}{\partial \ddot{a}_i})] = 0 \tag{18.a}$$

where

$$\Xi = \frac{\partial l}{\partial \phi} - \frac{\partial}{\partial t}(\frac{\partial l}{\partial \phi_t}) - \frac{\partial}{\partial x}(\frac{\partial l}{\partial \phi_x}) \tag{18.b}$$

which is no longer a simple projection. Following this idea, a simple ansatz that can be introduced for the breather problem is:

$$\phi(x, y(t), k(t)) = 4arctan(\frac{\sinh[y(1+\dot{k})]}{\cosh kx}) \tag{19}$$

While this ansatz is rather arbitrary, it is very appropriate to make our point for the removal of the divergence in the equation for k. It is not also unreasonable since any shape oscillation can cause a small extension or contraction of the breather. When y is large the $\dot{k}$ term leads to a wobbling kink antikink pair. The reason for putting y as a factor is to avoid a forced time dependance of k. In the absence of perturbations $\dot{k} = 0$ so that the pure breather or kink antikink solutions still exist. For all these reasons it is hoped that this ansatz will lead to a successful quantitative comparison with the solution of the perturbed pde. The evolution equations are currently being derived.

For the ϕ^4 problem, an ansatz of the form

$$\phi(x, x_0(t), y_0(t)) = 1 - \tanh[\frac{y_0(x - x_0(1+\dot{y}_0))}{\sqrt{2}}] + \tanh[\frac{y_0(x + x_0(1+\dot{y}_0))}{\sqrt{2}}] \tag{20}$$

eliminates the singularity for $x_0 = 0$. Again, it is not unreasonable to assume that the separation of the kink antikink pair depends on the shape variable y_0. It is hoped using this ansatz to obtain a quantitative agreement with the solution of the partial differential equation.

V. Discussion and summary

The breather dynamics and its transition to a kink-antikink pair is an important source of chaos in the perturbed sine-Gordon equation. This has been shown by solving directly the partial differential equation. Such a direct approach involves long computational times and does not lead to a quantitative mechanism for the transition to chaos for sine-Gordon or the complicated resonance structure for the unperturbed but non-integrable ϕ^4 system. Several authors therefore used the collective coordinate approach [5,8,9]. In all these cases the choice of the ansatz introduced mathematical singularities.
It has been shown in section III that the source of the singularity lies on the projection on a mode that vanishes at one point in the evolution of the system. This was established by writing the Lagrange equations. At this point it should be remarked that the problem is attached to only one of the two coordinates. Introducing a relativistic effect on the problem free coordinate will not remedy the situation. For example in the (y,k) problem considered introducing $\dot{y}$ in the ansatz is useless because it is the effective mass connected with $\dot{k}^2$ in the Lagrangian that vanishes. Section IV showed the essential mathematical ingredient that a new ansatz must have. For the breather a phenomenological $\dot{k}$ dependance was introduced in the ansatz leading to a Lagrangian containing a $\ddot{k}$ dependance and a fourth order non singular evolution equation through a second order variational equation.
In the case of ϕ^4, a phenomenological dependance of the kink antikink separation was introduced. It is then clear that the evolution equation of the shape mode is not singular. The same goal could have been achieved by keeping the complete relativistic ansatz introduced in [9]. This would have introduced very complicated integrals over x the value of which could not have been calculated analytically. Following a different approach, Fei and Vazquez [10] took the Lagrangian obtained from the ansatz (15) and reduced it by some rather severe approximations to the one of a particel coupled with a harmonic oscillator. Using one adjustable parameter they were able to get a remarkable semi-quantitative agreement with the pde simulations of [11]. Using our approach we hope to get a quantitative agreement with the pde without an adjustable parameter. The price to pay is that the evolution equations will be much more complex than the ones of [11]. On the other hand it is hoped to get more insight on the collision process.

As a general conclusion we think it is important before going into the lengthy calculations for the evolution of the collective coordinates to check that the projection is not done on a null mode or on colinear modes. The calculations shown in section III despite of their formal appearance reveal a lot on the physics of the problem.

REFERENCES

[1] R. K. Dodd, J. C. Eilbeck, J. D. Gibbon and H. C. Morris
"Solitons and Nonlinear wave equations"
Academic Press (1982)

[2] A. C. Scott and D. Mclaughlin
"Perturbation analysis of fluxon dynamics"

Phys. Rev. A **18**, nb. 4 (1978)
and references 17 and 18 therein.

[3] M. J. Rice and E. J. Mele
"Phenomenelogical theory of soliton formation in lihgtly doped polyacetylene"
Solid State Comm. **35**, p. 487 (1980)

[4] G. Reinisch, J. C. Fernandez, N. Flytzanis, M. Taki and S. Pnevmatikos
"Phase lock of a weakly biased inhomogeneous long Josephson junction to an external microwave source"
Phys. Rev. B **38**, nb. 16, p. 11284 (1988)

[5] O. Legrand
"Kink-antikink dissociation and annihilation: a collective-coordinate description"
Phys. Rev. A **36**, 5068 (1987)

[6] J. G. Caputo and N. Flytzanis
"Kink-antikink collisions in sine-Gordon and ϕ^4 models: problems in the variational approach
submitted to Phys. Rev. A, December 1991

[7] L. Vazquez
"A conservative scheme for lagrangian systems"
In preparation.

[8] R. Flesch
"Collective coordinates for non-linear Klein-Gordon field theory"
PhD doctoral dissertation, USC, (1987)

[9] S. Jeyadev and J. R. Schrieffer
"Collective coordinate description of soliton dynamics in trans-polyacetylene-like systems"
Synthetic Metals **9**, p. 451 (1984)

[10] Z. Fei and L. Vazquez
"Resonance phenomena in a dynamical system with two degrees of freedom "
submitted to Physica D

[11] D. K. Campbell, J. F. Schonfeld and C. A. Wingate
"Resonance structure in kink-antikink interactions in ϕ^4 theory"
Physica **9D**, p. 451 (1983)

EXACT SOLUTION OF THE PERTURBED SINE-GORDON BREATHER PROBLEM

Erich F. MANN

Max-Planck-Institut für Metallforschung, Institut für Physik, Heisenbergstraße 1, D-7000 Stuttgart 80, Germany

Abstract

For the Cauchy problem of a sine-Gordon breather under the action of arbitrary, small perturbations an exact solution in form of quadratures is presented. No application of inverse scattering methods is made. Besides standard methods above all Bäcklund transformations are utilized. The adiabatic approximation originates in an integral form in quite a natural manner. The complete solution allows us to derive expressions for the radiation of energy. Examples for constant external force are considered in more detail.

1 Introduction

Solitons in physics have a long history [1-3]. Kinks and breathers as solitonic solutions of the sine-Gordon equation have for the first time been studied, in the framework of dislocation theory, by A. Seeger about forty years ago [4-6]. Since already in 1870 Enneper had derived this equation in connection with differential geometry (cf. [1]) , we prefer to refer to it as Enneper equation. During the past fifteen years much effort has been spent in investigating the perturbed Enneper equation [3]. For treating the perturbed breather several methods have been developed, but they all had to make use of inverse scattering theory. In the following a perturbation theory for the sine-Gordon (Enneper) breather is presented that is free of inverse scattering methods and that only uses classical concepts. All the information needed for describing the soliton properties is supplied by the Bäcklund transformation. The answer of the system to arbitrary, small perturbations may be exactly formulated in terms of quadratures. The so-called adiabatic approximation arises automatically as the discrete part of the complete solution.

2 Fundamentals and General Solution

The perturbed Enneper equation

$$u_{xx} - u_{tt} = \sin u - \epsilon P(x,t) \tag{1}$$

is thought to describe the influence of a small perturbation ϵP on the unperturbed breather solution u_b in the form $u = u_b + u_\epsilon$. For $|u_\epsilon| \ll 1$ u_ϵ obeys equation (1) linearized about u_b:

$$u_{\epsilon,xx} - u_{\epsilon,tt} = (\cos u_b)u_\epsilon - \epsilon P. \tag{2}$$

The unperturbed breather is given, with $\xi = x/\mathrm{ch}\sigma, \tau = (\mathrm{th}\sigma)t$, by

$$u_b(x,t) = -4\arctan\left(\frac{1}{\mathrm{sh}\sigma}\frac{\sin\tau}{\mathrm{ch}\xi}\right), \quad \cos u_b = 1 - \frac{8\mathrm{sh}^2\sigma\mathrm{ch}^2\xi\sin^2\tau}{(\mathrm{sh}^2\sigma\mathrm{ch}^2\xi + \sin^2\tau)^2}, \tag{3}$$

where σ is a parameter characterizing the amplitude, the frequency, and the extension of the breather. Here we restrict ourselves to the breather at rest. The running breather, to be obtained by a Lorentz transformation, will be dealt with elsewhere [7].

For initial conditions $u_\epsilon(x,0) = f_0(x), u_{\epsilon,t}(x,0) = f_1(x)$ the Cauchy problem for (2) is solved by $u_\epsilon = u_\epsilon^0 + u_\epsilon^1$, where u_ϵ^0 and u_ϵ^1 are solutions of the Cauchy problems, with $L \equiv \partial_{xx} - \partial_{tt} - \cos u_b$,

$$Lu_\epsilon^0 = 0; \quad u_\epsilon^0(x,0) = f_0(x), \quad u_{\epsilon,t}^0(x,0) = f_1(x) \tag{4}$$

$$Lu_\epsilon^1 = -\epsilon P; \quad u_\epsilon^1(x,0) = u_{\epsilon,t}^1(x,0) = 0. \tag{5a}$$

According to Duhamel's principle the solution of (5a) is given by

$$u_\epsilon^1(x,t) = \int_0^t \varphi(x,t;t')dt', \tag{5b}$$

where $\varphi(x,t;t')$ is the solution of the Cauchy problem

$$L\varphi = 0; \quad \varphi(x,t';t') = 0, \quad \varphi_t(x,t';t') = -\epsilon P(x,t'), \quad 0 \le t' \le t. \tag{5c}$$

We concentrate on the problem (5) with vanishing initial values and see that the inhomogeneous problem is reduced to the homogeneous problem (5c), where the perturbation appears in the inital conditions. The homogeneous equation

$$\varphi_{xx} - \varphi_{tt} = (\cos u_b)\varphi \tag{6}$$

describes perturbation solutions or excited solutions of the unperturbed Enneper equation (1) (with ϵ =0) in the vicinity of the breather state u_b. Its most general solution, needed for solving problems (4) and (5), is furnished by the Bäcklund transformation.

In the coordinates $p = (x-t)/2$ and $q = (x+t)/2$ the Enneper equation reads $\partial_{pq}u = \sin u$. The Bäcklund transformation (BT) $B_{\sigma_i}u_0 = u_i$, expressed by the pair of first-order differential equations

$$\begin{aligned}\partial_p(u_i - u_0)/2 &= (\cos\sigma_i)^{-1}(1+\sin\sigma_i)\sin[(u_i+u_0)/2]\\ \partial_q(u_i + u_0)/2 &= (\cos\sigma_i)^{-1}(1-\sin\sigma_i)\sin[(u_i-u_0)/2],\end{aligned} \tag{7}$$

generates a new solution u_i representing a soliton (kink) superimposed on the solution u_0 of the Enneper equation. For $u_0 = 0$ the pure kink solution

$$u_i^{(0)}(x,t) = 4 \arctan \exp\{(\cos\sigma_i)^{-1}(x - t\sin\sigma_i) + \alpha_i\} \tag{8}$$

results, where σ_i denotes the Bäcklund parameter and α_i an integration constant. Two subsequent BTs with different parameters σ_1, σ_2 generate a soliton pair

$$u_{p(air)} = B_{\sigma_2} u_1 = B_{\sigma_1} u_2, \quad u_1 = B_{\sigma_1} u_0, \quad u_2 = B_{\sigma_2} u_0. \tag{9}$$

According to the Bianchi theorem, for u_p also the algebraic form holds

$$u_p = u_0 + 4 \arctan\{\frac{\cos[(\sigma_1+\sigma_2)/2]}{\sin[(\sigma_1-\sigma_2)/2]} \tan\frac{u_1 - u_2}{4}\}. \tag{10}$$

The breather form (3) follows herefrom with $\alpha_i = u_0 = 0, \sigma_1 = \sigma_2^* = i\sigma$($\sigma$ real).

The most general variation of the pair solution u_p in the vicinity of the breather solution u_b is most simply obtained by a variation with respect to the 4 free parameters α_i, σ_i and the arbitrary solution u_0

$$\delta u_p = \sum_i \left(\frac{\partial u_p}{\partial \alpha_i}\right)_0 \delta\alpha_i + \sum_i \left(\frac{\partial u_p}{\partial \sigma_i}\right)_0 \delta\sigma_i + \left(\frac{\delta u_p}{\delta u_0}\right)_0 \delta u_0, \quad i = 1,2. \tag{11}$$

Here $()_0$ means taking the breather values $\alpha_i = u_0 = 0, \sigma_1 = \sigma_2^* = i\sigma$. The variation δu_0 as an excited state of the vacuum state $u_0 = 0$ obeys the Klein-Gordon equation with plane waves as solutions

$$(\partial_{xx} - \partial_{tt} - 1)\delta u_0 = 0, \quad \delta u_0 \sim \exp\{i(kx \pm \omega t)\}, \quad \omega = +\sqrt{k^2+1}. \tag{12}$$

The variations $\delta u_i/\delta u_0$ entering (11) may be obtained from the system (7) linearized by replacing $u_i \to u_i^{(0)} + \delta u_i$, $u_0 \to \delta u_0$ with $u_i^{(0)}$ from (8) or directly from integrating $(\partial_{xx} - \partial_{tt} - \cos u_i^{(0)})\delta u_i = 0$ [8].

The aforementioned procedure provides us with independent solutions forming a complete basis in terms of which the general solution of (6) may be expanded [8]

$$\varphi(x,t) = \sum_{\nu=1}^{4} A_\nu \varphi_\nu(x,t) + \int_{-\infty}^{+\infty} dk[A(k)\phi(x,t;k,\omega) + B(k)\phi(x,t;k,-\omega)],$$

$$\begin{gathered} \varphi_\nu = N^{-1}\psi_\nu, \quad N = \mathrm{sh}^2\sigma\,\mathrm{ch}^2\xi + \sin^2\tau, \quad \xi = x/\mathrm{ch}\sigma, \quad \tau = (\mathrm{th}\sigma)t \\ \psi_1 = \mathrm{sh}\xi\sin\tau, \quad \psi_3 = \tau\,\mathrm{sh}\xi\sin\tau - \mathrm{sh}^2\sigma\,\xi\,\mathrm{ch}\xi\cos\tau \\ \psi_2 = \mathrm{ch}\xi\cos\tau, \quad \psi_4 = \mathrm{ch}^2\sigma\,\mathrm{ch}\xi\sin\tau - \tau\,\mathrm{ch}\xi\cos\tau - \mathrm{sh}^2\sigma\,\xi\,\mathrm{sh}\xi\sin\tau \end{gathered} \tag{13}$$

$$\phi(x,t;k,\omega) = e^{i(kx+\omega t)}\left(1 + \frac{2(\sin^2\tau - \mathrm{ch}^2\xi) + i(k\,\mathrm{ch}\sigma\,\mathrm{sh}2\xi + \omega\,\mathrm{cth}\sigma\sin 2\tau)}{(\mathrm{sh}\sigma)^{-2}(1+\mathrm{ch}^2\sigma k^2)N}\right).$$

(In [8] the discrete solutions φ_3 and φ_4 containing terms linear in x and t had been disregarded. As will be shown below, the terms linear in t do not constitute secularities.)

In order to determine the coefficients A_ν and $A(k), B(k)$ by means of the initial conditions (5c) the form (13) is not well suited. The key for a simple procedure, hereby making inverse scattering methods superfluous, is the observation that the solution $\varphi(x,t)$ of (6) satisfies the linear combination, with $N = \mathrm{sh}^2\sigma\mathrm{ch}^2\xi + \sin^2\tau$,

$$L[\varphi] \equiv \frac{\partial^2\varphi}{\partial\xi^2} + \frac{\mathrm{sh}^2\sigma}{N}\left(\mathrm{sh}2\xi\frac{\partial\varphi}{\partial\xi} + \sin 2\tau\frac{\partial\varphi}{\partial\tau}\right) + \left(1 + \frac{2\mathrm{ch}^2\sigma\sin^2\tau(N-2\sin^2\tau)}{N^2}\right)\varphi$$

$$= -\int_{-\infty}^{+\infty}(1+\mathrm{ch}^2\sigma k^2)[A(k)e^{i(kx+\omega t)} + B(k)e^{i(kx-\omega t)}]dk. \qquad (14)$$

This relation means that through such a combination the contributions from the discrete terms in (13) disappear and that from the integrand in (13) only the pure plane wave terms are retained. Equation (14) can be deduced by repeated use of the Bäcklund transformations (9) and their linearized versions [7]. For the much simpler case of a perturbed kink a relation corresponding to (14), but with first derivatives only, had been found by Seeger already in 1951 [5].

A second relation is obtained by differentiating (14) with respect to t

$$\frac{\partial}{\partial t}L[\varphi] = -i\int_{-\infty}^{+\infty}\omega(1+\mathrm{ch}^2\sigma k^2)[A(k)e^{i(kx+\omega t)} - B(k)e^{i(kx-\omega t)}]dk. \qquad (15)$$

Now we consider equations (14) and (15) at $t = t'$ and introduce the initial conditons (5c) by setting

$$\varphi(x,t') = \varphi_x(x,t') = \varphi_{xx}(x,t') = 0; \quad \varphi_t(x,t') = -\epsilon P(x,t'). \qquad (16)$$

The advantage of the formulations (14) and (15) is that by inverse Fourier transformation in the variable x these equations can be directly solved for the coefficients A and B, which now become functions of t'. The integrals containing first and second derivatives of $P(x,t')$ with respect to x may be integrated by parts. For $P(x,t')$ restricted at infinity one obtains save for terms vanishing with the subsequent k-integration

$$A(k,\omega;t') = \frac{\epsilon}{4\pi i\omega}\int_{-\infty}^{+\infty}\phi^*(x',t';k,\omega)P(x',t')dx'; \quad B(k,\omega;t') = A(k,-\omega,t'). \qquad (17)$$

This remarkable result, namely that the expansion coefficients A, B may be expressed in the basis functions $\phi(x,t;k,\omega)$ as given in (13), has also been stated by McLaughlin and Scott [9], deduced, however, with inverse scattering methods.

The coefficients A_ν in (13) are determined by inserting in $\varphi(x,t)$ the expressions (17) and requiring the initial condition $\varphi(x,t';t')=0$. The appearing Fourier integrals in k can be solved exactly. The result is

$$A_1(t')=\mathrm{sh}^3\sigma\cos\tau' I_1-\mathrm{sh}\sigma\tau'\sin\tau' I_3,\quad A_3(t')=\mathrm{sh}\sigma\sin\tau' I_3$$

$$A_2(t')=\mathrm{sh}^3\sigma\sin\tau' I_2-\mathrm{sh}\sigma(\mathrm{ch}^2\sigma\sin\tau'-\tau'\cos\tau')I_4,\quad A_4(t')=\mathrm{sh}\sigma\cos\tau' I_4$$

$$I_\nu(t')=\int_{-\infty}^{+\infty}\frac{f_\nu(\xi')}{N'}\epsilon P(x',t')dx',\quad f_1=\xi'\mathrm{ch}\xi',\, f_2=\xi'\mathrm{sh}\xi',\, f_3=\mathrm{sh}\xi',\, f_4=\mathrm{ch}\xi'$$

$$N'=\mathrm{sh}^2\sigma\mathrm{ch}^2\xi'+\sin^2\tau',\quad \xi'=x'/\mathrm{ch}\sigma,\quad \tau'=(\mathrm{th}\sigma)t'. \tag{18}$$

The discrete part of (13) can therefore in compact form be written as

$$\varphi_d(x,t;t')=\sum_{\nu=1}^{4}A_\nu(t')\varphi_\nu(x,t)=\mathrm{sh}\sigma\int_{-\infty}^{+\infty}(\varphi_1'\varphi_3-\varphi_3'\varphi_1+\varphi_2'\varphi_4-\varphi_4'\varphi_2)\epsilon P(x',t')dx'$$

$$=\frac{\mathrm{sh}\sigma}{N}\int_{-\infty}^{+\infty}\frac{\epsilon P(x',t')}{N'}[\mathrm{sh}^2\sigma(\mathrm{ch}\xi'\cos\tau'\mathrm{sh}\xi\sin\tau+\mathrm{sh}\xi'\sin\tau'\mathrm{ch}\xi\cos\tau)(\xi'-\xi)$$

$$+(\mathrm{ch}\xi'\cos\tau'\mathrm{ch}\xi\cos\tau-\mathrm{sh}\xi'\sin\tau'\mathrm{sh}\xi\sin\tau)(\tau'-\tau)$$

$$+\mathrm{ch}^2\sigma(\mathrm{ch}\xi'\cos\tau'\mathrm{ch}\xi\sin\tau-\mathrm{ch}\xi'\sin\tau'\mathrm{ch}\xi\cos\tau)]dx'. \tag{19}$$

The occurrence of the linear factor $t-t'$ is not of real significance, since in the subsequent integration (5b) it may be made to disappear by an integration by parts

$$\int_0^t(t-t')g(t')dt'=\int_0^t\int_0^{t'}g(t'')dt''dt'. \tag{20}$$

This shows that finally no terms linear in t are retained.

By means of the equations (5b), (13), (17), and (19) the Cauchy problem (5a) with vanishing initial values is exactly solved in terms of quadratures. In the case of non-vanishing initial values, the much simpler problem (4) has to be solved additionally along similar lines. It should be emphasized that the present method yields exact results exclusively in form of integrals. Since, as is shown below, the discrete part φ_d of (13), given explicitly in (19), corresponds to the so-called adiabatic approximation, that means that also the adiabatic approximation for an arbitrary perturbation arises in an integral form, not in a differential form as in other theories. The differential forms of the adiabatic approximation for any perturbations of the breather have been integrated here once for all.

3 Constant External Force

a) Adiabatic approximation

For a constant external force, such as a constant elastic stress in dislocation theory, $\epsilon P = s = const$. The new rest position for a straight dislocation initally at $u = 0$ is now $u = \arcsin s \approx s$. As initial values for the perturbed breather we choose $u(x,0) = u_s(x,0) = s$ and $u_t(x,0) = u_{b,t}(x,0)$ or $u_{s,t}(x,0) = 0$, where $u(x,t) = u_b(x,t) + u_s(x,t)$ and u_b denotes the unperturbed breather solution (3). In this case we have also a solution u_s^0 from (4), but as can easily be seen, there are no contributions of the form $\sum A_\nu \varphi_\nu$ of (13). The solution u_s^1 of (5), however, contributes such discrete terms, and by means of (19) and (20) these give rise to

$$u_s^{\mathrm{ad}}(x,t) = \frac{\pi \mathrm{ch}\sigma s}{N} \int_0^t \left\{ \left[(\mathrm{sh}^2\sigma - 1)\mathrm{Arsh}\, a' - \mathrm{ch}^2\sigma \frac{a'}{w'} \right] \mathrm{ch}\xi \cos\tau \right.$$

$$\left. + \frac{\cos\tau'}{\mathrm{sh}\sigma\, w'} (\mathrm{ch}^2\sigma \mathrm{ch}\xi - \mathrm{sh}^2\sigma\, \xi \mathrm{sh}\xi) \sin\tau \right\} dt', \tag{21}$$

where $a' = (\sin\tau')/\mathrm{sh}\sigma$, $w' = (1 + a'^2)^{1/2}$, and the other symbols as in (13), (18). The result (21) can be interpreted as the variation of the breather solution (3) with independently varying parameters $\mathrm{sh}\sigma, \mathrm{th}\sigma, \mathrm{ch}\sigma$. Thus it corresponds and, in fact, is equal to the adiabatic approximation as given by Karpman, Maslov, and Solov'ev [10] and integrated by Eßlinger [11].

Alternatively, we may write $u_s(x,t) = s + u_s'(x,t)$. Then u_s' satisfies equation (2) with $\epsilon P = s(1 - \cos u_b)$. In this case only the problem (5) needs to be solved. The discrete part φ_d in (19) contributes

$$u_s'^{ad}(x,t) = \frac{\pi \mathrm{ch}\sigma s}{N} \int_0^t \left\{ \left[(\mathrm{sh}^2\sigma - 1)(\mathrm{Arsh}\, a' - \frac{a'}{w'^3}) - \mathrm{ch}^2\sigma \frac{a'^3(4 + a'^2)}{w'^5} \right] \mathrm{ch}\xi \cos\tau \right.$$

$$\left. + \frac{\cos\tau'}{\mathrm{sh}\sigma} \frac{a'^2(4 + a'^2)}{w'^5} (\mathrm{ch}^2\sigma \mathrm{ch}\xi - \mathrm{sh}^2\sigma\, \xi \mathrm{sh}\xi) \sin\tau \right\} dt'. \tag{22}$$

This expression is equal to the adiabatic approximation as derived by Döttling [12,13] by means of a Hamiltonian formalism and integrated by Blüher [14]. Compared with (21) we see that the same physical problem may lead to different expressions for the adiabatic approximation. The total solutions u_s, of course, have to be the same in both cases.

b) Energy radiation

The total energy of a solution $u(x,t)$ of (1) with $\epsilon P = s$ is given by

$$E = \int_{-\infty}^{+\infty} \mathcal{E}(x,t)dx = \int_{-\infty}^{+\infty} [(u_x^2 + u_t^2)/2 + 1 - \cos u - su]dx, \tag{23}$$

where $E = const$. By means of (1) one may derive $\partial\mathcal{E}/\partial t = \partial(u_x u_t)/\partial x$. Therefore the quantity

$$\dot{E}_r = -[u_x u_t]_{x=a \gg \mathrm{ch}\sigma} \tag{24}$$

denotes the rate of energy increase in the region $a < x < \infty$, where a is a distance from the centre larger than the breather extension chσ. In other words, $\dot{E}_r$ represents the rate of energy radiated into the region $x > a$. Contributions to the asymptotic solution u are $u_s^{(0)} = s\cos t$ and $u_s^{(1)}$ of (5) with the integral part of (13) only. For $\mathrm{ch}(a/\mathrm{ch}\sigma) \gg 1$ and $\mathrm{sh}\sigma\mathrm{ch}(a/\mathrm{ch}\sigma) \gg 1$ the quantity ϕ in (13) simplifies to

$$\phi \to \phi_a(x,t;k,\omega) = e^{i(kx+\omega t)}\left(1 - 2\frac{1 - i\mathrm{ch}\sigma k}{1 + \mathrm{ch}^2\sigma k^2}\right). \tag{25}$$

By means of "Faltungen" (convolutions) one arrives at the expression

$$\varphi_a(x,t;t') = \frac{s}{2\pi}\int_{-\infty}^{+\infty} dx'\{I(x-x',t-t') + 2\int_0^{\infty} d\xi'' e^{-\xi''}[I(x-x'-x'',t-t')(\xi''-1$$
$$-\frac{\mathrm{ch}^2\sigma\sin^2\tau'\xi'' + \mathrm{sh}^2\sigma\mathrm{ch}\xi'\mathrm{sh}\xi'(\xi''-1)}{\mathrm{sh}^2\sigma\mathrm{ch}^2\xi' + \sin^2\tau'})$$
$$+\frac{\partial}{\partial t}I(x-x'-x'',t-t')\frac{\mathrm{sh}\sigma\mathrm{ch}\sigma\sin\tau'\cos\tau'\xi''}{\mathrm{sh}^2\sigma\mathrm{ch}^2\xi' + \sin^2\tau'}]\} \tag{26}$$

with

$$I(\tilde{x},\tilde{t}) = \int_{-\infty}^{+\infty} dk\exp(ik\tilde{x})\frac{\sin\omega\tilde{t}}{\omega} = \begin{cases}\pi J_0(\sqrt{\tilde{t}^2 - \tilde{x}^2}), & 0 < |\tilde{x}| < \tilde{t}\\ 0, & |\tilde{x}| > \tilde{t} > 0,\end{cases}$$

where $\omega = (k^2+1)^{1/2}$ and J_0 denotes the Bessel function of zeroth order. The derivative $\partial_t I = -\partial_{t'} I$ is meant in the sense that an integration by parts should follow. The first term in (26) gives rise to a contribution $s(1-\cos t)$ to $u_s^{(1)}$. The other terms in general cannot be integrated in closed form.

As an example we consider the low-amplitude breather characterized by $\sigma \gg 1$. Then only the terms $\sim \sin^2\tau'$ and $\sim \sin 2\tau'$ contribute and the integrations are readily performed. This leads to the asymptotic solution

$$u_s^a = s + s\pi\exp\left[-\sqrt{3}\pi\mathrm{ch}\sigma/2\right]\sin(\sqrt{3}x - 2t). \tag{27}$$

This represents an outgoing wave with the wave number $k = \sqrt{3}$ and the frequency $\omega = 2$. The rate of energy radiation (24) then becomes

$$\dot{E}_r = 2\sqrt{3}\pi^2 s^2\exp\left[-\sqrt{3}\pi\mathrm{ch}\sigma\right]\cos^2(\sqrt{3}x - 2t). \tag{28}$$

On an average, the total emission power (to both sides) of the breather perturbed by a constant external force s therefore is

$$W = 2\sqrt{3}\pi^2 s^2\exp\left[-\sqrt{3}\pi\mathrm{ch}\sigma\right], \tag{29}$$

in accordance with Malomed [15].

Acknowledgements

The author greatly appreciates the stimulating interest of Professor A. Seeger. Thanks are also due to Drs. W. Lay and A. Stahlhofen and Dipl.-Phys. U. Blüher and R. Beutler for many discussions.

References

1. A. Seeger: Solitons in Crystals, in: Continuum Models of Discrete Systems, Eds. E. Kröner, K.H. Anthony. University of Waterloo, Waterloo (Ontario) 1980, p. 253.
2. A.R. Bishop, J.A. Krumhansl, and S.E. Trullinger, Physica **1D**, 1 (1980).
3. Yu. S. Kivshar and B.A. Malomed, Rev. Mod. Phys. **61**, 763 (1989).
4. A. Kochendörfer and A. Seeger, Z. Phys. **127**, 533 (1950).
5. A. Seeger and A. Kochendörfer, Z. Phys. **130**, 321 (1951).
6. A. Seeger, H. Donth, and A. Kochendörfer, Z. Phys. **134**, 173 (1953).
7. E. Mann, to be published.
8. E. Mann, phys. stat. sol. (b) **144**, 115 (1987).
9. D.W. McLaughlin and A.C. Scott, Phys. Rev. **A18**, 1652 (1978).
10. V.I. Karpman, E.M. Maslov, and V.V. Solov'ev, Sov. Phys.- JETP **57**, 167 (1983).
11. J. Eßlinger, Diplomarbeit, Universität Stuttgart 1988.
12. R. Döttling, Diplomarbeit, Universität Stuttgart 1989.
13. R. Döttling, J. Eßlinger, W. Lay, and A. Seeger, in: Nonlinear Coherent Structures, Lecture Notes in Physics, Vol. 353, Eds. M. Barthes, J. Léon. Springer, Berlin 1990, p. 193.
14. U. Blüher, Diplomarbeit, Universität Stuttgart 1990.
15. B.A. Malomed, Physica **27D**, 113 (1987); cf. the first part of eq. (7.6).

This article was processed using the LaTeX macro package with ICM style

NUMERICAL RESULTS CONCERNING THE GENERALIZED ZAKHAROV SYSTEM

Hichem Hadouaj, Gérard A. Maugin and Boris A. Malomed
Université Pierre-et-Marie Curie,
Laboratoire de Modélisation en Mécanique associé au C.N.R.S.,
Tour 66, 4 Place Jussieu, 75252 Paris Cédex 05, France.

Abstract. A generalization of the well known Zakharov system of ion-acoustic waves (Langmuir solitons) has been obtained while studying the coupling between shear-horizontal surface waves and Rayleigh surface waves propagating on top of a structure made of a nonlinear elastic substrate and a superimposed thin elastic film. The generalization consists in a nearly integrable system made of a nonlinear Schrödinger equation (thus including self-interactions) coupled to two wave equations for the secondary acoustic system (Rayleigh mode). Here we present essentially the numerical simulations pertaining to the uncoupled case (pure SH mode) and the coupled case (influence of viscous dissipation in the Rayleigh subsystem, collision of solitons).

1. General problem

The problem considered consists in studying the possible propagation of *surface* solitary waves, eventually solitons, of the surface-wave type (amplitude decreasing in the substrate) in a structure made of a *nonlinear* elastic isotropic *substrate* (half-space $X_2 > 0$) and a superimposed *linear* elastic isotropic *thin film*, the latter being perfectly bonded to the former (Figure 1). The *nonlinearity* originates thus from the substrate while *dispersion* is induced by the film which plays the role of a *wave guide*. In the mathematical description, the thin film is reduced to an *interface* of vanishing thickness which, however, still carries a mass density (hence inertia) and membrane elasticity in agreement with a general continuum approach [1]. A general surface wave problem in this structure involves both an SH (shear horizontal) elastic component (polarized along X_3) and a Rayleigh two-component displacement polarized parallel to the so-called sagittal plane P_s [2]. The complete coupled nonlinear wave problem is a tedious one which is shown to be tractable in several steps. First in the linear approximation an SH *dispersive* surface mode of the type of Murdoch [3], and a classical Rayleigh (nondispersive) mode propagate independently as a consequence of the assumed isotropy of the materials. At the next order, both modes couple through the nonlinearity [4]. However, if the primary signal entered in the system through a transducer is of the SH type and is $O(\varepsilon)$, then the Rayleigh subsystem will develop an $O(\varepsilon^2)$ component. This nonlinear mutual coupling [4] is neglected in the first instance

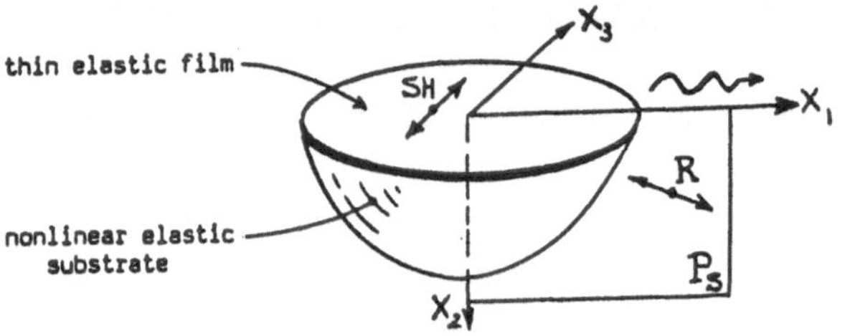

Figure 1 : Setting of the surface elastic-wave problem

and the pure *nonlinear* SH mode is shown to be governed by a single cubic Schrödinger equation *at the interface* for modulated signals with slowly varying envelope [5]. Then the problem accounting for the nonlinear coupling with the Rayleigh components is shown to be reducible to the announced generalized Zakharov system [6] when the main field still is of the SH type. Here essentially numerical simulations are presented, analytical results being found in other publications [5], [7].

2. The pure SH surface-wave problem

In this simplified case, using the boundary condition provided by the theory of material interfaces [1], we find that the initial mechanical (two space dimensions) problem is governed by the following set of equations in nondimensional units [5] :

$$\beta^2 U_{tt} - (U_{xx} + U_{yy}) = \beta^2 \Delta \left\{ \left[U_x \left(U_x^2 + U_y^2 \right) \right]_x + \left[U_y \left(U_x^2 + U_y^2 \right) \right]_y \right\} \text{ for } X_2 = y > 0 ,$$

$$\hat{U}_{tt} - \hat{U}_{xx} = U_y \left\{ 1 + \beta^2 \, \Delta \, \left(U_x^2 + U_y^2 \right) \right\} , \quad U = \hat{U} , \quad \text{at } X_2 = y = 0 \qquad , (2.1)$$

$$U \, (x, \, y \to \infty \, , \, t) = 0$$

where subscripts indicate space (x and y) and time (t) differentiation, Δ is the *nonlinearity* parameter, and β is the *dispersion* parameter. In the absence of nonlinearity ($\Delta = 0$) the above system yields *Murdoch's linear surface waves* [3] ; in the absence of dispersion (zero left-hand side in $(2.1)_2$), it yields *Mozhaev's nonlinear surface waves* [8]. The full system possesses all good ingredients to exhibit *solitary* waves of the surface wave type (so as to satisfy the last of (2.1)). This is proven analytically by using the Whitham-Newell [9] technique of treatment of nonlinear dispersive small amplitude, almost monochromatic waves [5]. In the process "*wave action*" conservation laws and "dispersive" nonlinear dispersion relations are established for this type of surface waves that could also be approached by using Whitham's averaged Lagrangian technique as modified by Hayes to account for the transverse modal behavior [10]. The analysis [5] is conducted simultaneously in the bulk (y > 0) and at the interface (y = 0). Combining the two at the interface results in a single nonlinear

Schrödinger (NS) equation for the envelope of complex amplitude a (in reduced coordinates) :

$$i\, a_t + p\, a_{xx} + q\, |a|^2\, a = 0 \tag{2.2}$$

where p and q are real and depend on the working regime $(\omega_0\ ,\ k_0)$ along the linear dispersion relation of Murdoch's waves. Explicitly,

$$p\ (\omega_0\ ,\ k_0) = \frac{1}{2}\,\omega_0'' \quad ,\quad q\ (\omega_0\ ,\ k_0) = \frac{3}{8}\,\Delta\,\beta^4\,\omega_0\ \frac{\left(\beta^2\ \omega_0^2\ -\ 2\ k_0^2\right)}{\beta^2\ +\ 2\ \left(\omega_0^2\ -\ k_0^2\right)}\ , \tag{2.3}$$

where ω_0'' is the curvature of the linear dispersion relation. The NS equation (2.2) is *exactly integrable* [11] and admits *bright* and *dark* envelope (true) solitons depending on the sign of the product pq. If the nonlinear material making up the substrate is known (e.g., $LiNbO_3$ [4] for which $\Delta > 0$), then this criterion allows one to select the thin film material to guarantee the existence of the desired stable surface solitary wave. In the present case with $\Delta > 0$, $\frac{1}{2} < \beta^2 < 1$ (film of aluminum) and $\beta^2 < \frac{1}{2}$ (film of gold) provide stable bright and dark solitons, respectively [5]. The analytical solutions thus obtained are used as initial-boundary value conditions in direct numerical simulations performed on the original (obviously *non* exactly integrable) two-space-dimension system (2.1). Explicit and implicit numerical finite-difference methods in three-dimensional.

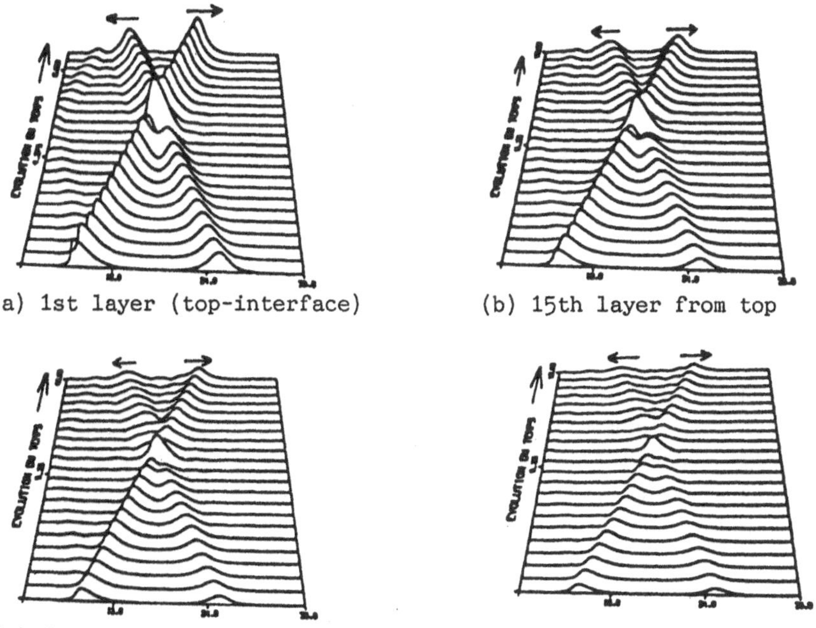

(a) 1st layer (top-interface)
(b) 15th layer from top
(c) 25th layer from top
(d) 35th layer from top

Figure 2 : Collision of unsymmetric envelope- solitons of the Surface-wave type in system (2.1).

Euclidian space-time grids were used for this (see [12] for technical details). The surface waves indeed propagate as *solitary* waves along the $X_1 = x$ direction with a nice exponential decrease with depth (along $X_2 = y$). A lack of accuracy in numerical solutions in depth will show up at the interface sooner or later. A predominance of nonlinear effects over dispersion may yield the formation of *surface shock waves after* a typical steepening [12]. While (2.2) obviously exhibits a true *solitonic* behavior in soliton interactions, the *rather* pure solitonic behavior of system (2.1) for small amplitudes, can only be checked numerically. This is indeed practically the case as shown in Figures 2 which exhibit the interaction of two colliding unequal solitons in the SH system at different depths in the substrate (fifty layers are accounted for in the computation along depth). At this point it should be noted that there is no difficulty to account for the *viscosity* in the pure SH system and the subsequent alteration in (2.2).

3. The coupled SH-Rayleigh problem

In agreement with Section 1, the main displacement field $O(\varepsilon)$ is the SH component, in which case the Rayleigh components are $O(\varepsilon^2)$. Considering slowly-varying envelope solutions for the SH components, a long asymptotic evaluation [6] allows one to show that, after integration along the transverse coordinate y, and appropriate scaling, the whole problem is governed at $y = 0$ by the following system of equations for the complex amplitude a of the SH mode and the real components v and w of elastic displacements along x and y, respectively, parallely to the sagittal plane P_s (Figure 1).

$$
\begin{aligned}
&i\, a_t + a_{xx} \pm 2\lambda\, |a|^2 a + 2a\,(\alpha_L n_1 + \alpha_T n_2) = 0 , \\
&(n_1)_{tt} - c_L^2 (n_1)_{xx} - \eta_L (n_1)_{xxt} = -\mu_L (|a|^2)_{xx} , \\
&(n_2)_{tt} - c_T^2 (n_2)_{xx} - \eta_T (n_2)_{xxt} = -\mu_T (|a|^2)_{xx} ,
\end{aligned}
\tag{3.1}
$$

where $n_1 = v_x$, $n_2 = w_x$, and viscosity has been introduced for the Rayleigh components only (on account of the last remark in Section 2). System (3.1), a *nearly integrable system* only, is a system which generalizes the system of Zakharov [13]- for which $\lambda = 0$, $w \equiv 0$, $\alpha_T = \mu_T = 0$ - that appears in ion-acoustic systems in plasmas (Langmuir solitons). This system has been extensively studied analytically. The general system (3.1) obviously is richer and presents many interesting features. Two of these are especially examined below.

4. Dissipation-induced evolution of solitons

We consider the evolution of envelope solitary waves in the (SH) a-system of (3.1) under the influence of dissipation (viscosities η_L and η_T) in the (v , w) Rayleigh systems. The two are coupled through the coupling coefficients μ_L and μ_T. In spite of its appearance, system (3.1) conserves the *number of surface phonons* (or *wave action*)

$$N = \int_{-\infty}^{+\infty} |a|^2 \, dx \tag{4.1}$$

In the analytical treatment $[7]_a$, which applies the *balance-equation analysis* to the *slow* dissipation-induced evolution of the exact one-soliton solution of the Zakharov system (for the sake of simplicity $w = 0$, $\mu_T = 0$ in (3.1) ; this system is *not* exactly integrable, three different scenarii of evolution are shown to be possible : (i) adiabatic (slow) transformation of a moving *subsonic* soliton into the stable quiescent one, (ii) complete adiabatic decay of a *transsonic* soliton with a small amplitude, and (iii) coming of the *transsonic* soliton with a large amplitude into a critical state, from which a further adiabatic evolution is not possible. In the latter case a numerical investigation of the further evolution of the soliton is particularly enligthening. In a general case, it is shown that it abruptly splits into the stable quiescent soliton, the slowly decaying small-amplitude transsonic one, and a pair of left and right-traveling acoustic pulses slowly fading under the action of the weak dissipation. This is exhibited in the numerical simulations in Figures 3 and 4. The abrupt splitting seems to be a new type of *inelastic* process for a soliton induced by small perturbations (see the review given in Ref.[14]). This concludes our brief excursion in the evolution of *one* soliton in the *damped* generalized Zakharov system.

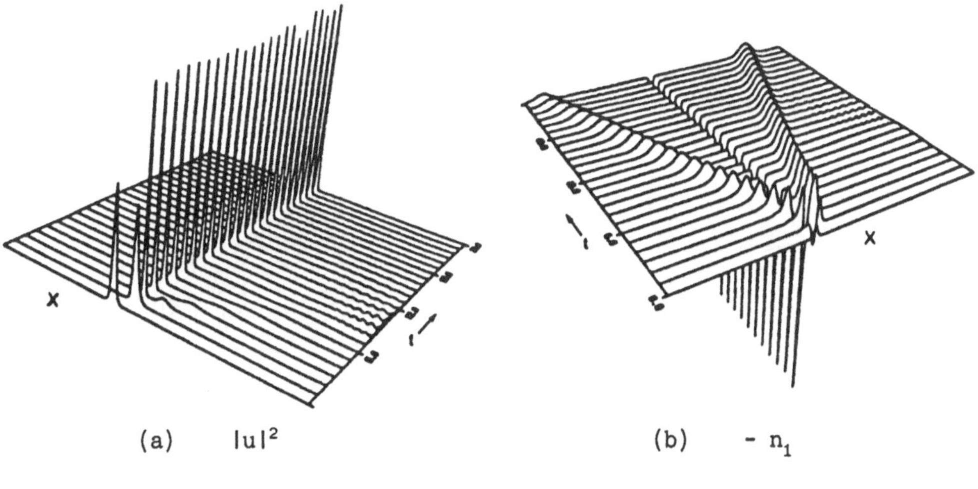

Figure 3 : Dissipation-induced evolution of the exact one-soliton solution of Zakharov's system : abrupt split into three pulses in the n_1-system (large wave action, large velocity)

5. Soliton-soliton collision in the generalized Zakharov System.

As seen above the generalized Zakharov system (3.1) in the absence of viscosity in the Rayleigh subsystem admits both *subsonic* and *transsonic* one-soliton solutions. The question naturally arises of the interaction (collision) of such solitons, for instance in the symmetric soliton-soliton collisions. In the analytical study $[7]_b$ the collision-induced emission of acoustic waves (in the Rayleigh subsystem) was treated for soliton velocities much larger than their amplitudes. In particular, it was shown that the acoustic losses are exponentially small unless the solitons' velocities are much larger than the characteristic sound velocity in the Rayleigh subsystem. The numerical simulation of the head-on soliton-soliton collision brings up two basic phenomena : (i) the collision of *subsonic* solitons always leads to their *fusion into a breather*, provided the system is sufficiently far from the integrable limit ; (ii) the collision between *transsonic* solitons gives rise to a multiple production of solitons (both sub- and transsonic solitons are produced), and the *quasi-elastic* character of

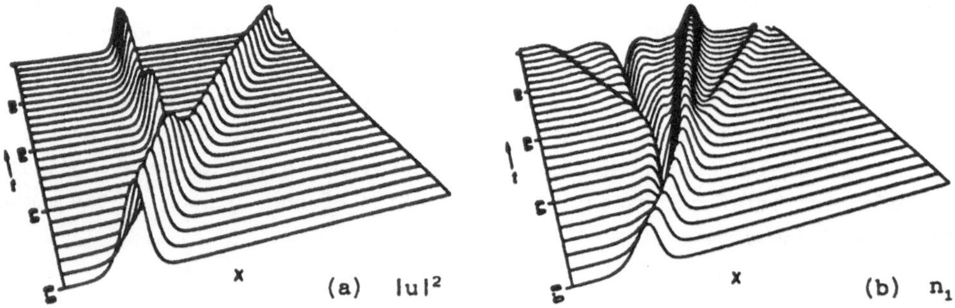

Figure 4 : Dissipation-induced evolution of the exact one-soliton solution of Zakharov's system : rearrangement of the soliton in the intermediate case. (Smaller values of wave action and velocity than in Figure 4).

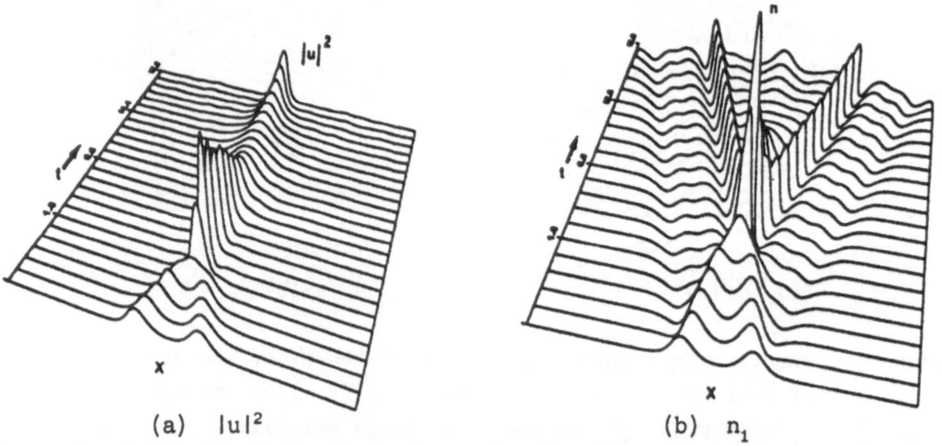

Figure 5 : Collision-induced fusion of *subsonic* solitons into a breather with acoustic emission in the Rayleigh subsystem

the collision is recovered in the limit of large velocities. This is illustrated in Figures 5, 6 and 7.

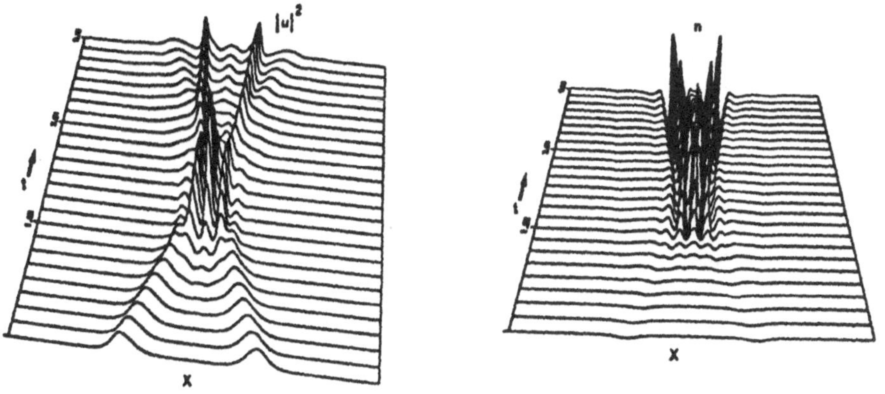

(a) $|u|^2$ (b) n_1

Figure 6 : Collision of two transsonic solitons at moderate velocities

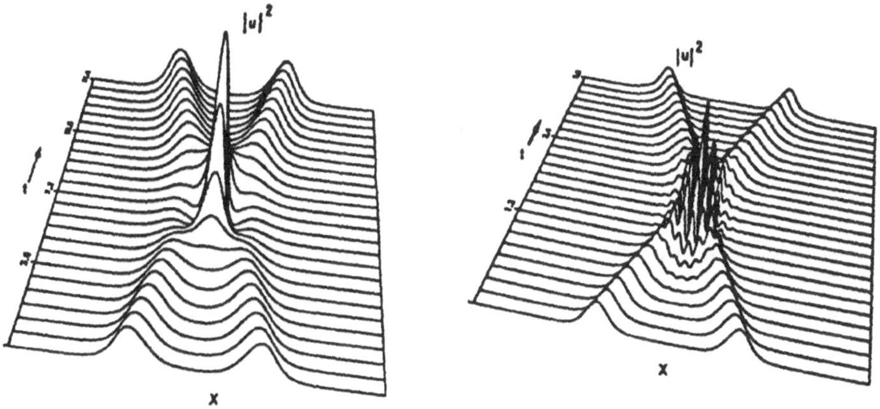

(a) (b)

Figure 7 : Comparison between the soliton-soliton collision for system (3.1) close to the NS equation (a) and the collision of two *transsonic* solitons at high velocities in the generalized Zakharov system (b).

6. Conclusion

It appears that the initial, purely mechanical, surface-wave problem considered yields, on the one hand, a very interesting physical application which may be of interest in signal processing (we have a mechanical analog of light solitons in optical fibers; compare [15]) and, on the other hand, a class of paradigmatic problems in soliton theory for *nearly integrable* systems made of an exactly integrable equation coupled nonlinearly to d'Alembert equations. The *sine-Gordon-d'Alembert* systems introduced previously by Maugin and Pouget [16] in a different physical context belong to the same class. The *modified-Boussinesq-d'Alembert* system introduced recently by Maugin and Cadet [17] in martensitic alloys appears to be even more difficult to deal with.

References

[1] N. Daher and G.A. Maugin, *Acta Mechanica*, 60 (1986), 217.

[2] G.A. Maugin, in : *Advances in applied Mechanics*, ed. J.W. Hutchinson, Vol.23, pp 373-434, Academic Press, New York (1983).

[3] A.I. Murdoch, *J. Mech. Phys. Solids*, 24 (1976), 137.

[4] G.A. Maugin, *Nonlinear Electromechanical Effects and Applications*, World Scientific, Singapore (1985), pp. 36-44.

[5] H. Hadouaj and G.A. Maugin, *C.R.Acad. Sci. Paris*, II-309 (1989), 1877 ; G.A. Maugin and H. Hadouaj, *Phys. Review, B* (1991), in the press.

[6] H. Hadouaj , G.A. Maugin, and B.A. Malomed, *Phys. Review , B* (1991).

[7] H. Hadouaj, B.A. Malomed, and G.A. Maugin, *Phys. Review, A* (1991 a,b).

[8] V.G. Mozhaev, *Physics letters*, A139 (1989), 333.

[9] D.J. Benney and A.C. Newell, *J. Math. and Phys.*, 46 (1967), 133 ; A.C. Newell, *Solitons in Mathematics and Physics*, S.I.A.M, Phil.(1985).

[10] G.B. Whitham, *Linear and Nonlinear Waves*, J. Wiley, New York (1974) ; W.D. Hayes, *Proc. Roy. Soc. London*, A320 (1970), 187.

[11] V.E. Zakharov and A.B. Shabat, *Sov. Phys. JETP.* , 34 (1972), 62 ; 37 (1973), 823.

[12] H. Hadouaj and G.A. Maugin, in : *Mathematical and Numerical Aspects of Wave Propagation*, S.I.A.M, Philadelphia (1991); *Wave Motion* (Special issue on " Nonlinear Waves in Deformable Solids ". I.C.I.A.M'91, Washington, D.C), 14 ,(1991) in the press.

[13] V.E. Zakharov, *Sov. Phys. JETP*, 35 (1972), 908 ; E. Infeld and G. Rowlands, *Nonlinear Waves, Solitons and Chaos*, Cambridge University Press, U.K (1990), pp. 319-321.

[14] Yu. S. Kivshar and B.A. Malomed, *Rev. Mod. Phys.*, 61 (1989), 763.

[15] A. Hasegawa , *Optical Solitons in Fibers*, Springer , Berlin (1989).

[16] G.A. Maugin and A. Miled, *Phys. Rev., B33* (1986), 4830 ; J.Pouget and G.A. Maugin, *Phys. Rev., B30* (1984), 5306 ; *ibid, B31* (1985), 4633.

[17] G.A. Maugin and S. Cadet, *Int. J. Engng. Sci*, 29 (1991), 243.

Resonances in nonlinear Klein-Gordon kink scattering by impurities

Yuri S. Kivshar,[1,2] *Angel Sánchez*[1] *and Luis Vázquez*[1]

[1] Departamento de Física Teórica I, Facultad de Ciencias Físicas Universidad Complutense, E-28040 Madrid, Spain
[2] On leave from: Institute for Low Temperature Physics and Engineering, 47 Lenin Avenue, 310164 Kharkov, U.S.S.R

Abstract. The scattering of topological kinks by point-like impurities is numerically studied in the framework of the sine-Gordon and ϕ^4 models. In the first case, we show that previous approximate analytical results are indeed applicable. Thus, for low velocities, the reflection coefficient depends oscillatorily on the distance between impurities, i.e., resonant scattering takes place. On the other hand, this effect disappears for higher velocities. This result is found to hold also for the ϕ^4 model nonlinearly coupled to the impurities, whereas the linear coupling for the same model gives rise to a different behaviour.

1 Introduction.

In recent years, it has become clear the importance of the interplay between disorder and nonlinearity in many physical contexts, which is at the root of many novel and striking phenomena and their corresponding unsolved mathematical problems [1,2]. Among these, the question as to whether solitons, supported by nonlinearity, and Anderson localization effects influence each other, changing the transmission properties of systems, deserves careful study in view of its practical consequences. As a first step in this direction, the scattering properties of solitons in a system with few impurities must be understood. Even more, as there are three main types of solitons (see [3,4] and references therein), their scattering features are different, and a separate analysis is required. The scattering of non-topological solitons by one and two impurities has been already studied numerically by Li *et al.* [5]: both kink-like and envelope solitons were considered in the simple model of a nonlinear atomic chain with nearest-neighbor interparticle interactions. A part of the results obtained there [5] has been explained analytically [6] by means of the perturbation theory for solitons based on the inverse scattering transform (IST), [7] (see [8] for a comprehensive review of this technique). On the contrary, the current knowledge about the scattering properties of the other kind of solitons, topological kinks similar to those beared by Nonlinear Klein-Gordon (NKG) equations (like the sine-Gordon [sG] or ϕ^4 ones), is quite limited.

As it was previously pointed out [6,7], interference phenomena are the most remarkable effects to study wave properties of solitons. Interference may arise when the characteristic distance between impurities becomes commensurable with the characteristic length of the linear waves emitted by the soliton. As a consequence, the simplest way to observe interference is to consider a system with two point impurities [6,7]. Solitons emit a rather wide spectrum of linear waves, and the problem is far more difficult than the scattering of a monochromatic plane wave. As a matter of fact, the above mentioned condition that the two characteristic lengths must be commensurate has to be valid for *an averaged* spectral structure of the soliton emission. When a soliton has an internal frequency (like, e.g., an envelope soliton), resonant scattering is naturally expected [6,7], while for topological kinks this phenomenon does not seem to be possible. However, resonant scattering has been predicted [6] for a slowly moving sG kink when the spectral density of the emission generated by it has a narrow maximum: interference effects should appear as oscillations of the soliton reflection coefficient dependence on the distance between impurities. It is not at all a trivial matter to ask if this prediction is actually useful, for if the kink is too slow, it may be pinned or reflected by attractive or repulsive impurities, respectively (see [8] and references therein). As these possible effects were neglected in the analytical calculation [7] of the reflection coefficient, it is crucial to determine whether they become as relevant as to inhibit the interference effects by forbidding kink propagation. This is the problem we address here, namely the resonant scattering of sG and ϕ^4 kinks: we study it numerically as to compare the so obtained results to the analytical predictions, thus establishing their validity range, if any.

2 Model, predictions and numerical procedure

The model we deal with is an inhomogeneous NKG system which, in dimensionless units, is described by the equation

$$\phi_{tt} - \phi_{xx} + V'(\phi) + \epsilon[\delta(x) + \delta(x - D)]\, V_{imp}(\phi) = 0; \tag{1}$$

In particular, we will consider the following choices for potentials and perturbations:

$$V'(\phi) = V_{imp}(\phi) = \sin\phi, \tag{2}$$

$$V'(\phi) = -\phi + \phi^3, \; V_{imp}(\phi) = V'(\phi), \tag{3}$$

$$V'(\phi) = -\phi + \phi^3, \; V_{imp}(\phi) = -\phi, \tag{4}$$

Equation (2) corresponds to a sG model with two point-like impurities, whereas equations (3) and (4) are similarly perturbed ϕ^4 models in which the impurities are coupled to the wavefield either nonlinearly, equation (3), or linearly, equation (4). All of these systems are very well-known in the unperturbed case, i.e., $\epsilon 00$, and their properties have been widely described, but the perturbed problems are rather difficult and cannot be solved exactly. Nevertheless, recalling that the homogeneous sG equation is integrable, some theoretical analysis of the problem (2) is possible through perturbation theory for solitons based on IST. This analysis has been recently carried out by Kivshar *et al.* [6], who were interested in the influence of the parameter D on the scattering properties of sG kinks, or more precisely, in their reflection coefficient. This coefficient, R, may be defined as $R \equiv E_{em}^{(-)}/E_k$, where $E_k = 8/\sqrt{1-v^2}$ is the energy of a sG kink with velocity v far away from the inhomogeneous region, and $E_{em}^{(-)}$ is the energy the kink emits backwards as radiation, due to its interaction with an impurity. The emitted energy $E_{em}^{(-)}$, and, consequently, the reflection coefficient R can be calculated analytically by means of the aforementioned IST perturbation theory, the main restriction of it being the necessary assumption that the kink velocity does not change during the scattering (the so called Born approximation). By this means, the emitted spectral density can be shown [6] (see similar computations in [7,8]) to be given by

$$\varepsilon_2(k) = 4\varepsilon_1(k)\cos^2\left\{\frac{D}{2v}[kv - \omega(k)]\right\}, \tag{5}$$

$$\varepsilon_1(k) = \frac{\pi\epsilon^2}{8v^6}(1-v^2)^2\,\frac{[k-\omega(k)]^2}{\cosh^2[\pi\sqrt{1-v^2}\omega(k)/2v]}, \tag{6}$$

where $\omega(k) \equiv \sqrt{1+k^2}$, and $n = 1,2$ stand for the case when one and two impurities are present in the system, respectively. Having these expressions in mind, it is straightforward to obtain the corresponding reflection coefficients, which turn out to be

$$R_n = \frac{1}{8}\sqrt{1-v^2}\int_0^\infty dk\,\varepsilon_n(-k), \tag{7}$$

where $n = 1,2$ stands for the case with one or two impurities, respectively.

It can be seen from equations (5) and (6) that for small v, $v^2 << 1$, the spectral density $\varepsilon_1(k)$ has a single maximum at $k = 0$, with a quite narrow peak of width of order $2v/\pi \sim v$. As a consequence, such maximum will provide the main contribution to the emitted energy and should give rise to a resonant dependence: it is possible to obtain from equations (5)–(7) an approximate estimation for the value $R_2/2R_1$ when $v^2 << 1$, which turns out to oscillate as

$$\frac{R_2}{2R_1} \simeq 1 + \frac{1}{(1+D^2/\pi^2)^{1/4}}\cos\left[\frac{D}{v} - \frac{1}{4}\tan^{-1}\left(\frac{D}{\pi}\right)\right]. \tag{8}$$

On the other hand, if the speed is large, there are two maxima at $\pm k_m$, $k_m = 2v/\pi\sqrt{1-v^2} \simeq (1-v^2)^{-1/2}$, and the function $\varepsilon_1(k)$ is not exponentially small in the region $|k| < k_m$. Hence, after averaging over all wave numbers there is no leading contribution, the oscillatory dependence dissapears, and resonant scattering is not to be expected.

It must be noticed that, of course, equation (8) makes sense only from a theoretical viewpoint, because due to total reflection the kink velocity can not be less than a certain threshold, $v_{thr} \equiv \sqrt{\epsilon/2}$, below which the kink is reflected by the impurity. This is the fundamental reason for the necessity of numerical simulations: to see whether velocities over the threshold still are well accountd for by the perturbative prediction. The v_{thr} value can be obtained thinking of the kink as a point-like particle moving in an effective potential originated by the impurities (see, e.g., [8]). Finally, let us insist that this result applies only to sG systems; the non-integrability of the ϕ^4 system does not allow to use the same technique, though it is possible to obtain some results which we will describe elsewhere [13].

The numerical procedure we use to simulate the kink scattering is the finite difference scheme of Strauss and Vázquez [9] for nonlinear Klein-Gordon equations. This scheme has been succesfully employed to study a number of different perturbed nonlinear Klein-Gordon problems (for instance, see [10,11]; see also references therein). Moreover, its most important property is that it exactly conserves the system energy in the unperturbed evolution, which is relevant to the accuracy of the computation we intend to do; notice that all we must evaluate is the energy content at the left, between and at the right of the impurities. Further details on the scheme can be found in the literature [9,11].

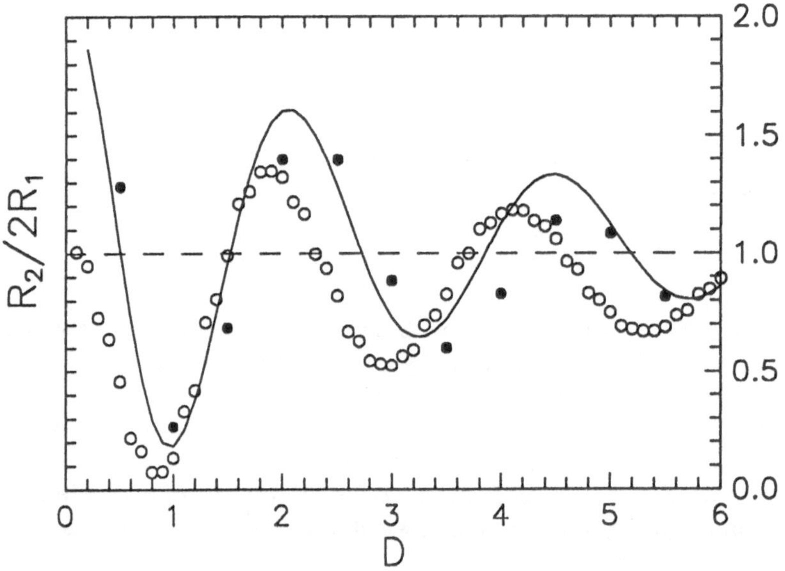

Fig. 1. Reflection coefficient for a $v = 0.4$ sG kink vs distance D between impurities.

3 Results and discussion

The numerical results for two values of the initial kink velocity in the sG model (2) are shown in figures 1 and 2. We plot the ratio $R_2/2R_1$ which not only can be directly compared to the prediction (8), but also is a suitable quantity to search for interference effects. Indeed, when the impurities are far from each other, i.e., when $D \rightarrow \infty$, the reflection coefficient has to coincide with $2R_1$, and the above mentioned quotient must go to unity; any difference from unity may be treated as coming from interference. In both figures, the full lines correspond to the analytically computed dependence given by Eqs.(5)–(7). In addition, for $v = 0.4$

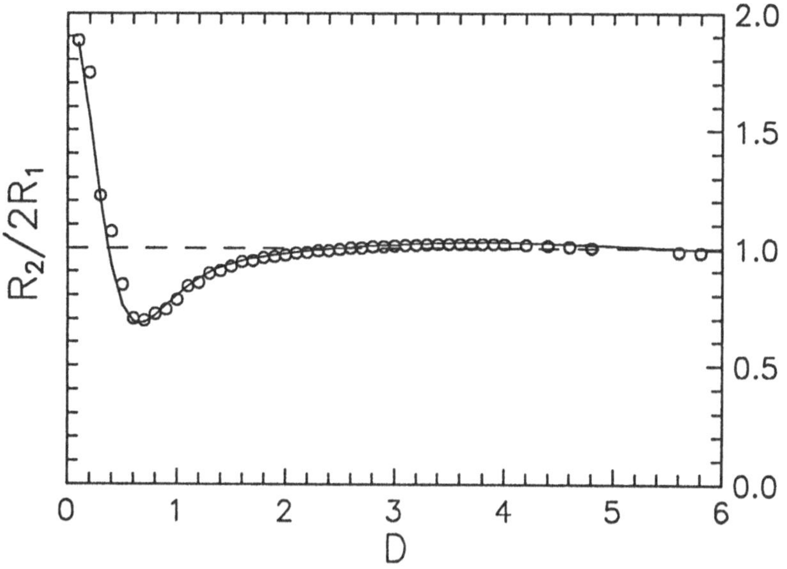

Fig. 2. Reflection coefficient for a $v = 0.9$ sG kink vs distance D between impurities.

we performed some simulations for attractive delta-functions, $\epsilon = -0.1$, because predictions do not depend on the sign of ϵ, cf. equation (6; these are shown as full circles. It comes out that the agreement between analytical and numerical curves is fairly good. Therefore, in view of our previous considerations, we conclude that the numerical simulations establish that the resonant scattering of the sG kink is stipulated by the spectral properties of its emission, according to the above discussed perturbative predictions.

The main differences arise at $v = 0.4$ (figure 1); in particular, the asymptotic behaviors of the analytical and the numerical curves are not the same. This disagreement, that in principle should not be expected, can be explained in a natural way. Recall that the analytical results for two impurities were obtained under the assumption that the kink *does not change* its velocity during the scattering (Born approximation). However, as a matter of fact, after the first scattering the kink loses some part of its kinetic energy, so that it interacts with the second impurity at a *smaller* velocity, say $v - \Delta v$. Hence, the ratio $R_2/2R_1$ does not go to 1 when the distance between deltas, go to infinity; rather well, it verifies

$$\frac{R_2}{2R_1} \to \frac{[E_{em}^{(-)}(v) + E_{em}^{(-)}(v - \Delta v)]}{2E_{em}^{(-)}(v)}, \tag{9}$$

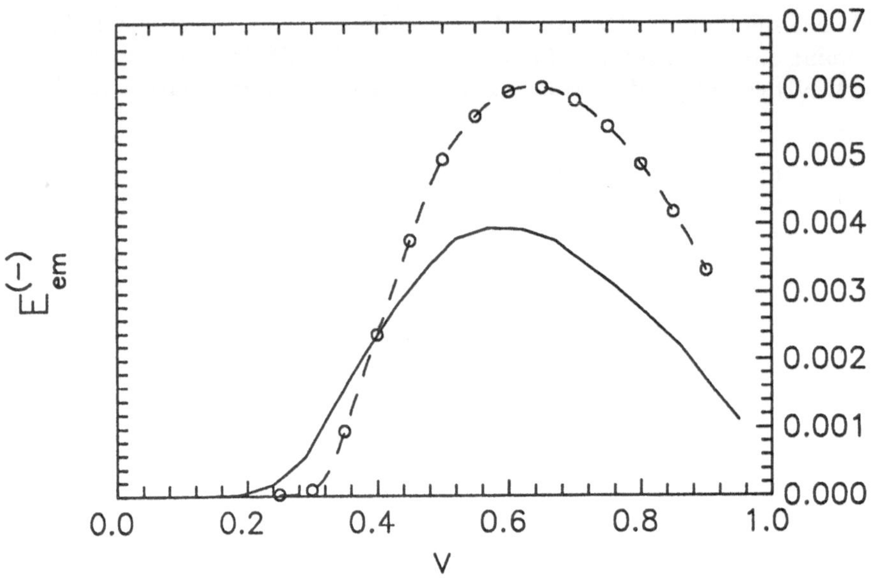

Fig. 3. Emitted energy by a sG kink vs kink velocity.

$E_{em}^{(-)}$ being the energy reflected by a single impurity. To understand the difference between $E_{em}^{(-)}(v)$ and $E_{em}^{(-)}(v - \Delta v)$, we have analyzed the emitted energy for a single impurity versus the kink velocity. Figure 3 presents both numerical and analytical dependences.[1] From this plot, it turns out that when the initial velocity is $v = 0.4$, $E_{em}^{(-)}(v) > E_{em}^{(-)}(v - \Delta v)$, so that the asymptotics of $R_2/2R_1$ computed numerically is always *smaller* than the analytical predictions. This ceases to be true when v becomes larger than $v_{cr} \simeq 0.6$ because in that range $E_{em}^{(-)}(v)$ decreases and hence $E_{em}^{(-)}(v - \Delta v) > E_{em}^{(-)}(v)$. Thus, in figure 2, the numerical asymptotics is slightly above the line $R_2/2R_1 = 1$, the difference being small due to the short time that this kink takes to cross the distance D between, much shorter than in the other case. The remaining, little discrepancy for attractive impurities is due to the fact that, in this case, the linearized sG model supports the so-called impurity mode (see, e.g., [12]), which is excited by the kink, giving an additional contribution to the radiated energy as computed in the region $x < 0$. Detailed analysis of the impurity mode excitation during the kink scattering and also a quasi-resonant behavior originated from an energy-exchange mechanism between the kink and this impurity mode will be presented elsewhere [13,14].

After this proof of the validity of the approximate perturbative results (which, besides, is also a new checking of the Strauss-Vázquez procedure) we have ca-

[1] The analytical results, that are below the numerical ones, assumed that the velocity of the kink does not change, but, in fact, the change in velocity will produce an additional emission. This effect can explain the amount of emission observed in the simulations, larger than the predicted.

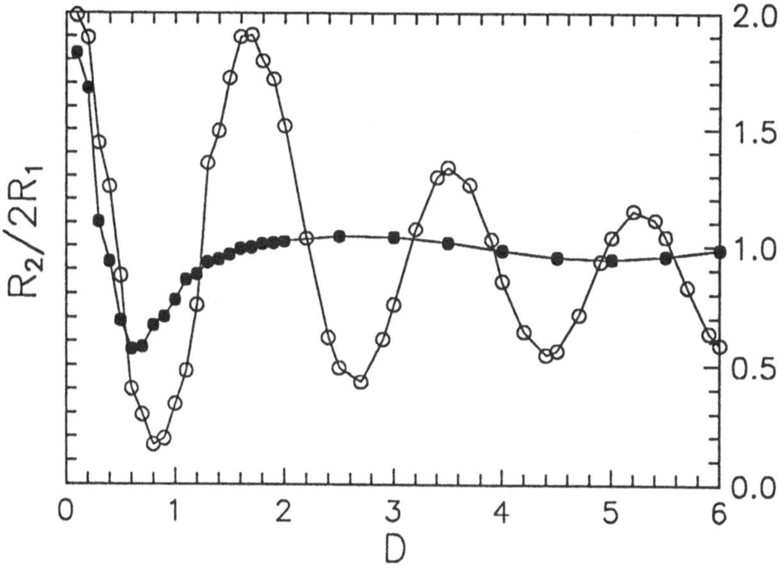

Fig. 4. Reflection coefficient for ϕ^4 kinks with nonlinear coupling [model (3)] vs distance D between impurities.

rried out identical simulations for the two ϕ^4 models. In the one with nonlinear coupling, everything is very much like the sG system, as is shown in figure 4: resonant scattering arises again for slow kinks, and attractive impurities exhibit also the signature of their mode (described in [15]). This strongly suggests that the radiation emission by the ϕ^4 kink is quite the same as that of the sG kink. On the contrary, the phenomena that take place for the ϕ^4 model linearly coupled to the impurities (figure 5) are different, and cannot be simply explained in terms of the spectral content of the emission. Let us stress that the reflection coefficient is never very small, and then we cannot speak of resonant scattering. Moreover, kinks with $v = 0.4$ are reflected by the joint action of the impurities if $D < 1.4$. Finally, we have observed that the role played by the localized impurity mode is much more important in this system, seemingly because of the linear nature of the coupling; notice that now the kink tails interact with the impurities and hence they are not a ground state of the model anymore. Further research on this model is needed to describe properly these cooperative scattering effects.

Acknowledgments

This work has been supported in part by the Dirección General de Investigación Científica y Técnica (D.G.I.C. y T.) of Spain, through project MAT90-0544. The stay of Yu. S. Kivshar in Madrid has been supported by the Universidad Complutense de Madrid through the Programa de Sabáticos Complutense.

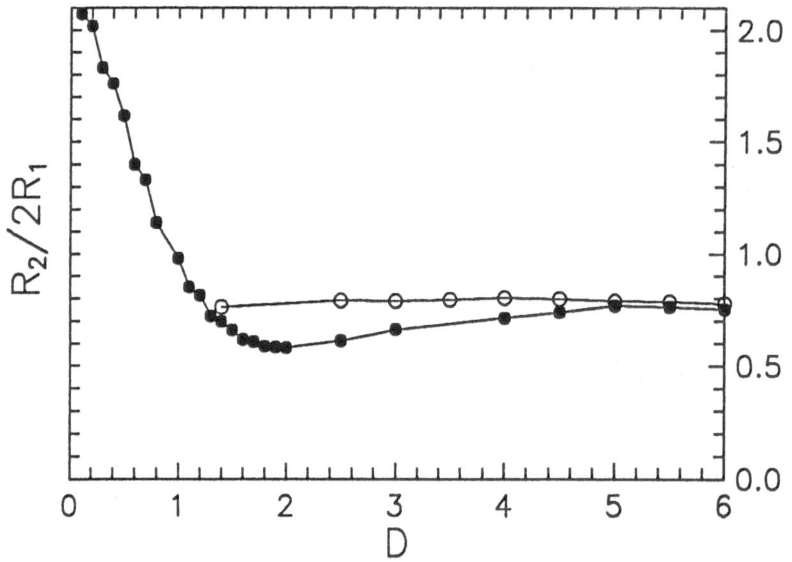

Fig. 5. Reflection coefficient for ϕ^4 kinks with linear coupling [model (4)] vs distance D between impurities.

References

1. Bishop, A. R., Campbell, D. K., Pnevmatikos, St. (eds.): *Disorder and Nonlinearity* (Springer Proceedings in Physics no.39, Springer, Berlin-Heidelberg, 1989).
2. Abdullaev, F. Kh., Bishop, A. R., Pnevmatikos, St. (eds.): *Nonlinearity with Disorder* (Springer Proceedings in Physics, Springer, Berlin-Heidelberg, in press, 1991).
3. Kivshar, Yu. S.: in Reference [2].
4. Sánchez, A. S., Vázquez, L.: Int. J. Mod. Phys. B, in press (1991).
5. Qiming Li, Pnevmatikos, St., Economou, E. N., Soukoulis, C. M.: Phys. Rev. B **37**, 3534 (1988).
6. Kivshar, Yu. S., Kosevich, A. M.,Chubykalo, O. A.: to be published.
7. Kivshar, Yu. S., Kosevich, A. M.,Chubykalo, O. A.: Phys. Lett. A **125**, 35 (1987).
8. Kivshar, Yu. S., Malomed, B. A.: Rev. Mod. Phys. **61**, 763 (1989).
9. Strauss, W., Vázquez, L.: J. Comp. Phys. **28**, 271 (1978).
10. Sánchez, A., Vázquez, L.: Phys. Lett. A **152**, 184 (1991).
11. Sánchez, A., Vázquez, L., Konotop, V. V.: Phys. Rev. A **43**, in press (1991).
12. Braun, O. M., Kivshar, Yu. S.: Phys. Rev. B **43**, 1060 (1991).
13. Kivshar, Yu. S., Sánchez, A., Vázquez, L.: to be published.
14. Kivshar, Yu. S., Zhang Fei, Vázquez, L.: to be published.
15. Fraggis, T., Pnevmatikos, St., Economou, E. N.: Phys. Lett. A **142**, 361 (1989).

This article was processed using the LaTeX macro package with ICM style

RESONANT KINK–IMPURITY INTERACTIONS

Zhang Fei, Yuri S. Kivshar†, and Luis Vázquez
Departamento de Física Teórica I, Facultad de Ciencias Físicas,
Universidad Complutense, E-28040 Madrid, Spain
† On leave from: Institute for Low Temperature Physics and Engineering,
UkrSSR Academy of Sciences, 47 Lenin Avenue, Kharkov 310164, USSR

ABSTRACT We demonstrate that the impurity mode plays an important role in the kink–impurity interactions, and a kink may be totally *reflected by an attractive impurity* if its initial velocity lies in some resonance "windows". This effect is quite similar to the resonance phenomena in kink-antikink collisions in some nonlinear Klein-Gordon equations, and it can be explained by a resonant energy exchange mechanism. Taking the sine-Gordon and the ϕ^4 models as examples, we find a number of resonance windows by numerical simulations, and develop a collective-coordinate approach to describe the interactions analytically.

1 Introduction

It is well known that nonlinearity may drastically change transport properties of disordered systems when it contributes to create soliton pulses [1,2]. As a first step to understand the soliton transmission through disordered media, one has to study the soliton scattering by a single impurity. The kink-impurity interactions have been explained, for a long time, by the well-known model in which a kink moving in an inhomogeneous medium is considered as an extended classical particle obeying Newton's Law of Motion (see Refs. [3]-[5] and references therein). In particular, it was shown by Malomed [6] that a sine-Gordon kink may be trapped by an attractive impurity due to radiative losses, and a threshold velocity was found analytically. However, the previous theoretical studies [3]-[6] totally ignored the fact that the underlying nonlinear system supports a localized impurity mode which may be *excited* due to the kink scattering. Recently, the importance of the impurity mode have been noticed in our papers [7]-[9]. We have found that a kink may be totally reflected by an attractive impurity due to resonance energy exchange between the kink translational mode and the impurity mode. This effect is quite similar to the resonance phenomena in kink-antikink collisions in some nonlinear Klein-Gordon equations [10]–[13]; and it cannot be predicted by the previous theoretical approach, which took into account only radiative losses [3]-[6].

In the present paper we briefly review our recent numerical and analytical results related to the kink-impurity interactions in two well-know kink-bearing systems, namely the sine-Gordon (SG) and the ϕ^4 models. In section 2 we describe the resonance phenomena in kink-impurity interactions in the SG model, and develop a collective-coordinate approach to explain the phenomena analytically. Section 3 is devoted to the similar problem in the ϕ^4 model. We draw conclusions in section 4.

2 Kink-impurity interactions in the SG model

Firstly, we consider the sine-Gordon system including a local inhomogeneity

$$u_{tt} - u_{xx} + [1 - \epsilon\delta(x)] \sin u = 0, \tag{1}$$

where $\delta(x)$ is the Dirac δ-function. When the perturbation is absent ($\epsilon = 0$), the SG model (1) supports a topological soliton, the so-called kink which is given by

$$u_k(x,t) = u_k(z) = 4\tan^{-1}\exp(\sigma z), \tag{2}$$

where $z = (x - X)/\sqrt{1 - V^2}, X = Vt + X_0$ is the kink coordinate, $\dot{X} = V$ is its velocity, and $\sigma = \pm 1$ is the kink polarity (without loss of generality we assume that $\sigma = +1$).

To describe the motion of the kink (2) in the presence of a localized inhomogeneity, the so-called adiabatic perturbation theory for solitons was usually used [3]–[5]. In the framework of this perturbation theory, the kink coordinate X is considered as a collective variable, and its evolution is described by a simple motion equation of a classical particle with mass $m = 8$ placed in the effective potential

$$U(X) = -2\epsilon/\cosh^2 X. \tag{3}$$

For $\epsilon > 0$, the impurity in Eq.(1) gives rise to an attractive potential $U(X)$ to the kink. Since the particle (kink) conserves its energy, it can not be trapped by the potential well if it has a non-zero velocity at infinity. However, according to the result of Malomed [6], the kink may be trapped by an attractive impurity due to radiative losses, and there exists a threshold velocity $V_{thr}(\epsilon)$, such that if the kink initial velocity is larger than $V_{thr}(\epsilon)$, it will pass through the impurity and escape to infinity, otherwise it will be trapped by the impurity [6]. In this consideration the reflection of the kink is *impossible.*

We have studied the kink-impurity interactions by numerical simulations [7]. We use a conservative scheme [14] to integrate Eq.(1), and carry out the simulations in the spatial interval $(-40, 40)$ with discrete stepsizes $\Delta x = 2\Delta t = 0.04$. When handling the Dirac δ–function, we take its value equal to $1/\Delta x$ at $x = 0$, and zero otherwise. The initial conditions are always taken as a kink centered at $X = -6$, moving toward the impurity with a given velocity $V_i > 0$. We have made intensive numerical simulations of the problem for $\epsilon > 0$ (attractive impurity), and here we will describe the results for the case $\epsilon = 0.7$ in detail.

In the numerical simulations, we find that there are three different regions of initial kink incoming velocity, namely, region of pass, of capture, and of reflection (see Fig.1); and a critical velocity $V_c \approx 0.2678$ (for $\epsilon = 0.7$) exists, such that if the incoming velocity of the kink is larger than V_c, the kink will pass the impurity inelastically and escape to the positive direction, losing part of its kinetic energy through radiation and excitation of an impurity mode. In this case, there is a linear relationship between the squares of the kink initial velocity V_i and its final velocity V_f : $V_f^2 = \alpha(V_i^2 - V_c^2)$, $\alpha \approx 0.887$ being constant.

If the incoming velocity of the kink is smaller than V_c , the kink cannot escape to infinity from the impurity after the first interaction, but will stop at a certain distance and return back, due to the attracting force of the impurity, to interact with the impurity again. For most of the velocities, the kink will lose energy again in the second interaction and finally get trapped by the impurity (see Fig. 1). However, for some special incoming velocities, the kink may escape to the *negative* infinity after the second interaction, i.e., the kink may be totally *reflected* by the impurity (see Fig. 1 and Fig. 2). This effect is quite similar to the resonance phenomena in kink-antikink collisions[10]–[13]. The reflection is possible only if the kink initial velocity is situated in some resonance windows. By numerical simulation, we have found eleven such windows. The detailed results are presented in Table I.

In order to understand the resonance structure, we define the center of the kink, $X(t)$, as the point at which the field function $u(x,t)$ is equal to π. We define T_{12} as the time between the first and the second interaction. It is clear that the attractive potential caused by the impurity falls off exponentially, so that using the same arguments as in Ref.[10] [see Eq.(3.6)], we obtain an approximate formula to estimate $T_{12}(V)$

$$T_{12}(V) = \frac{a}{\sqrt{V_c^2 - V^2}} + b \tag{4}$$

where V is the kink initial velocity, a and b are two constants. For $\epsilon = 0.7$, the parameters are empirically determined by numerical data: $a \approx 3.31893, b \approx 1.93$. We have found that formula (4) is very accurate for the velocities over the interval $(0.10, 0.267)$.

On the other hand, as we have observed that the first kink-impurity interaction always results in exciting the impurity mode, and the resonant reflection of the kink after the second interaction is just a reverse process, i.e., to extinguish the impurity mode (see Fig.2), when the timing is right, to restore enough of the lost kinetic energy and to escape from the impurity to infinity. Favorable timing in this case means that the occasion of the second interaction coincides with the passage of the impurity oscillation through some phase angle characteristics of the impurity mode extinction. Thus, the condition for restoration of the kink kinetic energy after the second interaction ought to be of the form

$$T_{12}(V) = nT + \tau. \tag{5}$$

where T_{12} is the time between the first and the second interaction, T is the period of the impurity mode oscillation, τ is an offset phase, and n is an integer. From numerical data we find that $\tau \approx 2.3$ (for the case $\epsilon = 0.7$)

Combining Eqs.(4) and (5), we may obtain a formula to predict the centers of the resonance windows,

$$V_n^2 = V_c^2 - \frac{11.0153}{(nT + 0.3)^2}, \quad n = 2, 3, ... \tag{6}$$

Similar formulas have been derived for kink-antikink collisions [10]–[13]. In Table I, we show the centers of the resonance windows predicted by Eq.(6), where $T = 2\pi/\Omega \approx 6.707$, Ω is determined by Eq.(8) with $\epsilon = 0.7$. Numerical $T_{12}(V_n)$ is defined as the time between the first and the second interactions. Note that $T_{12}(V_{n+1}) - T_{12}(V_n) \approx 6.7$ is just another expression of the resonance condition (5). From Table I we see that formula (6) can give very good predictions of the resonance windows.

To analyze the kink-impurity interactions analytically, first, we note that the nonlinear system (1) supports a localized mode. By linearizing Eq.(1) in small u, the shape of the impurity mode can be found analytically

$$u_{im}(x,t) = a(t)e^{-\epsilon|x|/2}, \tag{7}$$

where $a(t) = a_0 \cos(\Omega t + \theta_0)$, Ω is the frequency of the impurity mode,

$$\Omega = \sqrt{1 - \epsilon^2/4} \tag{8}$$

and θ_0 is an initial phase. As a matter of fact, the impurity mode (7) can be considered as a small-amplitude breather trapped by the impurity, with energy [8]

$$E_{im} = \frac{1}{2}\int_{-\infty}^{\infty} [(\frac{\partial u_{im}}{\partial t})^2 + (\frac{\partial u_{im}}{\partial x})^2 + (1 - \epsilon\delta(x))u_{im}^2] = \Omega^2 a_0^2/\epsilon. \tag{9}$$

Now we analyze the kink-impurity interactions by collective-coordinate method taking into account two dynamical variables, namely the kink coordinate $X(t)$ [see Eq. (2)] and the amplitude of the impurity mode oscillation $a(t)$ [see Eq. (7)]. Substituting the ansatz

$$u = u_k + u_{im} = 4\tan^{-1}\exp[x - X(t)] + a(t)e^{-\epsilon|x|/2} \tag{10}$$

into the Lagrangian of the system,

$$L = \int_{-\infty}^{\infty} [\frac{1}{2}u_t^2 - \frac{1}{2}u_x^2 - (1 - \epsilon\delta(x))(1 - \cos u)], \tag{11}$$

and assuming a and ϵ are small enough so that the higher-order terms can be neglected, we may derive the following (reduced) effective Lagrangian

$$L_{eff} = 4\dot{X}^2 + \frac{1}{\epsilon}(\dot{a}^2 - \Omega^2 a^2) - U(X) - aF(X), \tag{12}$$

where $U(X)$ is given by Eq.(3), and $F(X) = -2\epsilon \tanh X/\cosh X$. The equations of motion for the two dynamical variables are

$$\begin{cases} 8\ddot{X} + U'(X) + aF'(X) = 0, \\ \ddot{a} + \Omega^2 a + (\epsilon/2)F(X) = 0. \end{cases} \tag{13}$$

The system (13) describes a particle (kink) with coordinate $X(t)$ and mass 8 placed in an attractive potential $U(X)$ ($\epsilon > 0$), and "weakly" coupled with a harmonic oscillator $a(t)$ (the impurity mode). Here we say "weakly" because the coupling term $aF(X)$ is of order $O(\epsilon)$ and it falls off exponentially. The system (13) is a generalization of the well-known equation $8\ddot{X} = -U'(X)$ describing the kink-impurity interactions in the adiabatic approximation (see,e.g., Ref. [5]).

We find that the dynamical system (13) can describe all features of the kink-impurity interactions. Firstly, it may be used to calculate the threshold velocity of kink capture. By using an energy transfer argument, we have found the threshold analytically [4]

$$V_{thr} = \frac{\pi\epsilon}{\sqrt{2}} \frac{\sinh[\Omega Z(V_{thr})/2V_{thr}]}{\cosh(\Omega\pi/2V_{thr})}, \tag{14}$$

where $Z(V) = \cos^{-1}[(2V^2 - \epsilon)/(2V^2 + \epsilon)]$. For a given $\epsilon > 0$, this equation can be solved by Newton iteration to obtain the threshold velocity $V_{thr}(\epsilon)$. Comparing the analytical results with the direct numerical simulations of Eq.(1), we find that the perturbation theory used in Ref.[6] is valid only for very small ϵ, ($\epsilon \leq 0.05$), while formula (14) gives good estimations of $V_{thr}(\epsilon)$ for ϵ over the region (0.2, 0.7).

Furthermore, Eqs. (13) can be easily solved numerically. We perform the numerical simulations under initial conditions $X(0) = -7$, $\dot{X}(0) = V_i > 0$, $a(0) = 0$, $\dot{a}(0) = 0$. We find that, for a given $\epsilon > 0$, there exists a threshold velocity $V_{thr}(\epsilon)$ such that if the initial velocity of the particle is larger than $V_{thr}(\epsilon)$, then the particle will pass the potential well $U(X)$ and escape to $+\infty$, with final velocity $V_f < V_i$ because part of its kinetic energy is transferred to the oscillator.

Below the threshold velocity, the behaviour of the particle is very interesting. More precisely, if the initial velocity of the particle is smaller than the threshold velocity, the particle can not escape to $+\infty$ after the first interaction with the oscillator, but it will return to interact with the oscillator again. Usually the particle can be trapped by the potential well, which corresponds to a kink trapped by the attractive impurity. However, for some special initial velocities, the second interaction may cause the particle to escape to $-\infty$ with final velocity $V_f < 0$. This resonance phenomenon can be explained by the mechanism of resonant energy exchange between the particle and the oscillator. The resonant reflection of the particle by the potential well corresponds to the reflection of the kink by the attractive impurity. Therefore, the collective-coordinate approach can give a qualitative explanation of the resonance effects in the kink-impurity interactions in the SG model (1).

3 Kink-impurity interactions in the ϕ^4 model

Now let's consider the kink-impurity interactions in the ϕ^4 model

$$\phi_{tt} - \phi_{xx} + [1 - \epsilon\delta(x)](-\phi + \phi^3) = 0 \tag{15}$$

The inelastic interaction of a kink with an attractive impurity ($\epsilon > 0$) was briefly discussed by Belova and Kudryavtsev [15]. They analyzed the problem by collective-coordinate approach using two dynamical variables: the kink translational mode and its internal mode. By numerical simulation of the collective-coordinate dynamical system, they predicted that the kink may be reflected by the impurity due to energy exchange with its internal mode. However, they totally ignored the impurity mode. As we have observed that, although the SG kink does not have internal mode, it still can be reflected by an attractive impurity due to resonant energy exchange between the kink translational mode and the impurity mode, so we have reason to believe that the impurity mode may also play an important role in the ϕ^4 kink-impurity interactions.

We have studied the kink-impurity interactions in the ϕ^4 model [9], and extended the previous work in three directions. Firstly, by numerical simulation we have confirmed the previous claim that a kink can be reflected by an attractive impurity, meanwhile, we have observed more resonance windows of kink reflection. For example, at $\epsilon = 0.5$ we have found seven resonance windows below the threshold velocity of kink capture. Secondly, and the most importantly, we have observed that both the the impurity mode and the kink internal mode take part in the interactions, and the resonance window structure cannot be predicted by supposing that there exists only one localized mode. In particular, we have found that due to the joint effect of the impurity and the kink internal mode oscillation, some resonance windows may disappear. Finally, we have developed a collective–coordinate approach taking into account three dynamical variables: the kink coordinate, the amplitude of the impurity mode and that of the kink internal mode. Our collective–coordinate approach can give a qualitative description of the resonance phenomena in the kink-impurity interactions. The detailed results is reported Ref.[9].

4 Conclusion

We have briefly reviewed our recent numerical and analytical results related to the kink-impurity interactions in the sine-Gordon and the ϕ^4 models. In particular, we have demonstrated that a kink can be totally reflected by an attractive impurity *if* its initial velocity is situated in some well-defined resonance windows. This effect can be explained by a mechanism of resonant energy exchange between the kink translational mode and the impurity mode (for the SG system), as well as the kink internal mode (for the ϕ^4 model).

This work is partially supported by the Direccion General de Investigacion Cientifica y Tecnica (Spain) under Grant No. TIC 73/89. One of us (Zhang Fei) is also supported by the Ministry of Education and Science of Spain. Yu.S. Kivshar acknowledges the financial support of Complutense University through a sabbatical programm.

References

[1] *Disorder and Nonlinearity.* Eds. A.R. Bishop, D.K. Campbell, and St. Pnevmatikos (Springer-Verlag, Berlin, 1989).

[2] Yu.S. Kivshar, in *Nonlinearity with Disorder*, Eds. F.Kh. Abdullaev, A.R. Bishop and St. Pnevmatikos (Springer-Verlag, Berlin, 1991) in press.

[3] J. F.Currie, S.E. Trullinger, A.R. Bishop, and J.A. Krumhansl, Phys. Rev. B **15**, 5567 (1977).

[4] D. W.Maclaughlin and A.C. Scott, Phys. Rev. A **18**, 1652 (1978).

[5] Yu.S. Kivshar and B. A. Malomed, Rev. Mod. Phys. **61**, 763 (1989).

[6] B.A. Malomed, Physica D **15**, 385 (1985).

[7] Yu. S. Kivshar, Zhang Fei, and L. Vázquez, Phys. Rev. Lett. (1991) submitted.

[8] Zhang Fei, Yu.S. Kivshar, B.A. Malomed, and L.Vazquez, Phys. Lett. A (1991) submitted.

[9] Zhang Fei, Yu.S. Kivshar, and L.Vazquez, (to be published).

[10] D. K. Campbell, J. F. Schonfeld, and C.A. Wingate, Physica D **9**, 1 (1983).

[11] M. Peyrard and D. K. Campbell, Physica D **9**, 33 (1983).

[12] D. K. Campbell, M. Peyrard, and P. Sodano, Physica D **19**, 165 (1986).

[13] D. K. Campbell, Zhang Fei, L.Vazquez, and R.J. Flesch , (1991) to be published.

[14] Zhang Fei and L. Vázquez, Appl. Math. Comput., (1991) in press.

[15] T.I.Belova and A.E. Kudryavtsev, (preprint, 1986) unpublished.

TABLE I Resonance windows of the kink-impurity interactions in the SG model

n	V_n predicted by Eq.(6)	Numerical $T_{12}(V_n)$	Resonance Windows
6	0.25498	42.5	(0.2548, 0.25505)
7	0.25842	49.2	(0.25825, 0.2585)
8	0.26064	56.2	(0.2605, 0.2607)
9	0.26215	62.8	(0.26205, 0.26222)
10	0.26323	69.5	(0.26315, 0.26327)
11	0.26403	75.9	(0.26395, 0.26408)
12	0.26463	82.8	(0.26461, 0.264635)
13	0.26510	89.6	(0.26510, 0.26512)
14	0.26547	97.1	(0.26546, 0.26547)
15	0.26577	103.3	(0.26577, 0.26579)
16	0.26602	109.9	(0.26600,0.26602)

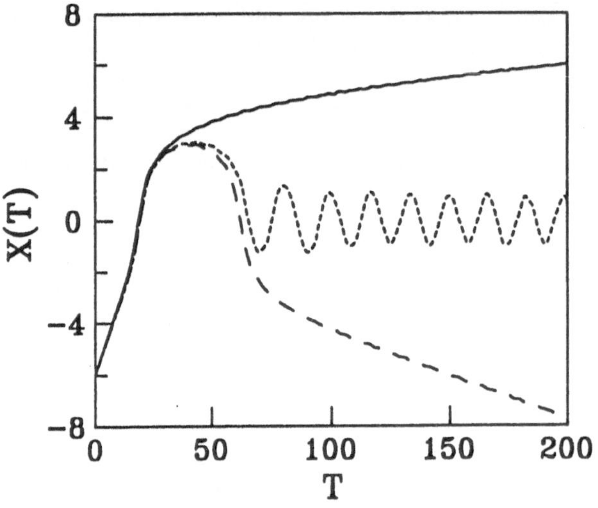

Fig.1 The kink coordinate $X(t)$ vs time for initial velocities V_i situated in three diffferent regions: the region of pass (solid line, $V_i = 0.268$), of capture (dotted line, $V_i = 0.257$), and of reflection (dashed line, $V_i = 0.255$).

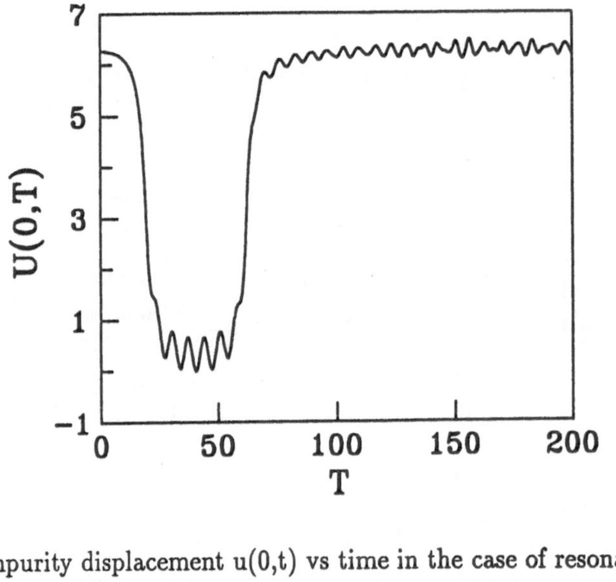

Fig.2 The impurity displacement u(0,t) vs time in the case of resonance ($V_i = 0.255$). Note that between the two interactions there are four small bumps which show the impurity mode oscillation, and after the second interaction the energy of the impurity mode is resonantly transferred back to the kink.

LOCALIZED SELF-SIMILAR STRUCTURES FOR A COUPLED NLS EQUATION: AN APPROXIMATE ANALYSIS

L. Gagnon
Centre d'Optique Photonique et Laser, Département de physique
Université Laval, Ste-Foy (Québec), Canada, G1K 7P4

ABSTRACT

We perform an approximate analysis of some particular self-similar solutions of the (2+1)-dimensional coupled nonlinear Schrödinger equation. These solutions are invariant under a point-symmetry subgroup of the model that involves the Schrödinger conformal symmetry. We use a variational approach to classify them and to determine their approximate structures.

Many physical systems deal with the propagation of two waves that interact nonlinearly. In this paper, we concentrate on the model

$$\begin{aligned} &\sqrt{-1}\,\psi_z^{(1)} + \psi_{xx}^{(1)} + \psi_{yy}^{(1)} + \eta\left[|\psi^{(1)}|^2 + (1+h)|\psi^{(2)}|^2\right]\psi^{(1)} = 0, \\ &\varepsilon\sqrt{-1}\,\psi_z^{(2)} + \psi_{xx}^{(2)} + \psi_{yy}^{(2)} + \eta\left[|\psi^{(2)}|^2 + (1+h)|\psi^{(1)}|^2\right]\psi^{(2)} = 0, \end{aligned} \tag{1}$$

where $\psi^{(i)}(x,y,z)$ are complex functions (throughout the text, $i = 1,2$), h is a real parameter, $\eta = \pm 1$ and $\varepsilon = \pm 1$. The above (2+1)-dimensional coupled nonlinear Schrödinger equation is of particular interest in the field of transverse effects in nonlinear optics (a review and extensive bibliography on the subject can be found in Ref. 1). In fact, it is the basic model that describes the time-independent copropagation[2] ($\varepsilon = 1$) and counterpropagation[3,4] ($\varepsilon = -1$) of two waves in a self-focusing ($\eta = 1$) or self-defocusing ($\eta = -1$) media.

Equation (1) is not completely integrable in the sense that it cannot be solved by inverse scattering techniques. However, it has the property of being invariant under a point-symmetry group for which the corresponding algebra is the direct sum between the 9-dimensional Schrödinger algebra sch(2) and a change of phase generator. The whole set of generators then includes one conformal symmetry C, one dilation D, one rotation J, two Galilean boosts K_x and K_y, three coordinate translations P_x, P_y and P_z and two constant change of phase $M^{(1)}$ and $M^{(2)}$. Thus, equation (1) is very appropriate for the application of the symmetry reduction method.

Here we concentrate on the specific 2-dimensional symmetry subalgebra

$$X_1 = J + a_1 M^{(1)} + a_2 M^{(2)} \ , \ X_2 = C + P_z + b_1 M^{(1)} + b_2 M^{(2)} \ , \tag{2}$$

where a_i and b_i are real parameters and

$$\begin{aligned} &J = y\,\partial_x - x\,\partial_y \ , \qquad M^{(i)} = -\varepsilon^{i-1}\sqrt{-1}\left(\psi^{(i)}\partial_{\psi^{(i)}} - c.c.\right), \qquad P_z = \partial_z \ , \\ &C = z\left(z\,\partial_z + x\,\partial_x + y\,\partial_y\right) - \frac{1}{4}\left(x^2+y^2\right)\left(M^{(1)} - M^{(2)}\right) - \sqrt{-1}\,z\left(M^{(1)} + M^{(2)}\right) . \end{aligned} \tag{3}$$

Following the standard symmetry reduction procedure[5], we calculate the invariants of the subgroups by solving the equation

$$X_i\,Q\left(x,y,z,\psi^{(1)},\psi^{(1)*},\psi^{(2)},\psi^{(2)*}\right) = 0 \ , \tag{4}$$

where Q is an auxiliary function. Since subalgebra (2) has generic orbits of codimension 1 in the space of independent variables and 4 in the space of dependent variables, the solution of Eq. (4) leads to five invariants ξ, $f^{(i)}(\xi)$ and $f^{(i)*}(\xi)$ satisfying

$$\begin{aligned} &\psi^{(i)} = \left(1+z^2\right)^{-1/2} f^{(i)}(\xi)\exp\left[\sqrt{-1}\,\varepsilon^{i-1}\left(\frac{1}{4}z\,\xi^2 + a_i\,\theta - b_i \arctan z\right)\right], \\ &\xi^2 = r^2\left(1+z^2\right)^{-1} , \end{aligned} \tag{5}$$

where $r^2 = x^2 + y^2$, $\theta = \arctan(y/x)$ and a_i are chosen as integers. Substituting relations (5) in Eq. (1) yields the reduced coupled ordinary differential equations

$$f^{(i)}_{\xi\xi} + \frac{1}{\xi} f^{(i)}_{\xi} + \left(b_i - \frac{1}{4}\xi^2 - \frac{a_i^2}{\xi^2}\right) f^{(i)} + \eta\left[\left|f^{(i)}\right|^2 + (1+h)\left|f^{(3-i)}\right|^2\right] f^{(i)} = 0 \ . \tag{6}$$

The task of solving Eq. (6) exactly is quite difficult. Usually, one restricts the analysis to the determination of conditions under which the reduced equations are of Painlevé type, that is, when none of their solutions have movable critical points. The method is well adapted for single equations since a large classification of second and third order Painlevé type equations exists[6,7]. This is not so easy for coupled systems. In any case, one can show for the uncoupled case $h = -1$, that Eq. (6) does not even have the Painlevé property. For all these reasons, we restrict our analysis of Eq. (6) to the determination of approximate solutions. The method we use is based on a variational principle and is well described elsewhere[8,9]. In the following, we will only summarize the main steps of the calculations.

Equation (6) can be reformulated as a variational problem with the Lagrangian

$$V = V^{(1)} + V^{(2)} - \eta\,(1+h)\,|f^{(1)}|^2\,|f^{(2)}|^2 \quad , \tag{7}$$

where

$$V^{(i)} = \left|f_\xi^{(i)}\right|^2 - b_i\,|f^{(i)}|^2 + \frac{1}{4}\,\xi^2\,|f^{(i)}|^2 + \frac{a_i^2}{\xi^2}\,|f^{(i)}|^2 - \frac{1}{2}\,\eta\,|f^{(i)}|^4 . \tag{8}$$

Equation (6) is then derived from the cylindrical Euler equations

$$\frac{\partial}{\partial\xi}\left[\frac{\partial V}{\partial f_\xi^{(i)*}}\right] + \frac{1}{\xi}\,\frac{\partial V}{\partial f_\xi^{(i)*}} - \frac{\partial V}{\partial f^{(i)*}} = 0 \ . \tag{9}$$

The essence of the variational approach lies in the choice of the most appropriate trial functions that describe, as faithfully as possible, the exact solutions behaviour. On the other hand, since we want to obtain simple analytical results, we have to restrict our choice to a generic one. We found that a good compromise between simplicity and accuracy is given by the trial functions

$$f^{(i)} = A_i\,L^{(i)}\!\left(\frac{\xi}{W_i}\right) \ , \tag{10}$$

where A_i and W_i are real parameters and

$$L_1^{(i)} = \exp\left[-\zeta_i^2\right] \qquad a_i = 0$$

$$L_2^{(i)} = \zeta_i \exp\left[-\zeta_i^2\right] \qquad a_i = \pm 1$$

$$L_3^{(i)} = \left(1 - 2\zeta_i^2\right) \exp\left[-\zeta_i^2\right] \qquad a_i = 0 \tag{11}$$

$$L_4^{(i)} = \zeta_i^2 \exp\left[-\zeta_i^2\right] \qquad a_i = \pm 2$$

The choice of real functions $L^{(i)}(\zeta_i)$, $\zeta_i = \xi / W_i$, is based on the form of the exact localized solutions of Eq. (6) in the linear limit $\eta = 0$, which are the well known Laguerre-Gauss modes. Relation (11) gives the expression of the first four modes.

Substituting the ansatz (10) in the Lagrangian (7) and integrating the ξ-variable from 0 to infinity yield a reduced Lagrangian denoted $\langle V \rangle$ that is independent of ξ. Thus, solving the corresponding reduced Euler equations

$$\frac{\partial}{\partial \xi}\left[\frac{\partial \langle V \rangle}{\partial y_{i\xi}}\right] + \frac{1}{\xi}\frac{\partial \langle V \rangle}{\partial y_{i\xi}} - \frac{\partial \langle V \rangle}{\partial y_i} = \frac{\partial \langle V \rangle}{\partial y_i} = 0 \ , \quad y_i \equiv A_i \text{ and } W_i \ , \tag{12}$$

lead to the four relations

$$E_i + \frac{(1+h)}{\alpha_{5i}\, W_{3-i}^2}\left[2\,\alpha_6 - W_i \frac{d\alpha_6}{dW_i}\right] E_{3-i} = \frac{\eta}{\alpha_{5i}}\left[2\,\alpha_{3i} + 2\,a_i^2\,\alpha_{4i} - \frac{1}{2}\,W_i^4\,\alpha_{2i}\right] , \tag{13}$$

and

$$b_i = \frac{1}{\alpha_{1i}\, W_i^2}\left[\alpha_{3i} + a_i^2\,\alpha_{4i} + \frac{1}{4}\,W_i^4 \alpha_{2i} - \eta\, E_i\,\alpha_{5i} - \eta\,(1+h)\frac{E_{3-i}}{W_{3-i}}\,\alpha_6\right] . \tag{14}$$

The constant $E_i = A_i^2\, W_i^2$ are proportional to the energy Σ_i in each wave through

$$\Sigma_i = \int_0^{2\pi} \int_0^{\infty} |\psi^{(i)}|^2 \, r \, dr \, d\theta = 2\pi \int_0^{\infty} |U^{(i)}|^2 \, \xi \, d\xi = 2\pi \, \alpha_{1i} E_i \, . \tag{15}$$

The parameters α_{ki} (k=1,...,5) and α_6 are given by

$$\alpha_{1i} = \int_0^{\infty} [L^{(i)}]^2 \, \zeta_i \, d\zeta_i \, , \qquad \alpha_{2i} = \int_0^{\infty} [L^{(i)}]^2 \, \zeta_i^3 \, d\zeta_i \, ,$$

$$\alpha_{3i} = \int_0^{\infty} \left[\frac{dL^{(i)}}{d\zeta_i}\right]^2 \zeta_i \, d\zeta_i \, , \qquad \alpha_{4i} = \int_0^{\infty} [L^{(i)}]^2 \, \zeta_i^{-1} \, d\zeta_i \, , \tag{16}$$

$$\alpha_{5i} = \int_0^{\infty} [L^{(i)}]^4 \, \zeta_i \, d\zeta_i \, , \qquad \alpha_6 = \int_0^{\infty} \left[L^{(1)}\left(\frac{\xi}{W_1}\right)\right]^2 \left[L^{(2)}\left(\frac{\xi}{W_2}\right)\right]^2 \xi \, d\xi$$

and can be evaluated analytically.

Relations (13) and (14) are parametric equations that give E_i and b_i as function of the widths W_1 and W_2 of the approximate localized solutions (10). We solved them for various values of W_1 and W_2.

For instance, figure 1 shows the normalized energy $S = (2 + h)\,\Sigma_i$ in each wave as function of b for two identical beams, i.e. $A_1 = A_2$, $W_1 = W_2$ and $b_1 = b_2 = b$. The curve numbers refer to the mode number in Eq. (11). The "+" and "–" signs refer to a self-focusing medium ($\eta = 1$) and a self-defocusing medium ($\eta = -1$) respectively. The points b = 1,2,3... and $S = 0$ correspond to the linear limit $\eta \to 0$. The case described by curve 1^+ was studied in Ref. 10. The energy values at $b \to -\infty$ are 4π, 16π, 24π, 32π and correspond approximately (within 5%) to the energy of the first self-trapping solutions of Eq. (1) (no z-dependence in the amplitudes)[11,12]. The fundamental self-trapping solution is known to be unstable and to eventually collapse under a self-focusing process (a review and extensive bibliography can be found in Ref. 13). We suspect a similar behaviour for the higher-order self-trapping solutions.

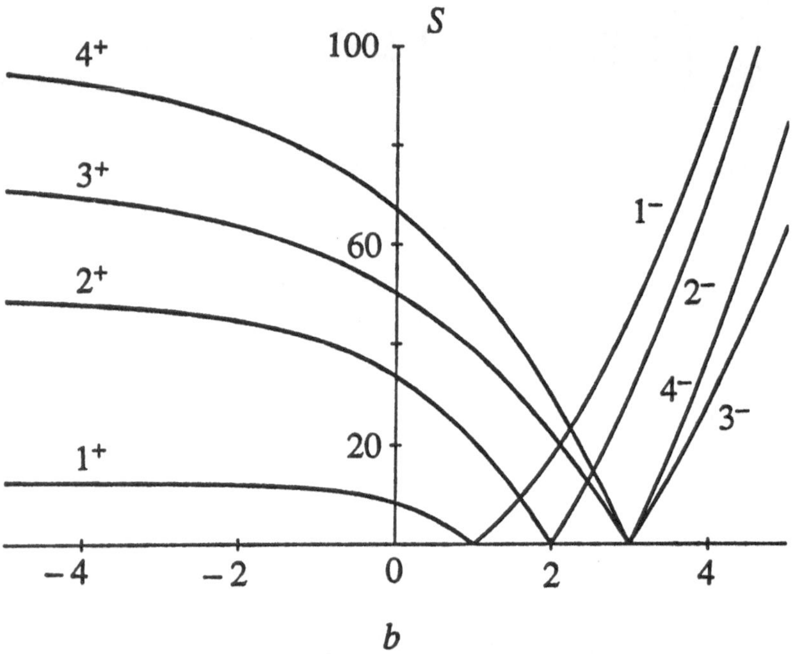

Figure 1: Normalized energy $S = (2 + h)\Sigma_i$ in each wave for two identical modes. Numbers refer to $L_1,\ldots, L_4$ and "+" and "–" signs stand for $\eta = 1$ and $\eta = -1$.

In addition to the above identical localized self-similar solutions, there is a large set of solutions where $W_1 \neq W_2$. For example, solid curves on figures 2 show the energies Σ_1 and Σ_2 as function of b_1 and b_2 for two beams in the fundamental mode with $h = 1$ and $\eta = 1$. For comparison, coarse dashed curve give the energy for the case $W_1 = W_2$ and fine dashed curve gives the energy for $h = -1$ (no nonlinear coupling). In figure 2, we have chosen $W_1 = 0.928$ which leads to $0.695 \leq W_2 \leq 1.209$ and provides the possible self-similar solutions centered around $b_1 = b_2 = -2$. The points on the solid curves correspond to $W_2 = 1.209$.

The most significative result of our analysis is the possible coexistence of self-similar solutions having different mode profiles. Our calculations show that this "nonlinear superposition" seems to be always possible within a certain parameter range. Specific examples of a such behaviour will be reported elsewhere.

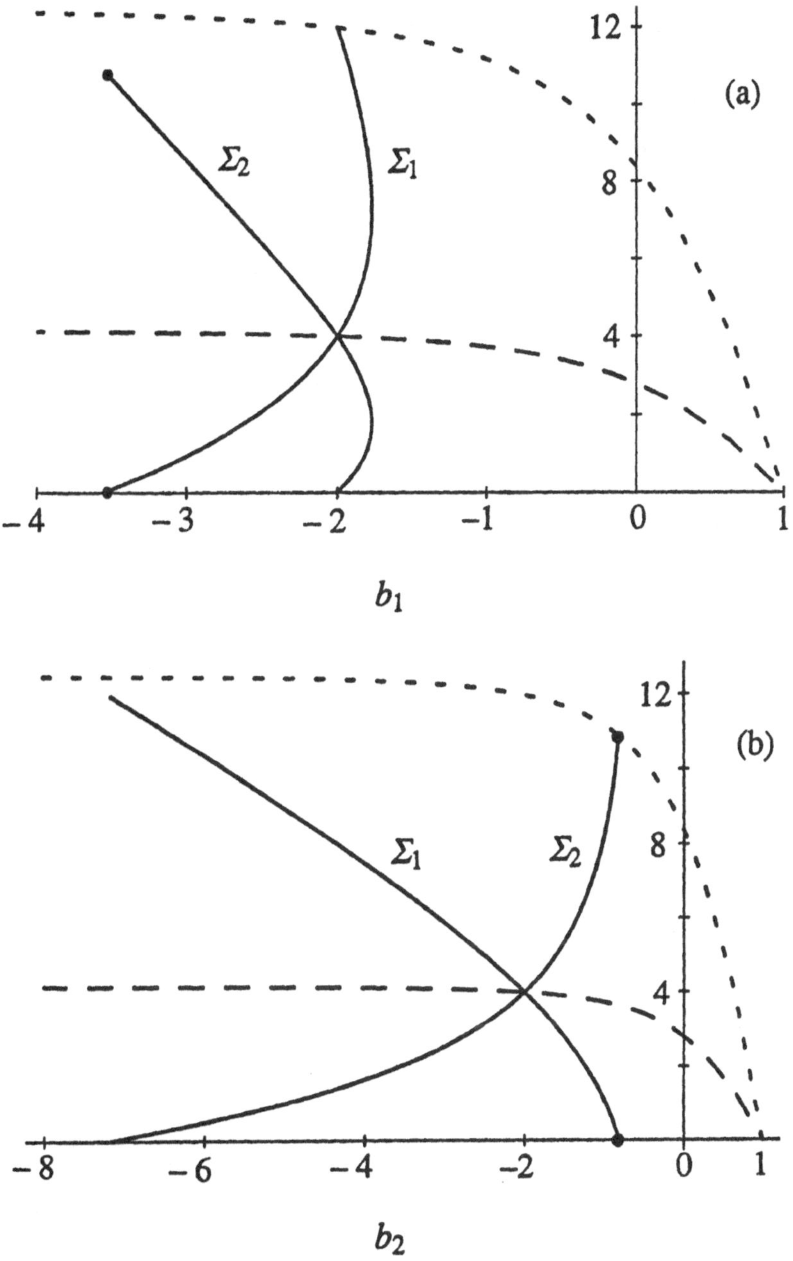

Figure 2: Energies Σ_1 and Σ_2 as function of b_1 (a) and b_2 (b) for two waves in the fundamental mode with $\eta = 1$, $h = 1$, $W_1 = 0.928$ and $0.695 \leq W_2 \leq 1.209$.

In conclusion, one can say that the results of this study are indicative of a large set of self-similar nonlinear coherent structures predicted by the model (1). The coexistence of these self-similar coupled waves can be of interest in the study of various nonlinear systems and in particular for the counterpropagation of two Gaussian optical beams in a Kerr media[4]. In that case, the question of temporal stability of such modes has to be addressed. We plan to go back to that issue in a near future.

ACKNOWLEDGEMENTS

The author is grateful to Prof. W. Firth from University of Strathclide (UK) for his suggestions. He also thanks the scientific committee of the 7th International Workshop on Nonlinear Coherent Structures in Physics and Biology for a financial support.

REFERENCES

1. N. B. Abraham and W. J. Firth, J. Opt. Soc. Am. B **7**, 951 (1990)
2. P. D. Maker and R. W. Terhune, Phys.Rev. A **137**, 801 (1965)
3. A. E. Kaplan, Opt. Lett. **8**, 560 (1983)
4. W. J. Firth, A. Fitzgerald and C. Paré, J. Opt. Soc. Am B **7**, 1087 (1990)
5. P. Olver, *Applications of Lie Groups to Differential Equations* (Springer, Berlin, 1986)
6. E. L. Ince, *Ordinary Differential Equations* (Dover, New York, 1956)
7. F. J. Bureau, Ann. Mat. Pura Appl. IV **41**, 163 (1972)
8. D. Anderson, Phys. Scr. **18**, 35 (1978)
9. L. Gagnon and C. Paré, J. Opt. Soc. Am. A **8**, 601 (1991)
10. J. H. Marburger and F. S. Felber, Phys. Rev. A **17**, 335 (1978)
11. R. Y. Chiao, E. Garmire and C. H. Townes, Phys. Rev. Lett **13**, 479 (1964)
12. D. Pohl, Opt. Commun. **2**, 305 (1970)
13. J. J. Rasmussen and K. Rypdal, Physica Scripta **33**, 481 (1986); K. Rypdal and J. J. Rasmussen, Physica Scripta **33**, 498 (1986)

SEARCHING FOR SOLITONS WITH A DIRECT BINARY OPERATOR METHOD.

F. Lambert and R. Willox*.
Vrije Universiteit Brussel,
Dienst Theoretische Natuurkunde,
Pleinlaan 2, B-1050 Brussel.

*Research Assistant, National Foundation for Scientific Research, Belgium.

1. Introduction.

Soliton equations in 1+1 dimensions with sech-squared solitons and corresponding two-soliton solutions:

$$u_2(x,t) = 2\partial_x^2 \ln f_2, \quad f_2 = 1 + \exp\theta_1 + \exp\theta_2 + A_{12}\ exp(\theta_1 + \theta_2),$$
$$\theta_i = -k_i x + \omega_i t + \tau_i \tag{1}$$

are known to appear in infinite families[1] with a characteristic coupling factor A_{12}. Hirota's binary operators[2] can be used in a direct way for the construction of sequences of candidate soliton equations which admit two-soliton solutions with the same coupling. The possibility of obtaining in this way an infinite sequence of candidate soliton solutions, for a particular class of dispersion relations and an appropriate choice of A_{12}, may often hint at the existence of an integrable hierarchy, especially when the lower members of the sequence are also found to admit three-soliton solutions.
Here we develop this idea into an algorithmic procedure based on the construction of multidimensional bilinear equations. The method applies to first order equations with respect to time and is shown to produce a new soliton equation related to the Sawada-Kotera (SK) hierarchy[3].

2. Key-properties of the D-operators.

Hirota's D-operators are defined by their action on a pair of functions:

$$D_{x_1}^{p_1} \dots D_{x_n}^{p_n} f.g = (\partial_{x_1} - \partial_{x_1'})^{p_1} \dots (\partial_{x_n} - \partial_{x_n'})^{p_n} f(x_1, \dots x_n) g(x_1', \dots x_n')\Big|_{x_i'=x_i} \tag{2}$$

Three key-properties of these operators can be derived from this definition[4]. They are related to their action on exponentials

Property 1.

If $\Theta = \sum_{i=1}^n \alpha_i x_i + \tau$ and $\Theta' = \sum_{i=1}^n \beta_i x_i + \tau'$ with $\alpha_i, \beta_i, \tau, \tau' =$ constant (3)

then: $$D_{x_1}^{p_1} \dots D_{x_n}^{p_n} \exp\Theta . \exp\Theta' = [\prod_{i=1}^n (\alpha_i - \beta_i)^{p_i}] \exp(\Theta + \Theta'). \tag{4}$$

Property 2.

Let $(p_1,\dots p_n;m)$ be a set of integers with $p_1+\dots+p_n=2m$.
If $q_{p_1x_1,\dots p_nx_n}=\partial_{x_1}^{p_1}\dots\partial_{x_n}^{p_n}q(x_1,\dots x_n)$, then:

$$f^{-1}D_{x_1}^{p_1}\dots D_{x_n}^{p_n}f.f\Big|_{f=\exp q/2}=q_{p_1x_1,\dots p_nx_n}+Q_m(p_1,\dots p_n;q), \tag{5}$$

where $Q_m(p_1,\dots p_n;q)$ stands for the "even part-partitional nonlinearity" which involves even-order partial derivatives of q and which contains as many different terms of degree p $(2\le p\le m)$:

$$C(r_{ij})\prod_{i=1}^{p}q_{r_{i1}x_1,\dots r_{in}x_n},\quad \sum_{i=1}^{p}r_{ij}=p_j,\quad \sum_{j=1}^{n}r_{ij}=2m_i,$$
$$m_i=\text{ integer},\quad r_{ij}=\text{ integer or zero}, \tag{6}$$

as there are different ways of dividing $2m$ elements, of which p_i are of type x_i, among p indistinguishable non-empty boxes with the condition that each box must contain an even number of elements, each coefficient $C(r_{ij})$ being equal to the combinatorial weight of the corresponding partition (number of such partitions if all elements can be distinguished).
This second property can be compared with the partitional property of the ordinary derivative:

$$\exp(-q)\partial_x^n\exp(q)\equiv\sum[c_1!\dots c_n!(2!)^{c_2}\dots(n!)^{c_n}]^{-1}n!q_x^{c_1}q_{2x}^{c_2}\dots q_{nx}^{c_n} \tag{7}$$

where the sum is taken over all partitions of n which are written as follows:

$$n=c_1+2c_2+\dots+nc_n,\quad c_i=\text{ number of parts equal to } i.$$

Property 3.

Let $P(\partial_{x_1},\dots\partial_{x_n})$ be a polynomial differential operator satisfying the conditions:

$$P(0,\dots 0)=0\quad\text{and}\quad P(-\partial_{x_1},\dots-\partial_{x_n})=P(\partial_{x_1},\dots\partial_{x_n}). \tag{8}$$

and let F_2 be an expression of the form:

$$F_2=1+\exp\Theta_1+\exp\Theta_2+A(\alpha_{ij})\exp(\Theta_1+\Theta_2),\quad \Theta_i=\sum_{j=1}^{n}\alpha_{ij}x_j+\tau_i \tag{9}$$

with parameters α_{ij} satisfying:

$$P(\alpha_{i1},\dots\alpha_{in})\equiv\exp(-\Theta_i)P(\partial_{x_1},\dots\partial_{x_n})\exp\Theta_i=0 \tag{10}$$

It then follows from property 1 that:

$$P(D_{x_1},\dots D_{x_n})F_2.F_2=2\{P(\alpha_{11}-\alpha_{21},\dots\alpha_{1n}-\alpha_{2n})$$
$$+A(\alpha_{ij})P(\alpha_{11}+\alpha_{21},\dots\alpha_{1n}+\alpha_{2n})\}\exp(\Theta_1+\Theta_2). \tag{11}$$

Property 3, taken with $n=2,\ x_1=x,\ x_2=t$ is known as the Hirota theorem[2].

3. Direct bilinear method in 1 + 1 dimensions.

Starting from a sequence of linear polynomial differential operators with respect to x and t one may construct bilinear equations which, on account of the above properties, give rise to candidate soliton equations "in primary form".
This construction can be illustrated with the following class of bilinear operators:

$$P_m(\partial_x, \partial_t) = \partial_x \left(\partial_x^{2m+1} + \partial_t\right), \quad m = \text{ integer.} \tag{12}$$

i) At $m = 1$ one has: $\quad P_1(k_i, \omega_i) \equiv \exp(-\Theta_i) P_1(\partial_x, \partial_t) \exp \Theta_i \equiv k_i(k_i^3 - \omega_i)$ (13)
Taking $\omega_i = k_i^3$ it follows from property 3 that:

$$D_x(D_x^3 + D_t) f_2.f_2 \Big|_{\substack{\Theta_i = -k_i x + k_i^3 t + r_i \\ A_{12} = A_{12}^{KdV}}} = 0, \quad A_{12}^{KdV} = \left(\frac{k_1 - k_2}{k_1 + k_2}\right)^2. \tag{14}$$

Application of property 2 shows that the following equation for $q(x,t)$:

$$f^{-2} D_x \left(D_x^3 + D_t\right) f.f \Big|_{f = \exp q/2} \equiv q_{xt} + q_{4x} + 3q_{2x}^2 = 0, \tag{15}$$

with the partitional nonlinearity $Q_2(4;q) = 3q_{2x}^2$ (determined by the even part partitions of 4 : $4 = 2 + 2$, $C(2,2) = \frac{1}{2!}\binom{4}{2,2} = 3$), admits two-soliton generating solutions $q_2 = 2 \ln f_2$ with $A_{12} = A_{12}^{KdV}$.
Setting $u = q_{2x}$ one finds that $u(x,t)$ satisfies the equation:

$$u_t + u_{3x} + 6uu_x = 0, \tag{16}$$

which is recognized as the KdV equation (the well-known integrability of this equation is here strongly suggested by the fact that the corresponding bilinear form satisfies the N-soliton test[5)]).

ii) At $m = 2$ one has: $\quad P_2(k_i, \omega_i) \equiv \exp(-\Theta_i) P_2(\partial_x, \partial_t) \exp \Theta_i \equiv k_i(k_i^5 - \omega_i)$ (17)
Taking $\omega_i = k_i^5$ it follows from rel.(11) that:

$$D_x(D_x^5 + D_t) f_2.f_2 \Big|_{\substack{\Theta_i = -k_i x + k_i^5 t + r_i \\ A_{12} = A_{12}^{SK}}} = 0, \quad A_{12}^{SK} = \left(\frac{k_1^2 - k_1 k_2 + k_2^2}{k_1^2 + k_1 k_2 + k_2^2}\right) A_{12}^{KdV}. \tag{18}$$

This means that the evolution equation:

$$f^{-2} D_x \left(D_x^5 + D_t\right) f.f \Big|_{f = \exp q/2} \equiv q_{xt} + q_{6x} + 15 q_{2x} q_{4x} + 15 q_{2x}^3 = 0, \tag{19}$$

with the partitional nonlinearity $Q_3(6;q) = 15 q_{2x} q_{4x} + 15 q_{2x}^3$ (according to the even part partitions of 6 : $\;6 = 4 + 2$, with $\binom{6}{4,2} = 15$ and $6 = 2 + 2 + 2$ with

$\frac{1}{3!}\binom{6}{2,2,2} = 15$) has two-soliton generating solutions $q_2 = 2\ln f_2$ with $A_{12} = A_{12}^{SK}$. This equation is the primary version of the fifth order SK equation. It is a straightforward matter to verify that the corresponding bilinear form passes the N-soliton test.

4. Multidimensional bilinear equations.

Further members of the KdV hierarchy (and the SK hierarchy) can also be derived at $m = 2,3$ and 4 ($m = 3$ and 5) from two-soliton considerations. One considers multidimensional expressions involving binary operators $D_{2p+1} \equiv D_{t_{2p+1}}$ with respect to appropriate auxiliary time variables t_{2p+1}:

$$B_{2m+2}(D_x, D_t, \ldots D_{2p+1} \ldots; \alpha_p) = D_x\left(D_x^{2m+1} + D_t\right) + \sum_p \alpha_p \left(D_x^{2p+1} + D_{2p+1}\right) F_{2m-2p+1}(D_{2r+1}), \tag{20}$$

where α_p = constant, $1 \leq p \leq m-1$, and where $F_{2m-2p+1}(D_{2r+1})$ denotes a product of auxiliary D-operators, of weight level $2m - 2p + 1$.
The integers p, r and the constants α_p are to be chosen such that:

$$B_{2m+2}(D_x, D_t, \ldots D_{2p+1} \ldots; \alpha_p) f_2.f_2\Big|_{\substack{\Theta_i = -k_i x + k_i^{2m+1} t + \sum_p k_i^{2p+1} t_{2p+1} \\ A_{12} = A_{12}^{KdV}\left(A_{12}^{SK}\right)}} = 0 \tag{21}$$

and such that all auxiliary variables can be eliminated, eventually, from the corresponding multidimensional evolution equation:

$$f^{-2} B_{2m+2}(D_x, D_t, \ldots D_{2p+1} \ldots; \alpha_p) f.f\Big|_{f = \exp q/2} = 0 \tag{22}$$

by means of available lower-order equations in the hierarchy.
This construction, followed by the appropriate dimensional reduction, is most easily illustrated at $m = 2$ with $A_{12} = A_{12}^{KdV}$.
Applying once more the property 3, one sees that:

$$D_x\left(D_x^5 + D_t\right) f_2.f_2\Big|_{\substack{\Theta_i = -k_i x + k_i^5 t + \tau_i \\ A_{12} = A_{12}^{KdV}}} = 20 k_1^2 k_2^2 (k_1 - k_2)^2 \exp(\Theta_1 + \Theta_2) \tag{23}$$

and that the introduction of an auxiliary t-variable t_3, by setting $\tau_i = k_i^3 t_3$, enables us to construct the "counter-term":

$$D_x^3\left(D_x^3 + D_3\right) f_2.f_2\Big|_{\substack{\Theta_i = -k_i x + k_i^5 t + k_i^3 t_3 \\ A_{12} = A_{12}^{KdV}}} = 24 k_1^2 k_2^2 (k_1 - k_2)^2 \exp(\Theta_1 + \Theta_2) \tag{24}$$

Thus $$\left[D_x\left(D_x^5+D_t\right)-\tfrac{5}{6}D_x^3\left(D_x^3+D_3\right)\right]\Bigg|_{\substack{\Theta_i=-k_ix+k_i^5t+k_i^3t_3\\ A_{12}=A_{12}^{KdV}}}=0. \tag{25}$$

It follows from property 2 that the 3-dimensional evolution equation:

$$\begin{aligned} &f^{-2}\left[D_x\left(D_x^5+D_t\right)-\frac{5}{6}D_x^3\left(D_x^3+D_3\right)\right]f.f\Bigg|_{f=\exp q/2} \\ &\equiv q_{xt}+\frac{1}{6}\left(q_{6x}+15q_{2x}q_{4x}+15q_{2x}^3\right)-\frac{5}{6}\left(q_{3x,t_3}+3q_{2x}q_{x,t_3}\right)=0, \end{aligned} \tag{26}$$

has two-soliton generating solutions $q_2(x_1,t,t_3)=2\ln f_2(x,t,t_3)$.
The auxiliary variable t_3 can be eliminated with the help of the primary KdV equ.(3) (in which t has been replaced by t_3) according to which:

$$q_{x,t_3}=-q_{4x}-3q_{2x}^2 \quad \text{and} \quad q_{3x,t_3}=-q_{6x}-6q_{2x}q_{4x}-6q_{3x}^2. \tag{27}$$

The resulting two-dimensional equation:

$$q_{xt}+q_{6x}+10q_{2x}q_{4x}+5q_{3x}^2+10q_{2x}^3=0 \tag{28}$$

is the primary form of the fifth-order KdV equation (its known integrability is also suggested by the fact that the corresponding bilinear forms pass the N-soliton test). Introducing at $m=3$ the auxiliary fifth-order SK-time t_5 one verifies in the same way that:

$$\left[D_x\left(D_x^7+D_t\right)-\frac{7}{10}D_x^3\left(D_x^5+D_5\right)\right]f_2.f_2\Bigg|_{\substack{\Theta_i=-k_ix+k_i^7t+k_i^5t_5\\ A_{12}=A_{12}^{SK}}}=0. \tag{29}$$

Dimensional reduction of the corresponding evolution equation:

$$f^{-2}\left[D_x\left(D_x^7+D_t\right)-\frac{7}{10}D_x^3\left(D_x^5+D_5\right)\right]f.f\Bigg|_{f=\exp q/2}=0 \tag{30}$$

by means of equ.(19) produces the following two-dimensional candidate soliton equation:

$$q_{xt}+q_{8x}+21q_{2x}q_{6x}+21q_{3x}q_{5x}+21q_{4x}^2+126q_{2x}^2q_{4x}+63q_{2x}q_{3x}^2+63q_{2x}^4=0 \tag{31}$$

which is recognized as the primary version of the seventh-order SK equation.

5. Soliton equations with higher-order regularized long wave dispersion.

Let us now consider a family of linear operators associated with higher-order regularized long wave dispersion:

$$L_m\left(\partial_x,\partial_t\right)=\partial_x\left(\partial_t+\partial_x-\partial_x^{2m}\partial_t\right), \quad m=\text{ integer}. \tag{32}$$

At $m = 1$, it is found on account of property 3 that:

$$D_x\left(D_t + D_x - D_x^2 D_t\right) f_2.f_2\Big|_{\substack{\Theta_i = -k_i x + k_i\left(1-k_i^2\right)^{-1} t + r_i \\ A_{12} = A_{12}^{KdV}}} = -\frac{4k_1^2 k_2^2 \left(k_1 - k_2\right)^2}{\left(1-k_1^2\right)\left(1-k_2^2\right)} \exp\left(\Theta_1 + \Theta_2\right) \tag{33}$$

whereas:

$$D_t\left(D_x^3 + D_3\right) f_2.f_2\Big|_{\substack{\Theta_i = -k_i x + k_i^3 t_3 + r_i \\ A_{12} = A_{12}^{KdV}}} = \frac{12k_1^2 k_2^2 \left(k_1 - k_2\right)^2}{\left(1-k_1^2\right)\left(1-k_2^2\right)} \exp\left(\Theta_1 + \Theta_2\right) \tag{34}$$

It follows that:

$$\left[D_x\left(D_t + D_x - D_x^2 D_t\right) + \frac{1}{3} D_t\left(D_x^3 + D_3\right)\right] f_2.f_2\Big|_{\substack{\Theta_i = -k_i x + k_i\left(1-k_i^2\right)^{-1} t + k_i^3 t_3 \\ A_{12} = A_{12}^{KdV}}} = 0 \tag{35}$$

Dimensional reduction of the corresponding evolution equation:

$$f^{-2}\left[D_x\left(D_t + D_x - D_x^2 D_t\right) + \frac{1}{3} D_t\left(D_x^3 + D_3\right)\right] f.f\Big|_{f = \exp q/2} = 0 \tag{36}$$

by means of the primary KdV equation (14) leads to the two-dimensional equation:

$$\left(q_t + q_x - q_{2x,t}\right)_{xx} - 2q_{3x}q_{xt} - 4q_{2x}q_{2x,t} = 0 \tag{37}$$

which is the primary form of the AKNS equation. This equation belongs to an integrable family[6)] higher-order members of which can be derived (at $m = 2$ and 3) in a similar way.
Introducing SK-equations as auxiliary equations (instead of KdV equations) at $m = 3$ it is found that:

$$\left[D_x\left(D_t + D_x - D_x^6 D_t\right) + \tfrac{1}{7} D_t\left(D_x^7 + D_7\right) + \tfrac{3}{5} D_t D_x^2\left(D_x^5 + D_5\right)\right] f_2.f_2\Big|_{\substack{\Theta_i = -k_i x + k_i\left(1-k_i^6\right)^{-1} t + k_i^5 t_5 + k_i^7 t_7 \\ A_{12} = A_{12}^{SK}}} = 0 \tag{38}$$

Dimensional reduction of the corresponding evolution equation:

$$f^{-2}\left[D_x\left(D_t + D_x - D_x^6 D_t\right) + \frac{1}{7} D_t\left(D_x^7 + D_7\right) + \frac{3}{5} D_t D_x^2\left(D_x^5 + D_5\right)\right] f.f\Big|_{f = \exp q/2} = 0 \tag{39}$$

by means of the equ.(19) and (31) produces the two-dimensional equation:

$$\begin{aligned}q_{2x,t} + q_{3x} - q_{8x,t} - 18q_{2x}q_{6x,t} - 15q_{2x,t}q_{6x} - 27q_{3x}q_{5x,t} - 30q_{3x,t}q_{5x} \\ - 33q_{4x}q_{4x,t} - 3q_{7x}q_{xt} - 144q_{2x}q_{4x}q_{2x,t} - 81q_{2x}^2q_{4x,t} \\ - 180q_{2x}q_{3x}q_{3x,t} - 45q_{3x}q_{4x}q_{xt} - 45q_{2x}q_{5x}q_{xt} \\ - 63q_{3x}^2q_{2x,t} - 108q_{2x}^3q_{2x,t} - 135q_{2x}^2q_{3x}q_{xt} \\ + 9q_{3x}\int_x^{+\infty} \partial_t\left[q_{2y}q_{4y} + q_{2y}^3\right] dy = 0\end{aligned} \tag{40}$$

The integrability of this question, suggested by the fact that the corresponding bilinear forms pass the 4-soliton test, was recently confirmed by the construction of a Lax-pair.

References.

1) A. Newell, "Solitons in Mathematics and Physics", SIAM, Philadelphia (1985), 61.
2) Y. Matsuno, "Bilinear Transformation Method", Acad. Press, Orlando (1984), 14.
3) J. Weiss, J. Math. Phys. 25 (1984), 13.
4) F. Lambert, R. Willox, J. Phys. Soc. Japan 58 (1989), 1860.
5) A. Newell, Zeng Yumbo, J. Math. Phys. 27 (1986), 2016.
6) F. Calogero, A. Degasperis, "Solitons and the Spectral Transform I", North Holland, Amsterdam (1982), 126.

THE INVERSE PROBLEM OF DYNAMICS FOR THE NONLINEAR KLEIN-GORDON EQUATION. PULSONS AND BUBBLES IN THE MODELS WITH LOGARITHMIC NONLINEARITIES

Eugene M. Maslov
Fondation Louis de Broglie, 23 quai de Conti, 75006 Paris, France,
and
The USSR Academy of Sciences,
IZMIRAN, Troitsk, Moscow Region, 142092, USSR*)

We review the results on the inverse problem of dynamics for the nonlinear Klein-Gordon equation in n+1 dimensions [1-3].

First, we formulate a theorem, which gives the procedure of reconstruction of nonlinearities from a given rotationally-invariant scalar field distribution. As an illustration, all power nonlinearities corresponding to a certain class of spatially-localized time-dependent algebraic solutions are constructed. Some other wave equations having spherically-symmetric soliton solutions with finite energy are also obtained.

Further, we demonstrate that the logarithmic nonlinearity is the only one, which admits the real solutions in the form of solitons with oscillating amplitude. The structure of these multidimensional pulsating solitons (pulsons) is investigated. We also construct wave equations having solutions in the form of expanding bubbles.

Finally, the results of the numerical experiments [4] on collisions of the pulsons are discussed.

References

[1] E. M. Maslov, Preprint IZMIRAN n°10 (836) (Moscow, 1988).
[2] E. M. Maslov, Phys. Lett., 151A (1990) 47.
[3] E. M. Maslov, Inverse Problems, 7 (1991) L1.
[4] A. G. Shagalov, unpublished.

*) Permanent address.

Lecture Notes in Physics

For information about Vols. 1–365
please contact your bookseller or Springer-Verlag

Vol. 366: M.-J. Goupil, J.-P Zahn (Eds.), Rotation and Mixing in Stellar Interiors. Proceedings, 1989. XIII, 183 pages. 1990.

Vol. 367: Y Osaki, H. Shibahashi (Eds.), Progress of Seismology of the Sun and Stars. Proceedings, 1989. XIII, 467 pages. 1990.

Vol. 368: L. Garrido (Ed.), Statistical Mechanics of Neural Networks. Proceedings. VI, 477 pages. 1990.

Vol. 369: A. Cassatella, R. Viotti (Eds.), Physics of Classical Novae. Proceedings, 1989. XII, 462 pages. 1990.

Vol. 370: H.-D. Doebner, J.-D. Hennig (Eds.), Quantum Groups. Proceedings, 1989. X, 434 pages. 1990.

Vol. 371: K.W. Morton (Ed.), Twelfth International Conference on Numerical Methods in Fluid Dynamics. Proceedings, 1990. XIV, 562 pages. 1990.

Vol. 372: F. Dobran, Theory of Structured Multiphase Mixtures. IX, 223 pages. 1991.

Vol. 373: C. B. de Loore (Ed.), Late Stages of Stellar Evolution. Computational Methods in Astrophysical Hydrodynamics. Proceedings, 1989. VIII, 390 pages. 1991.

Vol. 374: L. Ting, R. Klein, Viscous Vortical Flows. V, 222 pages. 1991.

Vol. 375: C. Bartocci, U. Bruzzo, R. Cianci (Eds.), Differential Geometric Methods in Theoretical Physics. Proceedings, 1990. XIX, 401 pages. 1991.

Vol. 376: D. Berényi, G. Hock (Eds.), High-Energy Ion-Atom Collisions. Proceedings, 1990. IX, 364 pages. 1991.

Vol. 377: W. J. Duschl, S.J. Wagner, M. Camenzind (Eds.), Variability of Active Galaxies. Proceedings, 1990. XII, 312 pages. 1991.

Vol. 378: C. Bendjaballah, 0. Hirota, S. Reynaud (Eds.), Quantum Aspects of Optical Communications. Proceedings 1990. VII, 389 pages. 1991.

Vol. 379: J. D. Hennig, W. Lücke, J. Tolar (Eds.), Differential Geometry, Group Representations, and Quantization. XI, 280 pages. 1991.

Vol. 380: I. Tuominen, D. Moss, G. Rüdiger (Eds.), The Sun and Cool Stars: activity, magnetism, dynamos. Proceedings, 1990. X, 530 pages. 1991.

Vol. 381: J. Casas-Vazquez, D. Jou (Eds.), Rheological Modelling: Thermodynamical and Statistical Approaches. Proceedings, 1990. VII, 378 pages. 1991.

Vol. 382: V.V. Dodonov, V. I. Man'ko (Eds.), Group Theoretical Methods in Physics. Proceedings, 1990. XVII, 601 pages. 1991.

Vol. 384: M. D. Smooke (Ed.), Reduced Kinetic Mechanisms and Asymptotic Approximations for Methane-Air Flames. V, 245 pages. 1991.

Vol. 385: A. Treves, G. C. Perola, L. Stella (Eds.), Iron Line Diagnostics in X-Ray Sources. Proceedings, Como, Italy 1990. IX, 312 pages. 1991.

Vol. 386: G. Pétré, A. Sanfeld (Eds.), Capillarity Today. Proceedings, Belgium 1990. XI, 384 pages. 1991.

Vol. 387: Y. Uchida, R. C. Canfield, T. Watanabe, E. Hiei (Eds.), Flare Physics in Solar Activity Maximum 22. Proceedings, 1990. X, 360 pages. 1991.

Vol. 388: D. Gough, J. Toomre (Eds.), Challenges to Theories of the Structure of Moderate-Mass Stars. Proceedings, 1990. VII, 414 pages. 1991.

Vol. 389: J. C. Miller, R. F. Haglund (Eds.), Laser Ablation-Mechanisms and Applications. Proceedings. IX, 362 pages, 1991.

Vol. 390: J. Heidmann, M. J. Klein (Eds.), Bioastronomy - The Search for Extraterrestrial Life. Proceedings, 1990. XVII, 413 pages. 1991.

Vol. 391: A. Zdziarski, M. Sikora (Eds.), Ralativistic Hadrons in Cosmic Compact Objects. Proceedings, 1990. XII, 182 pages. 1991.

Vol. 392: J.-D. Fournier, P.-L. Sulem (Eds.), Large-Scale Structures in Nonlinear Physics. Proceedings. VIII, 353 pages. 1991.

Vol. 393: M. Remoissenet, M.Peyrard (Eds.), Nonlinear Coherent Structures in Physics and Biology. Proceedings. XII, 398 pages. 1991.

New Series m: Monographs

Vol. m 1: H. Hora, Plasmas at High Temperature and Density. VIII, 442 pages. 1991.

Vol. m 2: P. Busch, P. J. Lahti, P. Mittelstaedt, The Quantum Theory of Measurement. XIII, 165 pages. 1991.

Vol. m 3: A. Heck, J. M. Perdang (Eds.), Applying Fractals in Astronomy. IX, 210 pages. 1991.

Vol. m 4: R. K. Zeytounian, Mécanique des fluides fondamentale. XV, 615 pages, 1991.

Vol. m 5: R. K. Zeytounian, Meteorological Fluid Dynamics. XI, 346 pages. 1991.